AF393810

Quantum Field Theory and String Theory

NATO ASI Series

Advanced Science Institutes Series

A series presenting the results of activities sponsored by the NATO Science Committee, which aims at the dissemination of advanced scientific and technological knowledge, with a view to strengthening links between scientific communities.

The series is published by an international board of publishers in conjunction with the NATO Scientific Affairs Division

A	Life Sciences	Plenum Publishing Corporation
B	Physics	New York and London
C	Mathematical and Physical Sciences	Kluwer Academic Publishers
D	Behavioral and Social Sciences	Dordrecht, Boston, and London
E	Applied Sciences	
F	Computer and Systems Sciences	Springer-Verlag
G	Ecological Sciences	Berlin, Heidelberg, New York, London,
H	Cell Biology	Paris, Tokyo, Hong Kong, and Barcelona
I	Global Environmental Change	

Recent Volumes in this Series

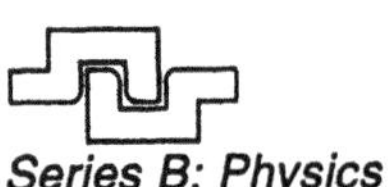

Series B: Physics

Quantum Field Theory and String Theory

Edited by

Laurent Baulieu and Vladimir Dotsenko
Université Pierre et Marie Curie (Paris VI)
Paris, France

Vladimir Kazakov
Université Pierre et Marie Curie (Paris VI)
and École Normale Supérieure
Paris, France

and

Paul Windey
Université Pierre et Marie Curie (Paris VI)
Paris, France

Springer Science+Business Media, LLC

Proceedings of a NATO Advanced Research Workshop on
New Developments in String Theory, Conformal Models, and Topological Field Theory,
held May 10–21, 1993,
in Cargèse, France

NATO-PCO-DATA BASE

The electronic index to the NATO ASI Series provides full bibliographical references (with keywords and/or abstracts) to more than 30,000 contributions from international scientists published in all sections of the NATO ASI Series. Access to the NATO-PCO-DATA BASE is possible in two ways:

—via online FILE 128 (NATO-PCO-DATA BASE) hosted by ESRIN, Via Galileo Galilei, I-00044 Frascati, Italy

—via CD-ROM "NATO Science and Technology Disk" with user-friendly retrieval software in English, French, and German (©WTV GmbH and DATAWARE Technologies, Inc. 1989). The CD-ROM also contains the AGARD Aerospace Database.

The CD-ROM can be ordered through any member of the Board of Publishers or through NATO-PCO, Overijse, Belgium.

Library of Congress Cataloging-in-Publication Data

Quantum field theory and string theory / edited by Laurent Baulieu ...
[et al.].
       p.   cm. -- (NATO ASI series. B, physics ; v. 328)
    "Published in cooperation with NATO Scientific Affairs Division".
    "Proceedings of a NATO Advanced Research Workshop on New
Developments in String Theory, Conformal Models, and Topological
Field Theory, held May 10-21, 1993, in Cargèse, France"--CIP t.p.
verso.
    Includes bibliographical references and index.
    ISBN 978-1-4613-5735-3   ISBN 978-1-4615-1819-8 (eBook)
    DOI 10.1007/978-1-4615-1819-8
    1. Quantum field theory--Congresses. 2. String models-
-Congresses. 3. Quantum gravity--Congresses.   I. Baulieu, Laurent.
II. North Atlantic Treaty Organization. Scientific Affairs
Division. III. NATO Advanced Research Workshop on New Developments
in String Theory, Conformal Models, and Topological Field Theory
(1993 : Cargèse, France) IV. Series: NATO ASI series. Series B,
Physics ; v. 328.
QC174.45.A1Q36265 1994
530.1'43--dc20   94-44814
                                                          CIP

Additional material to this book can be downloaded from http://extra.springer.com.

ISBN 978-1-4613-5735-3

©1995 Springer Science+Business Media New York
Originally published by Plenum Press, New York in 1995
Softcover reprint of the hardcover 1st edition 1995

10 9 8 7 6 5 4 3 2 1

All rights reserved

No part of this book may be reproduced, stored in a retrieval system, or transmitted in any form or by any means, electronic, mechanical, photocopying, microfilming, recording, or otherwise, without written permission from the Publisher

LIST OF SPEAKERS OF THE WORKSHOP CARGESE-93

I.Antoniadis
C. Bachas
M. Bellon
D. Bernard
M. Bershadsky
D. Boulatov
P. Bouwknegt
E. Brezin
S. Dalley
J. -M. Daul
R. Dijkgraaf
M. Douglas
G. Felder
J. -L. Gervais
D. Gross
G. Harris
P. Horava

C. Itzykson
V. Kazakov
A. Kirillov
I. Kostov
E. Martinec
A. Migdal
T. Miva
S. Mukhi
H. Nicolai
E. Rabinovici
F. Ravanini
V. Ogievetski
H. Ooguri
A. Schwimmer
N. Warner
P. Wiegmann
Al. Zamolodchikov

PREFACE

The Cargèse Workshop "Quantum Field Theory and String Theory" was held from May 10 to May 21, 1993.

The broad spectrum of the work presented at the Workshop was the reflection of a time of intensive search for new ways of solving some of the most fundamental problems in string theory, quantum gravity and non-perturbative field theory. A number of talks indicated the emergence of new promising domains of investigation. It is this very diversity of topics which, in our opinion, represents one of the most attractive features of the present volume which we hope will provide a good orientation in the abundant flow of ideas and publications in modern quantum field theory.

Many contributions to the present proceedings are concerned with two dimensional quantum field theory. The continuous advances in the domain of two dimensional integrable theories on the lattice as well as in the continuum, including conformal field theories, Liouville field theory and matrix models of two dimensional quantum gravity are very well represented. Other papers address physically realistic (and therefore very complicated) problems like developed turbulence, the Hofstadter problem, higher dimensional gravity and phenomenological strings. A new elegant class of topological field theories is presented. New ideas in the string representation of multicolor quantum chromodynamics were widely discussed at the Workshop, more particularly the example of the exactly solvable two dimensional case.

We would like to point out that some of the papers included in this volume contain, along with original developments, an extensive review of a particular topic and can serve as a convenient source of information even for the non-specialist or for students.

We would like to thank the many people who contributed to the successful organization of the conference and in particular Marie-France Hanseler for her constant attention to the well being of all the participants. She was once again one of the pillars of the Cargèse Institute.

We are greatly indebted to NATO Division for Scientific Affairs and to *Département de la Formation Permanente* of the Centre National de la Recherche Scientifique for their generous financial support. Partial support was also provided by the Science grant SCI*0394 from the European Commission.

Last, but not least, we would like to thank all the participants for helping create an excellent working atmosphere and especially the contributors to this volume for the quality of their presentation and also for significantly simplifying our task in editing this volume.

Laurent Baulieu
Vladimir Dotsenko
Vladimir Kazakov
Paul Windey

CONTENTS

FRACTAL STRUCTURE IN 4D GRAVITY

I. ANTONIADIS

Centre de Physique Théorique
Ecole Polytechnique
91128 Palaiseau, France

Abstract: We argue that classical general relativity is drastically modified at cosmological distance scales, due to the large quantum fluctuations of the conformal factor. The infrared dynamics is generated by an effective action induced by the trace anomaly, analogous to the Polyakov action in two dimensions, and it describes a scale invariant phase of quantum gravity in the far infrared. We derive scaling relations for the partition function and physical observables, which can in principle be tested in numerical simulations of simplicial four geometries with S^4 topology. In particular, we predict the form of the critical curve in the coupling constant plane, and determine the scaling of the Newtonian coupling with volume which permits a sensible continuum limit.

1   Introduction

At accessible distance scales, from centimeters to light years, gravitational phenomena are described quite accurately by the classical Einstein theory. It is well known that this theory is beset with ultraviolet divergences at the quantum level and is virtually certain to require drastic modification at the Planck scale. A corollary of this severe behavior at ultra-short distances would seem to be mild behavior in the infrared. One does not normally think of quantum fluctuations of the metric field as important at large distances, and indeed in perturbation theory around flat space there is no sign of infrared problems. However, gravitation is a long range force characterized by massless excitations that cannot be shielded. If the classical spacetime is curved on some characteristic distance scale, fluctuations with wavelengths longer than this scale need not remain small, a fact long ago pointed out in Newtonian theory by Jeans. Infrared behavior of fluctuations depends very much on the background geometry, and may force large modifications of the classical background.

A typical example of large long range fluctuations is provided by the dynamics of massless fields in rapidly expanding cosmological backgrounds, such as de Sitter space

Quantum Field Theory and String Theory, Edited by
L. Baulieu *et al.*, Plenum Press, New York, 1995

"

which is the natural maximally symmetric ground state of Einstein gravity with a positive cosmological term. In particular, the graviton propagator grows without bound at distances larger than the horizon length [1, 2], while the spin zero or conformal part of the propagator provides the dominant contribution [2]. This suggests that classical theory is only valid at intermediate distances, larger than the Planck scale but smaller than the horizon, and that in the far infrared regime quantum gravitational effects become important. However, since this region is expected to be insensitive to the ultraviolet structure of the theory, one may hope to say something sensible in the language of low energy effective Lagrangian, without knowledge of Planck scale phenomena.

In order to construct a low energy effective Lagrangian for gravity that incorporates infrared fluctuations correctly, one must decide first of all what is the relevant order parameter at large distance scales. Here observational cosmology comes to our aid to suggest that the FRW scale factor, or more generally, the conformal factor of the metric tensor should play an important role. In the classical Einstein theory this scalar part of the metric does not propagate. It is determined in terms of the matter sources and has no independent dynamics of its own. At the quantum level there is a trace anomaly in the energy-momentum tensor of conformally coupled matter fields. The existence of the conformal anomaly means that the classical constraints which fix the scalar part of the metric fluctuations in terms of matter sources cannot be maintained upon quantization. In other words, the conformal factor becomes *unconstrained* in the full quantum theory.

The low energy effective Lagrangian must be modified accordingly to take account of the trace anomaly and fluctuations of the conformal factor, and this modified theory reanalyzed to discover the correct infrared behavior of the quantum theory of gravity. In two dimensional quantum gravity, or equivalently non-critical string theory, the effective action induced by the trace anomaly modifies the dynamics of the conformal factor at all distance scales in dramatic fashion. Certainly nothing like the fractal behavior and scaling relations of random surfaces in 2D gravity can arise from the classical two dimensional Einstein-Hilbert action which (being a topological invariant) yields no dynamics whatsoever [11, 23].

In ref. [5] we obtained the effective Wess-Zumino action for the conformal factor of the metric induced by the trace anomaly of conformal matter in four dimensions. We analyzed this continuum effective theory showing that it possesses a non-trivial, infrared stable fixed point, characterized by certain anomalous scaling relations. We argued that this fixed point describing a scale invariant phase of 4D quantum gravity is the true ground state of the theory, approached at distance scales much larger than the horizon length of any given classical background. Conformal symmetry, apparently broken by the trace anomaly is restored dynamically by the large fluctuations in the conformal factor at these scales. An important consequence of restoration of scale invariance at large distances is the screening of the effective cosmological term as measured by the average scalar curvature in the ground state. In particular, the two-point correlator of Ricci scalars $\langle R(x)R(x') \rangle$ falls to zero with a certain universal power of the invariant distance between x and x' that depends only on the total number of massless fields in the theory through its effective central charge.

In principle, it should be possible to verify the existence of such a scale invariant phase of 4D quantum gravity by numerical simulations. In ref. [6] we suggested that a comparison of the continuum predictions can be made with simplicial simulations of four-geometries with the topology of S^4 [7, 8]. This topology includes in particular the physically interesting case of Euclidean de Sitter space.

2 The Effective Theory

Let us start with the conformal decomposition of the metric:

$$g_{ab}(x) = e^{2\sigma(x)}\bar{g}_{ab}(x) \, , \tag{1}$$

where $e^{\sigma(x)}$ is the conformal factor and $\bar{g}_{ab}(x)$ is a fixed fiducial metric. By consideration of the general form of the trace anomaly for conformal fields in four dimensions and taking into account the Wess-Zumino integrability condition, we determined the general form of the effective action whose σ variation is the trace anomaly [5, 9]. Treating this effective action as the fundamental quantum action for the σ field at large distances and requiring that general covariance be exactly preserved in the vacuum state of this σ theory, we found that the total trace anomaly of the full theory must vanish. In other words the absence of diffeomorphism anomalies in quantum gravity requires that the vacuum is a scale invariant conformal fixed point where the beta functions of all renormalized couplings are zero.

At the fixed point the effective Euclidean action for σ reads:

$$S_{eff} = \int d^4x \sqrt{\bar{g}} \left\{ \frac{Q^2}{(4\pi)^2}[\sigma \bar{\Delta}_4 \sigma + \tfrac{1}{2}(\bar{G} - \tfrac{2}{3}\bar{\Box}\bar{R})\sigma] - \frac{3}{\kappa}e^{2\alpha\sigma}[\alpha^2(\bar{\nabla}\sigma)^2 + \tfrac{\bar{R}}{6}] + \lambda e^{4\alpha\sigma} \right\} \tag{2}$$

where G is the Gauss-Bonnet integrand whose integral is the Euler number,

$$\chi_E = \frac{1}{32\pi^2} \int d^4x \sqrt{g}\, G \, , \tag{3}$$

and Δ_4 is the Weyl covariant fourth order operator acting upon scalars:

$$\Delta_4 = \Box^2 + 2R^{ab}\nabla_a\nabla_b - \tfrac{2}{3}R\Box + \tfrac{1}{3}(\nabla^a R)\nabla_a. \tag{4}$$

The quantity Q^2 plays the role of the central charge at the infrared fixed point and is proportional to the coefficient of the Gauss-Bonnet term in the quantum trace anomaly. For free fields, it is given by:

$$Q^2 = \frac{1}{180}(N_S + 11N_F + 62N_V - 28 + 1411), \tag{5}$$

where N_S, N_F and N_V stand for the number of massless scalars, Dirac fermions and vectors, respectively. The number -28 denotes the σ-contribution, while 1411 is the one-loop contribution of transverse graviton and reparametrization ghosts using Einstein theory [9].

The physical metric at the conformal fixed point becomes $g_{ab}(x) = e^{2\alpha\sigma(x)}\bar{g}_{ab}(x)$, with α determined by the condition that the Einstein-Hilbert action have its canonical scale dimension in this metric. This condition gives a quadratic equation which fixes α in terms of the central charge [5]:

$$\alpha = \frac{1 - \sqrt{1 - \frac{4}{Q^2}}}{\frac{2}{Q^2}}.$$ (6)

With α determined in this way the condition that the volume (cosmological) term have dimension 4 requires that the cosmological and Newtonian couplings satisfy the relation:

$$\lambda\kappa^2 = \frac{18\pi^2\alpha^2}{Q^2}\left(1 + \frac{4\alpha^2}{Q^2} + \frac{6\alpha^4}{Q^4}\right).$$ (7)

As mentioned above, all of these results were obtained by treating the metric $\bar{g}_{ab}$ as fixed. In other words the transverse, traceless sector of the theory containing the physical spin-2 gravitons was neglected completely. Our basic hypothesis is that these relations remain true in the infrared when the graviton modes are included, up to a possible renormalization of the value of Q^2. In fact, this is not unreasonable from a Wilsonian effective action point of view. Consider the functional integration over transverse gravitons as well as over matter fields, with both infrared and ultraviolet cut-offs, ℓ and a. At short distances, graviton effects presumably become uncontrollable due to the presence of the dimensionful Newtonian coupling κ, so that a cannot be taken to zero in the effective action without introducing higher dimension operators. Conversely, at large distance scales, the quantum effects of the specifically transverse, tracefree fluctuations of the metric are expected to become less important, so that the effective action should be determined by the coefficients of the relevant operators in the infrared. Yet, the conformal anomaly tells us that the metric fluctuations of the *scalar* part of the metric tensor induced by massless fields *never* decouple, and remain relevant at large distance scales. A block spin averaging procedure over larger and larger volumes should produce then a long distance effective action which is regular as the infrared cutoff ℓ is removed to ∞. If this is the case, then an infrared stable renormalization group fixed point of the effective low energy theory is approached as $\ell \to \infty$. Scale invariance at this fixed point requires that the low energy effective action must be of the form (2) when expanded up to four derivatives of σ.

In order to derive the scaling behavior of the partition function of the effective σ theory, let us subject the σ field to the constant shift [23]:

$$\sigma \to \sigma + \frac{\omega}{\alpha},$$ (8)

and use the translational invariance of the integration measure $[\mathcal{D}\sigma]$ to find:

$$Z(\kappa, \lambda) \equiv \int [\mathcal{D}\sigma] e^{-S_{eff}[\sigma]} = e^{-\frac{Q^2}{\alpha}\chi_E\omega} Z(\kappa e^{-2\omega}, \lambda e^{4\omega}).$$ (9)

In what follows we restrict to the topology S^4 for which the Euler number $\chi_E = 2$. It is convenient to define also the partition function at fixed volume

$$Z(\kappa; V) \equiv \int [\mathcal{D}\sigma] e^{\lambda V - S_{eff}[\sigma]} \delta\left(\int d^4x \sqrt{\bar{g}} e^{4\alpha\sigma} - V\right).$$ (10)

Then performing the translation (8) above, we obtain [6]:

$$
\begin{aligned}
Z(\kappa; V) &= e^{-2(\frac{Q^2}{\alpha}+2)\omega} Z(\kappa e^{-2\omega}, e^{-4\omega}V) \\
&= V^{-\frac{Q^2}{2\alpha}-1} \tilde{Z}(\kappa V^{-\frac{1}{2}}) ,
\end{aligned}
\tag{11}
$$

where in the second line we put $e^{4\omega} \propto V$.

3 Numerical Simulations

3.1 Simplicial Gravity

The numerical method that has proven most fruitful up until now is that of "dynamical triangulation", a variation of Regge calculus in which geometries are constructed by gluing together fundamental simplices of fixed volume [10]. The four-simplices share common faces with their neighbors which are regular tetrahedra of edge length a. The volume of a fundamental J-simplex is:

$$
V_J = a^J \Omega_J = \frac{a^J}{J!}\sqrt{\frac{J+1}{2^J}},
\tag{12}
$$

where a is the lattice spacing, so that the total volume of the simplicial manifold is

$$
\int d^4x \, \sqrt{g} \to N_4 V_4
\tag{13}
$$

with N_4 the total number of 4-simplices in the configuration.

As in the Regge approach, curvature is residing on the two-dimensional hinges, $i.e.$ equilateral triangles with the same fixed edge length. If n_i is the number of 4-simplices sharing a given equilateral triangle i, then the deficit angle δ_i is given by

$$
\delta_i = 2\pi - n_i\theta ,
\tag{14}
$$

where $\theta = \arccos(1/4) \simeq 1.3$ is the angle between two tetrahedra sharing a triangle. Then, the Einstein-Hilbert action takes the value:

$$
\int d^4x \, \sqrt{g} \, R \to \sum_i \delta_i V_2 = (2\pi N_2 - 10\theta N_4)V_2,
\tag{15}
$$

where we used $\sum_i n_i = 10N_4$ (since each 4-simplex has 10 triangles in its boundary).

By inspection of Eqs. (15) and (13), one considers the lattice action:

$$
S(\mathcal{T}) = -k_2 N_2(\mathcal{T}) + k_4 N_4(\mathcal{T}) ,
\tag{16}
$$

for each simplicial triangulation $\mathcal{T}$. Dynamics is then specified by summing over all triangulations weighted by this action with each triangulation otherwise having equal statistical weight. In this approach there is no diffeomorphism invariance at the lattice level, and no gauge-fixing problems. As a corollary, neither is it self evident that the statistical model(s) constructed by the dynamical triangulation procedure lie in the same universality class as quantum gravity in four dimensions. However, naive comparison of (16) with the Einstein-Hilbert action, using the substitutions (13), (15) and (12), leads to the following identification of the bare simplicial action parameters with the (unrenormalized) parameters of the continuum theory:

$$k_2 = \frac{\sqrt{3}\pi}{4}\frac{a^2}{\kappa}$$

$$k_4 = \frac{5\sqrt{3}\theta}{4}\frac{a^2}{\kappa} + \frac{\sqrt{5}}{96}\lambda a^4 \; . \tag{17}$$

A continuum infrared fixed point of the lattice theory can be obtained, in principle, by block averaging over larger and larger sublattices, thereby determining the mapping from the bare lattice action with parameters (k_2, k_4) to the renormalized parameters of the effective action at larger and larger distance scales. The fixed point of this mapping, if it exists, is the infrared fixed point of the continuum theory. In practice, in the dynamical triangulation approach one does not need to perform the block averaging procedure explicitly, since by taking more and more four-simplices (*i.e.* $N_4 \to \infty$), one is effectively averaging over larger and larger sublattices in any case. Since by definition, at the fixed point the parameters of the effective action do not change under further block averaging, the bare parameters (k_2, k_4) should become *identical* to the renormalized parameters in this limit of infinite lattice size, $N_4 \to \infty$. For this reason, at the infrared fixed point we shall not distinguish between the bare parameters of the lattice action (16) and the renormalized or dressed parameters of the effective action in the large volume, scaling limit.

We do not add any higher derivative couplings to the action, since these should correspond to irrelevant operators in the infrared, with vanishing coefficients at the fixed point of the infrared renormalization group. Indeed, in the continuum theory, the coefficients of possible R^2 and Weyl-squared terms in the action vanish at the conformal fixed point [5, 9]. The trace anomaly induced action (2) should not be added to the bare lattice action (16) either. It is *nonlocal* in the full metric (1) and should be generated *dynamically* by the quantum fluctuations of the simplicial geometries, in analogy with the situation in the two dimensional case.

3.2 The Critical Curve

Because the number of triangulations with fixed S^4 topology which can be made from a given number N_4 of 4-simplices is exponentially bounded with respect to N_4 [7, 8], the partition function of the dynamical triangulation approach,

$$Z_{DT}(k_2, k_4) \equiv \sum_{T} e^{-S(T)} = \sum_{N_4} Z(k_2; N_4)e^{-k_4 N_4} \lesssim \sum_{N_4} e^{-[k_4 - k_4^c(k_2)]N_4} \tag{18}$$

must exist in a region of the coupling constant plane $k_4 > k_4^c(k_2)$. By approaching the boundary of this region from above one can hope to arrive at a continuum limit in which physical correlation lengths go to infinity when expressed in lattice units. In this precise sense, one is searching for a "critical" curve in the (k_2, k_4) plane corresponding to the existence of a non-trivial infinite volume limit of the lattice theory defined by (16) and (18). Only on this curve, and as we shall see below, only at one point on this curve can one hope to find a second order phase transition where correlation lengths go to infinity in units of the lattice spacing and the continuum limit of the theory may be defined.

A prediction for the critical curve can be obtained by combining the two equations (17) so that the lattice spacing is removed:

$$
\begin{aligned}
k_4^c(k_2) &= \frac{5\theta}{\pi}k_2 + \frac{\sqrt{5}}{18\pi^2}(\lambda\kappa^2)k_2^2 \\
&\simeq 2.1k_2 + 0.03(\lambda\kappa^2)k_2^2 .
\end{aligned}
\tag{19}
$$

We emphasize that this relation should hold for the *bare* lattice parameters only in the strictly *infinite* volume continuum limit, where the fixed point identification of bare and renormalized parameters is valid. In any realistic finite volume simulation we must expect that there will be unknown finite volume corrections to the form of the curve, so that the practical utility of the relation (19) depends crucially on the ability to do simulations with very large lattices where the finite volume effects are both controllable and small. We are encouraged by the preliminary results of the simulations with $N_4 \sim 10^4$, which seem to indicate a critical curve $k_4^c(k_2)$ which is approximately linear with a slope slightly more than 2 [11]. The relation (19) has a quadratic term as well which could be in principle used to determine Q^2 using the fixed point relation (7). However, this term seems to be suppressed and in practice high enough statistics is required.

3.3 Continuum Limit

The main prediction of infrared conformal dominance is provided by the finite volume scaling relation (11). Translating this continuum relation to the lattice by using (17), we obtain the following scaling relation for the fixed volume partition function at large volumes [6]:

$$
Z(k_2; N_4)e^{-k_4^c(k_2)N_4} \sim \tilde{Z}(\tilde{k}_2)N_4^{\gamma-3} ,
\tag{20}
$$

where

$$
\tilde{k}_2 = k_2\sqrt{N_4} ,
\tag{21}
$$

and

$$
\gamma(Q^2) = 2 - \frac{Q^2}{2\alpha} .
\tag{22}
$$

The content of the scaling relation (20) is that, when k_2 is scaled to zero with the square root of the volume as in (21) keeping $\tilde{k}_2$ fixed, the prefactor $\tilde{Z}$ becomes volume independent and the entropy exponent γ depends only on the effective central charge Q. Therefore, a clear test of this scaling relation is that γ must be independent of the parameter $\tilde{k}_2$, if our hypothesis of infrared conformal dominance is correct. Otherwise, γ would acquire non-trivial $\tilde{k}_2$ dependence. If γ is indeed independent of $\tilde{k}_2$, a measurement of γ would provide a non-perturbative way to compute the graviton contribution to the central charge. Note also that the entropy exponent goes to $-\infty$ in the semiclassical limit $Q^2 \to \infty$, while $\gamma = 1$ for $Q^2 = 4$. In the latter case one would expect logarithmic behavior in analogy with the $c = 1$ case in two dimensions. For $Q^2 < 4$ the exponent α in (6) becomes complex and the theory could exhibit a phase transition with qualitatively new phenomena. Perturbative calculations of the graviton contributions [9] lead to the value $\gamma \sim -1.3$.

We are proposing therefore that the procedure for finding the infrared fixed point of quantum gravity in the continuum limit from the dynamical triangulation starting point is to first locate the critical curve in the (k_2, k_4) plane which should approximate (19) for large enough volumes. Then one should attempt to run simulations with even larger volumes, moving along the critical curve towards the origin by rescaling $k_2 \to 0$ as $N_4 \to \infty$ in accordance with (21), keeping $\tilde{k}_2$ fixed. In this double scaling limit one can then test the validity of the scaling behavior (20) and measure γ for pure quantum gravity at its infrared fixed point.

In addition to the scaling of the finite volume partition function, one may consider a variety of physical observables as well. The simplest observable with nice scaling behavior is the average curvature $\langle R \rangle$ defined by [7]:

$$\langle R \rangle \equiv \left\langle \frac{\int d^4 x \sqrt{g} R}{\int d^4 x \sqrt{g}} \right\rangle = \frac{2\pi V_2}{V_4} \left[\frac{1}{N_4} \frac{\partial}{\partial k_2} \ln Z(k_2; N_4) - \frac{5\theta}{\pi} \right] , \tag{23}$$

where the last equality holds for fixed volume and we used the relations (13), (15) and (16). Now inserting the scaling behavior (20) and the expression (19) one obtains:

$$\begin{aligned}
\langle R \rangle &= \frac{2\pi V_2}{V_4} \left[2\sqrt{5} f(Q^2) k_2 + \frac{1}{\sqrt{N_4}} \frac{\partial}{\partial \tilde{k}_2} \ln \tilde{Z}(\tilde{k}_2) \right] \\
&\sim \frac{1}{\sqrt{N_4}} \to 0 ,
\end{aligned} \tag{24}$$

where the proportionality factor depends only on Q^2 and the rescaled $\tilde{k}_2$. This shows that the average curvature scales to zero with the square root of the volume.

Hence scaling $k_2 \to 0$ with $\tilde{k}_2$ fixed yields a large volume continuum limit consistent with naive dimensional analysis. This removes the main obstruction to the interpretation of the numerical simulations, mentioned in ref. [7], since from naive scaling, $\langle R \rangle_{\text{lattice}} = a^2 \langle R \rangle_{\text{continuum}}$ should vanish in the continuum limit. Yet this does not seem to be the case in the simulations for any non-zero value of k_2^c, such as that suggested by the apparent divergence of the specific heat. For that reason, previous attempts to identify an *ultraviolet* fixed point of the lattice theory at $k_2^c \neq 0$ failed. The numerical evidence to date suggests that $\langle R \rangle_{\text{lattice}}$ vanishes *only* at $k_2 = 0$, scaling linearly to zero with k_2 [7]. This is completely consistent with the scaling $k_2 \sim N_4^{-1/2} \to 0$, and provides an independent indication of the correctness of the double scaling limit proposed in (21) to reach the *infrared* fixed point.

References

[1] L.H. Ford, Phys. Rev. D31 (1985) 710; I. Antoniadis, J. Iliopoulos, T.N. Tomaras, Phys. Rev. Lett. 56 (1986) 1319; B. Allen and M. Turyn, Nucl. Phys. B292 (1987) 813; E.G. Floratos, J. Iliopoulos, and T.N. Tomaras, Phys. Lett. B197 (1987) 373.

[2] I. Antoniadis and E. Mottola, Jour. Math. Phys. 32 (1991) 1037.

[3] A. M. Polyakov, Phys. Lett. B103 (1981) 207; V. G. Knizhnik, A. M. Polyakov, and A. B. Zamolodchikov, Mod. Phys. Lett. A3 (1988) 819.

[4] F. David, Mod. Phys. Lett. A3 (1988) 1651; J. Distler and H. Kawai, Nucl. Phys. B321 (1989) 509.

[5] I. Antoniadis and E. Mottola, Phys. Rev. D45 (1992) 2013.

[6] I. Antoniadis, P. O. Mazur and E. Mottola, Ecole Polytechnique preprint CPTH-A214.1292.

[7] J. Ambjørn and J. Jurkiewicz, Phys. Lett. B278 (1992) 42; J. Ambjørn, Z. Burda, J. Jurkiewicz, and C. Kristjansen, Acta Phys. Polon. B23 (1992) 991; J. Ambjørn, J. Jurkiewicz, and C. Kristjansen, Nucl. Phys. B393 (1993) 601.

[8] M. E. Agishtein and A. A. Migdal, Mod. Phys. Lett. A7 (1992) 1039; Nucl. Phys. B385 (1992) 395.

[9] I. Antoniadis, P. O. Mazur and E. Mottola, Nucl. Phys. B388 (1992) 627.

[10] J. Ambjørn, B. Durhuus, J. Frohlich and P. Orland, Nucl. Phys. B270 (1986) 457; A. Billoire and F. David, Nucl. Phys. B275 (1986) 617; D. V. Boulatov, V. A. Kazakov, I. K. Kostov and A. A. Migdal, Nucl. Phys. B275 (1986) 641.

[11] J. Ambjørn, private communication; B. Brügmann, Phys. Rev. D47 (1993) 3330 [*cf.* Fig. 1].

A ONE DIMENSIONAL IDEAL GAS OF SPINONS,
OR SOME EXACT RESULTS ON THE XXX SPIN CHAIN
WITH LONG RANGE INTERACTION

D. BERNARD, V. PASQUIER and D. SERBAN

Service de Physique Théorique de Saclay [1]
F-91191, Gif-sur-Yvette, France

Abstract: We describe a few properties of the XXX spin chain with long range interaction. The plan of these notes is :
1 — The Hamiltonian.
2 — Symmetry of the model.
3 — The irreducible multiplets.
4 — The spectrum.
5 — Wave functions and statistics.
6 — The spinon description.
7 — The thermodynamics.

Introduction. The XXX spin chain with long range interaction is a variant of the spin half Heisenberg chain, with exchange inversely proportional to the square distance between the spins. It possesses the remarkable properties that its spectrum is additive and that the elementary excitations are spin half objects obeying a half-fractional statistics intermediate between bosons and fermions. In this sense, it gives a model for an ideal gas of particles with fractional statistics . The model is gapless; its low energy properties belong to the same universality class as the Heisenberg model, and are described by the level one $su(2)$ WZW conformal field theory.

1   The Hamiltonian.

The Hamiltonian of the trigonometric isotropic spin chain with long range interaction is given by [1, 2] :

$$H = \left(\frac{\pi}{N}\right)^2 \sum_{i \neq j} \frac{(P_{ij} - 1)}{\left(2 \sin \frac{\pi(i-j)}{2N}\right)^2} \tag{1}$$

Quantum Field Theory and String Theory, Edited by
L. Baulieu *et al.*, Plenum Press, New York, 1995

where P_{ij} is the operator which exchange the spins at the sites i and j. We restrict ourselves to the $su(2)$ case, in which case the spin variables can only take two values: $\sigma_i = \pm$. The sum is over all the distinct pairs of sites labeled by integers $i, j, \cdots$ ranging from 1 to N.

The spectrum of (1), which has been conjectured by Haldane [3], possesses a remarkable additivity property as well as a rich degeneracy. It can be described as follow. To each eigenstate multiplet is associated a set of rapidities $\{m_p\}$ which are non-consecutive integers ranging from 1 to $(N-1)$. The energy of an eigenstate $|\{m_p\}\rangle$ with rapidities $\{m_p\}$ is:

$$H|\{m_p\}\rangle = \left(\sum_p \epsilon(m_p)\right)|\{m_p\}\rangle \quad \text{with} \quad \epsilon(m) = \left(\frac{\pi}{N}\right)^2 m(m-N) \qquad (2)$$

The degeneracy of the multiplet with rapidities $\{m_p\}$ is described by its $su(2)$ representation content as follows. Encode the rapidities in a sequence of $(N-1)$ labels 0 or 1 in which the 1's indicate the position of the rapidities; add two 0's at both extremities of the sequence which now has length $(N+1)$. Since the rapidities are never equal nor differ by a unit, two labels 1 cannot be adjacent. A sequence can be decomposed into the product of elementary motifs. A motif is a series of Q consecutive 0's, and it corresponds to a spin $\frac{Q-1}{2}$ representation of $su(2)$. The representation content of a sequence is then the tensor product of its motifs.

The degeneracy of the spectrum can also be described in terms of path, in a way surprisingly similar to the path description of the six-vertex corner transfer matrix [4].

2 Symmetry of the Model

The symmetry algebra responsible for the degeneracy of the model was identified as the $su(2)$ Yangian [5]. A Yangian is an infinite dimensional associative algebra generated by elements T_n^{ab}, with n a positive integer, and $a, b = \pm$ in the $su(2)$ case. These generators satisfy quadratic relations which can be arranged into an Yang-Baxter equation by introducing the transfer matrix $T(x)$, with matrix elements, $T^{ab}(x) = \delta^{ab} + \sum_{n \geq 0} x^{-n-1} T_n^{ab}$. The commutation relations then take the following form [6, 7] :

$$R(x-y)\,(1 \otimes T(x))\,(T(y) \otimes 1) = (T(y) \otimes 1)\,(1 \otimes T(x))\,R(x-y) \qquad (3)$$

The matrix $R(x)$ is the solution of the Yang-Baxter equation given by: $R(x) = x + P$, where P is the permutation operator which exchanges the two auxiliary spaces. The transfer matrix was constructed in [8] . Its expression is :

$$T^{ab}(x) = \delta^{ab} + \sum_{i,j=1}^{N} X_i^{ab}\left(\frac{1}{x - L}\right)_{ij} \qquad (4)$$

with $L_{ij} = (1 - \delta_{ij})\theta_{ij}P_{ij}$, $\theta_{ij} = z_i/z_{ij}$ with $z_{ij} = z_i - z_j$, and X_i^{ab} is the canonical matrix $|a\rangle\langle b|$ acting on the ith spin only. The transfer matrix (4) form a representation of the exchange algebra (3) for any values of the complex numbers z_j. The center of the

$su(2)$ Yangian algebra (3) is generated by the so-called quantum determinant $Det_q T(x)$ defined by [9]:

$$Det_q T(x) = T_{--}(x-1)T_{++}(x) - T_{-+}(x-1)T_{+-}(x) \tag{5}$$

In the representation (4) , the quantum determinant is a pure number for any values of the z_j's given by :

$$Det_q\, T(x) \;=\; 1 + \sum_{i,j=1}^{N} \Big(\frac{1}{x-\Theta}\Big)_{ij} \;=\; \frac{\Delta_N(x+1)}{\Delta_N(x)} \tag{6}$$

with $\Delta_N(x)$ the characteristic polynomial of the $N \times N$ matrix Θ with entries θ_{ij}: $\Delta_N(x) = \det(x - \Theta)$.

The trigonometric spin chain corresponds to $z_j = \omega^j$ with ω a primitive N^{th} root of the unity. For these values of z_j, the transfer matrix (4) commutes with the Hamiltonian (1). For $z_j = \omega^j$, the matrix Θ can be diagonalized giving the following expression for Δ_N :

$$\Delta_N(x) = \prod_{j=1}^{N}\Big(x + \frac{N+1}{2} - j\Big) \tag{7}$$

3 The Irreducible Multiplets

Solving the model consists in finding all the irreducible components of the Yangian symmetry algebra and computing the energy in each of these blocks. For the values of the z_j's induced by the spin chain, $z_j = \omega^j$, the representation (4) is reducible. It is completely reducible since the transfer matrix is hermitic: $t_n^{ab\,\dagger} = t_n^{ba}$. Each irreducible sub-representation possesses a unique highest weight (h.w.) vector $|\Lambda\rangle$ which is annihilated by $T_{+-}(x)$ and which is an eigenvector of the diagonal components $T_{\pm\pm}(x)$ of the transfer matrix :

$$T(x)|\Lambda\rangle = \begin{pmatrix} t_{++}(x) & 0 \\ \star & t_{--}(x) \end{pmatrix} |\Lambda\rangle$$

Here, $t_{\pm\pm}(x)$ are rational functions in x, but not operators. Since the quantum determinant (4) take the same value in any of the irreducible block, these two functions are related by :

$$\frac{\Delta_N(x+1)}{\Delta_N(x)} = t_{--}(x-1)t_{++}(x)$$

Hence, only one of them, say $t_{--}(x)$, is independent. It uniquely characterizes the $su(2)$ Yangian representation. We therefore have to compute all the functions $t_{--}(x)$ arising from the decomposition of the Yangian representation induced by the spin chain, but also to identify the h.w. vectors in order to be able to compute the energy spectrum.

Obviously, the ferromagnetic vacuum $|\Omega\rangle = |++\cdots++\rangle$ is a h.w. vector: the corresponding $t_{--}(x)$ is one, and the energy is zero. The h.w. vectors in the one-magnon sector are $|m\rangle = \sum_j \omega^{mj}\sigma_j^-|\Omega\rangle$, with, $1 \le m \le (N-1)$: the corresponding eigenvalue is $t_{--}(x) = \frac{P_1(x+1)}{P_1(x)}$, with $P_1(x) = (x + \frac{N+1}{2} - m)$, and the one-magnon energy is $\epsilon(m) = \left(\frac{\pi}{N}\right)^2 m(m-N)$.

In order to determine all the highest weight vectors, we decompose the hilbert space into subspaces of fixed magnon number. A M-magnon state $|\Psi\rangle$ has M spin reversed :

$$|\Psi\rangle = \sum_{n_1,\cdots,n_M} \psi_{n_1,\cdots,n_M} \, \sigma_{n_1}^- \cdots \sigma_{n_M}^- |\Omega\rangle \tag{8}$$

where σ_n^a denote the Pauli matrices acting on the spin located on the site n. By construction, the coefficients $\psi_{n_1,\cdots,n_M}$ of the M-magnon wave functions are symmetric in their indices. The wave function coefficients are unspecified for two coincident indices $\psi_{\cdots,n,\cdots,n,\cdots}$. By convention, we choose these coefficients to be zero.

Since these indices range from 1 to N, to any M-magnon state is associated a symmetric polynomial $\Psi(z_1,\cdots,z_M)$ in M variables of degree less than $(N-1)$ such that $\Psi(\omega^{n_1},\cdots,\omega^{n_M}) = \psi_{n_1,\cdots,n_M}$. In the following, we restrict ourselves to the class of magnon states deriving from polynomials of the following form :

$$\Psi(z_1,\cdots,z_M) = \prod_{p<q}(z_p - z_q)^2 R(z_1,\cdots,z_M) \tag{9}$$

with $R(z_1,\cdots,z_M)$ a symmetric polynomial of degree less than $(N - 2M + 1)$. This class of states does not include all the states of the spin chain but, as we will see, all the highest weight vectors are in this class.

As explained in the Appendix, the operators $T_{--}(x)$ and $T_{+-}(x)$ act on this class of states. Therefore, the action of these operators on these magnon states induces an action on the polynomials. As shown in the Appendix, we find :

$$T_{--}(x)\,\Psi(z.) = \left(1 + \sum_{p=1}^{M} \frac{1}{x + \frac{N+1}{2} - D_p}\right)\Psi(z.) \tag{10}$$

and

$$T_{+-}(x)\,\Psi(z.) = \left(1 + \sum_{p=1}^{M} \frac{1}{x + \frac{N+1}{2} - D_p}\right)\Psi(z_1 = 0, z.) \tag{11}$$

Here we have introduced differential operators D_p which have recently been proved useful in the Calogero-Sutherland model [10, 11, 12, 8]. For the following, we also need another set of differential operators $\widehat{D}_p$. Both are defined by :

$$\begin{aligned}
D_p &= z_p \partial_{z_p} + \sum_{p \neq q} \theta_{pq} K_{pq} \\
\widehat{D}_p &= z_p \partial_{z_p} + \sum_{q>p} \theta_{pq} K_{pq} - \sum_{q<p} \theta_{qp} K_{pq}
\end{aligned} \tag{12}$$

Here, $\theta_{pq} = z_p/z_{pq}$ and, the operator K_{pq} exchanges the positions: $K_{pq} z_p = z_q K_{pq}$. The differential D_p are covariant under permutation of the positions, $K_{pq} D_p = D_q K_{pq}$, whereas the $\widehat{D}_p$'s are not. On the other hand, the differentials $\widehat{D}_p$ commute, $\left[\widehat{D}_p, \widehat{D}_q\right] = 0$, but the differentials D_p's does not: $\left[D_p, D_q\right] = (D_p - D_q)K_{pq}$. The sum of the mth

powers of both differentials form two sets of commuting operators. However, they are not independent thanks to the following relation:

$$1 + \sum_{p=1}^{M} \frac{1}{x - D_p} = \left(1 + \frac{1}{x - \widehat{D}_1}\right) \cdots \left(1 + \frac{1}{x - \widehat{D}_M}\right) \tag{13}$$

In Eq. (13) it is understood that the operators are acting on functions symmetric in their arguments.

¿From eq. (11), we learn that the highest weight vectors correspond to polynomials $\Psi(z)$ with $R(z)$ given by :

$$R(z_1, \cdots, z_M) = (\prod_{p=1}^{M} z_p)\phi(z_1, \cdots, z_M) \tag{14}$$

with $\phi(z_1, \cdots, z_M)$ a symmetric polynomial of degree less than $(N - 2M)$. Thus, in the M-magnon sector, there are $\frac{(N-M)!}{M!(N-2M)!}$ independent highest weight vectors.

¿From eq. (10), we learn that the eigenfunctions of $T_{--}(x)$ are the eigenvectors of the commuting hamiltonians of the Calogero-Sutherland model, or equivalently of the commuting operators $\widehat{D}_p$. The symmetric eigenfunctions $\Psi^{\{m_p\}}(z.)$ with,

$$\sum_p (\widehat{D}_p)^n \; \Psi^{\{m_p\}}(z.) = \sum_p m_p^n \; \Psi^{\{m_p\}}(z.),$$

are polynomials with degree between 1 and $(N - 1)$ if $1 \leq m_p \leq (N - 1)$. Hence, the M-magnon highest weight vectors $|\{m_p\}\rangle$, with wave function given by $\Psi^{\{m_p\}}(z.)$, are labeled by sets of M integers $\{m_1, \cdots, m_M\}$. Due to the Vandermond prefactor in eq.(9), these integers never coincide nor differ by a unit. Using the factorisation relation (13), one find that the value of $t_{--}(x)$ for these highest weight vectors are :

$$t_{--}(x) = \frac{P_1(x+1)}{P_1(x)} \quad \text{with} \quad P_1(x) = \prod_{p=1}^{M}(x + \frac{N+1}{2} - m_p) \tag{15}$$

The dimension of the irreducible multiplets are encoded in the transfer matrix eigenvalues. The eigenvalues $t_{--}(x)$ are given by eq. (15). The remaining eigenvalues $t_{++}(x)$ are computed from the relation (4). They can also be written in a product form. The result is :

$$T(x)|\{m_p\}\rangle = \frac{P_1(x+1)}{P_1(x)} \begin{pmatrix} \frac{P_0(x+1)}{P_0(x)} & 0 \\ \star & 1 \end{pmatrix} |\{m_p\}\rangle \tag{16}$$

with $P_1(x)$ given in eq.(15), and $P_0(x)$ and $P_1(x)$ factorize $\Delta_N(x)$:

$$\Delta_N(x) = P_0(x) \; P_1(x)P_1(x - 1). \tag{17}$$

One can check that the factorization equation (17) admits solutions only if the roots of $P_1(x)$ are not adjacent. This provides one way to recover the rapidities selection rules.

Let us decompose the sequence of rapidities $\{m_p\}$ in elementary motifs as explained in Section 1. To each motif of length Q, we associate a canonical transfer matrix defined by :

$$T^{ab}_{motif}(x) = \delta^{ab} + \frac{S^{ab}}{x - x_0} \tag{18}$$

where S^{ab} are the matrices forming the spin $\frac{Q-1}{2}$ representation of $su(2)$ and x_0 is the position of the most left label 0 of the motif. It is easy to check that the matrix (18) satisfy the commutation relations (3) . The representation induced by the transfer matrix (16) is then seen to be equivalent to the irreducible tensor product of the transfer matrices associated to each motifs:

$$T(x) \cong \bigotimes_{motifs} T_{motif}(x + \frac{N+1}{2}) \tag{19}$$

This is proved by comparing the eigenvalues of the diagonal transfer matrix elements on the h.w. vector. Therefore, we find that the multiplet of a rapidity sequence is the tensor product of each of its motifs.

We have found one (and only one) irreducible representation for each rapidity sequence. Their direct sum is a vector space of dimension 2^N. Therefore, it fills the Hilbert space of the spin chain, and there is no other irreducible multiplet.

4 The Spectrum

Since all the irreducible multiplets are now identified, finding the spectrum consists in computing the action of the Hamiltonian on the highest weight vectors. The Hamiltonian (1) is $su(2)$ invariant, hence it acts on the M-magnon subspace. In the magnon basis, this action is :

$$\begin{aligned}
(H\psi)_{n_1,\cdots,n_M} = \ & - \ 2\left(\frac{\pi}{N}\right)^2 \sum_p \sum_{k_p \neq n_p} \frac{\omega^{k_p}\omega^{n_p}}{(\omega^{k_p} - \omega^{n_p})^2} \left(\psi_{n_1,\cdots,k_p,\cdots,n_M} - \psi_{n_1,\cdots,n_p,\cdots,n_M}\right) \\
& - \left(\frac{\pi}{N}\right)^2 \sum_{pq} \frac{\omega^{n_p}\omega^{n_q}}{(\omega^{n_p} - \omega^{n_q})^2} \psi_{n_1,\cdots,n_M}
\end{aligned}$$

The Hamiltonian act on the state of the form (9). Using eq. (43) given in the Appendix, one realizes that the action induced on the polynomials $\Psi(z)$ is :

$$\begin{aligned}
(H_M\Psi)(z) &= \left(\frac{\pi}{N}\right)^2 \left(\sum_{p=1}^M z_p \partial_{z_p}(z_p \partial_{z_p} - N) + 4 \sum_{p<q} \frac{z_p z_q}{z_{pq} z_{qp}} \right) \Psi(z) \\
&= \left(\frac{\pi}{N}\right)^2 \sum_{p=1}^M \widehat{D}_p(\widehat{D}_p - N)\Psi(z) \tag{20}
\end{aligned}$$

In Eq. (20) one recognizes the Calogero-Sutherland Hamiltonian at a special value of the coupling constant. In other words, the spin chain in the M-magnon sector has been mapped on the M-body Calogero problem. The last equality in (20), gives the energy of a multiplet $\{m_p\}$:

$$E(\{m_p\}) = \sum_p \left(\frac{\pi}{N}\right)^2 m_p(m_p - N)$$

This completes the proof of the spectrum.

16

5 Wave Functions and Statistics

Only the wave functions of the h.w. vectors are relevant since those of their descendents are obtained by recursive action of the transfer matrix. We now show how recent results on the Calogero models can be used to find explicit expressions for these wave functions. The latter are based on the construction of operators, which we denote by Λ_M in the M-magnon sector, intertwining the Calogero Hamiltonian and the free Hamiltonian:

$$H\,\Lambda_M = \Lambda_M \Delta \quad \text{with} \quad \Delta = \sum_{p=1}^{M} z_p \partial_{z_p}(z_p \partial_{z_p} - N) \tag{21}$$

These intertwiners were defined in [10, 13]. One of their definitions is the following Vandermond determinant of the operators D_p :

$$\Lambda_M = \sum_{\sigma \text{ perm.}} \text{sign}(\sigma)\, D_{\sigma_{M-1}}^{M-1} \cdots D_{\sigma_2}^2 D_{\sigma_1}$$

In this formula, it is understood that Λ_M is acting on antisymmetric functions. For example, for two magnons: $\Lambda_2 = z_1\partial_{z_1} - z_2\partial_{z_2} - \frac{z_1 + z_2}{z_1 - z_2}$.

The operators Λ_M are antisymmetric. Therefore, the symmetric wave functions are obtained by acting first with the antisymmetrizer on the plane waves $z_1^{m_1} \cdots z_M^{m_M}$, and then with Λ_M:

$$\Psi(z_1, \ldots, z_M) = \Lambda_M \left(\text{Det}(z_p^{m_q})_{pq} \right) \tag{22}$$

It is easy to check that the wave functions (22) are symmetric polynomials vanishing at coincident points. If the rapidities are such that $1 \leq m_p \leq (N-1)$, these polynomials have degree less than $(N-1)$ and satisfy the condition (14). They thus are the wave functions of the h.w. vectors. In other words, since the plane waves z^m are the wave functions of the one-magnon h.w. vectors, the operator $\mathcal{D} = \Lambda_M \circ \text{Det}$ map tensor products of M one-magnon states into M-magnon states:

$$|m_1\rangle \otimes \cdots \otimes |m_M\rangle \xrightarrow{\mathcal{D}} |\{m_1, \cdots, m_M\}\rangle \tag{23}$$

This map is a generalization of the Slater determinant in the sense that it implements the rapidity selection rules: if two of the rapidities m_p and m_q are either equal or differ by a unit, then the resulting wave function vanish identically. The fact that the result vanish if two of the rapidities coincide is obvious from the definition, while the fact that it vanishes if they differ by a unit results from an explicit check on the two-magnon case (which has all generality thanks to the symmetry of the wave functions and the recursive definition of Λ_M given in [13]).

6 The Spinon Description

The magnons are the excitations over the ferromagnetic vacuum; the excitations over the antiferromagnetic vacuum are conveniently described in terms of spinons .

For N even, the antiferromagnetic vacuum corresponds to the alternating sequence of symbols $010101\cdots010$. Its rapidities sequence is $\{m_j^0 = 2j - 1\}_{j=1}^{N/2}$. The excitations are obtained by flipping and moving the symbols 0 and 1. We classify the sequence by their number M of rapidities. The spinon number N_{sp} of a sequence is then defined by $M = \frac{N - N_{sp}}{2}$. Since M is an integer, $(N - N_{sp})$ is always even. The spin S_z of the Yangian highest weight vector of the sequence is $S_z = \frac{1}{2}(N - 2M) = \frac{N_{sp}}{2}$.

A sequence of rapidities $\{m_j; j = 1, \cdots, M = \frac{N - N_{sp}}{2}\}$, in the N_{sp} sector, can be decomposed into $(M + 1)$ elementary motifs. As we defined them in Sect.1, an elementary motif is a series of consecutive 0. We will think about the elementary motifs as the possible orbitals for spin half objects, which are called spinons. At fixed N_{sp}, there are $(1 + \frac{N - N_{sp}}{2})$ orbitals. To a spinon in the jth orbital, with $j = 0, \cdots, \frac{N - N_{sp}}{2}$, we assign a momentum $k = \frac{2\pi j}{N}$. Thus, the spinon momenta vary from zero to $k_0 = \frac{2\pi}{N}\left(\frac{N - N_{sp}}{2}\right)$. By convention, a sequence of rapidities $\{m_j\}$ corresponds to the filling of the $(1 + \frac{N - N_{sp}}{2})$ orbitals with respective occupation numbers $n_j = n_j^+ + n_j^- = (m_{j+1} - m_j - 2)$, with $m_0 = -1$ and $m_{M+1} = N + 1$. The length of the jth elementary motif is then $Q_j = n_j + 1$. By construction, the total occupation number is the spinon number :

$$\sum_{j=0}^{\frac{N - N_{sp}}{2}} (n_j^+ + n_j^-) = N_{sp} \tag{24}$$

Since an elementary motif of length Q corresponds to a spin $\frac{Q-1}{2}$ representation of $su(2)$, the full degeneracy of the sequences is then recovored by assuming that, at fixed spinon number, the spinon behaves as bosons. Notice that this property is specific to the $su(2)$ spin chain.

The spinons are not bosons but "semions" since the number of available orbitals varies with the total occupation number [14]. In particular, the spinons are always created by pairs.

The energy of a collection of $N_{sp} = N - 2M$ spinons is :

$$E - E_{vac} = E_0(M) \;+\; \sum_j \sum_\sigma 2(M - j)(M - \frac{N}{2} + j)n_j^\sigma \tag{25}$$
$$+ \; \sum_{j,j'} \sum_{\sigma\sigma'} (M - \sup(j, j'))n_j^\sigma n_{j'}^{\sigma'}$$

with $E_0(M) = \frac{1}{3}M(M - 1)(4M + 1) - (N - 1)M^2$.

The low energy, low temperature, behavior is classified [15] as the level one $su(2)$ WZW conformal field theory. In the spinon description the states consist of semi-infinite sequences of symbols 0 and 1. The two primary states, which correspond to the two integrable representations of the $su(2)$ current algebra at level one, are the vacuum, with sequence $010101\cdots$ and the spin half primary, with sequence $0010101\cdots$. The excited states are given by finite rearrangement of the primary sequence. The Virasoro generator acts as $L_0 = \sum_j (m_j^0 - m_j)$, where $\{m_j^0\}$ are the primary sequences. Its spinon representation is :

$$L_0 = \left(\frac{N_{sp}}{2}\right)^2 + \sum_{j\geq 0} j(n_j^+ + n_j^-) \tag{26}$$

with $N_{sp} = \sum_j (n_j^+ + n_j^-)$, the total spinon number. The excitations over the vacuum possess an even number of spinons, whereas those over the spin half primary contain an odd number of spinons.

7 The Thermodynamics

Following ref.[3], the thermodynamics can be derived from the spinon description using methods similar to the thermodynamic Bethe ansatz [9].

First, we consider the system with a fixed spinon density $D_{sp} = \frac{N_{sp}}{N}$. In the $N \to \infty$ limit, the spinon orbitals are labeled by momenta k continuously varying from zero to $k_0 = \frac{2\pi}{N}\left(\frac{N - N_{sp}}{2}\right) = \pi(1 - D_{sp})$. Let $n_\pm(k)$ be the spinon occupation numbers of the kth orbital. By definition, they satisfy :

$$\sum_\sigma \int_0^{k_0} \frac{dk}{2\pi} \, n_\sigma(k) = D_{sp} \tag{27}$$

In the continuum limit, the energy per unit of length is :

$$\frac{E}{N} \equiv \mathcal{E}(D_{sp}; n_\pm(k)) = \quad \mathcal{E}_0(D_{sp}) + \sum_\sigma \int_0^{k_0} \frac{dk}{2\pi} \, \epsilon_0(k) n_\sigma(k)$$
$$+ \sum_{\sigma\sigma'} \int_0^{k_0} \frac{dk dk'}{(2\pi)^2} V(k, k') n_\sigma(k) n_{\sigma'}(k') \tag{28}$$

with $\mathcal{E}_0(D_{sp}) = (\frac{\pi}{2})^2(1 - D_{sp})^2(\frac{1 + 2D_{sp}}{3})$, $\epsilon_0(k) = \frac{1}{2}(k_0 - k)(k_0 + k - \pi)$ and $V(k, k') = \frac{\pi}{2}(k_0 - \sup(k, k'))$. In each orbital the spinons behave as bosons, therefore the entropy of a configuration of $n_\pm(k)$ spinons is :

$$\frac{S}{N} \equiv \mathcal{S}(D_{sp}; n_\pm(k)) = \sum_\sigma \int_0^{k_0} \frac{dk}{2\pi} \, \left((n_\sigma(k) + 1)\log(n_\sigma(k) + 1) - n_\sigma(k)\log n_\sigma(k)\right) \tag{29}$$

The free energy per unit of length is :

$$\frac{F}{N} = \mathcal{F} = \mathcal{E} - T\mathcal{S} - h\mathcal{M} \tag{30}$$

with h the exterior magnetic field and $\mathcal{M} = \int_0^{k_0} \frac{dk}{2\pi}(n_+(k) - n_-(k))$ the magnetization.

At fixed spinon density, the thermodynamic equilibrium is determined by minimizing (30) with respect to the variation of the spinon occupation numbers subject to the constraint (27). This gives :

$$n_\pm(k) = \frac{1}{e^{\beta(\epsilon(k) \mp h - A)} - 1} \tag{31}$$

where A is the Lagrange multiplier and $\epsilon(k)$ is the dressed energy defined by :

$$\epsilon(k) = 2\pi \frac{\delta\mathcal{E}}{\delta n_\sigma(k)} = \epsilon_0(k) + 2\sum_{\sigma'} \int_0^{k_0} \frac{dk'}{2\pi} V(k, k') n_{\sigma'}(k') \tag{32}$$

At fixed density, the Lagrange multiplier is determined from the constraint (27). This completely specifies the thermodynamics of the spinon gas.

The spin chain corresponds to a spinon gas of arbitrary density; i.e. the spinon chemical potential $\mu = \frac{\partial \mathcal{F}}{\partial D_{sp}}$ is zero. Minimizing the free energy with respect to the density fixes A to be $\beta A = -\log(2\cosh(\beta h))$. The constraint (27) then gives the mean density $\overline{D}_{sp}$.

The coupled Eqs. (31,32) can be solved. Deriving twice eq.(32) with respect to k gives :

$$\frac{\partial^2 \epsilon(k)}{\partial k^2} = -\left(1 + \frac{1}{2}\sum_\sigma n_\sigma(k)\right) \tag{33}$$

with the boundary conditions: $\epsilon(k_0) = \epsilon(0) = 0$, and $\epsilon'(0) = -\epsilon'(k_0) = \frac{\pi}{2}$. Eq.(33) is integrated by introducing the dressed momenta $p = \frac{d\epsilon}{dk}$. It varies from $-\frac{\pi}{2}$ to $\frac{\pi}{2}$, and it satisfies $\frac{dp}{dk} = \frac{d^2\epsilon}{dk^2} = \frac{d(p^2/2)}{d\epsilon}$. As function of p, the occupation number are then given by :

$$n_\pm = \exp\left[\beta(\eta \pm \epsilon_{dr} \pm h)\right] \tag{34}$$

with

$$\eta(p) = \frac{1}{2}\left[p^2 - \left(\frac{\pi}{2}\right)^2\right] \quad \text{and} \quad \exp(\beta\eta) = \frac{\sinh(\beta\epsilon_{dr})}{\sinh(\beta h)} \tag{35}$$

In the limit $h \to 0$, the dressed energy is $\epsilon_{dr} \sim he^{\beta\eta}$.

The free energy is found by integrating the thermodynamical relations :

$$\left(\frac{\partial \mathcal{F}}{\partial T}\right)_h = -\mathcal{S} \quad \text{and} \quad \left(\frac{\partial \mathcal{F}}{\partial h}\right)_T = -\mathcal{M} \tag{36}$$

The result is the following simple answer :

$$\mathcal{F} = -T\int_{-\frac{\pi}{2}}^{\frac{\pi}{2}} \frac{dp}{\pi} \, \log\left[\frac{\sinh(\beta(\epsilon_{dr} + h))}{\sinh(\beta h)}\right] \tag{37}$$

Notice that this is the free energy of a gas of non-interacting particles which, in the limit $h \to 0$, have energies given by $\eta(p)$.

Appendix

In this Appendix we prove the Eqs.(10,11). First we compute the action of $T_{--}(x)$ on these magnon states. We recall that $T_{--}(x)$ can be written as $T_{--}(x) = 1 + \sum_{i,j}\left(\frac{1}{x-L}\right)_{ij} P_j^-$, where P_j^- is the projector on the spin ($\sigma_j = -$) acting on the jth spin only. The projector P_j^- on the M-magnon states (8) gives states with one spin down marked of the form :

$$P_j^-|\Psi\rangle \equiv |\Psi_j\rangle = M \sum_{n_2,\cdots,n_M} \psi_{j;n_2,\cdots,n_M} \sigma_j^- \sigma_{n_2}^- \cdots \sigma_{n_M}^- |\Omega\rangle \tag{38}$$

They corresponds to polynomials $\Psi(z_1|z_2,\cdots,z_M)$ symmetric in $z_2,\cdots,z_M$ with the point z_1 distinguished. To evaluate the action of $\sum_{i,j}(L^n)_{ij}P_j^-$, we remark that on this class of states, L_{ij} acts as follows :

$$\sum_k L_{jk}|\Psi_k\rangle = |(L\Psi)_j\rangle \tag{39}$$

with,

$$(L\Psi)_{j;n_2,\cdots,n_M} = \sum_k \widehat{\theta}_{jk}\psi_{k;n_2,\cdots,n_M} + \sum_{q=1}^{M} \widehat{\theta}_{jn_q}(K_{1q}\psi)_{j;n_2,\cdots,n_M} \tag{40}$$

where $\widehat{\theta}_{jk} = \omega^j/(\omega^j - \omega^k)$, and K_{1q} permutes the indices in position 1 and q. Therefore, $|(L\Psi)_i\rangle = \sum_j (L^n)_{ij} P_j^-|\Psi\rangle$ can be recursively computed. Then,

$$\sum_{i,j}(L^n)_{ij} P_j|\Psi\rangle = M \sum_i |(L^n\Psi)_i\rangle \tag{41}$$

is obtained by symmetrizing the wave function coefficients of $|(L^n\Psi)_i\rangle$ in all its indices. I.e. its wave function coefficients, denoted $(L^n P\Psi)_{n_1,\cdots,n_M}$ are :

$$(L^n P\Psi)_{n_1,\cdots,n_M} = \left(1 + \sum_{p\neq 1} K_{1p}\right)(L^n\Psi)_{n_1;n_2,\cdots,n_M} \tag{42}$$

We now translate this action on the wave function coefficients into an action on the polynomials. We recall that in the basis of polynomials Q_k in one variable of degree $(N-1)$ specified by $Q_k(\omega^n) = \delta_k^n$, the matrix elements of the derivatives are:

$$(z\partial_z - \frac{N+1}{2})Q_j(z) = -\sum_{k\neq j}\frac{\omega^j}{\omega^j - \omega^k}\, Q_k(z) \tag{43}$$

$$z\partial_z(z\partial_z - N)Q_j(z) = -2\sum_{k\neq j}\frac{\omega^j\omega^k}{(\omega^j - \omega^k)^2}(Q_k(z) - Q_j(z))$$

The differentials D_p, introduced in eq.(12), acts on polynomials of the form (9); i.e. they preserve the form of these polynomials. Moreover, by comparing the formula, we recognize in eq.(40) the operator $\left(D_1 - \frac{N+1}{2}\right)$. Since the operators D_p are covariant by permutation of the coordinates, the polynomial of $(L^n P\Psi)$ is given by acting with $\sum_p \left(D_p - \frac{N+1}{2}\right)^n$ on Ψ. Resuming all the contributions, we obtain :

$$T_{--}(x)\,\Psi(z.) = \left(1 + \sum_{p=1}^{M}\frac{1}{x + \frac{N+1}{2} - D_p}\right)\Psi(z.)$$

The action of $T_{-+}(x)$ on these subclass of magnon states can be computed using the same methods. Its action is given by eq.(11). It is remarkable that the differentials D_p appear naturally in the study of the spin transfer matrix.

Acknowledgements: It is a pleasure to thank Olivier Babelon, Michel Gaudin and Duncan Haldane for stimulating discussions.

References

[1] F.D. Haldane, Phys.Rev.Lett. 60(1988)635.

[2] B.S. Shastry, Phys.Rev.Lett. 60(1988)639.

[3] F.D. Haldane, Phys.Rev.Lett. 66 (1991) 1529.

[4] E. Date, M. Jimbo, A. Kuniba, T. Miwa and M. Okado, Lett. Math. Phys. 17 (1989) 69.

[5] F.D. Haldane, Z.N. Ha, J.C. Talstra, D. Bernard and V. Pasquier, Phys.Rev.Lett. 69 (1992) 2021.

[6] V.G. Drinfel'd, "Quantum Groups.", Proc. of the ICM, Berkeley (1987)798.

[7] E.K. Sklyanin, Funct. Anal. Appl. 16 (1983) 263.

[8] D. Bernard, M. Gaudin, F.D. Haldane and V. Pasquier, J. Phys. A. , to be published.

[9] A.G. Izergin and V. Korepin, Sov.Phys.Dokl. 26 (1981) 653.

[10] E.M. Opdam, Invent. Math. 98 (1989) 1;
G.J. Heckman, Invent. Math. 103 (1991) 341.

[11] A.P. Polychronakos, Phys.Rev.Lett. 69 (1992) 703.

[12] I. Cherednik, Invent. Math. 106 (1991) 411; and Publ. RIMS 27 (1991) 727; and reference therin.

[13] O.Chalykh and A. Veselov, Comm. Math. Phys. 152 (1993) 29.

[14] F.D. Haldane, Phys.Rev.Lett. 67(1991)937.

[15] H.W.Blöte, J.L. Cardy, and Nightingale, Phys.Rev.Lett. 56 (1986) 742;
I. Affleck, Phys.Rev.Lett. 56 (1986) 746.

[16] C.N.Yang and C.P.Yang, Phys.Rev. 10(1969)1115.

KODAIRA-SPENCER THEORY OF GRAVITY

Michael BERSHADSKY

Harvard University
Lyman Laboratory
Cambridge MA 02138, USA

Abstract: We briefly review the topological model on Calabi-Yau 3-fold coupled to gravity. We discuss the Kodaira-Spencer Theory of Gravity which is equivalent to topological B-model on Calabi-Yau 3-fold and may be viewed as the closed string analog of Chern-Simons Theory.

1   Introduction

In this paper we discuss a field theory on Calabi-Yau space which is closely related to topological string theory . In a particular realization of the critical topological strings, the classical limit of the string field theory turns out to describe the classical deformation of the complex structure of Calabi–Yau manifolds (and the related variation of Hodge structure), i.e., the Kodaira–Spencer theory. This relation can be summarized by writing an action whose classical solution corresponds to all possible deformation of the complex structure of the Calabi–Yau manifold. We call this field theory the *Kodaira–Spencer (KS) theory of gravity*. It is a gravitational theory in 6 real dimensions with vacua being Calabi–Yau 3–folds and which gauges the complex structure of the manifold. The *Kodaira–Spencer theory can be viewed as the closed string field theory for the critical topological string on a Calabi–Yau*. This is a rather simple realization of a closed string field theory which may be helpful for further understanding of closed string field theory in more general cases. One can also consider the quantum Kodaira–Spencer theory, i.e., the higher loops on it which are the same as the partition function at higher genus of topological strings. In particular at one–loop the partition function can be related to an appropriate combination of determinants of various operators which turns out to be related to the Ray–Singer holomorphic torsion.

The case of Calabi–Yau 3–fold as a string theory has already been studied in [1] for both open and closed strings. In particular it was discovered there that in the case

Quantum Field Theory and String Theory, Edited by
L. Baulieu *et al.*, Plenum Press, New York, 1995

of the open string theory, the target space physics is equivalent to three dimensional Chern–Simons theory.

The topological string theory in question can be obtained by twisting an $N = 2$ superconformal σ model on a Calabi-Yau space. There are two ways of twisting $N = 2$ superconformal theory: either twisting the fermion number or the axial fermion number. These two choices of twisting correspond to either (c, c) ring or (a, c) ring as being the topological one. In the first case of (c, c) twisting, the topological correlation functions are sensitive to the Kähler class of manifold, this twisting is called the *A-twisting* or the *Kähler twisting*. On the other hand, in the case of (a, c) twisting, the topological correlation functions are only sensitive to the complex structure of the manifold. This twisting is called the *B–twisting* or the *complex twisting*. This paper is devoted to the later case.

The topological model is easier to study in the large volume case. In this limit the the Hilbert space of B-model can be identified with antiholomorphic forms wedged with holomorphic vectors, i.e.,

$$\mathcal{H} = \bigoplus_{p,q} \wedge^p \overline{T}_M^* \otimes \wedge^q T_M \ , \tag{1}$$

where in here and in the following $T_M, \overline{T}_M$ denote the holomorphic and anti-holomorphic tangent bundles respectively and $T_M^*, \overline{T}_M^*$ denote the holomorphic and anti-holomorphic cotangent bundles. This Hilbert space can be mapped on Hilbert space of forms by contracting the vector indices with holomorphic three form. This converts the (q, p) sector above to a differential form of degree $(3 - q, p)$. On this Hilbert space $\mathcal{H}'$ the dictionary for the supercharges turn out to be

$$G^+ = \frac{1}{2}(\partial^\dagger + \overline{\partial})$$

$$\overline{G}^- = \frac{1}{2}(\overline{\partial} - \partial^\dagger) \ .$$

Note that $Q_2 = G^+ + \overline{G}^- \leftrightarrow \overline{\partial}$ and the observables can be identified with the $\overline{\partial}$ cohomology elements of M [1]. The physical states with charge $(1_L, 1_R)$ are called marginal operators and are in one-to-one correspondence with deformations of complex structure of Calabi-Yau 3-fold.

The computations of topological B–model before coupling to gravity, can be related to classical questions in variation of Hodge structure, i.e., the complex structure of Calabi-Yau and how it varies. In the language of sigma models this is related to the fact that the B–model topological theory is independent of the volume of the manifold. Rescaling the volume to infinity implies that in the topological B–theory, not coupled to gravity, the path–integral configurations are dominated by constant maps, thus leading to classical geometry questions, and in particular the questions of variation of Hodge structure of Calabi–Yau 3-folds. As we will discuss later in the section this is essentially true (modulo a crucial subtlety) even after we couple to 2d-gravity.

---

[1] For convenience we will flip the convention for rightmoving $U(1)$ charge and replace $\overline{G}^-$ by $\overline{G}^+$

2 Coupling to Gravity

Bosonic string theory is in many ways like a twisted $N = 2$ theory [2] [3]. It has a *scalar* supercharge $Q_{BRST} = Q + \overline{Q}$, which is the usual BRST operator. It has anti-ghosts, $b, \overline{b}$ of spin (2,0) and (0,2), with the property

$$Q^2 = b_0^2 = 0$$

$$\{Q, b_0\} = H_L \ .$$

and it has two $U(1)$'s, $G, \overline{G}$ corresponding to the left and right ghost numbers. Identifying

$$2j_{BRST} \leftrightarrow G^+$$

$$b \leftrightarrow G^-$$

$$bc \leftrightarrow J \ .$$

and similarly for right-movers. Thus the notion of a physical state in the bosonic string becomes exactly the same as that of a chiral state in the twisted theory. Thus we can define coupling of twisted $N = 2$ theory to gravity by integrating correlation functions of chiral fields over moduli space of Riemann surface, with the insertion of G^-'s folded with $3g - 3$ Beltrami differentials. In particular the partition function of the twisted $N = 2$ theory coupled to gravity at genus $g > 1$, F_g, can be defined by

$$F_g = \int_{\mathcal{M}_g} \Big\langle \prod_{k=1}^{3g-3} (\int G^- \mu_k)(\int \overline{G}^- \overline{\mu}_k) \Big\rangle \ ,$$

where μ_i denote the Beltrami differentials, and $\mathcal{M}_g$ denotes the moduli space of genus g Riemann surfaces. For genus one the similar expression can be written not for the partition function but for the one point correlation function. The 0,1, and 2 point functions on the sphere are equal to zero, and the three point function is given by $\langle \phi_i \phi_j \phi_k \rangle = C_{ijk}$. The chiral fields ϕ_i, which after twisting have dimension zero, play the same role as $c\overline{c}V_i$ in the usual bosonic string theory, and

$$\phi_i^{(2)} = [G_{-1}^-[\overline{G}_{-1}^- \phi_i]\,] = \oint G^- \oint \overline{G}^- \phi_i$$

plays the same role as vertex operator V_i. The field $\phi_i^{(2)}$ has dimension $(1,1)$ and charge $(q_L - 1, q_R - 1)$, where $q_{L,R}$ are left and right charges correspondingly. For $q_L = q_R = 1$ the fields $\phi_i^{(2)}$ are chargeless and thus they are possible perturbations of the Lagrangian. One also has to remember that original $N = 2$ conformal theory has an hermitian structure that maps $G^+ \leftrightarrow G^-$ and chiral fields ϕ_i on antichiral ϕ_i^+. It is natural to consider the perturbed topological conformal field theory, where the perturbations respect the $N = 2$ hermitian structure

$$S_{pert} = S_{TCFT} + t^i \int d^2 z [G_{-1}^-[\overline{G}_{-1}^- \phi_i]\,] + \overline{t}^i \int d^2 z [Q[\overline{Q} \phi_i^+]\,]$$

It is a rather nice property of twisted *unitary* $N = 2$ theories that F_g is finite and well defined. The only potential divergence would have come from the regions near the boundary of moduli space of Riemann surfaces. But in such cases, the fact that the propagator annihilates the massless modes, implies that only the massive modes propagate and thus the integrand in F_g is exponentially small in these regions (the coefficient of exponent being fixed by the first non-vanishing eigenvalue of $L_0 = \overline{L}_0$).

A multipoint correlation functions of the $N = 2$ twisted theory coupled to gravity at genus g can be defined in a similar way

$$C^g_{i_1 i_2 \ldots i_n} = \int_{\mathcal{M}_g} \left\langle \int \phi^{(2)}_{i_1} \cdots \int \phi^{(2)}_{i_n} \prod_{k=1}^{3g-3} (\int G^- \mu_k)(\int \overline{G}^- \overline{\mu}_k) \right\rangle. \tag{2}$$

Despite an almost complete parallel between bosonic string and twisted $N = 2$ theories coupled to gravity, there are two notable differences. The first one is that the ghost number violation in bosonic string at genus g is universal and is given by $3g - 3$, whereas for twisted $N = 2$ theories it is given by $\hat{c}(g-1)$. In particular we see that $\hat{c} = 3$ is a critical case in that it gives the same degree of charge violation as bosonic string. So in particular this suggests that Calabi–Yau 3-folds are a specially interesting class to consider [4]. Note that only for $\hat{c} = 3$ the F_g has a chance to be non-zero for $g > 1$, by $U(1)$ charge conservation. For all the other values of $\hat{c}$, the only way to get a non-zero result is by introducing other correlators. The correlators involving chiral fields may be used to prevent vanishing of correlation functions only for $1 < \hat{c}$; For $\hat{c} \leq 1$ the charges of all $\phi^{(2)}_i$ are negative (the maximum being $\hat{c} - 1$) and so cannot be used to balance charges. In these cases, which happen to be intensively studied in connection with matrix models, one needs to include the full topological gravity multiplet and construct gravitational descendants which give rise to non-vanishing correlation functions [4] [5] [6].

It is convenient to collect all correlation functions on all genera in one object – generating function. To specify a theory let us fix a base point $P(t,\bar{t})$ in the moduli space of complex structures of Calabi Yau 3-fold. The marginal operators ϕ_i can be identified with vectors in the tangent space. Let x^i be the coordinates in the tangent space at base point $P(t,\bar{t})$ and λ be the string coupling constant. Then consider the generating function

$$W(\lambda, x^i, t, \bar{t}) = \sum_g \lambda^{2-2g} \sum \frac{1}{n!} x^{i_1} \ldots x^{i_n} C^{(g)}_{i_1 \ldots i_n}(t, \bar{t}) + (\frac{\chi}{24} - 1)\log \lambda , \tag{3}$$

where $C^{(g)}_{i_1 \ldots i_n} = 0$ for $2 - 2g + n \leq 0$. This generating function satisfies anomaly equation derived in [7]. To write the anomaly equation let us first introduce some notations. Let K be the Kähler potential on the moduli space of Calabi Yau 3-folds, $G_{\bar{a}b}$ – Kähler metric and $C^{bc}_{\bar{a}} = \overline{C}_{\bar{a}\bar{b}\bar{c}} G^{b\bar{b}} G^{c\bar{c}} e^{2K}$ is Yukawa coupling with two upper indices. The generating function $W(\lambda, x^i, t, \bar{t})$ satisfies the following equation

$$\frac{\partial}{\partial \bar{t}^a} e^W = \left[\frac{\lambda^2}{2} C^{bc}_{\bar{a}} \frac{\partial^2}{\partial x^b \partial x^c} - G_{a\bar{b}} x^b (\lambda \partial_\lambda + x^k \frac{\partial}{\partial x^k}) \right] e^W .$$

Expanding the above equation in series in λ and x^i and comparing coefficients in front of monomials one gets the anomaly equation for individual correlation functions. The generating function defined above turns out to be an effective action for massless modes of Kodaira-Spencer theory on Calabi Yau 3-fold, that will be discussed in the next sections.

3 Deformations of Complex Structure

The observables in topological B–model are in one-to-one correspondence with cohomology elements $H^p(\wedge^q T_M)$, where T_M is the holomorphic tangent bundle. The two forms $\phi_A^{(2)}$ are possible perturbations of the Lagrangian. In case $p = 1$, $q = 1$ operators $\phi_A^{(2)}$ for $A \in H^{(0,1)}(T_M)$ correspond to marginal deformations of the B–model and are in one to correspondence with deformations of complex structure of Calabi–Yau 3–fold M. In the spirit of string theory one expects that $A \in \Omega^{(0,1)}(T_M)$ should be the basic field in the field theory in question. This field theory is closely related to the mathematical theory of deformations of complex structures. Before proceeding further we must first review some elements of this theory.

The complex structure on manifold M is determined by the $\bar{\partial}$ operator. To the first order the change of complex structure is described by deformation of $\bar{\partial}$ operator $\bar{\partial} \to \bar{\partial} + A^i \partial_i$ [8]. This is a deformation of $\bar{\partial}$ operator acting on functions. One can describe not only the infinitesimal deformations of complex structure but a finite one. The new complex structure is described by requiring that functions satisfying the condition

$$(\bar{\partial} + A^i \partial_i) f = 0 , \tag{4}$$

are holomorphic in the new complex structure. In other words the kernel of the deformed $\bar{\partial}$ coincides with with kernel of (4). The integrability condition

$$\bar{\partial}(\bar{\partial} f + A^i \partial_i f) = (\bar{\partial} A^j + A^i \partial_i A^j) \partial_j f = 0$$

is equivalent to the Kodaira–Spencer (KS) equation [8]

$$\bar{\partial} A + \frac{1}{2}[A, A] = 0 . \tag{5}$$

Once again A is $(1, 0)$ vector field with coefficients in $(0, 1)$ forms and the brackets $[,]$ mean the commutator of two vector fields and wedging. Two solutions of (5) correspond to the same complex structure if they differ by a diffeomorphism. In the linear approximation Kodaira Spencer equation reduces to $\bar{\partial} A = 0$. The solution is defined modulo diffeomorphisms generated by vector fields $A \to A + \bar{\partial}\epsilon$, and thus A has to be a cohomology element. The ambiguity in the choice of cohomology representative is promoted to the ambiguity in the solution of Kodaira Spencer equation.

Before fixing the ambiguity in question let us mention that for any Calabi–Yau manifold there is an isomorphism

$${}' : \quad \Omega^{(0,p)}(\wedge^q T_M) \to \Omega^{(3-q,p)}(M) \tag{6}$$

given by the product with the holomorphic $(3, 0)$ form. Without lack of generality we impose the constraint

$$\partial A' = 0 . \tag{7}$$

To fix the ambiguity, $A \to A + \bar{\partial}\epsilon$, we impose the gauge condition

$$\bar{\partial}^\dagger A' = 0 . \tag{8}$$

This gauge condition requires the choice of metric on the Calabi–Yau manifold. It will be clear later that these conditions fix the solution uniquely.

Let A, B be vector fields with the coefficients in $(0,1)$ forms which satisfy the gauge condition $\partial A' = \partial B' = 0$. It was proven by Tian [9] that

$$[A, B]' = \partial(A \wedge B)' , \tag{9}$$

Later we will need the generalization of Tian's Lemma where A, B belong to $\Omega^p(\wedge^q T_M)$ [10]. Using this Lemma we can rewrite the KS equation in Tian form

$$\overline{\partial} A' + \frac{1}{2}\partial(A \wedge A)' = 0.$$

The tangent space to the moduli space of complex structures is given by $H^{(0,1)}(T_M)$. Let A_1 be an infinitesimal deformation of complex structure satisfying conditions (8), (7). Then for any A_1 one can "exponentiate" the deformation of complex structure by constructing the solution to the KS equation

$$A = \sum_{n=1}^{\infty} \epsilon^n A_n,$$

where ϵ is a formal expansion parameter (we put $\epsilon = 1$ later). We will show that it is possible to get a unique solution of the KS equation satisfying the gauge condition $\overline{\partial}^\dagger A' = 0$ such that A'_n is ∂–exact for $n > 1$. Note that this latter condition automatically implies that we can use Tian's form of the KS equation. This choice means that A'_1 is a harmonic form, which we will call *massless*, and A'_n for $n > 1$ can be written as a linear combination of eigenstates of Laplacian with positive eigenvalue. We will call these states the *massive* states.

Let us see how we can construct the solution recursively (following the work of [9], [11]) making sure that at each stage $\overline{\partial} A'_n = 0$ and that A'_n is ∂–exact for $n > 1$. Let A_1 satisfy the gauge condition (8) together with constraint (7). Thanks to Tian's Lemma the equation for A'_2 becomes

$$\overline{\partial} A'_2 + \frac{1}{2}\partial(A_1 \wedge A_1)' = 0.$$

Note that the solution to this equation for A'_2 is unique up to addition of $\overline{\partial}\nu$. In order to get rid of this ambiguity we will consider the gauge condition $\overline{\partial}^\dagger A'_2 = 0$. Then the solution can be written as

$$A'_2 = -\overline{\partial}^\dagger \frac{1}{\Delta}\partial(A_1 \wedge A_1)' . \tag{10}$$

where

$$\Delta = 2[\overline{\partial}\overline{\partial}^\dagger + \overline{\partial}^\dagger\overline{\partial}]$$

is the Laplacian. To see that the above is a solution, first note that it is well defined, because ∂ annihilates the kernel of Δ. Then acting by $\overline{\partial}$ and using the fact that $\overline{\partial}(A_1 \wedge A_1)' = 0$ (because A_1 is $\overline{\partial}$ closed and $\overline{\partial}$ commutes with the operation ′ since Ω is holomorphic) one checks that it is a solution to the equation. It also satisfies the conditions of being ∂–exact (because ∂ and $\overline{\partial}^\dagger$ cohomologies anticommute for a

Kähler manifold) and $\overline{\partial}^\dagger$ closed. The fact that there is always a solution to the above equation is also known as $\partial\overline{\partial}$-Lemma [12] [2]. This in particular means that with the gauge condition we have chosen

$$\frac{-1}{2\overline{\partial}}\partial \equiv -\overline{\partial}^\dagger\frac{1}{\Delta}\partial \, , \tag{11}$$

and it can be viewed as a propagator for massive modes. The equation (10) describes interaction between two massless modes and a massive one and then further propagation of the massive state.

The equation for the next iteration becomes $\overline{\partial}A_3' + \partial(A_2 \wedge A_1)' = 0$. The second term in this equation is $\overline{\partial}$ closed $\overline{\partial}\partial(A_2 \wedge A_1)' \sim \partial([A_1, A_1] \wedge A_1)' \sim [[A_1, A_1], A_1] = 0$ and therefore one may use the above propagator again.

$$A_3' = 2\overline{\partial}^\dagger\frac{1}{\Delta}\partial(A_1 \wedge (\overline{\partial}^\dagger\frac{1}{\Delta}\partial(A_1 \wedge A_1)')^\vee)'$$

where $(A')^\vee = A$. Note that this solution satisfies the required conditions. Again this contribution has a clear interpretation. Two massless states go to a massive one (as before), but now the propagator receives corrections due to the coupling with the massless state in the background.

It is already clear that $\partial\overline{\partial}^\dagger/\Delta$ is a propagator for the massive states for the field theory in question. The massless modes play the role of the background. It is quite remarkable that the KS equation reproduces the perturbation series of a ϕ^3 theory.

At $n-$th iteration step all $A_1, ... A_{n-1}$ satisfy the conditions $\partial A_1' = \cdots = \partial A_{n-1}' = 0$ and the KS equation becomes

$$\overline{\partial}A_n' + \frac{1}{2}\sum_{i=1}^{n}\partial(A_{n-i} \wedge A_i)' = 0 \, . \tag{12}$$

The second term of this equation is $\overline{\partial}$ closed. This follows from the equations satisfied for $\overline{\partial}A_i'$ dictated by induction and the Jacobi identity for the vector fields with coefficients in $(0, 1)$ forms and Tian's Lemma

$$\partial([A, B] \wedge C)' + \partial([C, A] \wedge B)' + \partial([B, C] \wedge A)' = 0 \, . \tag{13}$$

It follows from the above arguments therefore that equation (12) has a solution and it is ∂–exact. The perturbation theory described above is convergent in some open neighborhood of the origin [9] .

We just proved that for any $x \in H^{(0,1)}(T_M)$ there is a map $x \to A[x]$ given by the solution of the KS equation, with $A_1 = x$. This map can basically be viewed as shifting the complex structure of the Calabi-Yau labeled by $(t, \overline{t}) \to (t + x, \overline{t})$. For later convenience we will write $A[x] \to x + A(x)$. Decomposition into x and $A(x)$ is quite natural. A cohomology element x represent a *massless* mode while $A(x) = \sum_{n=2}^{\infty} A_n$ contains the *massive* modes of the field.

Under the deformation of complex structure, the holomorphic $(3, 0)$ form get changed. For infinitesimal deformation the deformed holomorphic form

---

[2]The $\partial\overline{\partial}$–Lemma reads: if ω is any $\overline{\partial}$ closed form and ω is also ∂ exact, then $\omega = \partial\overline{\partial}\phi$.

is equal to $\Omega_0 + x'$. For the finite deformations the holomorphic $(3,0)$ form mixes with $(2,1)$, $(1,2)$ and $(0,3)$ and it satisfies the equation

$$\bar{\partial}\Omega + \frac{1}{2}\partial(\Omega^\vee \wedge A)' = 0 \;,$$

where prime and check are defined with respect to the fixed holomorphic three form Ω_0. It follows from Tian's Lemma that the deformed holomorphic $(3,0)$ form is given as follows [11]

$$\Omega = \Omega_0 + A' + (A \wedge A)' + (A \wedge A \wedge A)' \;. \tag{14}$$

Coordinates in $H^{(0,1)}(T_M)$, denoted by x, may serve as affine coordinates on some open neighborhood of the moduli space of complex structures (see also [13]) thanks to Tian's mapping. These coordinates are in fact very special (not to be confused with special coordinates except for the particular case of base point at infinity) and we call them canonical coordinates. The defining property of canonical coordinates is that covariant holomorphic derivatives at the origin coincides with the ordinary derivatives

$$D_i D_j ... D_k F = \partial_i \partial_j ... \partial_k F \;.$$

In fact x^i in (3) are nothing else but the canonical coordinates around point $(t, \bar{t})$.

4 Kodaira–Spencer Theory as String Field Theory

So far we have discussed what seems to be a perturbative field theory which describes the perturbation of complex structure of Calabi–Yau manifolds starting from a base-point. Since the B–model describes the deformation of the complex structure, the effective string field theory of the B–models must be this underlying field theory, which we shall call the Kodaira–Spencer theory of gravity. We have two options in writing this field theory: We can either use the Kodaira-Spencer equation in the Tian gauge to write the action giving rise to these equations, or directly use the rules for constructing closed string field theory along general lines discussed in the literature (see [17] for a thorough review of the literature). We will follow the first line and see why it is the same as the second.

To write an action we first need to fix some data: the point P (which we sometimes denote also by $(t_0^i, \bar{t}_0^i)$) in the moduli space of complex structures (background) and a cohomology element $x \in H^{(0,1)}(T_M)$. The physical field A in the KS theory is a vector field with coefficients in $(0,1)$ forms which is also constrained to satisfy condition $\partial A' = 0$. For reasons that will be clear in a moment we assume that A includes only massive modes. This means that A lies in the subspace $\mathcal{H} \subset \Omega^{(0,1)}(T_M)$ orthogonal to $H_{\bar{\partial}}^{(0,1)}(T_M)$, or in other words

$$\int_M A' \wedge \bar{z}' = 0$$

for any $\bar{z} \in H_{\partial}^{(1,0)}(T^*)$. Thanks to constraint (7), this definition is independent of the choice of representative in .

The Kodaira–Spencer action is given as follows

$$\lambda^2 S(A, x|P) = \frac{1}{2} \int_M A' \frac{1}{\partial} \overline{\partial} A' + \frac{1}{6} \int_M ((x + A) \wedge (x + A))'(x + A)' , \qquad (15)$$

where λ^2 is the coupling constant. In spite of the non-local kinetic term, this action is well defined. Indeed, it follows from the $\partial\overline{\partial}$-Lemma that $\overline{\partial} A' = \partial\overline{\partial} v$ and therefore $\partial^{-1}\overline{\partial} A' = \overline{\partial} v + \partial\rho + \overline{z}$, where ρ and z summarize the ambiguities and $z \in H^{(1,0)}_{\partial}(T^*)$. The condition that A' is massive together with the constraint it satisfies implies that ρ and z do not contribute to the action which therefore is well defined. Note that to define the action we did not use the metric on Calabi–Yau manifold. We just used its complex structure[3]. This is just like the Chern–Simons theory. Thus the KS theory is a topological theory (or more properly it could be called a holomorphic topological theory in the sense that it does depend on the complex structure of the Calabi-Yau). Varying the KS action with respect to A we recover the Kodaira–Spencer equation in Tian's form

$$\overline{\partial} A' + \frac{1}{2}\partial((x + A) \wedge (x + A))' = 0 . \qquad (16)$$

The existence of this action explains the fact that in the perturbation expansion for $A(x)$, discussed before, one naturally gets Feynman rules of some field theory. In fact, they are nothing but the tree level diagrams of KS theory. Note that the propagator for KS action $\overline{\partial}^{-1}\partial$ is given by (11) in the appropriate gauge.

We now wish to see why the action (15) is the same as what we would have gotten from the target space theory of the B-model. For this, we employ the arguments of Witten [1] .

He used the fact that volume perturbation for the Calabi–Yau is BRST trivial in the B-model set up, to take the infinite volume limit. In this case, the worldsheet configurations *for a fixed worldsheet modulus* is dominated by constant maps. But as noted in [1], this is not the full story. The reason is that we are discussing a theory of 2d gravity which means we are integrating over the moduli of Riemann surfaces. No matter how large a volume of Calabi–Yau we choose, if we go close enough to the boundary of the moduli space we can get finite action. In other words the worldsheets which will have finite action are the ones concentrated in long thin tubes, which means that we are going to end up with an ordinary field theory as an *exact* field theory of string model (i.e., all the stringy massive modes are irrelevant because of topological triviality of these modes). Indeed this argument applies even taking into account potential anomalies, because there is no anomaly for the decoupling of the Kähler-moduli in the B-model.

So to fix the string field we have to recall that the field in question should have charge $(1,1)$, which in our case translates to the fact that A' should belong to $\overline{T}^*_M \wedge T_M$. Let us also recall the dictionary: In the large volume limit operator $\overline{\partial}$ is identified with BRST operator $\overline{\partial} = Q = G^+_0 + \overline{G}^+_0$, while $\partial = G^-_0 - \overline{G}^-_0 = b^-_0$. The string field A' should satisfy two constraints

$$\partial A' = b^-_0 A' = 0 \quad \text{and} \quad (L_0 - \overline{L}_0)A' = (\Delta - \overline{\Delta})A' = 0 . \qquad (17)$$

---

[3]To see that the action is well defined and independent of the choice of metric on M, we can also use the ∂ constraint to write $A' = \partial\phi$ and substitute it in the action to get a local action for ϕ.

In the case of the KS theory, the second constraint is a trivial consequence of Kählerian geometry and amazingly the first condition is precisely Tian's condition which led to the simplification and proof of integrability of the KS equation in the case of Calabi–Yau 3-fold. In order to borrow the machinery of closed string field theory, we need to find an expression for $c_0^- = c_0 - \bar{c}_0$. However there is no such object just because the b-cohomology is not trivial. What is true instead is that on the *massive* states of the theory, we can in fact define a

$$c_0^- = \frac{1}{\partial} = \frac{\partial^\dagger}{\Delta}$$

which satisfies

$$\{c_0^-, b_0^-\} = 1 \ ,$$

and we are thus forced to write down the action *only for the massive modes*. Therefore, the kinetic piece of the KS action coincides with the free part of the standard string field theory action

$$\frac{1}{2} \int A' \frac{1}{\partial} \bar{\partial} A' = \frac{1}{2}(A', c_0^- Q A') \ .$$

The gauge $\bar{\partial}^\dagger A' = 0$ is nothing else but the Siegel gauge in which both $b_0^- = \partial$ and $b_0^+ = \bar{\partial}^\dagger$ annihilates the physical fields. In this gauge the propagator takes the familiar form

$$\frac{b_0^+ b_0^-}{(L_0 + \bar{L}_0)} = \frac{\bar{\partial}^\dagger \partial}{\Delta}$$

Magically enough this is identical with the Kodaira–Spencer kinetic term and the propagator. The cubic interaction term is quite standard and gives rise to the interaction term of the Kodaira–Spencer action.

Thus the KS action is nothing else but the closed string field theory action, at least up to cubic order. One of the main difficulties of the closed string theory is the absence of a decomposition of the moduli space of Riemann surfaces compatible with Feynman rules. To avoid this problem one should introduce higher string vertices and as a result the closed string field theory becomes non–polynomial (see [17] and references there). The contribution to these higher string vertices comes entirely from the internal domains of the moduli space of Riemann surfaces. Quantized KS theory is defined as the large volume limit of topological sigma–model and as a topological theory it gets contribution entirely from the boundary of moduli space of Riemann surfaces. Therefore, the higher vertices should be absent in quantized KS theory. It is quite satisfactory that we thus end up with precisely the KS theory as the string field theory of the B-model. This is further confirmed in the next section where we will find that the KS theory, with the ghost fields added, already satisfies the BV master equation and needs no further corrections.

Let us now discuss the gauge symmetries of Kodaira–Spencer theory. As a string field theory we certainly expect it to have such symmetries. Being a theory of gravity the Kodaira–Spencer theory should be invariant under diffeomorphisms (we will make this statement precise in a moment). Put differently, the fact that the variation of $\bar{\partial}$ can also be affected by diffeomorphisms, and we do not wish to take this as a physical variation, we need to consider the theory as a gauge theory with respect to diffeomorphism group. The kinetic part of the action is clearly invariant under the shift of A by $\overline{\partial}$-exact term which means $\delta A = \overline{\partial}\epsilon = Q\epsilon$. This linearized gauge transformation can be extended to a full non-linear gauge transformation which turns out to be nothing else but an Ω-*preserving* diffeomorphism

$$z^i \longrightarrow z^i + \epsilon^i(z,\overline{z}) \ .$$

The condition that ϵ is Ω preserving diffeomorphism means that it satisfies the constraint $\partial \epsilon' = 0$. The full gauge transformation of the Kodaira-Spencer field A, which can be deduced from the variation of $\overline{\partial}$ under the diffeomorphism, is given as follows

$$\delta A = \overline{\partial}\epsilon - [\epsilon, (x + A)] \ ,$$

and using the Tian's Lemma it can be rewritten in a more familiar form $\delta A' = \overline{\partial}\epsilon' - \partial(\epsilon \wedge (x + A))'$. One can verify that this transformation is a symmetry of the action. Indeed the variation of the action is equal to

$$\lambda^2 \delta S = \ - \int_M A' \overline{\partial}((x + A) \wedge \epsilon)' + \tfrac{1}{2} \int_M ((x + A) \wedge (x + A))' \overline{\partial}\epsilon' \qquad (18)$$
$$- \tfrac{1}{2} \int_M ((x + A) \wedge (x + A))' \partial((x + A) \wedge \epsilon)' \ .$$

The first two terms cancel each other, as can be seen by integrating by parts. The vanishing of the third term follows from the Jacobi identity. Indeed, the last term can be rewritten as follows

$$\int_M ([(x + A), \epsilon] \wedge \ (x + A))'(x + A)' = \tfrac{1}{2} \int_M ([(x + A), (x + A)] \wedge \epsilon)'(x + A)' = \qquad (19)$$
$$\tfrac{1}{2} \int_M ([(x + A), (x + A)] \wedge (x + A))'\epsilon' = 0$$

To formulate the KS theory we fixed some data: point in the moduli space P and the cohomology element x. Note that the fact that x cannot be written as part of the kinetic term is because of the ∂^{-1} in the kinetic term, which renders the appearance of x meaningless. So the KS theory *does not have the degree of freedom to shift the complex structure* as a dynamical field in the theory. Instead the existence of the coupling with x *as a background field* in the interaction term is there to take care of this. One may ask how the theory changes if we choose a different base point P. We parametrize the position of the base point P in canonical coordinates $P = P(t, \overline{t})$. Ignoring the holomorphic anomaly the KS action depends only on t and is independent of $\overline{t}$. The shift in t coordinate can be achieved by shifting the field A by the solution of the KS equation (let $A_0(x)$ be the solution of KS equation). Then, consider the following identity

$$\lambda^2 S(A + A_0(x), x | t, \overline{t}) = \int_M A' \tfrac{1}{\partial}\left(\overline{\partial}A_0' + \tfrac{1}{2}\partial(x + A_0) \wedge (x + A_0))'\right) + \qquad (20)$$
$$\tfrac{1}{2} \int_M A_0' \tfrac{1}{\partial} \overline{\partial}A_0' + \tfrac{1}{6} \int_M ((x + A_0) \wedge (x + A_0))'(x + A_0)' +$$
$$\tfrac{1}{2} \int_M A' \tfrac{1}{\partial}\left[\overline{\partial}A' + \partial((x + A_0) \wedge A)'\right] + \tfrac{1}{6} \int_M (A \wedge A)'A' \ .$$

The first term vanishes due to the equation of motion. The second and third terms are naturally combined into the classical KS action evaluated on the solution of KS equation. The two remaining terms have an interpretation as the KS action around the *new background*. Indeed the combination in the square brackets coincides with the deformed $\bar{\partial}$ operator around the new background. There is still one subtlety: the *prime* operation is defined with respect to old background. In the new background the prime operation should be defined by contraction with the deformed holomorphic 3–form given by (14). Noticing, that only projection on $(3,0)$–forms contributes to the action, one can replace the prime operation around the old background by the prime operation around the new background. As a result of this formal manipulation we obtain the relation

$$S(A + A_0(x), x|t, \bar{t}) = S(A_0(x), x|t, \bar{t}) + S(A, 0|t + x, \bar{t}) . \tag{21}$$

In the original definition of the KS theory, t and $\bar{t}$ are complex conjugate to each other. Without the holomorphic anomaly, the KS action is independent of $\bar{t}$ and one can replace $S(\ |t + x, \bar{t})$ by $S(\ |t + x, \bar{t} + \bar{x})$. If such arguments were true they imply the background independence of the KS theory or background independence of the corresponding closed string field theory. The dependence of the KS action on $\bar{t}$ destroys background independence. In other words the holomorphic anomaly governs the background dependence of the KS action (see also discussion in [20]). In the presence of holomorphic anomaly, relation (21) may serve as the definition of the KS action where the condition $t = \bar{t}^*$ is relaxed.

We now come to a puzzle raised by Witten in his study of this theory [1]. It was pointed out in [1] that the fact that three point function C_{ijk} is not zero seems to be at odds with the fact that there is no obstruction to deforming by the marginal operators. The resolution of this puzzle in the context of the KS theory is simply that the massless fields, i.e., the string modes, *are not dynamical fields* and so there is no reason for the classical value of action to be independent of their expectation value (as we will discuss in more detail below). Thus the fact that the kinetic term cannot be defined unless we delete the massless modes means in particular that C_{ijk} may be non-zero even if the massless modes can be given arbitrary expectation value.

Being a quantum theory in six dimension it is not easy to explicitly compute higher loop amplitudes in the KS theory. In particular this 6-dimensional field theory looks highly non-renormalizable from the simple power counting argument. It is quite remarkable that topological string theory of the B–model provides a prescription to quantize the Kodaira-Spencer theory. The properly regularized Kodaira-Spencer theory should satisfy the property

$$e^{W(\lambda, x|t, \bar{t})} = \int DA e^{S(x, A|t, \bar{t})} , \tag{22}$$

where $W(\lambda, x|t, \bar{t})$ is the generating function for topological $N = 2$ CFT. We also introduce the notation $x = x^i \mu_i$, where μ_i is some basis in $H^{(0,1)}(T)$. Even though the r.h.s. of this equation is to be properly defined at higher loop, it is well defined as it stands for the tree level. Let us prove this relation at least at the tree level. In fact this relation also continues to hold at one–loop.

At the tree level, the contribution of the path-integral simply gives rise to the classical action evaluated for solutions to the field equations. Let us denote this action by $S_0(x, A_0|t, \bar{t})$ where $A_0(x)$ is such that $A_0(x) + x$ satisfies the KS equation (expanded about the base point $(t, \bar{t})$). Thus we need to show

$$W_0(x|t, \bar{t}) = \lambda^2 S_0(x, A_0(x)|t, \bar{t}) \, , \tag{23}$$

where W_0 is the tree level contribution to W (i.e., the coefficient of λ^{-2}). Note that in the x-coordinate, which is a canonical one, W_0 is defined by the condition

$$\partial_i \partial_j \partial_k W_0 = C_{ijk}(x) = \sum_{n=0}^{\infty} \frac{1}{n!} C^0_{ijks_1...s_n} x^{s_1}...x^{s_n} \, ,$$

and also W_0 has no linear or quadratic dependence on x. We see simply from the definition of S_0 that up to $O(x^3)$ they are thus equal. We need to show that it holds to all orders. Let us compare the third derivatives of both sides of (23). The third derivative of the classical action is given as follows

$$\frac{d^3 S_0}{dx^i dx^j dx^k} = [(\delta_A S)\partial_i \partial_j \partial_k A] + [(\delta_A^2 S)\partial_i A + (\delta_A \partial_i S)]\partial_j \partial_k A + \tag{24}$$
$$[(\delta_A^3 S)\partial_i A \partial_j A \partial_k A + \quad 3(\delta_A^2 \partial_i S)\partial_j A \partial A_k + 3(\delta_A \partial_i \partial_k S)\partial A_k + \partial_i \partial_j \partial_k S]$$

where δ_A is variational derivative with respect to A and $\partial_i = \partial/\partial x^i$ and symmetrization with respect to ijk is implicit. The first two terms vanish: the first one vanishes because $\delta_A S = 0$ by the equations of motion which is the definition of $A_0(x)$. The second term vanishes by taking derivative of $\delta_A S$, along the classical solution, with respect to x_i and expanding to the third order term. Finally the last term can be rewritten as

$$\frac{d^3 S_0}{dx^i dx^j dx^k} = \int ((\mu_i + \partial_i A_0) \wedge (\mu_j + \partial_j A_0))' \wedge (\mu_k + \partial_k A_0)' = C_{ijk}(x)$$

where the last equality follows from the alternative definition of Yukawa coupling discussed. This proves the equation (23).

$W_0(x|t, \bar{t})$ may be viewed as the effective action for the massless modes x having integrated out the massive modes. It is quite amazing that integrating the massive modes has only the effect of taking derivatives of the Yukawa coupling. One can use this fact to estimate the behavior of the partition function at genus g of the KS theory to all loops, as we approach the boundary of moduli space (see [7]).

The one-loop partition function of the KS theory is given by the product of determinants which coincides with appropriate combination of Ray-Singer torsions [14]

$$I(V) = \prod (\det' \Delta_V^{(p)})^{p(-)^p} \, ,$$

where $\Delta_V^{(p)} = \bar{\partial}_V^\dagger \bar{\partial}_V$ and $\bar{\partial}_V$ is the del-bar operator coupled to a vector bundle V. Namely, the one-loop partition function is equal to

$$F_1 = \frac{1}{2} \sum q(-1)^q \log I(\wedge^q T^*) = \frac{1}{2} \sum pq(-1)^{p+q} \log \det(\Delta_{p,q}) \tag{25}$$

Now, using the Quillen anomaly discussed in [17] one can deduce that (25) satisfies exactly the same anomaly equation as one-loop partition function of topological $N = 2$ CFT [16].

The relation between tree level amplitudes as well as one-loop coincidence present enough evidence to support the conjecture (22).

5 BV Formalism and Closed String Field Theory

In this section we quantize the KS action using the BV formalism which is particularly well suited to string theory. The interpretation of the KS theory as string field theory turns out to be very useful. In this interpretation the KS field A' is identified with the string field. But in string theory there are 'ghost' states, which mean that we are not restricted to ghost number $(1,1)$. Translated to the geometry of Calabi-Yau, this means that we should broaden the range of A so that $A \in \Omega^{(0,p)}(\wedge^q T_M)$; the ghost counting coincides with the fermion counting and is equal to $F_L + F_R = (p+q-3)$. The original KS field A' has the ghost number 2.

The consistent scheme for quantization string field theory is given by Batalin–Vilkovisky (BV) formalism [18] (for review see also [19]). In the Batalin-Vilkovisky formalism onc has to relax the condition for the ghost numbers of string field and include all possible fields with arbitrary ghost numbers. The fields A with ghost numbers $q(A) \leq 2$ are called fields, while the fields A^* with ghost numbers $q(A) > 2$ are called antifields. The space of functionals of fields–antifields is equipped with odd antibracket $\{\ ,\ \}$. The BRST symmetry is a canonical transformation in the antibracket. The BRST variations of the fields are given as follows

$$\delta_{BRST} \mathcal{A} = \{\mathcal{A}, S\} \ .$$

The original action is replaced by full action which depends on both fields and antifields. The full action satisfies two conditions. When all antifields are set to zero, the full action reduces to the original one. The full action also satisfies Batalin-Vilkovisky master equation

$$\{S, S\} = \hbar \Delta S \ , \tag{26}$$

where Δ is the natural Laplacian on the space of fields–antifields. The r.h.s. of (26) is contribution coming from the path integral measure. At the classical level ($\hbar = 0$), the Batalin-Vilkovisky equation is nothing else but the condition that full action is gauge invariant. The gauged fixed action is determined by an odd functional $\Psi(A)$ and is given by $S_\Psi(A) = S(A, A^* = \delta\Psi/\delta A)$.

In the case of the KS theory the full space of fields is a subspace $\mathcal{H}$ of $\oplus_{p,q} \Omega^{(0,p)}(\wedge^q T_M)$ satisfying the constraints (17). The space

$$\oplus_{p+q\leq 2} \Omega^{(0,p)}(\wedge^q T_M)$$

is the space of fields, while

$$\oplus_{p+q>2} \Omega^{(0,p)}(\wedge^q T_M)$$

is the space of antifields. Note that not all (p,q) are allowed, and the projection of $\mathcal{H}$ on $\oplus\Omega^{(0,p)}(\wedge^3 T_M)$ is empty. Taking into account that both fields and antifields satisfy constraints (17), we get exactly the same number of fields and antifields. Fields and antifields are paired with each other

$$A \in \Omega^{(0,p)}(\wedge^q T_M) \longleftrightarrow A^* \in \Omega^{(0,3-p)}(\wedge^{(2-q)} T_M) \ ,$$

and obey opposite statistics. The odd bracket structure on the space of field-antifields is given by

$$\{A_p^q(z), A_{\widetilde{p}}^{\widetilde{q}*}(w)\} = \delta_{p+\widetilde{p},3}\delta_{q+\widetilde{q},2}\Omega^{-1}\partial\delta(z,w)\overline{\Omega} \ ,$$

where $\delta(z,w)$ is the delta function on the Calabi–Yau manifold, defined as follows

$$\int_M \delta(x,y)\Omega(x) \wedge \overline{\Omega}(x) = 1 \ .$$

This structure is promoted to a canonical antibracket on the space of functionals and formally may be written as follows

$$\{F,L\} = \sum \int \left(\partial\left(\frac{\delta F}{\delta A}\right)^{\vee}\frac{\delta L}{\delta A^*} - \frac{\delta F}{\delta A^*}\partial\left(\frac{\delta L}{\delta A}\right)^{\vee}\right)^{\vee} \ .$$

It is quite remarkable that the full KS action is given by the same expression as the original KS action, but without any restrictions on the ghost numbers. Indeed, the ghost number conservation requires that either all fields in the action are elements of $\Omega^{(0,1)}(T_M)$, or at least one field has ghost number greater than 2 and therefore this field is an antifield. When all antifields are put to zero, the only contribution to the action comes from the original field $A \in \Omega^{(0,1)}(T_M)$. It is a tedious but straightforward check that the full action is invariant under the nonlinear gauge transformation. The proof is based on generalized Tian's Lemma (9) for arbitrary (p,q) forms and the generalized Jacobi identity (13).

The naive definition of the Laplacian turns out to be the correct one:

$$\Delta = \int \left(\frac{\delta}{\delta A^*}\partial\left(\frac{\delta}{\delta A}\right)^{\vee}\right)^{\vee} \ .$$

To verify that this definition is indeed covariant one has to take into account that $\delta A_p^q(x)/\delta A_r^s(y) = \delta_{p,r}\delta_{q,s}\delta(x,y)\Omega \wedge \overline{\Omega}$. Now we can check whether the full Kodaira Spencer action $S(A,A^*)$ satisfies the master equation (26). The gauge invariance of the full action implies that l.h.s of (26) is equal to zero. The r.h.s can be easily computed and it is equal

$$\Delta S \sim \int \partial(\Omega A_0^1) \wedge \overline{\Omega} = 0 \ .$$

Indeed, $\partial(\Omega A_0^1) = \partial(A_0^1)' = 0$ due to constraint (17). The above discussion implies that quantum corrections are not needed for maintaining the gauge invariance of the KS theory.

References

[1] E. Witten, *Chern–Simons Gauge Theory as a String Theory*, IASSNS-HEP-92/45, hep-th/9207094

[2] B. Gato-Rivera and A.M. Semikhatov, Phys. Lett. B293 (1992) 72-80

[3] M. Bershadsky, W. Lerche, D. Nemeschansky, N.P. Warner, Nucl.. Phys. B401 (1993) 304

[4] E. Witten, Nucl. Phys. B340 (1990) 281
R. Dijkgraaf and E. Witten, Nucl. Phys. B342 (1990) 486

[5] E. Verlinde and H. Verlinde, Nucl. Phys. B348 (1991) 457

[6] K. Li, Nucl. Phys. B354 (1991) 725-739

[7] M. Bershadsky, S. Cecotti, H. Ooguri and C. Vafa *Kodaira-Spencer Theory of Gravity and Exact Results for Quantum String Amplitudes*, HUTP-93/A025

[8] K. Kodaira and D.C. Spencer, Annals of Math. 67 (1958) 328
K. Kodaira, L. Niremberg and D.C. Spencer, Annals of Math. 68 (1958) 450
K. Kodaira and D.C. Spencer, Acta Math. 100 (1958) 281
K. Kodaira and D.C. Spencer, Annals of Math. 71 (1960) 43

[9] G. Tian, in *Essays on Mirror manifolds, ed. by S. T. Yau*, International Press, 1992
G. Tian, in *Mathematical aspects of String theory, ed. by S.T. Yau*, World Scientific, Singapore, 1987

[10] G. Tian, private communication

[11] A.N. Todorov, Comm. Math. Phys. 126 (1989) 325

[12] P. Griffiths and J. Harris *Principles of Algebraic Geometry* New York, Wiley, 1978

[13] M. Kuranishi, Annals of Math 75 (1962) 536

[14] D.B. Ray and I.M. Singer, Ann. Math. 98 (1973) 154

[15] J.M. Bismut and D.S. Freed, Comm. Math. Phys. 106 (1986) 159; 107 (1986) 103
J.M. Bismut, H. Gillet and C. Soule, Comm. Math. Phys. 115 (1988) 49, 79, 301

[16] M. Bershadsky, S. Cecotti, H. Ooguri and C. Vafa, Nicl. Phys. B405 (1993) 279

[17] B. Zwiebach, *Closed String Field Theory: Quantum Action and the B-V Master equation*, IASSNS-HEP-92/41, MIT-CTP-2102, hep-th/9206084

[18] I. A. Batalin and G. A. Vilkovisky, Phys. Rev D28 (1983) 2567

[19] M. Henneaux, *Lectures on the antifield-BRST formalism for gauge theories*, Proc. of XXII GIFT Meeting

[20] E. Witten, *Quantum Background Independence in String Theory*

3D GRAVITY AND GAUGE THEORIES

Dimitri BOULATOV

The Niels Bohr Institute
University of Copenhagen
Blegdamsvej 17,
DK-2100 Copenhagen Ø, DENMARK

Abstract: I argue that the complete partition function of 3D quantum gravity is given by a path integral over gauge-inequivalent manifolds times the Chern-Simons partition function. In a discrete version, it gives a sum over simplicial complexes weighted with the Turaev-Viro invariant. Then, I discuss how this invariant can be included in the general framework of lattice gauge theory (qQCD$_3$). To make sense of it, one needs a quantum analog of the Peter-Weyl theorem and an invariant measure, which are introduced explicitly. The consideration here is limited to the simplest and most interesting case of $SL_q(2)$, $q = e^{i\frac{2\pi}{k+2}}$. At the end, I dwell on 3D generalizations of matrix models.

1 Introduction

During the last few years, considerable progress has been made in our understanding of 2D quantum gravity and string theory (see review [8] and references therein). What helped greatly to fight the problem was the fortunate interplay of the methods of conformal field theory and the computational power of matrix models. For those who tries to think of quantum gravity seriously the next step has naturally been the path integral over 3D manifolds. It is not a priori doomed-to-fail enterprise. Indeed, although the problem is really a hard one, some interesting results have already been obtained. As well as in the 2D case, there are essentially two approaches. The first starts with a continuous formulation trying to make sense of a path integral over metrics. The main achievement on this way has been the connection with the Chern-Simons theory established by E.Witten [5]. The second approach is based completely on lattice experience. Here the path integral is substituted by a sum over all simplicial (or another kind of) complexes. The gained advantage is the finiteness of all involved quantities and relative simplicity, which allows for numerical investigations. However, the main problem, native to all lattice models, is what kind of continuum limit (if any) can be reached in every particular case?

In the present paper, I try to establish a connection between these two approaches, paying more attention to the second one, however.

Quantum Field Theory and String Theory, Edited by
L. Baulieu *et al.*, Plenum Press, New York, 1995

In Section 2 I remind a reader the basic notions of 3D general relativity and describe its connection with the Chern-Simons theory.

Section 3 is devoted to 3D simplicial gravity. I formulate the model and review some results of numerical investigations.

In Section 4 I define qQCD$_3$ and show that its weak-coupling limit is related to the Turaev-Viro invariant.

In Section 5 a model which can be regarded as a 3D generalization of the one-matrix model are introduced.

Section 6 contains some general remarks.

2   3d Gravity and Chern-Simons Interpretation

The partition function in Euclidean quantum gravity is intuitively defined as a sum (path integral) over all manifolds weighted with the exponential of a reparametrization invariant action

$$\mathcal{P} = \sum_{\mathcal{M}^3} e^S \tag{1}$$

By definition, the manifold is a topological space which can be globally covered with local coordinate systems. In other words, every its point has an open vicinity allowing for a continuous one-to-one map into R^3. On a manifold, one can define functions, vector fields, forms and tensors. To make sense of the partition function (1), a metric tensor, a volume form and an affine connection are needed. The metric tensor is a scalar bilinear-linear symmetric function on vectors, $i.e.$, a second-rank contravariant-variant tensor g_{ij}. The matrix g_{ij} has to be invertible: $g_{ij}g^{jk} = \delta_i^k$. If $g_{ij} \equiv 0$ on some sub-manifold, it should be regarded as non-compactness. Without metric one cannot define the functional integral measure.

The volume form is some fixed 3-form[a], $\tilde{V}$. It is always convenient to make it compatible with the metric. Then, in coordinates,

$$\tilde{V} = \sqrt{g}dx^1 \wedge dx^2 \wedge dx^3 \tag{2}$$

To choose a coordinate basis means to fix 3 mutually commutative vector fields

$$\hat{\partial}_1, \ \hat{\partial}_2, \ \hat{\partial}_3 : \ [\hat{\partial}_i, \hat{\partial}_j] = 0.$$

Sometimes, it is convenient to have a non-coordinate basis $\{\hat{e}_a\}$:

$$[\hat{e}_a, \hat{e}_b] = C_{ab}^c \hat{e}_c \tag{3}$$

Let me choose it such that

$$g_{ij}e_a^i e_b^j = \delta_{ab} \tag{4}$$

---

[a]In what follows, for convenience, I denote forms with the tilde and vector fields with the hat, $e.g.$, $\hat{\partial}_i \equiv \frac{\partial}{\partial x^i}$.

where e^i_a are the components: $\widehat{e}_a = e^i_a \widehat{\partial}_i$. I shall refer to them as the dreibein.

To introduce the Riemann tensor one needs the notion of an affine connection, which defines rules of a parallel transport of vectors: $\nabla_{\widehat{e}_a} \widehat{e}_b = \omega^c_{ba} \widehat{e}_c$.

If one uses forms[b], $\widetilde{e}^a = e^a_i dx^i$ and $\widetilde{\omega}^a_b = \omega^a_{bi} dx^i$, one can introduce the Riemann tensor $\widetilde{R}^a_b = \frac{1}{2} R^a_{b,ij} dx^i \wedge dx^j$ and the torsion $\widetilde{T}^a = \frac{1}{2} T^a_{ij} dx^i \wedge dx^j$ in the most elegant way (Cartan's structural equations)

$$\widetilde{R}^a_b = \mathrm{d}\widetilde{\omega}^a_b - \widetilde{\omega}^a_c \wedge \widetilde{\omega}^c_b \tag{5}$$

$$\widetilde{T}^a = \mathrm{d}\widetilde{e}^a - \widetilde{\omega}^a_b \wedge \widetilde{e}^b \tag{6}$$

If the torsion vanishes, the connection is said to be symmetric. In this case, it is determined by the commutator (3)

$$\omega^c_{ab} = \frac{1}{2}(C^a_{cb} + C^b_{ca} - C^c_{ab}) \tag{7}$$

The Einstein-Hilbert action can be written in the form

$$S = \lambda \int \epsilon_{abc} \widetilde{R}^a_b \wedge \widetilde{e}^c + \beta \int \widetilde{e}^1 \wedge \widetilde{e}^2 \wedge \widetilde{e}^3 \tag{8}$$

Witten suggested to consider the dreibein and the Levi-Civita connection as the gauge variable:

$$A_i = e^a_i P_a + \omega^a_{bi} J_{ab} \tag{9}$$

taking values in the $ISO(3)$ Lie algebra (if the signature is Euclidean and the cosmological constant is zero):

$$[J_{ab}, J_{cd}] = \delta_{ac} J_{bd} + \delta_{bd} J_{ac} - \delta_{bc} J_{ad} - \delta_{ad} J_{bc}$$

$$[J_{ab}, P_c] = P_a \delta_{bc} - P_b \delta_{ac} \tag{10}$$

$$[P_a, P_b] = 0$$

with the invariant metric on the algebra: $\langle P_a, P_b \rangle = \langle J_{ab}, J_{cd} \rangle = 0$, $\langle P_a, J_{bc} \rangle = \epsilon_{abc}$.

The obvious problem here is: what meaning do we give to the generators? If P_a's are to represent vector fields, $\widehat{P}_a$, forming a coordinate basis (they commute), then $[e^a_i \widehat{P}_a, e^b_j \widehat{P}_b]$ is not zero except for the case when the space is flat and the dreibein appears as a coordinate transformation: $e^a_i = \frac{\partial y^a}{\partial x^i}$. In a curved space, the Lie algebra generators have an indefinite meaning.

However, the construction is not so restrictive as might seem from this consideration.

Let $\widehat{v} = v^i \widehat{\partial}_i = \gamma^a \widehat{e}_a$ be an infinitesimal vector field. The variation of the basis $\widehat{\partial}_i$ under the diffeomorphism generated by $\widehat{v}$ is given by the Lie derivative

$$\pounds_{\widehat{v}} \widehat{\partial}_i \equiv [\widehat{v}, \widehat{\partial}_i] = \nabla_{\widehat{v}} \widehat{\partial}_i - \nabla_{\widehat{\partial}_i} \widehat{v} = (-\nabla_i \gamma^a + \omega^a_{bj} v^j e^b_i + v^j e^a_{i,j}) \widehat{e}_a \tag{11}$$

---

[b]The groups of indices abc and ijk belong to different bases!

where the comma means the derivative with respect to x^j. The first equality holds if the torsion tensor identically vanishes. So, from the view-point of the fixed non-coordinate basis $\hat{e}_a$, the variation of the basis $\hat{\partial}_i$ consists of (i) a "gauge transformation" $\nabla_i \gamma^a$, (ii) a "Lorentz rotation" $\omega^a_{bj} v^j e^b_i = \tau^a_b e^b_i$ and (iii) a coordinate shift $v^j e^a_{i,j} \approx e^a_i(x+v) - e^a_i(x)$. The last term can be removed by "pulling back" e^a_i to its initial point in the x-frame as a scalar function.

Of course, it just repeats the famous Witten's argument [5] that diffeomorphisms generated by vector fields can be regarded on-shell as gauge transformations of the field (9). Maybe, it should be stressed here that one is restricted to *reparametrizations*, which are not the most general transformations. In particular, they do not affect the commutators (3).

The complete algebra of vector fields is infinite dimensional, since the structure constants C^a_{bc} are arbitrary functions of coordinates. Hence, it cannot as a whole be reduced to any finite dimensional symmetry, if one insists on the interpretation of its generators as vector fields. However, we have seen that reparametrizations can be regarded as the gauge transformations. One can, in principle, get rid of them by fixing a gauge and pulling out of the path integral a volume they produce. For compact manifolds, this volume gives a topological invariant (up to some trivial (but maybe infinite) factor). Indeed, one can choose an arbitrary background metric, and the most convenient choice is a solution to the Einstein equation (classical vacuum). In this case, one finds an integral over flat connections. Witten has noticed that, if one considers the dreibein and connection as independent variables, the Riemann-Hilbert action (8) takes the form of the Chern-Simons one for the group $SO(4)$. In this case, the vanishing torsion and the Einstein equation are implied by the equations of motion and one finds that the gauge volume should be given by the Chern-Simons partition function. Off-shell, of course, any equivalence between diffeomorphisms and the gauge transformations disappears.

The non-renormalizability of $3D$ gravity means that one should work within a regularization scheme, the choice of which can be crucial (*i.e.*, answers will vary from scheme to scheme drastically).

3 Quantum Regge Calculus

The heuristic consideration of the previous section serves to support the following substitution for the path integral over all $3D$ geometries:

$$P = \sum_{\text{topologies}} I_{TV} \sum_C e^S \tag{12}$$

where the first sum goes over all topologies; $\sum_C$ is the sum over all simplicial complexes of a given topology; I_{TV} is the Turaev-Viro invariant [3]; S is a lattice action, which can be taken in the form

$$S = \alpha N_1 - \beta N_3 \tag{13}$$

(N_k is the number of k-dimensional simplexes in a complex).

The Turaev-Viro invariant is the most reasonable substitution for the gauge volume. For a negative cosmological constant, one finds $SO(4)$ Chern-Simons theory and, as $SO(4) = SU(2) \times SU(2)$, I_{TV} seems to be the most appropriate candidate [4]. Its "classical" limit was investigated long ago by Ponzano and Regge [5] in the framework of the Regge calculus [6]. Provided a triangulation is fixed, it describes an integral over lengths of all links with a weight equal to an exponential of the discretized Einstein-Hilbert action. The Turaev-Viro invariant in this context may be regarded simply as a regularization of the Ponzano-Regge construction.

As all lengths are included in I_{TV}, we can choose every tetrahedron in $\sum_C$ to be equilateral. This sum serves as a natural regularization of the path integral over classes of gauge-inequivalent manifolds.

Using reparametrizations, one can make lengths of the dreibein vectors equal to unity, three remaining local degrees of freedom being angles between them. As usual, on a lattice, one should work with a group rather than an algebra. It means that, instead of dreibeins, their integral curves have to be considered. In a discrete version, one fixes a finite number of the curves going from every vertex and associate them with links of a lattice. It is convenient to make lengths of all links equal to one another. In simplicial complexes, angles between them are quantized, which leads to a quantized total curvature. The Regge calculus gives the expression for it

$$\int d^3x \sqrt{g}R = a\left(2\pi N_1 - 6N_3 \arccos \frac{1}{3}\right) \tag{14}$$

a is a lattice spacing.

For manifolds, the Euler character vanishes $\chi = N_0 - N_1 + N_2 - N_3 = 0$. Together with the constraint $N_2 = 2N_3$, it implies that a natural action (linear in the numbers of simplexes) depends on two free parameters which should be related to bare cosmological and Newton constants.

To simulate all geometries, one has to sum over all possible complexes. Indeed, if a triangulation is fixed, commutators of lattice shifts (analogs of the structure constants in Eq. (3)) are fixed. *Fluctuating geometry assumes a fluctuating lattice.*

If a topology is fixed, the sum over simplicial complexes can be investigated numerically. Any two complexes of the same topology can be connected by a sequence of moves shown in Figure 1. The first move is called the triangle-link exchange: the common triangle of two tetrahedra on the left of Figure 1(a) is removed and three new triangles sharing the new link appear on the right. It increases (the inverse one decreases) the number of tetrahedra by 1. The second move consists in the subdivision of a tetrahedron: 4 new tetrahedra fill an old one. The inverse move is seldom possible. However, to perform it, one can always decrease the coordination number of a vertex by applying the triangle-link exchange.

Monte-Carlo simulations using these moves as basic "infinitesimal steps" appear to be quite efficient.

I do not intend to give a review of the numerical results here. An interested reader is referred to the original papers [7, 8, 9]. However, a few words should be said. All simulations so far have been carried out for the spherical topology of complexes. It appears that the number of spherical complexes of a given volume, N_3, is exponentially bounded as a function of N_3 for an arbitrary value of α. It means that the definition (12) is reasonable and P hopefully has an appropriate continuum limit.

One of the most interesting observations is the resolution of the problem of the unboundedness of the Riemann-Hilbert action within the discrete model. As Eq. (13) is linear in N_1, it seems that most probable configurations should be those having the maximum mean curvature, but they are surely lattice artifacts. However, it happened that, at N_3 and α fixed, the probability distribution for N_1 has roughly speaking the Gaussian shape

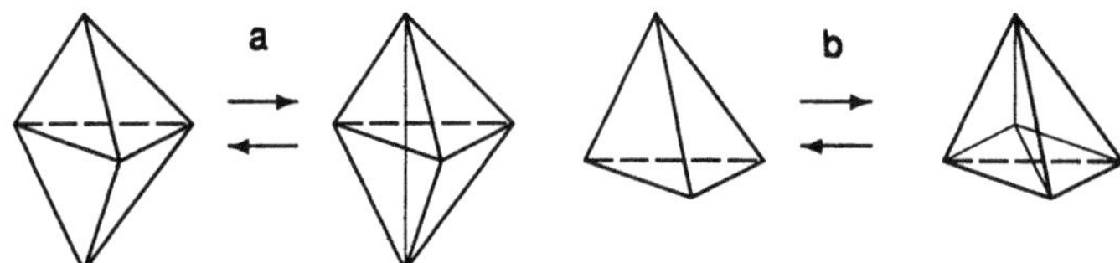

Figure 1. The triangle-link exchange (a) and the subdivision (b)

$$P_{N_3,\alpha}(N_1) \approx e^{-\frac{(N_1 - \langle N_1 \rangle)^2}{2\sigma^2}} \tag{15}$$

It means that, varying α, one just shifts a position of the maximum. Moreover, in Refs. [8], a first order phase transition was found at some critical value, α_c. In the "hot" phase ($\alpha < \alpha_c$), crumpled manifolds dominate the partition function. This phase is clearly unphysical. In the "cold" phase ($\alpha > \alpha_c$), it happens that

$$\langle N_1 \rangle(N_3) = c_1 N_3 + c_2 \tag{16}$$

is a linear function; c_1 is a constant smoothly depending on α. Hence, the mean curvature per unit volume makes sense in the large volume limit. However, after the naive rescaling, one finds that its value tends to the infinity in the continuum limit. But, the total curvature can not be regarded as an observable in quantum gravity. In the Einstein-Hilbert action, there are two terms with dimensionful coupling constants in fronts of them. After a regularization, one finds the action (13) (or similar) where the total curvature has lost its individuality and is mixed with the volume. Let us imagine a kind of renormalization group procedure: one increases a cut-off and integrate over fluctuations inside the blocks. The additive nature of the total curvature means that it should undergo an additive renormalization as well as a multiplicative one. It is natural to kill the first by shifting the cosmological constant. Therefore, the mean value of the total curvature is scheme-dependent and only fluctuations make sense.

Four dimensional numerical simulations show a similar picture [10]. The phase transition there is, presumably, of the second order which might be an evidence for graviton-like (*i.e.*, long-range) excitations in the system.

Here, the following comment is in order. One can easily obtain a continuous manifold from a simplicial complex by using a piece-wise linear approximation and then to smooth it. In three dimensions, any continuous manifold allows for a unique differentiable structure and *vice versa*. It means that the continuous and discrete models are hopefully equivalent. In four dimensions, the situation is much more complicated [11] and it is unclear whether an entropy of smooth manifolds can be correctly estimated within a lattice approximation. However, simplicial gravity is interesting in its own rights. One can simply say that, at the quantum level, the notion of the continuous manifold is more fundamental than of the smooth one.

4 q-Deformed Lattice Gauge Theory (qQCD$_3$)

In this section, I would like to show that the Turaev-Viro invariant can be interpreted as a lattice gauge model (although a not quite standard one). Let me start with reminding basic facts about lattice QCD [12].

Given a d-dimensional lattice, a gauge variable g_ℓ taking values in a compact group G is attached to each 1-dimensional link, ℓ, and the Boltzmann weight,

$$w_\beta(x_f) = \sum_R d_R \chi_R(x_f) e^{-\beta C_R}, \tag{17}$$

to each 2-dimensional face, f. The argument is a holonomy along the face, *i.e.*, the ordered product of gauge variables along a boundary, ∂f, of the face f:

$$x_f = \prod_{k \in \partial f} g_k \tag{18}$$

In Eq. (18), every factor is taken respecting an orientation of links and faces. The change of the orientation corresponds to the conjugation $g_k \to g_k^+$ (or $x_f \to x_f^+$).

By a lattice I mean a cell (polyhedral) decomposition of a d-dimensional manifold such that any cell can enter in a boundary of another one only once, and every two cells can border upon each other along only one less dimensional cell. Simplicial complexes and their duals obey this restriction by definition. In eq. (17), $\sum_R$ is the sum over all irreps of the gauge group G; $\chi_R(x_f)$ is the character of an irrep R; $d_R = \chi_R(I)$ is its dimension; C_R is a second Casimir and β is a number. The construction makes sense for compact groups when unitary finite dimensional irreps span the regular representation. Therefore, R is always a discreet index. The choice (17) provides that $w_\beta(x_f)$ becomes the group δ-function when $\beta \to 0$:

$$w_0(x_f) = \delta(x_f, I) \tag{19}$$

The partition function is defined as the integral over all field configurations:

$$Z_\beta = \int_G \prod_\ell dg_\ell \prod_f w_\beta \left(\prod_{k \in \partial f} g_k \right) \tag{20}$$

where dg is the Haar measure on the group G.

Now, we would like to make the gauge group quantum[c]. The simplest example of quantum group is $GL_q(2)$ elements of which can be defined as

$$g = \begin{pmatrix} a & b \\ c & d \end{pmatrix} \tag{21}$$

where

$$\begin{array}{lll} ab = qba & bd = qdb & bc = cb \\ ac = qca & cd = qdc & ad - da = (q - q^{-1})bc \end{array} \tag{22}$$

The matrices can be multiplied. If elements of both g_1 and g_2 obey Eq. (22) and are mutually commutative, the elements of the product obey (22) as well [13]. Therefore, matrices on different links of a lattice have to commute with one another in the tensor product (as well as with matrices performing gauge transformations).

The determinant

$$\mathrm{Det}_q \, g = ad - qbc \tag{23}$$

is central, therefore, one can put it equal to 1. In this way, one arrives at $SL_q(2)$, which has two real forms: $SU_q(2)$, for real q, and $SL_q(2, R)$, for $|q| = 1$ [13].

The relations (22) imply the existence of the R-matrix

$$R = \begin{pmatrix} q & 0 & 0 & 0 \\ 0 & 1 & 0 & 0 \\ 0 & q - q^{-1} & 1 & 0 \\ 0 & 0 & 0 & q \end{pmatrix} \tag{24}$$

and the $RTT = TTR$ equation

$$R g_1 \otimes g_2 = g_2 \otimes g_1 R \tag{25}$$

R itself obeys the Yang-Baxter equation

$$R_{12} R_{13} R_{23} = R_{23} R_{13} R_{12} \tag{26}$$

Indices show at which positions in the tensor cube $V \otimes V \otimes V$ acts the R-matrix.

For classical gauge groups the self-consistency of the model follows from the Peter-Weyl theorem stating that the algebra of regular functions on a compact group is isomorphic to the algebra of matrix elements of finite dimensional representations. The quantum version of this theorem was proven for real q's in Refs. [14]. In this case there is the one-to-one correspondence between representations of $SU_q(N)$ and $SU(N)$, and the notion of the matrix element is naturally generalized.

---

[c]In this context, the word "quantum" may be misleading, but it has already become standard having actually supplanted the term "q-deformed".

Therefore, by the space of functions, one may mean a vector space spanned by matrix elements of irreps, $T_{j,\alpha\beta}(g)$. Eq. (21) can be regarded as the fundamental representation. Matrix elements always obey the $RTT = TTR$ equation (25) and, by definition,

$$T_{j,\alpha\beta}(gh) = \sum_{\gamma=-j}^{j} T_{j,\alpha\gamma}(g)T_{j,\gamma\beta}(h) \tag{27}$$

The next ingredient is the integral, whose existence is postulated. It is defined simply as

$$\int dg\, T_j(g) = \delta_{j,0} \tag{28}$$

i.e., whatever is integrated the answer is always zero except for the trivial representation, which is just a constant.

If one has a product of functions, one can always re-expand products of matrix elements by using Clebsch-Gordan coefficients:

$$T_{j_1,\alpha\beta}(g)T_{j_2,\gamma\delta}(g) = \sum_{j_3=|j_1-j_2|}^{j_1+j_2} \sum_{\sigma,\varepsilon=-j_3}^{j_3} \langle j_1\alpha,\, j_2\gamma|j_3\sigma\rangle T_{j_3,\sigma\varepsilon}(g)\langle j_3\varepsilon|j_1\beta,\, j_2\delta\rangle \tag{29}$$

Applying this equation successively one can, in principle, reduce an arbitrary integral to the basic one (28).

The last ingredient is the character entering the definition of the weight (17). It should be said that this notion is missing for $SL_q(2)$. If one naively defines it as the quantum trace of a matrix element,

$$\chi_j(g) \overset{?}{=} \mathrm{Tr}_q T_j(g) = \sum_{\alpha=-j}^{j} q^{\alpha} T_{j,\alpha\alpha}(g) \tag{30}$$

then one finds that $[\chi_j(gh), \chi_i(gf)] \neq 0$. It seems to be impossible to q-deform the partition function (20) in a self-consistent way simply starting with this definition! It is a manifestation of the fact that qQCD$_D$ does not exist at arbitrary D.

On the other hand, the quantum dimension is equal to the quantum trace of the identity operator:

$$[2j+1] \equiv \sum_{\alpha=-j}^{j} q^{\alpha} = \frac{q^{j+\frac{1}{2}} - q^{-j-\frac{1}{2}}}{q^{\frac{1}{2}} - q^{-\frac{1}{2}}} \tag{31}$$

However, to define the partition function (20) in the quantum case, one does not actually need the notion of the character!

The profound correspondence between quantum groups and links of knots[d] suggests that the most adequate way to define qQCD$_3$ would be to connect all involved notions with certain geometric objects. After a projection onto a plane, the partition function can be given a meaning by putting into correspondence quantum-group quantities to all geometrical elements. If one takes another plane, quantum-group symmetries should

---

[d]I use the term "link" to denote 1-dimensional simplexes as well hoping it should not lead to misunderstanding.

provide the independence of the construction from a way of projection. I shall follow closely Ref. [15]. The basic notion is the tangle, which is defined as follows. One takes a spherical ball inside which there are a number of oriented loops and segments whose ends lie on the boundary of the ball. They are all colored with $SL_q(2)$ representations. One puts into correspondence to every tangle an operator O acting in the tensor product of representation spaces $V_{j_1} \otimes \ldots \otimes V_{j_n}$, (if there are n segments colored $j_1, \ldots, j_n$; their orientations show the direction of the action of O):

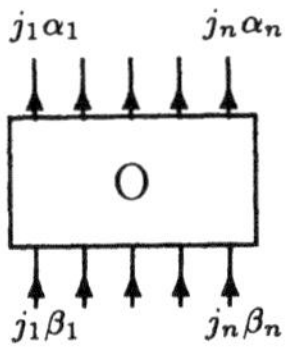

$$\equiv O_{j_1,\alpha_1\beta_1;\ldots;j_n,\alpha_n\beta_n} \tag{32}$$

For example, if there is only one segment and no loops, one finds the δ-function:

$$\equiv \delta_{\alpha,\beta} \tag{33}$$

The R-matrix distinguishes between under- and over-crossings:

$$\equiv R = \sum_i a_i \otimes b_i \tag{34}$$

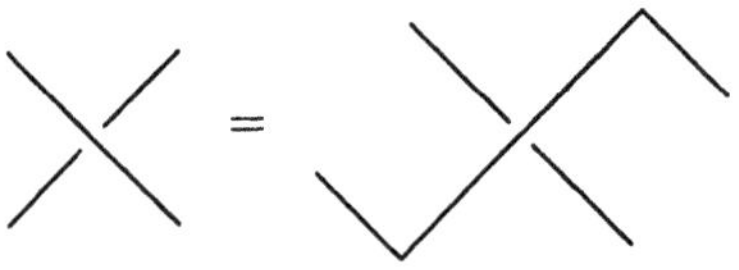

$$\equiv R^{-1} = \sum_i b_i \otimes s(a_i) \tag{35}$$

where s is the antipode in the $\mathcal{U}_q(sl(2))$ Hopf algebra. The Clebsch-Gordan coefficients are represented as the 3-valent vertices

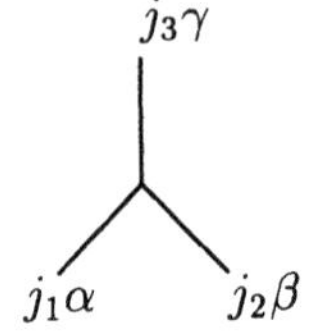

$$\equiv \langle j_3\gamma | j_1\alpha, j_2\beta \rangle \tag{36}$$

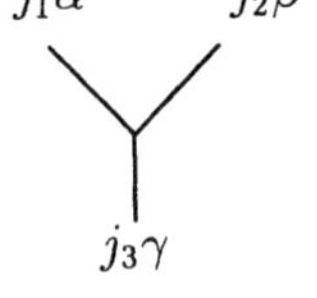

$$\equiv \langle j_1\alpha, j_2\beta | j_3\gamma \rangle \tag{37}$$

48

Matrix elements can be drawn as

$$\boxed{g} \equiv T_{j,\alpha\beta}(g) \tag{38}$$

The Yang-Baxter and $RTT = TTR$ equations take the familiar graphical forms

$$\tag{39}$$

and

$$\boxed{g_1}\boxed{g_2} = \boxed{g_2}\boxed{g_1} \tag{40}$$

One needs also the quantum trace of an operator, which is equivalent to the closure of a tangle

$$O \equiv \sum_{\alpha_1=-j_1}^{j_1} \cdots \sum_{\alpha_n=-j_n}^{j_n} \prod_{i=1}^{n} q^{\alpha_i}\, O_{j_1,\alpha_1\alpha_1;\ldots;j_n,\alpha_n\alpha_n} \tag{41}$$

To each link of the lattice, one puts into correspondence an integral of a product of matrix elements, the number of which is equal to the number of faces incident to the link. One can associate a tangle with every such integral. It means a cell decomposition of the manifold. The partition function can be constructed by connecting these tangles together or, equivalently, by gluing up the 3-cells. There appears an index loop going along a boundary of every face. In three dimensions, there is a natural cyclic order of faces sharing the same link. The index loops have to be ordered according to it. After that the partition function can be unambiguously defined.

If $q = e^{i\frac{2\pi}{k+2}}$, one has to restrict all indices to the fusion ring: $j = 0, \frac{1}{2}, 1, \ldots, \frac{k}{2}$. In this case the following tangle can serve as the definition of the matrix element

$$T_{j,\alpha\beta}(g) \equiv \tag{42}$$

The tensor product of matrix elements looks as

$$T_{j_1,\alpha_1\beta_1}(g)T_{j_2,\alpha_2\beta_2}(g) \equiv \quad \tag{43}$$

and the integral takes the form of the finite sum

$$\int dg\, T_{j,\alpha\beta}(g) \equiv d_0 \sum_{i=0}^{k/2} d_i \quad =$$

$$\frac{2\sin\frac{\pi}{k+2}}{k+2} \sum_{i=0}^{k/2} \sin\frac{\pi(2i+1)}{k+2}\ \frac{\sin\frac{\pi(2i+1)(2j+1)}{k+2}}{\sin\frac{\pi(2j+1)}{k+2}}\ \delta_{\alpha,\beta} = \delta_{j,0}\delta_{\alpha,0}\delta_{\beta,0} \tag{44}$$

where d_j is the quantum dimension conveniently normalized:

$$d_j = \sqrt{\frac{2}{k+2}}\sin\frac{\pi(2j+1)}{k+2} \tag{45}$$

To prove Eq. (44), I used results of Reshetikhin and Turaev [15]. My claim is that it can be regarded as the definition of the integral on the fusion ring of $SL_q(2)$, $q = e^{i\frac{2\pi}{k+2}}$.

In addition to the fusion ring irreps $\{V_j\}$, $j = 0, \frac{1}{2}, \ldots, \frac{k}{2}$; $\mathcal{U}_q(sl(2))$ has a number of representations having the vanishing quantum dimension [16]: $\{I_p\}$, $p = -\frac{1}{2}, 0, \frac{1}{2}, 1, \ldots, \frac{k+1}{2}$. Representations I_p for $0 \leq p \leq \frac{k+1}{2}$ although not irreducible are indecomposable.

The tensor product of two irreps from the fusion ring has the following decomposition

$$V_i \otimes V_j = \left(\bigoplus_{m=|i-j|}^{\min(i+j,k-i-j)} V_m \right) \oplus \left(\bigoplus_{\substack{-\frac{1}{2}\leq p \leq i+j-\frac{k+2}{2} \\ p=(i+j)\,\mathrm{mod}\,1}} I_p \right) \tag{46}$$

The set of representations $\{I_p\}$ forms an ideal

$$\{V_j\} \otimes \{I_p\} \subset \{I_p\} \qquad \{I_p\} \otimes \{I_p\} \subset \{I_p\} \tag{47}$$

As was proven by Reshetikhin and Turaev [15], the closure of any tangle vanishes if at least one representation of this type appears in it. More precisely, they have shown that any $\mathcal{U}_q(sl(2))$-linear operator acting in $\{I_p\}$ has the vanishing quantum trace. Obviously, it holds for operators obtained by cutting an internal line in an arbitrary closed tangle. As it takes place for any line, colors from the set $\{I_p\}$ never appear. Therefore, when all index loops are closed, these representations can be simply ignored. Thus, one has the following orthogonality property

$$\int dg\, T_{j_1,\alpha_1\beta_1}(g)T_{j_2,\alpha_2\beta_2}(g) \equiv d_0 \sum_{i=0}^{k/2} d_i \quad =$$

$$\sum_{m=|j_1-j_2|}^{\min(j_1+j_2,k-j_1-j_2)} d_0 \sum_{i=0}^{k/2} d_i \; \overset{j_1,\alpha_1 \; j_2,\alpha_2}{\underset{j_1,\beta_1 \; j_2,\beta_2}{\bowtie}} = \frac{d_0 \delta_{j_1,j_2}}{d_{j_1}} \; \overset{\alpha_1 \quad \alpha_2}{\underset{\beta_1 \quad \beta_2}{\cup\cap}} \tag{48}$$

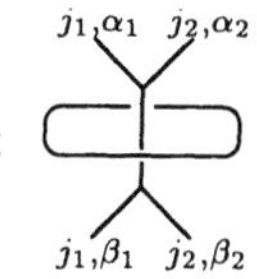

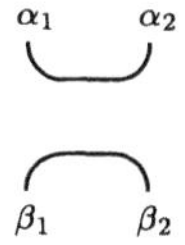

which allows for a Fourier decomposition of an arbitrary function spanned by matrix elements from the fusion ring. It gives an analog of the Peter-Weyl theorem. A reader must realize that Eqs. (44) and (48) do not hold for $\{I_p\}$ representations. For self-consistency, all Greek indices have to be summed over to form a link of 3-valent graphs and loops. *All equalities between tangles have to be understood as taking place after closing with an arbitrary tangle.*

So, we arrive at the following definition of qQCD$_3$ partition function on a 3-manifold $\mathcal{M}$ [17]:

$$Z_\beta(\mathcal{M}) = d_0^{N_3+N_0-2} \sum_{\{j_f\}} \prod_{f=1}^{N_2} [d_{j_f} c_{j_f}(\beta)] \sum_{\{j_\ell\}} \prod_{\ell=1}^{N_1} d_{j_\ell} \; J_{\{j_\ell\},\{j_f\}}(\mathcal{L}) \tag{49}$$

where $J_{\{j_\ell\},\{j_f\}}(\mathcal{L})$ is the Jones polynomial for a link $\mathcal{L}$ defined by a cell decomposition of $\mathcal{M}$. This link consists of N_1+N_2 unframed loops colored with sets of representations $\{j_\ell\}$ and $\{j_f\}$. Loops from the first set, $\{j_\ell\}$, go around 1-cells (links) pinching bunches of loops from the second set, $\{j_f\}$, which go along boundaries of 2-cells (faces); $c_j(\beta)$'s are numbers (weights). If all $c_j \equiv 1$, Z_0 is a topological invariant. In this case, the partition function (49) is obviously self-dual with respect to the Poincaré duality of complexes.

Eq. (49) is just a particular implementation of the general Reshetikhin-Turaev construction [15]. However, the link $\mathcal{L}$ here is not related to a surgery representation of the manifold.

In order to establish a connection with the Turaev-Viro invariant, let us consider lattices dual to simplicial complexes. Their 1-skeletons are 4-valent graphs and exactly 3 faces are incident to each link giving the integral of 3 matrix elements for every triangle in a simplicial complex:

$$\int dg \; T_{j_1,\alpha_1\beta_1}(g) T_{j_2,\alpha_2\beta_2}(g) T_{j_3,\alpha_3\beta_3}(g) = \frac{d_0}{d_{j_3}} \; \overset{j_1,\alpha_1 \; j_2,\alpha_2 \; j_3,\alpha_3}{\underset{j_1,\beta_1 \; j_2,\beta_2 \; j_3,\beta_3}{}} \tag{50}$$

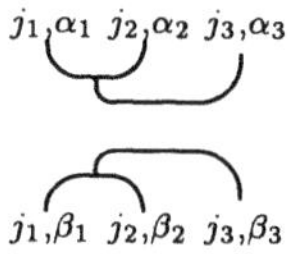

The right hand side of eq. (50) is the product of two 3-j symbols. Summing over lower indices one gets a Racah-Wigner 6-j symbol

$$\left\{ \begin{array}{ccc} j_1 & j_2 & j_3 \\ j_4 & j_5 & j_6 \end{array} \right\} = \frac{d_0^2}{d_{j_6}\sqrt{d_{j_2}d_{j_5}}} \; \tag{51}$$

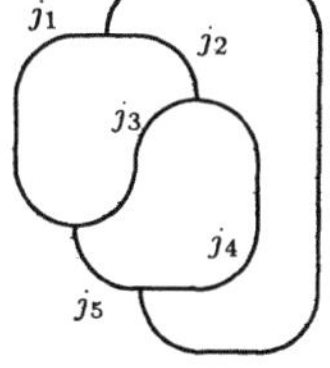

inside each tetrahedron of a simplicial complex. Representation indices, j_f, are attached to its 1-simplexes, f, (*i.e.* faces of the dual lattice). The partition function Z_0 can be written then in the Turaev-Viro form

$$Z_0 = d_0^{N_1 - N_2 - 2} \sum_{\{j_f\}} \prod_{f=1}^{N_2} d_{j_f} \prod_{t=1}^{N_0} \left\{ \begin{matrix} j_{t_1} & j_{t_2} & j_{t_3} \\ j_{t_4} & j_{t_5} & j_{t_6} \end{matrix} \right\} \tag{52}$$

where the indices $t_1, \ldots, t_6$ denote six edges of a t'th tetrahedron.

To prove that Z_0 is indeed a topological invariant it is sufficient to show that it is unchanged under the moves shown in Figure 1. However, the link representation (49) is more convenient in this respect. By using the analog of the group measure invariance

$$\int dg \ f(gh) \equiv d_0 \sum_{i=0}^{k/2} d_i \ \cdots = d_0 \sum_{i=0}^{k/2} d_i \ \cdots = \int dg \ f(g) \tag{53}$$

one can reduce the number of loops. The corresponding operations have a nice interpretation as topology preserving transformations of complexes [17].

For lattices, as they have been defined above, all loops are unframed. However, transforming complexes, one can obtain non-trivial framings and, in real calculations, has to follow them carefully. Practically, it is convenient to use the ribbon graph representation [15]. The framing of a ribbon loop is defined as a linking number of its edges. It is fixed by the condition that one of two sides of the ribbon is always turned toward the inside of a 2-cell which it encircles (or toward a 1-cell which it wraps).

The invariant is multiplicative with respect to the connected sum of complexes

$$Z_0(C_1 \# C_2) = Z_0(C_1) Z_0(C_2) \tag{54}$$

because for the sphere

$$Z_0(S^3) = 1 \tag{55}$$

Every oriented complex can be transformed into the canonical form, when there are single 0- and 3-dimensional cells and the equal number, ν, of 1- and 2-cells:

$$C = \sigma^0 \cup \left(\bigcup_{i=1}^{\nu} \sigma_i^1 \right) \cup \left(\bigcup_{j=1}^{\nu} \sigma_j^2 \right) \cup \sigma^3 \tag{56}$$

One can put into correspondence with each 1-cell, σ_i^1, a generator of the fundamental group $\gamma_i \in \pi_1(C)$. Each 2-cell, σ_j^2, gives a defining relation for $\pi_1(C)$:

$$\Gamma_j = \prod_{\sigma_k^1 \in \partial \sigma_j^2} \gamma_k = I \tag{57}$$

If the gauge group is a classical finite group G, the partition function Z_0 is well defined (after substituting the sum $\sum_{g \in G}$ for $\int dg$):

$$Z_0^{(G)} = \sum_{\{g_i\}} \prod_{j=1}^{\nu} \delta(\prod_{\sigma_k^1 \in \partial\sigma_j^2} g_k, I) \tag{58}$$

This expression equals the number of representations of the fundamental group by elements of the gauge one: $\pi_1(C) \to G$. Hence, it is an integer. In the quantum case, a similar interpretation exists. One have to consider an action of the fundamental group on the universal covering of a complex. It acts permuting cells of the covering, which can be regarded as a $\pi(C)$-module. The invariant can be said to be the "quantum analog" of Eq. (58), where the $\pi(C)$-action on the universal covering is represented by elements of a quantum gauge group. It is a real number.

As was proven by Turaev [18], the Turaev-Viro invariant is equal to the Reshetikhin-Turaev-Witten one modulo squared: $Z_0(\mathcal{M}) = |I(\mathcal{M})|^2$. Kohno [19] has shown that it is bounded from above as

$$Z_0(\mathcal{M}) \leq \left(\frac{1}{d_0}\right)^{2h} \tag{59}$$

where h is a Heegaar genus of $\mathcal{M}$, $i.e.$, the minimum genus of handlebodies appearing in Heegaar splittings of $\mathcal{M}$.

5 Generating Function for Simplicial Complexes

In Refs. [20, 17] the zero-dimensional field model generating all possible simplicial complexes weighted with the partition function (52) was suggested. Let $\phi(x,y,z)$ be a function on $G \otimes G \otimes G$ invariant under right shifts

$$\phi(x,y,z) = \phi(xu, yu, zu) \qquad \forall x,y,z,u \in G \tag{60}$$

and symmetric under even permutations. Odd ones are equivalent to the complex conjugation:

$$\phi(x,y,z) = \phi(y,z,x) = \phi(z,x,y) = \overline{\phi}(y,x,z) \tag{61}$$

It can be represented in terms of matrix elements as

$$\phi(x,y,z) = \sum_{\substack{j_1 j_2 j_3 \\ \{a_i,b_i\}}} \sqrt{d_{j_1} d_{j_2}}\, \varphi_{a_1 a_2 a_3}^{j_1 j_2 j_3} T_{j_1,a_1 b_1}(x) T_{j_2,a_2 b_2}(y) T_{j_3,a_3 b_3}(z) \begin{pmatrix} j_1 & j_2 & j_3 \\ b_1 & b_2 & b_3 \end{pmatrix}$$

$$= \sum_{\{j_i,a_i,b_i\}} \sqrt{\frac{d_{j_1} d_{j_2}}{d_{j_3}}} \quad \boxed{\varphi} \atop \boxed{x}\ \boxed{y}\ \boxed{z} \tag{62}$$

where $\begin{pmatrix} j_1 & j_2 & j_3 \\ b_1 & b_2 & b_3 \end{pmatrix}$ is the 3-j symbol; $\varphi_{a_1 a_2 a_3}^{j_1 j_2 j_3} = \overline{\varphi}_{a_2 a_1 a_3}^{j_2 j_1 j_3}$ and symmetric under cyclic permutations. This equation is a general Fourier decomposition of a function obeying (60).

The partition function is defined as the integral

$$P = \int \mathcal{D}\phi \, e^{-S} \tag{63}$$

where the action is taken in the form

$$S = \frac{1}{2} \int dx dy dz \, |\phi(x,y,z)|^2 -$$

$$\frac{\lambda}{12} \int dx dy dz du dv dw \, \phi(x,y,z)\phi(x,u,v)\phi(y,v,w)\phi(z,w,u) \tag{64}$$

The first term in eq. (64) can be imagined as two glued triangles and the second, as four triangles forming a tetrahedron. It is not surprising that, after the Fourier transformation, one finds a 6-j symbol associated with it:

$$S = \frac{1}{2} \sum_{j_1 j_2 j_3} \frac{1}{d_{j_3}} \quad - \quad \frac{\lambda}{12} \sum_{j_1 \ldots j_6} \frac{1}{d_{j_1}^2 d_{j_2} d_{j_3}} \tag{65}$$

The measure can be written in terms of Fourier coefficients

$$\mathcal{D}\phi = \prod_{\substack{j_1 j_2 j_3 \\ a_1 a_2 a_3}} d\varphi^{j_1 j_2 j_3}_{a_1 a_2 a_3} \tag{66}$$

If $q = e^{i \frac{2\pi}{k+2}}$, the product in eq. (66) runs over irreps from the fusion ring and, hence, is finite.

Practically, the partition function (63) has a meaning within the perturbation expansion in λ. Performing all possible Wick pairings, one gets in every order in λ all oriented simplicial complexes. For every 1-simplex in a simplicial complex, one has a loop carrying a representation index. It gives a corresponding quantum dimension. A 6-j symbol inside each tetrahedron has already appeared in eq. (65). Summing over all representations on links, one reproduces the Turaev-Viro partition function for a given simplicial complex.

Therefore, $\log P$ is a generating function of 3D simplicial complexes weighted with the Turaev-Viro invariant. Of course, P is only formally defined. However, this construction gives a framework for the strong coupling expansion in simplicial gravity, which can be carried out by iterating the Schwinger-Dyson equation for the partition function (63) [21].

A more down-to-earth model can be obtained by taking classical finite gauge group. Repeating all steps, one finds the sum over all simplicial complexes weighted with the invariant (58) times a volume dependent factor. To make a contact with the discrete action (13), one has to introduce a fugacity μ for the number of links as well. It can be done by adding three indices to ϕ:

$$\log P^{(G)} = \int \prod_{\{x_i \in G\}} \prod_{k_i=1}^{\mu} d\phi^{k_1 k_2 k_3}_{x_1 x_2 x_3} \exp\left\{ -\frac{1}{2} \sum_{\{x_i \in G\}} \sum_{k_i=1}^{\mu} |\phi^{k_1 k_2 k_3}_{x_1 x_2 x_3}|^2 + \right.$$

$$\frac{\lambda}{12} \sum_{\{x_i \in G\}} \sum_{k_i=1}^{\mu} \phi_{x_1 x_2 x_3}^{k_1 k_2 k_3} \phi_{x_1 x_4 x_5}^{k_1 k_4 k_5} \phi_{x_2 x_5 x_6}^{k_2 k_5 k_6} \phi_{x_3 x_6 x_4}^{k_3 k_6 k_4} =$$

$$\sum_{\{C\}} |G|^{N_0-1} \lambda^{N_3} \mu^{N_1} Z_0^{(G)}(C) \tag{67}$$

where $|G| = \sum_G 1$ is the rank of the group.

It can be easily seen that simplicial complexes have non-negative Euler characters

$$\chi = \sum_{i=1}^{N_0} p_i \geq 0 \tag{68}$$

where the sum runs over all vertices. Tetrahedra touching the i'th vertex form a 3D ball; p_i is the genus of its 2D boundary. By definition, a complex is a manifold *iff* $p_i = 0 \; \forall i$; *i.e.*, the vicinity of every point is a spherical ball.

After the rescaling, $\lambda = |G|\tilde{\lambda}$, $\mu = \frac{1}{|G|}\tilde{\mu}$, one obtains

$$|G| \log P^{(G)} = \sum_{\{C\}} \tilde{\lambda}^{N_3} \tilde{\mu}^{N_1} |G|^{\chi} Z_0^{(G)}(C) \tag{69}$$

and in the formal limit $|G| \to 0$ only manifolds for which $Z_0^{(G)}$ is finite contribute.

Any finite group can be embedded in the permutation group, S_n, for sufficiently large n; $|G| = n!$ in this case. It suggests that, at $\tilde{\lambda}$ and $\tilde{\mu}$ fixed, one should take n much bigger than the maximum rank of the fundamental group for typical complexes and try to continue analytically to $n! = 0$ (But how to do it practically?!). If the Poincaré hypothesis is true, only spheres should survive in this limit. Technically, it could mean a kind of double scaling.

Unfortunately, the model seems to be too complicated to be investigated analytically.

6 Conclusion

My aim in the present paper has been to draw attention to the quite promising problem of 3D quantum gravity. What one could learn from it concerns fundamental properties of the quantum vacuum. Non-renormalizability of gravity, non-boundedness of the Einstein-Hilbert action, topology changing processes, cosmological constant problem can be addressed within this simplified (comparing to $4D$ gravity) framework. Three dimensional geometry and topology possess a lot of beautiful mathematical structures. Many fundamental and long standing problems have not yet been solved. It is still a field of intensive research, which create an exciting atmosphere of a parallel rise of mathematical results and physical understanding.

Acknowledgments

I would like to express my gratitude to the organizers for the creative and friendly atmosphere at Cargese during the workshop. At different stages of work on problems touched in this paper, I have enjoyed the collaboration and discussions with M.Agistein, A.Alekseev, J.Ambjørn, C.Bachas, C.Itzykson, V.Kazakov, A.N.Kirillov, I.Kostov, A.Krzywicki, A.A.Migdal, M.Petropoulos and S.Piunikhin. I appreciate the financial support from the EEC grant CS1-D430-C.

References

[1] F.David, *"Simplicial quantum gravity and random lattices"*, Les Houches lectures, Session LVII (1992).

[2] E.Witten, *Nucl. Phys.* **B311** (1988/89) 46 and **B323** (1989) 113.

[3] V.G.Turaev and O.Y.Viro, *Topology* **31** (1992) 865.

[4] H.Ooguri and N.Sasakura, *Mod. Phys. Lett.* **A6** (1991) 3591;
F.Archer and R.M.Williams, *Phys.Lett.* **B273** (1991) 438.

[5] G.Ponzano and T. Regge, *in Spectroscopic and group theoretical methods in physics*, ed. F.Bloch (North-Holland, Amsterdam, 1968).

[6] T. Regge, *Nuovo Cimento* **19** (1961) 558.

[7] M.E. Agishtein and A.A. Migdal, *Mod. Phys. Lett.* **A6** (1991) 1863;
J.Ambjørn and S. Varsted, *Phys. Lett.* **B226** (1991) 258 and *Nucl. Phys.* **B373** (1992) 557.

[8] D.V. Boulatov and A. Krzywicki, *Mod. Phys. Lett.* **A6** (1991) 3005;
J.Ambjørn, D.V. Boulatov, A. Krzywicki and S. Varsted, *Phys. Lett.* **B276** (1992) 432.

[9] J.Ambjørn, Z.Burda, J.Jurkiewicz and C.F.Cristjansen, *Phys. Lett.* **B297** (1992) 253.

[10] M.Agishtein and A.A.Migdal, *Mod. Phys. Lett.* **A7** (1992) 1039;
J.Ambjørn and J.Jurkiewicz, *Phys. Lett.* **B278** (1992) 42.

[11] See for example, C.Nash, *Differential topology and quantum field theory*, Academic Press; chap. 1 and references therein.

[12] K.Wilson, *Phys. Rev.* **D10** (1975) 2445.

[13] L.D.Faddeev, N.Reshetikhin and L.Takhtajan, *Leningrad Math. J* **1** (1990) 193.

[14] S.L.Woronowicz, *Commun. Math. Phys.* **111** (1987) 613;
L.L.Vaksman and Ya.S.Soibelman, *Func. Anal. Appl.* **22** (1988) 170.

[15] N.Yu.Reshetikhin and V.G.Turaev, *Commun. Math. Phys.* **124** (1989) 307 and *Invent. Math.* **103** (1991) 547.

[16] P.Roche and D.Arnaucon, *Lett. Math. Phys.* **17** (1989) 295;
V.Pasquier and H.Saleur, *Nucl. Phys.* (1990);
G.Keller, *Lett. Math. Phys.* **21** (1991) 273.

[17] D.V. Boulatov, *Int. J. Mod. Phys.* **A8** (1993) 3139.

[18] V.G.Turaev, *C.R. Acad. Sci. Paris* **313** (1991) 395; *J. Diff. Geom.* **36** (1992) 35.

[19] T.Kohno, *Topology* **31** (1992) 203.

[20] D.V. Boulatov, *Mod. Phys. Lett.* **A7** (1992) 1629.

[21] H.Ooguri, *Prog. Theor. Phys.* **89** (1993) 1.

ON THE $\mathcal{W}$-GRAVITY SPECTRUM
AND ITS G-STRUCTURE

Peter Bouwknegt[1], Jim McCarthy[2] and Krzysztof Pilch[1]

[1] Department of Physics and Astronomy
University of Southern California
Los Angeles, CA 90089-0484, USA

[2] Department of Physics and Mathematical Physics
University of Adelaide
Adelaide, SA 5005, Australia

Abstract: We present results for the BRST cohomology of $\mathcal{W}[\mathfrak{g}]$ minimal models coupled to $\mathcal{W}[\mathfrak{g}]$ gravity, as well as scalar fields coupled to $\mathcal{W}[\mathfrak{g}]$ gravity. In the latter case we explore an intricate relation to the (twisted) $\mathfrak{g}$ cohomology of a product of two twisted Fock modules.

1 Introduction

The BRST quantization of two dimensional $\mathcal{W}[\mathfrak{g}]$ gravity coupled to $\mathcal{W}[\mathfrak{g}]$ matter poses the interesting mathematical problem of computing the semi-infinite cohomology of a $\mathcal{W}$-algebra with values in a tensor product of two (positive energy) $\mathcal{W}$-modules. In this note we study this cohomology both for free scalar fields as well as for $\mathcal{W}$ minimal models coupled to $\mathcal{W}$-gravity, *i.e.* we study the cohomology of the tensor products of two Fock spaces at irrational α_+^2, and of an irreducible $\mathcal{W}$-module with a Fock space. In these cases we give the complete results for the cohomologies. The work described in this paper is an extension of [1, 2], where we presented results for the case in which the 'Liouville' momentum takes values in one specific Weyl chamber. We refer to [1] for further references on the subject.

Strictly speaking, the relevant BRST operator has only been shown to exist, by explicit construction, for $\mathcal{W}_3 \equiv \mathcal{W}[sl(3)]$ [3, 4]. However, since our analysis is insensitive to the specific form of the BRST operator, pending the existence proof we have formulated our results for arbitrary simple, simply-laced Lie algebras $\mathfrak{g}$.

For irrational α_+^2, it turns out that there is an intimate connection between the $\mathcal{W}[\mathfrak{g}]$ cohomology of a tensor product of two $\mathcal{W}$ Fock spaces and the (twisted) $\mathfrak{g}$ cohomology

Quantum Field Theory and String Theory, Edited by
L. Baulieu *et al.*, Plenum Press, New York, 1995

of the product of two twisted $\mathfrak{g}$ Fock spaces, which is the finite-dimensional analogue of a G/G coset model. Closely related observations have been made in [5, 6].

This note is organized as follows. In section 2 we discuss the (twisted) $\mathfrak{g}$ cohomology of the product of two twisted Fock spaces. In appendix A we give the complete result for $\mathfrak{g} \cong sl(2)$ and $sl(3)$. Our results in this section mainly serve the purpose to formulate the results for the $\mathcal{W}$-cohomology through the correspondence alluded to above, but we believe they are also interesting in their own right. In section 3 we consider the $\mathcal{W}[\mathfrak{g}]$ cohomology of a tensor product of two Fock spaces and explain the correspondence with the (twisted) $\mathfrak{g}$ cohomology. Finally, in section 4, we present a complete result for the $\mathcal{W}[\mathfrak{g}]$ minimal models coupled to $\mathcal{W}[\mathfrak{g}]$ gravity. At the end we included a table of some of the states for explicitness. We compare our results to previously obtained results, in particular those of [7, 8], and find complete agreement.

2 G-cohomology of a product of two twisted Fock spaces

Let $\mathfrak{g}$ be a finite-dimensional simple Lie algebra. Fix a triangular decomposition $\mathfrak{g} \cong \mathfrak{n}_- \oplus \mathfrak{h} \oplus \mathfrak{n}_+$, and a corresponding Chevalley basis $\{e_{-\alpha}, h_i, e_\alpha\}$, $\alpha \in \Delta_+$, $i = 1, \ldots, \mathrm{rank}\,\mathfrak{g}$. For any $\mathfrak{g}$-module V in the BGG-category $\mathcal{O}$ [9] (loosely speaking, the category of modules with weights bounded from above), we can consider its 'twisted' cohomology $\mathrm{H}^i_{\mathrm{tw}}(\mathfrak{g}, V)$, which is the finite-dimensional analogue of the so-called 'semi-infinite' cohomology introduced by Feigin [10]. This cohomology is defined as follows (see [11, 12] for more details): Introduce a ghost system (b_A, c^A) for each generator e_A of $\mathfrak{g}$, with (anti-)commutators $\{b_A, c^B\} = \delta_A{}^B$, and denote the corresponding ghost Fock space F^{gh}. The (physical) ghost vacuum $|\mathrm{gh}\rangle$ satisfies

$$b_\alpha|\mathrm{gh}\rangle = b_i|\mathrm{gh}\rangle = c^{-\alpha}|\mathrm{gh}\rangle = 0\,, \qquad \alpha \in \Delta_+,\ i = 1, \ldots, \mathrm{rank}\,\mathfrak{g}\,. \tag{2.1}$$

The ghost Fock space is graded by ghost number, $\mathrm{gh}(c^A) = 1$, $\mathrm{gh}(b_A) = -1$, and is a $\mathfrak{g}$-module under the action

$$\pi^{\mathrm{gh}}(e_A) = -\sum_{B,C} f_{AB}{}^C c^B b_C\,. \tag{2.2}$$

Note that the highest weight of F^{gh} equals 2ρ, where ρ is the principal vector of $\mathfrak{g}$ $((\rho, \alpha) = 1, \forall \alpha \in \Delta_+)$, as is easily computed via

$$\pi^{\mathrm{gh}}(h_i)|\mathrm{gh}\rangle = -\sum_{\alpha \in \Delta_+} f_{i-\alpha}{}^{-\alpha} c^{-\alpha} b_{-\alpha}|\mathrm{gh}\rangle = \sum_{\alpha \in \Delta_+} (\alpha_i^\vee, \alpha)|\mathrm{gh}\rangle = (\alpha_i^\vee, 2\rho)|\mathrm{gh}\rangle\,. \tag{2.3}$$

The (twisted) cohomology $\mathrm{H}_{\mathrm{tw}}(\mathfrak{g}, V)$ is defined as the cohomology of the (BRST) operator

$$d = \sum_A c^A \left(\pi(e_A) + \tfrac{1}{2}\pi^{\mathrm{gh}}(e_A) \right)\,, \tag{2.4}$$

acting on the (graded) complex $V \otimes F^{\mathrm{gh}}$.

The twisted cohomology of a subalgebra of $\mathfrak{g}$ is defined similarly by restricting to the appropriate subset of generators. In particular we are interested in the cohomologies of (twisted) nilpotent subalgebras $\mathfrak{n}_+^w \equiv w \cdot \mathfrak{n}_+ \cdot w^{-1}$, corresponding to Weyl group elements $w \in W$. In this case the sums run over $\alpha \in w(\Delta_+)$.

To orient the discussion it is worth noting in the "untwisted" case ($w' = 1$) that the computation of $\mathrm{H}^i_{\mathrm{tw}}(\mathfrak{n}_+, V)_\lambda \cong \mathrm{H}^i(\mathfrak{n}_+, V)_\lambda \cong \mathrm{Ext}^i_{\mathcal{O}}(M_\lambda, V)$ for various modules $V \in \mathcal{O}$ is a classical problem in mathematics. In the case when V is a finite-dimensional irreducible module L_Λ the result is well-known [13] (see (2.7) below, where this result is derived as an illustration of standard techniques). For many other interesting modules, such as Verma modules, the problem has only been solved partially.

Besides Verma modules M_λ and contragredient Verma modules $\overline{M}_\lambda$ there exists a class of modules in $\mathcal{O}$, the so-called twisted Fock spaces F_λ^w (labelled by elements $w \in W$), that interpolate between M_λ and $\overline{M}_\lambda$. [Our conventions are such that $F_\lambda^1 \cong \overline{M}_\lambda$ and $F_\lambda^{w_0} \cong M_\lambda$.] These modules are, in a sense, finite-dimensional analogues of Wakimoto modules [14] and were introduced in [11] (see [15] for explicit realizations). They are uniquely characterized by the property that they are free over $\mathcal{U}(\mathfrak{n}_+^w \cap \mathfrak{n}_-)$, cofree over $\mathcal{U}(\mathfrak{n}_+^w \cap \mathfrak{n}_+)$ and have a unique highest weight vector (of weight λ).

The cohomology of a twisted Fock space F_λ^w with respect to the nilpotent subalgebra $\mathfrak{n}_+^w$ with the same twist w, is given by [11, 12]

$$\mathrm{H}^i_{\mathrm{tw}}(\mathfrak{n}_+^w, F_\lambda^w) \cong \delta^{i,0}\, \mathbb{C}_{\lambda+\rho-w\rho} \,. \tag{2.5}$$

In fact, (2.5) uniquely characterizes the module F_λ^w in the category $\mathcal{O}$.

For Λ an integral dominant weight, *i.e.* $\Lambda \in P_+$, there exist resolutions of the irreducible module L_Λ in terms of twisted Fock spaces F_λ^w (for any $w \in W$) with terms

$$C_w^i L_\Lambda \cong \bigoplus_{\{\sigma \in W \,|\, \ell_w(\sigma)=i\}} F_{\sigma*\Lambda}^w \,, \tag{2.6}$$

where $\ell_w(\sigma)$ is the twisted length of $\sigma \in W$, which can be expressed in terms of the usual length ℓ through $\ell_w(\sigma) = \ell(w^{-1}\sigma) - \ell(w^{-1})$ and $\sigma * \Lambda = \sigma(\Lambda + \rho) - \rho$ denotes a shifted action of the Weyl group.

To illustrate an application of (2.6) we reproduce the known result for $\mathrm{H}_{\mathrm{tw}}(\mathfrak{n}_+^w, L_\Lambda)$ as alluded to above. Simply take a resolution of L_Λ in terms of Fock spaces twisted by the same $w \in W$ and apply (2.5) to the resulting double complex. We find

$$\mathrm{H}^i_{\mathrm{tw}}(\mathfrak{n}_+^w, L_\Lambda) \cong \bigoplus_{\{\sigma \in W \,|\, \ell_w(\sigma)=i\}} \mathbb{C}_{\sigma*\Lambda+\rho-w\rho} \,. \tag{2.7}$$

We are interested in computing the twisted cohomology $\mathrm{H}^i_{\mathrm{tw}}(\mathfrak{g}, F_\lambda^w \otimes F_\mu^{w'w_0})$, or rather the cohomology relative to the Cartan subalgebra $\mathfrak{h}$, which we will denote by $\mathrm{H}^i_{\mathrm{tw}}(\mathfrak{g}, \mathfrak{h}; F_\lambda^w \otimes F_\mu^{w'w_0})$. [The Weyl group element w_0 denotes the unique element of longest length in W.] This cohomology corresponds to the physical states of the finite-dimensional analogue of the so-called G/G-model. Using the fact that $\mathfrak{g} \cong \mathfrak{n}_+^{w'} \oplus \mathfrak{h} \oplus \mathfrak{n}_+^{w'w_0}$, it can be related to a generalization of (2.5) by invoking a reduction theorem (see *e.g.* [12])

$$
\begin{aligned}
\mathrm{H}^i_{\mathrm{tw}}(\mathfrak{g}, \mathfrak{h}; F_\lambda^w \otimes F_\mu^{w'w_0}) &\cong \bigoplus_{p+q=i} \left(\mathrm{H}^p_{\mathrm{tw}}(\mathfrak{n}_+^{w'}, F_\lambda^w) \otimes \mathrm{H}^q_{\mathrm{tw}}(\mathfrak{n}_+^{w'w_0}, F_\mu^{w'w_0}) \right)_{\mathfrak{h}} \\
&\cong \mathrm{H}^i_{\mathrm{tw}}(\mathfrak{n}_+^{w'}, F_\lambda^w)_{-\mu-\rho-w'\rho} \,.
\end{aligned} \tag{2.8}
$$

One can show that nonzero cohomology can only arise if λ and μ can be parametrized
as

$$\lambda = \sigma * \Lambda, \qquad \mu = -\sigma' * \Lambda - 2\rho, \qquad (2.9)$$

for some dominant weight Λ and $\sigma \preceq \sigma'$, $\sigma, \sigma' \in W$. [We take the usual Bruhat ordering
"$\preceq$" on W [16]. In particular for $sl(2)$ and $sl(3)$, this is just the ordering of W by the
length $\ell(\cdot)$.] Moreover, if one restricts the discussion to dominant integral weights, $i.e.$
$\Lambda \in P_+$, then one can show that the cohomology does not depend on the particular
$\Lambda \in P_+$. We will henceforth restrict the discussion to $\Lambda \in P_+$ and adopt the shorthand
notations $F_\sigma^w = F_{\sigma*\Lambda}^w$ and $F_{-\sigma'}^{w'w_0} = F_{-\sigma'*\Lambda-2\rho}^{w'w_0}$.

In order to summarize the computation of the dimensions of these cohomology
groups, we introduce a set of polynomials by

$$\mathcal{P}_{\sigma,\sigma'}^{w,w'}(q) = \sum_i (-1)^{\ell_w(\sigma)+\ell_w(\sigma')+i} \, q^i \, \dim \mathrm{H}_{\mathrm{tw}}^i(\mathfrak{g},\mathfrak{h}; F_\sigma^w \otimes F_{-\sigma'}^{w'w_0}) . \qquad (2.10)$$

They satisfy the following basic relations:

1. $\mathcal{P}_{\sigma,\sigma'}^{w,w'}(q) = 0$ for $\sigma \succ \sigma'$.

2. (i) $\mathcal{P}_{\sigma,\sigma}^{w,w'}(q) = 1$,   (ii) $\mathcal{P}_{\sigma,\sigma'}^{w,w}(q) = \delta_{\sigma,\sigma'}$,   (iii) $\mathcal{P}_{\sigma,\sigma'}^{w,w'}(1) = \delta_{\sigma,\sigma'}$.

3. $\mathcal{P}_{\sigma,\sigma'}^{w,w'}(q) = \mathcal{P}_{\sigma'w_0,\sigma w_0}^{w'w_0,ww_0}(q)$ (reflection symmetry).

4. $\mathcal{P}_{\sigma,\sigma'}^{w,w'}(q) = \mathcal{P}_{\sigma,\sigma'}^{ww_0,w'w_0}(q^{-1})$ (Poincaré duality).

The identities in 2) follow from the fact that (i) for $\sigma = \sigma'$ there is only one state
in the complex, (ii) for $w = w'$ the cohomology is given by (2.5), and (iii) by applying
the Lefschetz principle. Identity 3) follows from the observation that in (2.9) we could
equally well have swapped the order of the two Fock spaces and chosen the weight $\Lambda' =
-w_0\Lambda$ to parametrize (λ, μ). Finally, identity 4) follows from the fact that the module
contragredient to F_λ^w is $F_\lambda^{ww_0}$. In Appendix A we list all polynomials for $\mathfrak{g} \cong sl(2)$ and
$sl(3)$.

In the particular case $(w, w') = (w_0, 1)$, where the polynomials correspond to the $\mathfrak{g}$
cohomology of a product of two Verma modules $M_\sigma \otimes M_{-\sigma'}$, it is known that $\mathcal{P}_{\sigma,\sigma'}^{w_0,1}(q) =
R_{\sigma,\sigma'}(q)$ for $\ell(\sigma') - \ell(\sigma) \leq 3$ [17]. Here, $R_{\sigma,\sigma'}(q)$ denote the Kazhdan-Lusztig R-
polynomials [18, 16].

The above discussion has a straightforward generalization to the affine Lie algebras, in which case one is interested in computing the relative semi-infinite cohomology
$\mathrm{H}^{\infty/2+i}(\widehat{\mathfrak{g}},\widehat{\mathfrak{h}}; F_\lambda^w \otimes F_\mu^{w'w_0})$ of the tensor product of two Wakimoto modules, twisted by
finite-dimensional Weyl group elements $w, w' \in W$ [12]. Once more the cohomology
can arise only for the weights satisfying the affine analogue of (2.9), with σ, σ' in the
affine Weyl group $\widehat{W}$.

It is known that for $w = w'$ (see, $e.g.$ sections 4 and 5 in [12])

$$\mathrm{H}^{\infty/2+i}(\widehat{\mathfrak{g}},\widehat{\mathfrak{h}}; F_\sigma^w \otimes F_{-\sigma'}^{ww_0}) \cong \delta_{\sigma,\sigma'}\delta^{i,0} \, \mathbb{C}, \qquad (2.11)$$

which is the analogue of (2.5). We expect that for general w and w' the set of polynomials as in (2.10) with $\sigma, \sigma' \in \widehat{W}$ will in fact be the same as in the finite-dimensional
case. For $\widehat{sl}(2)$ this can be verified explicitly from the results of [6]. The general case
appears to be an open problem.

3 $\mathcal{W}$-cohomology of Fock spaces at irrational α_+^2

Let $\mathcal{W}[\mathfrak{g}]$ be the $\mathcal{W}$-algebra associated to some simply-laced simple Lie algebra $\mathfrak{g}$ (see [19] for a review and a list of notations). In the remainder of this paper we will present some new results for the semi-infinite cohomology $H^i(\mathcal{W}[\mathfrak{g}], V^M \otimes V^L)$ of $\mathcal{W}[\mathfrak{g}]$ on the product of two positive energy $\mathcal{W}[\mathfrak{g}]$ modules V^M and V^L. Specifically, for the 'Liouville' module V^L, representing the $\mathcal{W}[\mathfrak{g}]$ gravity sector, we will take the Fock space of an appropriate set of free scalar fields, while for the matter module V^M we will take either a Fock space (section 3) or an irreducible module, $i.e.$ $\mathcal{W}[\mathfrak{g}]$ minimal model (section 4). Although, in general, the Cartan subalgebra of $\mathcal{W}[\mathfrak{g}]$ will not be diagonalizable on the modules V, there is still an analogue of the relative cohomology. One can show that the cohomology possesses a multiplet structure of 2^ℓ states (where $\ell = \operatorname{rank} \mathfrak{g}$), which is essentially due to the ghost zero modes [8, 1]. The lowest ghost number state in each multiplet will be called a prime state. Throughout this paper we will only formulate the results for the prime states in the cohomology.

To be precise, our results are only valid for $\mathfrak{g} \cong sl(2)$ and $sl(3)$, for these are the only cases for which the differential (BRST operator) has been constructed explicitly (see [3, 4] for the latter, and also [20] for some higher rank results). However, one expects that such a differential exists for the other $\mathcal{W}[\mathfrak{g}]$ algebras. In the discussion below we use only very generic properties of the differential, and our results should therefore be valid for the other $\mathcal{W}[\mathfrak{g}]$ algebras as well.

Let $F(\Lambda, \alpha_0)$ denote the Fock space of $\ell \equiv \operatorname{rank} \mathfrak{g}$ scalar fields $\phi^k(z)$, normalized such that $\phi^k(z)\phi^l(w) = -\delta^{kl}\ln(z - w)$, coupled to a background charge $\alpha_0\rho$. The Fock space vacuum $|\Lambda\rangle$ is labelled by a vector Λ in the weight space of $\mathfrak{g}$ such that $p^k|\Lambda\rangle = \Lambda^k|\Lambda\rangle$.

A realization of $\mathcal{W}[\mathfrak{g}]$ on the Fock space $F(\Lambda, \alpha_0)$ can be constructed by means of the Drinfel'd-Sokolov reduction. In particular, the stress energy tensor is given by

$$T(z) = -\tfrac{1}{2}(\partial\phi(z) \cdot \partial\phi(z)) - i\alpha_0\rho \cdot \partial^2\phi(z)\,. \tag{3.1}$$

It generates the Virasoro subalgebra of $\mathcal{W}[\mathfrak{g}]$. The central charge and conformal dimension of $F(\Lambda, \alpha_0)$ are given by

$$c = \ell - 12\alpha_0^2|\rho|^2\,, \qquad h(\Lambda) = \tfrac{1}{2}(\Lambda, \Lambda + 2\alpha_0\rho)\,. \tag{3.2}$$

Let us first consider the cohomology $H^i(\mathcal{W}[\mathfrak{g}], F(\Lambda^M, \alpha_0^M) \otimes F(\Lambda^L, \alpha_0^L))$. Imposing the condition that the total central charge vanishes in order for the differential to be nilpotent, leads to the following parametrization of α_0^M and α_0^L

$$\alpha_0^M = \alpha_+ + \alpha_-\,, \qquad -i\alpha_0^L = \alpha_+ - \alpha_-\,, \qquad \alpha_+\alpha_- = -1\,. \tag{3.3}$$

By standard arguments, based on the composition series of a Fock space, one can show that the cohomology is trivial ($i.e.$ contains at most tachyonic states) if either $F(\Lambda^M, \alpha_0^M)$ or $F(\Lambda^L, \alpha_0^L)$ is irreducible (see $e.g.$ [21]). Recall that a Fock space $F(\Lambda, \alpha_0)$ is reducible if and only if there exists a root $\alpha \in \Delta_+$ such that (see $e.g.$ [1])

$$(\Lambda + \alpha_0\rho, \alpha) \in \pm(\mathbb{N}\alpha_+ + \mathbb{N}\alpha_-)\,, \tag{3.4}$$

and 'completely degenerate' if (3.4) holds for all roots $\alpha \in \Delta_+$. Therefore, in the reducible case, it is convenient to parametrize Λ by

$$\Lambda + \alpha_0 \rho = w^{-1}(\alpha_+ \sigma(\Lambda^{(+)} + \rho) + \alpha_-(\Lambda^{(-)} + \rho)) \equiv \Lambda(w,\sigma) + \alpha_0 \rho, \qquad (3.5)$$

for some weights $\Lambda^{(+)}, \Lambda^{(-)}$ and $w, \sigma \in W$. If $F(\Lambda, \alpha_0)$ is reducible in the direction of $\alpha \in \Delta_+$, then we can choose $\Lambda^{(+)}$ and $\Lambda^{(-)}$ such that $(\Lambda^{(+)}, \alpha) \in \mathbf{N}$, $(\Lambda^{(-)}, \alpha) \in \mathbf{N}$. If $F(\Lambda, \alpha_0)$ is completely degenerate then one can choose $\Lambda^{(+)}, \Lambda^{(-)} \in P_+$ as well as $\sigma = 1$, and if moreover α_+^2 is irrational then this parametrization of Λ in terms of $\Lambda^{(+)}, \Lambda^{(-)} \in P_+$ and $w \in W$ is unique. In this paper we will only consider the completely degenerate case. In the case that $F(\Lambda, \alpha_0)$ is reducible only in certain directions the cohomology essentially reduces to the one of a smaller $\mathcal{W}$-algebra. It can be analysed similarly.

The highest weight vectors of all $F(\Lambda(w,1))$, $w \in W$, have the same eigenvalues with respect to the $\mathcal{W}[\mathfrak{g}]$ generators because these are invariant under $\Lambda + \alpha_0 \rho \to w(\Lambda + \alpha_0 \rho)$. Therefore, all $F(\Lambda(w,1))$ contain one and the same irreducible $\mathcal{W}[\mathfrak{g}]$ module $L(\Lambda)$. Thus, by arguing in the usual way, we obtain a family of resolutions $C_w^i L(\Lambda)$ of $L(\Lambda)$, parametrized by $w \in W$, that are based on $F(\Lambda(w,1))$. For $\alpha_+^2 \notin \mathbb{Q}$ these resolutions have a finite number of terms (see section 4 for $\alpha_+^2 \in \mathbb{Q}$). Specifically, for $\Lambda = \alpha_+ \Lambda^{(+)} + \alpha_- \Lambda^{(-)}$ we have

$$C_w^i L(\Lambda^{(+)}, \Lambda^{(-)}) \cong \bigoplus_{\{\sigma \in W | \ell_w(\sigma) = i\}} F(\Lambda(w,\sigma)). \qquad (3.6)$$

where $\Lambda(w,\sigma)$ is defined in (3.5). For $w = 1$ these resolutions were already constructed in [22, 23].

Let us now turn to the discussion of the cohomology $\mathrm{H}^i(\mathcal{W}[\mathfrak{g}], F(\Lambda^M, \alpha_0^M) \otimes F(\Lambda^L, \alpha_0^L))$. By analysis of the Kac-determinants for $\mathcal{W}[\mathfrak{g}]$ [1] and standard reasoning using composition series for F, one can argue that:

For Λ^M of the form (3.5), i.e.

$$\Lambda^M(w,\sigma) + \alpha_0^M \rho = w^{-1}\left(\alpha_+ \sigma(\Lambda^{(+)} + \rho) + \alpha_-(\Lambda^{(-)} + \rho)\right), \qquad (3.7)$$

the cohomology $\mathrm{H}^i(\mathcal{W}[\mathfrak{g}], F(\Lambda^M, \alpha_0^M) \otimes F(\Lambda^L, \alpha_0^L))$ can only be nontrivial if $\Lambda^L = \Lambda^L(w', \sigma')$ for some $w', \sigma' \in W$, where

$$-i(\Lambda^L(w',\sigma') + \alpha_0^L \rho) = w'^{-1}\left(\alpha_+ \sigma'(\Lambda^{(+)} + \rho) + \alpha_-(\Lambda^{(-)} + \rho)\right). \qquad (3.8)$$

At this point we note a remarkable similarity between the resolutions of (3.6) and (2.6), which suggests that the matter module $F(\Lambda^M(w,\sigma))$ behaves like the twisted Wakimoto module F_σ^w, while the Liouville module $F(\Lambda^L(w',\sigma'))$ behaves like the dual of the matter module, *i.e.* like $F_{-\sigma'}^{w'w_0}$. For irrational α_+^2 we therefore expect a close correspondence between the $\mathcal{W}[\mathfrak{g}]$ cohomology of a tensor product of two Fock spaces and the (twisted) $\mathfrak{g}$ cohomology of a tensor product of two twisted Wakimoto modules. More precisely, we assert that

$$\sum_i (-1)^{\ell_w(\sigma)+\ell_{w'}(\sigma')+i}\, q^i\, \dim \mathrm{H}^i(\mathcal{W}[\mathfrak{g}], F(\Lambda^M(w,\sigma)) \otimes F(\Lambda^L(w',\sigma'))) = \mathcal{P}^{w,w'}_{\sigma,\sigma'}(q)\,, \quad (3.9)$$

where the polynomials $\mathcal{P}^{w,w'}_{\sigma,\sigma'}(q)$ were defined in (2.10) (see Appendix A for an explicit list of all polynomials for $\mathfrak{g} \cong sl(2)$ and $sl(3)$).

For $\mathfrak{g} \cong sl(2)$ the result agrees with [21, 24]. For $sl(3)$ the result (3.9) is in complete agreement with the states explicitly constructed in [7] (in [7] only states corresponding to, in our conventions, $w = \sigma$, $w' = \sigma'$ of ghost number $\ell_{w'}(w') = -\ell(w')$ were considered). We have explicitly constructed some additional physical states. The results are consistent with (3.9).

Let us now examine the consequences of (3.9) for the (chiral) ground ring at irrational α_+^2 in the $\mathcal{W}[sl(3)]$ case. From the explicit results in Appendix A we conclude that for every $\Lambda^{(+)}, \Lambda^{(-)} \in P_+$ we have a ground ring element iff $(w, w') = (1, w_0)$, $(\sigma, \sigma') = (1, w_0)$. Moreover, this element is unique. Let us denote it by $\phi_{(\Lambda^{(+)}, \Lambda^{(-)})}$. The ground ring is freely generated by $\phi_{(\Lambda_1,0)}, \phi_{(\Lambda_2,0)}, \phi_{(0,\Lambda_1)}$ and $\phi_{(0,\Lambda_2)}$ (in the notation of [7] these correspond to $\gamma_2^0, \gamma_1^0, x_1$ and x_2, respectively). [Note that our present conventions differ from those in [1].]

It is reasonable to believe that an equation similar to (3.9) holds for rational α_+^2 if one replaces the $\mathfrak{g}$ Fock space F^w with its corresponding affinization $i.e.$ a (twisted) Wakimoto module of $\widehat{\mathfrak{g}}$, and let σ run over the affine Weyl group. In particular we claim (see (2.11)):

$\mathrm{H}^i(\mathcal{W}[\mathfrak{g}], F(\Lambda^M(w,\sigma)) \otimes F(\Lambda^L(w,\sigma')))$ *(i.e. $w = w'$) is nontrivial only if $\sigma = \sigma'$ in which case it is one-dimensional and concentrated in dimension $i = 0$ (tachyonic state).*

For $\widehat{sl}(2)$ this claim is known to be correct, while for other $\widehat{\mathfrak{g}}$ it is consistent with sample calculations. In particular, this assertion leads to the cohomology of $\mathcal{W}$ minimal models (section 4) where it does not contradict previous results. [In the affine Lie algebra case it corresponds to (2.11). For the analogous statement for (contragredient-) Verma modules see [1].]

For $\mathfrak{g} \cong sl(2)$ it is well-known that the dimensions of the cohomology groups $\mathrm{H}^i(\mathcal{W}[\mathfrak{g}], F(\Lambda^M(w,\sigma)) \otimes F(\Lambda^L(w',\sigma')))$ are insensitive as to whether α_+^2 is rational or irrational, as a consequence of an $SO(2,\mathbb{C})$ symmetry relating different values of α_+ [21]. This $SO(2,\mathbb{C})$ symmetry does not persist to higher rank $\mathcal{W}$-algebras, and in fact for $\mathfrak{g} = sl(3)$ it has been shown that the cohomology at $c^M = 2$ (corresponding to $\alpha_\pm = \pm 1$) is considerably larger than for irrational α_+^2 (see [2] for some results for $\mathcal{W}_3$ at $c^M = 2$). There are two, possibly related, reasons for this phenomenon. Firstly, for $\alpha_+^2 = \pm 1$ the parametrization (3.5) in terms of $\Lambda^{(+)}, \Lambda^{(-)} \in P_+$ is still highly redundant. Secondly, at $\alpha_\pm = \pm 1$, both the complex and the cohomology carry a representation of $\mathfrak{g}$, where the $\mathfrak{g}$-generators are given by the (zero modes of a) Frenkel-Kac-Segal vertex operator construction in the matter sector.

4 $\mathcal{W}$-cohomology of minimal models

In this section we will give a complete classification of physical states for a $\mathcal{W}[\mathfrak{g}]$ minimal model coupled to $\mathcal{W}[\mathfrak{g}]$ gravity.

The $\mathcal{W}[\mathfrak{g}]$ minimal models arise for $\alpha_+^2 = p/p' \in \mathbb{Q}$ (p and p' relatively prime integers) and are labelled by two integrable weights $\Lambda^{(+)} \in P_+^{p-h^\vee}$ and $\Lambda^{(-)} \in P_+^{p'-h^\vee}$ such that the highest weight is given by $\Lambda = \alpha_+ \Lambda^{(+)} + \alpha_- \Lambda^{(-)}$. Here, P_+^k denotes the set of integrable weights of $\widehat{\mathfrak{g}}$ at level k and $h^\vee$ is the dual Coxeter number.

We have a set of resolutions, parametrized by $w \in W$, of the minimal model $L(\Lambda^{(+)}, \Lambda^{(-)})$ in terms of twisted Fock spaces similar to (3.6)

$$C_w^i L(\Lambda^{(+)}, \Lambda^{(-)}) \cong \bigoplus_{\{\sigma \in \widehat{W} \mid \ell_w(\sigma) = i\}} F(\Lambda(w, \sigma)), \tag{4.1}$$

where $\Lambda(w, \sigma)$ is defined as in (3.5), but now the sum over σ runs over the affine Weyl group $\widehat{W}$. [We recall that any $w \in \widehat{W}$ can be written as $w = t_\beta \overline{w}$ for some $\overline{w} \in W$ and translation t_β such that $w\lambda = t_\beta \overline{w}\lambda = \overline{w}\lambda + k\beta$ for affine weights λ of level k. In the following ρ is regarded as an element of $P_+^{h^\vee}$.] The twisted length ℓ_w on the affine Weyl group $\widehat{W}$ is defined by (see [11, 15])

$$\ell_w(\sigma) = \lim_{N \to \infty} \left(\ell(t_{-Nw\rho}\sigma) - \ell(t_{-Nw\rho}) \right). \tag{4.2}$$

Since, for each $\sigma \in \widehat{W}$ there are only a finite number of possible cancellations between $t_{-Nw\rho} = wt_{-N\rho}w^{-1}$ and σ, the limit in (4.2) is in fact reached at a finite value of N. Note furthermore that for $\sigma \in W$ the length defined by (4.2) reduces to the usual twisted length $\ell_w(\sigma) = \ell(w^{-1}\sigma) - \ell(w^{-1})$.

Now, consider the cohomology $\mathrm{H}^i(\mathcal{W}[\mathfrak{g}], L(\Lambda^{(+)}, \Lambda^{(-)}) \otimes F(\Lambda^L, \alpha_0^L))$. By taking an arbitrary resolution $C_w^i L(\Lambda^{(+)}, \Lambda^{(-)})$ of $L(\Lambda^{(+)}, \Lambda^{(-)})$ one finds that the cohomology is nontrivial if and only if $\Lambda^L = \Lambda^L(w, \sigma)$ for some $w \in W$ and $\sigma \in \widehat{W}$ where

$$-i(\Lambda^L(w, \sigma) + \alpha_0^L \rho) = w^{-1}\left(\alpha_+ \sigma(\Lambda^{(+)} + \rho) + \alpha_-(\Lambda^{(-)} + \rho)\right). \tag{4.3}$$

Now, as discussed in section 3, $\mathrm{H}^i(\mathcal{W}[\mathfrak{g}], F(\Lambda^M(w, \sigma)) \otimes F(\Lambda^L(w, \sigma'))) \cong \delta_{\sigma, \sigma'}\delta^{i,0}\mathbb{C}$. Thus by taking, for any $\Lambda^L = \Lambda^L(w, \sigma)$, a resolution $C_w^i L(\Lambda^{(+)}, \Lambda^{(-)})$ of $L(\Lambda^{(+)}, \Lambda^{(-)})$ with the same twist $w \in W$, the same argument as in *e.g.* [24] immediately yields

1. The cohomology $\mathrm{H}^i(\mathcal{W}[\mathfrak{g}], L(\Lambda^{(+)}, \Lambda^{(-)}) \otimes F(\Lambda^L, \alpha_0^L))$ *is nontrivial iff* $\Lambda^L = \Lambda^L(w, \sigma)$ *for some* $w \in W$ *and* $\sigma \in \widehat{W}$.

2. For $\Lambda^L = \Lambda^L(w, \sigma)$ *there is precisely one (prime) state in the cohomology. Its ghost number is given by* $\ell_w(\sigma)$ *and its energy level by* $E = \frac{1}{2}|\alpha_+(\Lambda^{(+)} + \rho) + \alpha_-(\Lambda^{(-)} + \rho)|^2 - \frac{1}{2}|\Lambda^L(w, \sigma) + \alpha_0^L \rho|^2$.

In [1] we obtained this result for the particular case that $-i(\Lambda^L + \alpha_0^L \rho)$ is in the fundamental Weyl chamber D_+. This corresponds to those $\Lambda^L(w, \sigma)$ where for any $\sigma \in \widehat{W}$ the Weyl group element $w = w_\sigma$ is determined in such a way that $-i(\Lambda^L(w_\sigma, \sigma) + \alpha_0^L \rho) \in D_+$. The above result extends our previous work to all Weyl chambers. For $\mathfrak{g} \cong sl(2)$ it agrees with [21, 24].

For $\mathfrak{g} \cong sl(3)$ and the trivial module $L(\Lambda^{(+)}, \Lambda^{(-)}) \cong \mathbb{C}$ (*i.e.* $p = 3$, $p' = 4$ and $\Lambda^{(+)} = \Lambda^{(-)} = 0$) the results can be compared to those for the 'two-scalar $\mathcal{W}_3$ string' [8]. We find a complete agreement. For illustrative purposes we provide, in this case, a table of physical states for low ghost numbers. In each row the table lists, for given $\sigma \in \widehat{W}$ ($\ell(\sigma) \leq 4$), the values of the (energy-)level E and the ghost number $\ell_w(\sigma)$, $w \in W$ (see (4.2)), of the corresponding (prime) state in $\mathrm{H}^i(\mathcal{W}_3, \mathbb{C} \otimes F(\Lambda^L(w, \sigma)))$. The table can be compared to the results of [8].

Acknowledgements: P.B. would like to thank the organizers of the Cargèse workshop for the invitation and opportunity to present this talk. J.M. would like to thank U.S.C., and P.B. and K.P. the University of Adelaide, for hospitality. We enjoyed several discussions with Robert Perret on the $\mathfrak{g}$ cohomology. This work was supported by the Packard Foundation, by the U.S. Department of Energy under contract #DE-FG03-84ER-40168 and the Australian Research Council.

A Appendix

In this appendix we give some explicit polynomials $\mathcal{P}^{w,w'}_{\sigma,\sigma'}(q)$. In particular we present an exhaustive list for $\mathfrak{g} \cong sl(2)$ and $sl(3)$. The results have been verified by explicit computation.

$$\mathcal{P}^{w,w}_{\sigma,\sigma'}(q) = \begin{cases} 1 & \text{if } \sigma = \sigma' \\ 0 & \text{otherwise} \end{cases} \tag{A.4}$$

$$\mathcal{P}^{w,wr_i}_{\sigma,\sigma'}(q) = \begin{cases} 1 & \text{if } \sigma = \sigma' \\ q^{\pm 1} - 1 & \text{if } \sigma \prec r_{w\alpha_i}\sigma = \sigma', \ w\alpha_i \in \Delta_{\mp} \\ 0 & \text{otherwise} \end{cases} \tag{A.5}$$

For $i \neq j$

$$\mathcal{P}^{w,wr_ir_j}_{\sigma,\sigma'}(q) = \begin{cases} 1 & \text{if } \sigma = \sigma' \\ q^{\pm 1} - 1 & \text{if } \sigma \prec r_{w\alpha_i}\sigma = \sigma', \ w\alpha_i \in \Delta_{\mp} \\ & \text{or } \sigma \prec r_{wr_i\alpha_j}\sigma = \sigma', \ wr_i\alpha_j \in \Delta_{\mp} \\ (q^{\pm 1} - 1)^2 & \text{if } \sigma \prec r_{w\alpha_i}\sigma \prec r_{\tilde{w}\alpha_i}r_{w\alpha_j}\sigma = \sigma', \ w\alpha_i \in \Delta_{\mp}, \\ & wr_i\alpha_j \in \Delta_{\mp} \\ (q - 1)(q^{-1} - 1) & \text{if } \sigma \prec r_{w\alpha_i}\sigma \prec r_{w\alpha_i}r_{w\alpha_j}\sigma = \sigma', \ w\alpha_i \in \Delta_{\mp}, \\ & wr_i\alpha_j \in \Delta_{\pm} \end{cases} \tag{A.6}$$

Note that (A.6) can be summarized as

$$\mathcal{P}^{w,wr_ir_j}_{\sigma,\sigma'}(q) = \sum_{\sigma \preceq \sigma'' \preceq \sigma'} \mathcal{P}^{w,wr_i}_{\sigma,\sigma''}(q) \mathcal{P}^{wr_i,wr_ir_j}_{\sigma'',\sigma'}(q) . \tag{A.7}$$

The right hand side of (A.7) is in fact the polynomial associated to the first term in the spectral sequence of $\mathrm{H}^i_{\mathrm{tw}}(\mathfrak{g}, \mathfrak{h}; F^w_\sigma \otimes F^{wr_ir_jw_0}_{-\sigma'})$ with respect to the decomposition $\mathfrak{g} \cong \mathfrak{n}^{wr_i}_+ \oplus \mathfrak{h} \oplus \mathfrak{n}^{wr_iw_0}_+$. In this particular case this spectral sequence collapses at the first term, hence the result (A.7).

For $sl(3)$ the above determines all but the six polynomials $\mathcal{P}^{w,ww_0}_{\sigma,\sigma'}(q)$, which are listed below

Table 1. Physical states for the two-scalar $\mathcal{W}_3$ string

$\sigma \backslash w$	E	1	r_1	r_2	$r_2 r_1$	$r_1 r_2$	$r_1 r_2 r_1$
1	0	0	0	0	0	0	0
r_1	1	1	-1	1	1	-1	-1
r_2	1	1	1	-1	-1	1	-1
r_0	2	-1	-1	-1	1	1	1
$r_1 r_2$	3	2	0	2	0	-2	-2
$r_2 r_1$	3	2	2	0	-2	0	-2
$r_1 r_0$	4	0	-2	2	2	-2	0
$r_2 r_0$	4	0	2	-2	-2	2	0
$r_0 r_1$	5	-2	-2	0	2	0	2
$r_0 r_2$	5	-2	0	-2	0	2	2
$r_1 r_2 r_1$	4	3	1	1	-1	-1	-3
$r_0 r_1 r_0$	6	-1	-3	1	3	-1	1
$r_0 r_2 r_0$	6	-1	1	-3	-1	3	1
$r_1 r_2 r_0$	8	3	-1	3	1	-3	-3
$r_2 r_1 r_0$	8	3	3	-1	-3	1	-3
$r_1 r_0 r_2$	9	1	-3	3	3	-3	-1
$r_2 r_0 r_1$	9	1	3	-3	-3	3	-1
$r_0 r_1 r_2$	11	-3	-3	-1	3	1	3
$r_0 r_2 r_1$	11	-3	-1	-3	1	3	3
$r_1 r_2 r_1 r_0$	10	4	2	2	-2	-2	-4
$r_1 r_0 r_2 r_0$	11	2	-2	4	2	-4	-2
$r_2 r_0 r_1 r_0$	11	2	4	-2	-4	2	-2
$r_0 r_1 r_0 r_2$	13	-2	-4	2	4	-2	2
$r_0 r_2 r_0 r_1$	13	-2	2	-4	-2	4	2
$r_1 r_2 r_0 r_1$	14	4	0	4	0	-4	-4
$r_2 r_1 r_0 r_2$	14	4	4	0	-4	0	-4
$r_0 r_1 r_2 r_1$	14	-4	-2	-2	2	2	4
$r_1 r_0 r_2 r_1$	16	0	-4	4	4	-4	0
$r_2 r_0 r_1 r_2$	16	0	4	-4	-4	4	0
$r_0 r_1 r_2 r_0$	18	-4	-4	0	4	0	4
$r_0 r_2 r_1 r_0$	18	-4	0	-4	0	4	4

$$\mathcal{P}^{w_0,1}_{\sigma,\sigma'}(q) = \mathcal{P}^{1,w_0}_{\sigma,\sigma'}(q^{-1}) = \begin{cases} 1 & \text{if } \sigma = \sigma' \\ q - 1 & \text{if } \ell(\sigma') - \ell(\sigma) = 1 \\ (q-1)^2 & \text{if } \ell(\sigma') - \ell(\sigma) = 2 \\ (q^3 - 2q^2 + 2q - 1) & \text{if } \ell(\sigma') - \ell(\sigma) = 3 \end{cases} \tag{A.8}$$

For $i,j \in \{1,2\}$, $i \neq j$

$$\mathcal{P}^{r_i,r_jr_i}_{\sigma,\sigma'}(q) = \mathcal{P}^{r_jr_i,r_i}_{\sigma,\sigma'}(q^{-1}) = \begin{cases} 1 & \text{if } \sigma = \sigma' \\ q - 1 & \text{if } \sigma \prec r_i\sigma = \sigma' \\ & \text{or } \sigma \prec r_{r_j\alpha_i}\sigma = \sigma' \\ q^{-1} - 1 & \text{if } \sigma \prec r_j\sigma = \sigma' \\ (q-1)(q^{-1} - 1) & \text{if } \sigma \prec \sigma r_i \prec \sigma r_i r_j = \sigma' \end{cases} \tag{A.9}$$

References

[1] P. Bouwknegt, J. McCarthy and K. Pilch, USC-93/11, hep-th/9302086, Lett. Math. Phys. **29** (1993), to be published.

[2] P. Bouwknegt, J. McCarthy and K. Pilch, *On the BRST structure of $\mathcal{W}_3$ gravity coupled to $c = 2$ matter*, USC-93/14, hep-th/9303164.

[3] J. Thierry-Mieg, Phys. Lett. **B197** (1987) 368.

[4] M. Bershadsky, W. Lerche, D. Nemeschansky and N.P. Warner, Phys. Lett. **B292** (1992) 35. WB

[5] V. Sadov, *On the spectra of $\widehat{sl}(N)_k/\widehat{sl}(N)_k$-cosets and W_N gravities*, HUTP-92/A055, hep-th/9302060; *The hamiltonian reduction of the BRST complex and $N = 2$ SUSY*, HUTP-93/A006, hep-th/9304049.

[6] O. Aharony, O. Ganor, J. Sonnenschein and S. Yankielowicz, Phys. Lett. **B305** (1993) 35; and references therein.

[7] M. Bershadsky, W. Lerche, D. Nemeschansky and N.P. Warner, Nucl. Phys. **B401** (1993) 304.

[8] H. Lu, C.N. Pope, X.J. Wang and K.W. Wu, *The complete spectrum of the W_3 string*, CTP TAMU 50/93, hep-th/9309041, and references therein.

[9] I.N. Berstein, I.M. Gel'fand and S.I. Gel'fand, in: Proc. Summer School of the Bolyai János Math. Soc., ed. I.M. Gel'fand (New York, 1975).

[10] B.L. Feigin, Usp. Mat. Nauk **39** (1984) 195.

[11] B.L. Feigin and E.V. Frenkel, Comm. Math. Phys. **128** (1990) 161.

[12] P. Bouwknegt, J. McCarthy and K. Pilch, J. Geom. Phys. **11** (1993) 225.

[13] R. Bott, Ann. Math. **66** (1957) 203;
B. Kostant, Ann. Math. **74** (1961) 329.

[14] M. Wakimoto, Comm. Math. Phys. **104** (1986) 605.

[15] P. Bouwknegt, J. McCarthy and K. Pilch, Prog. Theor. Phys. Suppl. **102** (1990) 67.

[16] J.E. Humphreys, *Reflection groups and Coxeter groups*, Cambridge University Press (1990).

[17] O. Gabber and A. Joseph, Ann. Sci. Ec. Norm. Sup. **14** (1981) 261;
B.D. Boa, Contemp. Math. **139** (1991) 1;
K.J. Karlin, Trans. Amer. Math. Soc. **249** (1986) 29.

[18] D. Kazhdan and G. Lustig, Inv. Math. **53** (1979) 165.

[19] P. Bouwknegt and K. Schoutens, Phys. Rep. **223** (1993) 183.

[20] H. Lu, C.N. Pope and X.J. Wang, *On higher-spin generalizations of string theory*, CTP TAMU-22/93, `hep-th/9304115`;
K. Hornfeck, *Explicit construction of the BRST charge for* $\mathcal{W}_4$, DFTT-25/93, `hep-th/9306019`;
C.-J. Zhu, *The BRST quantization of the nonlinear* $_2$ *and* W_4 *algebras*, SISSA/77/93/EP, `hep-th/9306026`.

[21] B.H. Lian and G.J. Zuckerman, Phys. Lett. **B254** (1991) 417; Phys. Lett. **B266** (1991) 21; Comm. Math. Phys. **145** (1992) 561.

[22] E. Frenkel, $\mathcal{W}$*-algebras and Langlands-Drinfel'd correspondence*, in Proc. of the 1991 Cargèse workshop on "New Symmetry Principles in Quantum Field Theory," eds. J. Fröhlich et. al., Plenum Press (1992).

[23] M. Niedermaier, Comm. Math. Phys. **148** (1992) 249.

[24] P. Bouwknegt, J. McCarthy and K. Pilch, Comm. Math. Phys. **145** (1992) 541.

LIGHT-CONE QUANTISATION
OF MATRIX MODELS AT $c > 1$

Simon DALLEY

Joseph Henry Laboratories, Princeton University

Abstract: The technique of (discretised) light-cone quantisation, as applied to matrix models of relativistic strings, is reviewed. The case of the $c = 2$ non-critical bosonic string is discussed in some detail to clarify the nature of the continuum limit. Further applications for the technique are then outlined.

1 Introduction

Random surface problems appear in many branches of theoretical and mathematical physics. A number of them may be reformulated as, or arise from, matrix field theories [1, 2, 8, 4, 5, 6], planar diagrams modelling the fluctuating surfaces. The possible physical applications, together with relations to the mathematics of systems of integrable differential equations [7] and moduli space of Riemann surfaces [8, 9], make this a fascinating and novel (ab)use of quantum field theory. While condensed matter problems involve the thermal fluctuations of observable surfaces in 3D Euclidean space, for applications to high-energy physics the surfaces are worldsheets which enter indirectly through the weak coupling expansions of relativistic string theories purporting to describe quantum gravity and confining gauge theories. Matrix models of relativistic strings should be defined in Minkowski space, their spectrum, or a self-consistent truncation of it, comprising string excitations. While knowledge of the fractal geometry of perturbative diagrams is sometimes useful, numerical computation of critical random surface properties in Euclidean space is no substitute for direct calculation of the relativistic string observables; for example, the almost uninvestigated issues of strongly coupled string theories would be otherwise neglected.

I.Klebanov and the author have suggested [10] that discretised light-cone quantisation (DLCQ), introduced with a view to calculating the bound-state spectrum of gauge theories directly [11], may be very appropriately applied to matrix models also. This method, although primarily a numerical one, paints a clear physical picture of otherwise

Quantum Field Theory and String Theory, Edited by
L. Baulieu *et al.*, Plenum Press, New York, 1995

complex dynamics. In the case of string theories, this means the size, shape, and energy of strings. The unfortunate circumstance that the simplest bosonic string theory, to which one would add more structure for a physically realistic object, is (expected to be) tachyonic poses no particular problem to DLCQ of matrix models. On the contrary, one may analyse this pathology with some rigour and, as a result, set about curing it.

The following section contains an elementary review of matrix models, light-cone quantisation for a $c = 2$ model being carried out in section 3. Section 4 describes the critical behaviour of the discretised version, while future avenues of development are discussed in section 5.

2 Matrix Models of Random Surfaces

Consider an NxN hermitian matrix field $\phi_{ab}(x)$ in c dimensions subject to the following (Euclidean) action

$$S_E = \int d^c x \, \mathrm{Tr} \left(\frac{1}{2} (\partial_\alpha \phi)^2 + \frac{1}{2} \mu \phi^2 - \frac{1}{3\sqrt{N}} \lambda \phi^3 \right) . \tag{2.1}$$

The Feynman rules are those of ordinary ϕ^3 field theory except that worldlines are double lines, each line carrying a "colour" index a. Colour is conserved along the propagator and at vertices on account of the global $U(N)$ symmetry, $\phi \to \Omega^\dagger \phi \Omega$, of the action;

$$< \phi_{ab}(x) \phi_{cd}(y) > = \delta_{ad} \delta_{bc} \int \frac{d^c p}{(2\pi)^c} \frac{e^{ip \cdot (x-y)}}{p^2 + \mu} , \tag{2.2}$$

$$< \phi_{ab}(x) \phi_{cd}(x) \phi_{ef}(x) > = \delta_{bc} \delta_{de} \delta_{fa} N . \tag{2.3}$$

The $1/N$ expansion of the theory described by (2.1) is a topological expansion in surfaces [1], the Feynman diagrams being understood as fishnet approximations drawn on continuous 2-dimensional surfaces embedded in R^c;

$$\int \mathcal{D}\phi \, e^{-S_E} \sim \sum \lambda^v \left(\frac{1}{N} \right)^{-\chi} \int \text{embeddings} . \tag{2.4}$$

v is the number of vertices, χ the Euler number, and one integrates over all embeddings of the graph with rules (2.2)(2.3) . The discretised surface picture is formalised by considering the dual graphs [8, 4] that join centres of neighbouring loops, which for a ϕ^3 theory specifies a simplicial triangulation. Each triangle carries unit intrinsic area and its centre coincides with the Feynman vertex; in particular this leaves the angle between neighbouring triangles unspecified in this model. The idea is then to tune the coupling λ to a critical value λ_c at which surfaces of large intrinsic area v are favourable. At this point one may be able to take a continuum limit for surfaces.[a]

The observables are given by the Green's functions of the matrix field theory. A closed string *state* is a hole cut into the surfaces at some fixed time, t_0 say. Such states are given by the singlet operators in the matrix model;

$$\mathrm{Tr}[\phi(x_1) \ldots \phi(x_B)]_{t=t_0} \tag{2.5}$$

---

[a]There may be no scaling of the intrinsic geometry (curvatures, etc) even though the area scales – the $c \leq 1$ models are examples of this – while at $c > 1$ there is even a danger, commonly attributed to tachyons, of non-scaling area.

is a B-bit string, the bits being embedded at $x_1, \ldots, x_B$ respectively and forming a closed chain dual to the external legs of the Feynman diagrams. Also present are direct products of (2.5) (multi-string states) and non-singlets; the latter do not have an interpretation as closed strings and should be eliminated, either by performing a self-consistent truncation or effecting a dynamical decoupling (by gauging the $U(N)$ for example [12]). The critical point λ_c of large surfaces should manifest itself as singularities of Green's functions, such as the string propagator $\sim< \text{Tr}[\phi \cdots \phi]\text{Tr}[\phi \cdots \phi] >$, and thus in particular the spectrum of string excitations (2.5) should exhibit some sort of critical behaviour. Applying DLCQ to the matrix field theory in Minkowski space, one derives the spectrum as a function of λ and can search for such behaviour.

The simplest non-trivial example to consider is $c = 2$, for which (2.1) models, in its $1/N$ expansion, a discrete version of the worldsheet action for the $c = 2$ non-critical bosonic string [13],

$$A = \int d^2\sigma \sqrt{-\det g}(\Lambda + \nu R(\sigma) + T \sum_{\mu,\nu=1}^{2} \eta_{\mu\nu} g^{\alpha\beta} \partial_\alpha x^\mu \partial_\beta x^\nu + O[(\partial x)^4]) \ . \tag{2.6}$$

Λ, ν, and T are renormalised parameters associated with $\lambda, 1/N$, and μ respectively. Stretching energy of the worldsheet is governed by the propagator (2.2) , which specifies the probability amplitude for the separation of centres of neighbouring triangles. This exponential fall-off gives the Gaussian term in A plus non-renormalisable higher derivative terms. Naively the latter are irrelevant, but such reasoning assumes scaling of the worldsheet in some sense. The $c = 2$ theory (2.1) is rendered perturbatively finite by normal ordering, this being necessary in any case to eliminate graphs dual to pathological triangulations where two or more sides of the same triangle are identified. According to general arguments [14], the sum of graphs at a given order in $1/N$ in a UV finite theory is also finite for sufficiently small coupling constant; the $1/N$ expansion itself is only asymptotic. As $\lambda \to \lambda_c$ one then approaches the edge of the domain of convergence. The phenomenon is similar to, but certainly different from, the crossover to non-borel summability in the non-matrix field theory ($N = 1$).

3 Light-Cone Quantisation

Let us now rotate $x^0 \to ix^0$ and find the relativistic string spectrum by light-cone quantisation. Defining light-cone variables $x^\pm = (x^0 \pm x^1)/\sqrt{2}$, with $x^+ = x_-$, the light-cone energy and momentum $P^\pm = \int dx^- T^{+\pm}$ are

$$P^+(x^+) = \int dx^- \text{Tr}(\partial_- \phi)^2$$
$$P^-(x^+) = \int dx^- \text{Tr}(\tfrac{1}{2}\mu\phi^2 - \frac{\lambda}{3\sqrt{N}}\phi^3) \ . \tag{3.7}$$

The light-cone Hamiltonian P^- propagates a field configuration from one x^+ slice to another. Choosing a free field representation at $x^+ = 0$ say[b],

$$\phi_{ij} = \frac{1}{\sqrt{2\pi}} \int_0^\infty \frac{dk^+}{\sqrt{2k^+}}(a_{ij}(k^+)e^{-ik^+x^-} + a_{ji}^\dagger(k^+)e^{ik^+x^-}) \ , \tag{3.8}$$

---

[b]The symbol † is always understood to have purely quantum meaning and never acts on indices.

and imposing the canonical commmutation relations at equal x^+ lines,

$$[\phi_{ij}(x^-), \partial_-\phi_{kl}(\tilde{x}^-)] = \frac{i}{2}\delta(x^- - \tilde{x}^-)\delta_{il}\delta_{jk} \,, \qquad (3.9)$$

the modes a_{ij} satisfy standard commutators;

$$[a_{ij}(k^+), a^\dagger_{lk}(\tilde{k}^+)] = \delta(k^+ - \tilde{k}^+)\delta_{il}\delta_{jk} \,. \qquad (3.10)$$

An important feature is that longitudinal momentum k^+ is positive semi-definite for positive energy quanta[c] In the Fock space constructed using (3.10) , the orthonormal single closed-string states are the singlets

$$N^{-B/2}\mathrm{Tr}[a^\dagger(k_1^+)\cdots a^\dagger(k_B^+)]|0> \,, \quad \sum_{i=1}^{B} k_i^+ = P^+ \,. \qquad (3.11)$$

If we let $N \to \infty$ the multi-string states can be neglected, since $1/N$ is the string coupling constant, and P^- propagates without splitting or joining strings. Thus we solve for the free string spectrum. (Non-singlets states are discarded by hand.) P^+ and P^- can be simultaneously diagonalised and so eigenfunctions of P^- with given P^+ will be some superposition

$$\Psi = \sum_B \int_0^{P^+} dk_1 \cdots dk_B \delta(P^+ - \sum k_i) f_B(k_1, \ldots, k_B) N^{-B/2}\mathrm{Tr}[a^\dagger(k_1)\cdots a^\dagger(k_B)]|0> \,.$$
$$(3.12)$$

Explicitly one finds

$$: P^- := \tfrac{1}{2}\mu \int_0^\infty \frac{dk^+}{k^+} a^\dagger_{ij}(k^+)a_{ij}(k^+) - \frac{\lambda}{4\sqrt{N\pi}}$$
$$\times \int_0^\infty \frac{dk_1^+ dk_2^+}{\sqrt{k_1^+ k_2^+(k_1^+ + k_2^+)}} \left\{ a^\dagger_{ij}(k_1^+ + k_2^+)a_{ik}(k_2^+)a_{kj}(k_1^+) + a^\dagger_{ik}(k_1^+)a^\dagger_{kj}(k_2^+)a_{ij}(k_1^+ + k_2^+)\right\} \quad (3.13)$$

where repeated indices are summed over. For the free string theory each of the terms in (3.13) has a simple local action on the string. The mass term acts as a tensional energy $\sim \mu \sum_{i=1}^{B} 1/k_i$ and does not change the state. The cubic term can coalesce two neighbouring bits in the trace, or perform the inverse process, and changes both the length and momentum distribution (structure function) of the string. Since $c = 2$, there are no transverse oscillations in the target space. Unlike $c \le 1$ however, we are dealing with truely stringy degrees of freedom – an infinite number of particle fields – which supplement the centre of mass motion

Free string states will satisfy a relativistic dispersion relation $2P^+P^- = M^2$ and by diagonalising P^- in the basis of states of a fixed P^+ we can find the spectrum of masses M^2. This diagonalisation cannot be performed analytically for the full Hamiltonian in general, so we must introduce a cutoff [11, 15], rendering the number of states of momentum P^+ finite, and compute numerically. It is therefore important that we can neglect zero momentum modes $a^\dagger(0)$ since we could in principle include arbitrary numbers of them without changing P^+. Such modes have infinite energy for

---

[c]This positivity constraint results in the extra 1/2 in (3.9) , ensuring for example that $[P^+, \phi] = \partial_-\phi$.

$\mu \neq 0$, according to (3.13) ; however, they can be infinite in number and so a more careful analysis is required to lift this ambiguity. Indeed, it is believed by some that a proper constrained quantisation of these zero modes is necessary to describe spontaneous symmetry breaking and other non-perturbative effects in light-cone formalism. In those cases it is argued that the true vacuum, if one exists at all, is not the Fock one, : $P^- : |0> = 0$, but a condensate of $a^\dagger(0)$ modes. But one should recall that we are interested in (2.1) only insofar as it generates random surfaces through the $1/N$ expansion and perturbation theory. Therefore we should always take the $\phi = 0$ vacuum, for which there is no condensate of zero modes, since this is the one with respect to which the planar diagrams are defined. For the $c = 2$ model, at each order in $1/N$ we work with a convergent perturbation expansion, the non-perturbative effects presumably manifesting themselves through the e^{-N} corrections to the asymptotic expansion in $1/N$. The latter are non-perturbative effects of string theory and specifying their details is equivalent to stating how one is going to stabilise the unbounded ϕ^3 theory (without disturbing the $1/N$ expansion), a question that will not be considered here.

Another useful way of viewing these and other issues is to consider triangulations in light-cone perturbation theory. Indeed, for relativistic strings one could have set up the random surface expansion from this vantage point from the very beginning. Using Feynman's causality prescription on x^+ rather than x^0, the single-particle propagator becomes

$$\lim_{\epsilon \to 0} \left(\int_\epsilon^\infty \frac{dp^+}{4\pi i p^+} \theta(x^+) e^{-i(x^- p^+ + x^+ \mu/2p^+)} + \int_{-\infty}^{-\epsilon} \frac{dp^+}{4\pi i p^+} \theta(-x^+) e^{-i(x^- p^+ - x^+ \mu/2p^+)} \right) + \frac{\delta(x^+)}{2\pi\mu} . \tag{3.14}$$

The first two terms can be given the usual particle and anti-particle interpretation by viewing the negative energy (p^-) states as propagating backwards in time (x^+). The third term is a special contribution from $p^+ = 0$ and corresponds in the dual diagram to the propagation of a zero momentum string-bit; neglecting this case, all particles and anti-particles move forward in x^+ carrying positive p^+. In early work [16], the quantisation procedure used was the one adopted here, where it was proved not only that x^+-ordered and x^0-ordered perturbation theory are equivalent, but also that the third term eventually does not contribute in non-vacuum diagrams. This seems to indicate once again that, provided we restrict to perturbation theory of (2.1) , the zero modes can be ignored in computing the string propagator.

4 Discretisation and Critical Behaviour

The desired cut-off will be introduced by compactifying x^- and imposing periodic boundary conditions, $M_{ij}(x^-) = M_{ij}(x^- + L)$ [11]. Then the allowed momenta are labelled by positive integers n_i;

$$k_i^+ = \frac{2\pi n_i}{L} \ , \quad P^+ = \frac{2\pi K}{L} \ , \quad \sum_{i=1}^{B} n_i = K \ . \tag{4.15}$$

For fixed P^+, removing the cut-off $L \to \infty$ corresponds to sending $K \to \infty$. The "harmonic resolution" K represents the total number of momentum units available to the string. The longest string has K bits of one unit[d], the shortest one bit of K units, and in general the states can be labelled by the ordered partitions of K modulo cyclic permutations. Light-cone quantisation (3.8) - (3.13) may now be repeated for discrete variables $k^+ \to n$ and one finds the mass relation

$$\frac{2P^+P^-}{\mu} = K(V - xT) \; ; \; x = \frac{\lambda}{2\mu\sqrt{\pi}} \; . \tag{4.16}$$

V is the discrete version of the mass term in (3.13) while T is the cubic term. For finite K the r.h.s. is a finite-dimensional symmetric matrix with real dimensionless elements – μ is the quantity of dimension mass^2 which plays the role of string tension $1/\alpha'$ – which may be diagonalised as a function of the dimensionless parameter x; V is diagonal while T is off-diagonal.

The following picture of the critical behaviour, supported by the numerical results [10, 17], begins to emerge. In the light cone formalism of critical string theory the longitudinal momentum supported between two points on the string is proportional to the amount of σ-coordinate space between these points. We can adopt a similar co-ordinate system for the non-critical strings (2.5) . Indeed, fixing a particular bit as origin, we can define a positive scalar field on this σ-space by $X = \Delta b/\Delta\sigma$, where b is the distance, measured in number of bits, from the origin. For example, the zero mode $\int X d\sigma$ is the intrinsic length of the string. As we remove the cutoff on the longitudinal momentum allowed for bits ($K \to \infty$), and hence on discreteness of σ-space, the scalar field will generically take constant values almost everywhere in σ-space. We would like to be able to tune the theory to a critical point where the scalar field is in a long wavelength regime. In this case it would be somewhat similar to the Liouville mode of Poyakov's string [13]. What would this long wavelength regime mean for the spectrum? Firstly we would expect long string dominance; the expectation value of length (or length raised to some power) for eigenstates would typically diverge as $x \to x_c$ (and we have been assuming that this is not distinct from λ_c discussed earlier). Since, roughly speaking, each string-bit carries finite energy, we expect that $|M^2| \to \infty$ as a result. $M^2 \to \infty$ at $x = 0$ for infinitely long strings, but as $x \to x_c$, if only for consistency, we must see $M^2 \to -\infty$ for the low-lying eigenstates if they are long. Indeed this will tend to happen to the lowest eigenvalues of any real symmetric matrix as one increases its dimension for sufficiently large off-diagonal elements. Only if the Hamiltonian is an explicitly bounded operator combination ($H \sim O^\dagger O + \text{const}$) can this instability be avoided in general. We might also expect to see a continuous spectrum at $x = x_c$ if we compare with the Liouville theory results [18]

$$2P^+P^-\alpha' = p_\phi^2 + 4r - \frac{1}{6} \; , \tag{4.17}$$

but it has been difficult to confirm this numerically. Moreover the groundstate has finite negative mass squared in (4.17) , while the matrix model's is infinitely negative at $x = x_c$. It has been suggested [17] that perhaps μ should be renormalised to zero

---

[d]This sector alone represents what one might call "critical string theory" [8]

as a result, but a derivation of this requirement is still lacking. In any case we have
the first direct demonstration that the non-critical bosonic string with area action is
tachyonic above $c = 1$. The use of an unbounded ϕ^3 matrix potential does not *a priori*
spell tachyons at any order in $1/N$ in the expansion about $\phi = 0$, but the string theory
described at the critical coupling is nevertheless tachyonic.[e]

5 Future Directions

In order to identify the critical behaviour at $c = 2$ more precisely it is useful to investigate the effects of adding an explicit polymerisation term [19]

$$S = S_E - \int d^2x \, \frac{g}{N^2} (\text{Tr}[\phi^2])^2 \, . \tag{5.18}$$

For large g this worldsheet contact interaction seems to favour short strings. This is
quite unlike the critical point at $g = 0$ and, assuming that there is a phase transition
somewhere in between, casts doubt on branched polymer behaviour of the $c = 2$
matrix model in Minkowski space; recall that this behaviour was identified in $c > 1$
Euclidean dimensions from numerical simulations [20] and combinatorial estimates [21]
of dynamical triangulations at the critical point. Moreover simple polymerisation is not
the only possibility in the Euclidean game. The phase diagram needs to be investigated
in more detail before a clearer picture can be gained.

The tachyon we found is expected to persist at $c > 2$. To regulate the transverse
dimensions we can use a transverse lattice. To eliminate the zero modes associated with
$x_\perp$, which would otherwise obscure the stringy part of the spectrum, we can perform
the Eguchi-Kawai (EK) compactification [22] to single links in each direction. The
resulting field theories are much the same as the $c = 2$ one; namely, we deal with
UV finite two-dimensional field theories with convergent perturbation expansions at
given order in $1/N$. If we use Hermitian matrix models (2.1) the EK reduction induces
more general interactions $V_{\text{eff}}(\phi)$ in the effective $c = 2$ potential [23]. Unfortunately
$V_{\text{eff}}(\phi)$ is not known explicitly and can only be calculated as an expansion in powers
of the inverse transverse lattice spacing $1/a$. Truncating the expansion arbitrarily at
some order, the resulting model will exhibit polymeric behaviour for sufficiently small
a since the leading term is a contact interaction similar to that appended in (5.18) , the
induced g being $\sim 1/a^2$. At sufficiently large a the (unreduced) transverse lattice sites
are obviously uncoupled and we should recover copies of the $c = 2$ model at each site.

In order to deal with an exact EK reduced system, one can employ complex matrix
models similar to the old Weingarten model [2]. Complex matrices M live on the links
of a c-dimensional hypercubic lattice. For non-critical string theory we must use an
action given by traces around all (oriented) loops of length 4 on the lattice [6], which
comprises the standard plaquette action plus zero area loops. Expansion of this theory

---

[e]The use of an unbounded potential is rather a symptom of the divergence of the $1/N$
string perturbation expansion, as commented earlier.

in the coupling constant and $1/N$ reproduces the random surfaces of a dynamical quadrangulation in c dimensions, the the probability distribution for neighbouring vertices $\{x, y\}$ of the quadrangulation in this case being

$$\sum_{\widehat{\mu}} \delta(x - y - a\widehat{\mu}) + \delta(x - y + a\widehat{\mu}) , \qquad (5.19)$$

for orthonormal vectors $\widehat{\mu}$. For $c = 1$ it has been proven that this gives the same answers at the critical point as using Feynman propagators [6]. To perform DLCQ we must take the naive continuum limit for two of the dimensions, which seems to produce an intractable two-dimensional kinetic term unfortunately. Instead we could use the Feynman propagator for these two continuous dimensions and study actions like [8, 25]

$$\int d^2x \mathrm{Tr}[\partial_\alpha M \partial^\alpha M^\dagger + \mu M M^\dagger + \lambda_1 M M M^\dagger M^\dagger + \lambda_2 M M^\dagger M M^\dagger] . \qquad (5.20)$$

This is a $c = 3$ model with EK reduction of the transverse direction – a single complex matrix M lies on the periodic link – particularly simple since it has no standard plaquette term. One may now study the DLCQ as a function of a. At sufficiently small a there should be a roughening transition from the $c = 2$ to the $c = 3$ phase. It should be noted however that the computational accuracy diminishes significantly for $c > 2$ due to the increase in the number of degrees of freedom, each string bit now carrying at least one more Z_2 variable – e.g. the real and imaginary parts of M.

Even if reliable data on models such as (5.20) can be collected, we still expect tachyons. We must try other possibilities to eliminate them, such as introducing some dependence on extrinsic geometry, which is also of interest in condensed matter problems. This is certainly possible for the complex matrix models, at least on the lattice target space, since the orientation of simplices is rather manifest. At a more fundamental level, we should look for tachyon-free matrix models with long string dominance if we wish to describe interacting continuum strings. As indicated earlier, such would be a delicate theory; using a positive definite Hamiltonian tends to counteract precisely the requirement of the critical point – crossover to divergence of perturbation theory.

For other applications of light-cone matrix models, such as to confining gauge theories, continuity of strings may not be so important. For example the interesting Regge-like trajectories found in two-dimensional gauged matrix models [12] are most probably the result of restriction to sectors of fixed (discrete) string length^f. According to the suggestion made earlier, this freezes the Liouville-like zero mode and exposes the quasi-harmonic energy levels. While it is unrealistic to expect solution of the gauged models in higher dimensions, since they are more complicated than pure large-N QCD, the two-dimensional models may help us to understand more clearly the relationship between gluonic and fundamental strings. They describe a limit of higher dimensional gauge theory in which all transverse directions are very compact; $x_\perp$-independent transverse potentials $A_\perp(x^+, x^-)$ play the role of matter ϕ in the gauged $c = 2$ matrix model.

---

^{f}In ref.[12] low-lying mass eigenstates tended to consist of strings of some given length for mysterious dynamical reasons.

Clearly there are many interesting questions in applications of string theory which may be addressed by the light-cone matrix models through analytic and numerical techniques.

Acknowledgments

It is a pleasure to thank I.Klebanov and K.Demeterfi for interesting discussions. Financial support comes through S.E.R.C.(U.K.) post-doctoral fellowship RFO/B/91/9033.

References

[1] G.'t Hooft, *Nucl. Phys.* **B72** 461 (1974).

[2] D. Weingarten, *Phys. Lett.* **90** (1980) 280.

[3] F.David, *Nucl. Phys.* **B257** 45 (1985).

[4] V.A.Kazakov, I.K.Kostov, and A.A.Migdal, *Phys. Lett.* **B157** (1985) 295.

[5] V.A.Kazakov and A.A.Migdal, *Nucl. Phys.* **B311** (1989) 171.

[6] S. Dalley, *Mod. Phys. Lett.* **A7**, 1651 (1992).

[7] M.R.Douglas, *Phys. Lett.* **B238** (1990) 176.

[8] R.C.Penner, J. Diff. Geom. **27** (1988) 35.

[9] M.Kontsevich, Commun. Math. Phys. **147** (1992) 1.

[10] S.Dalley and I.R.Klebanov, *Phys. Lett.* **B298** 79 (1993).

[11] H.-C. Pauli and S. Brodsky, *Phys. Rev.* **D32** (1985) 1993 and 2001.

[12] S.Dalley and I.R.Klebanov, *Phys. Rev.* **D47** 2517 (1993).

[13] A.M.Polyakov, *Phys. Lett.* **B103** (1981) 207.

[14] J.Koplik, A.Neveu, and S.Nussinov, *Nucl. Phys.* **B123** (1977) 109.

[15] C. B. Thorn, *Phys. Lett.* **70B** (1977) 85; *Phys. Rev.* **D32** (1978) 1073.

[16] Chang and Ma, *Phys. Rev.* **180** (1969) 1506; J.Kogut and Soper, *Phys. Rev.* **D1** (1970) 2901.

[17] K.Demeterfi and I.R.Klebanov, PUPT–1370, hep-th/9301006, presented at the 7th Nishinomiya-Yukawa Memorial Symposium "Quantum Gravity", November 1992.

[18] T. L. Curtright and C. B. Thorn, *Phys. Rev. Lett.* **48**, 1309 (1982).

[19] S.Das, A.Dhar, A.Sengupta, and S.Wadia, *Mod. Phys. Lett.* **A5** (1990) 1041.

[20] D.V.Boulatov, V.A.Kazakov, I.K.Kostov, and A.A.Migdal, *Nucl. Phys.* **B257** (1985) 641.

[21] J.Ambjorn, B.Durhuus, and J.Frohlich, *Nucl. Phys.* **B257** (1985) 433.

[22] T.Eguchi and H.Kawai, *Phys. Rev. Lett.* **48** (1982) 1063.

[23] L.Alvarez-Gaumé, C.Crnkovic, and J.F.L.Barbòn, *Nucl. Phys.* **B394** (1993) 383.

[24] I. R. Klebanov and L. Susskind, *Nucl. Phys.* **B309**, 175 (1988).

[25] S. Dalley and T.R. Morris, *Int. Jour. Mod. Phys.* **A5**, 3929 (1990).

MULTICRITICAL POINTS of 2-MATRIX MODELS

Jean-Marc DAUL

Laboratoire de Physique Théorique de l'Ecole Normale Supérieure
(Unité propre du CNRS),
24 rue Lhomond, 75005 Paris, France

1 Introduction

We present results obtained with V.Kazakov and I.Kostov [1] about the two-matrix model: following the ideas of M.Douglas [2],we obtained explicit expressions for critical potentials corresponding to any rational conformal field theory, and for the one- and two-loop averages. We also show how to analyse the critical behaviour of these correlation functions.

We shall be interested in the partition function

$$Z = \int dX\, dY\, e^{-\frac{N}{\lambda}Tr[U(X)+V(Y)-XY]} \tag{1.1}$$

where X, Y are hermitian $N \times N$ matrices and U, V are polynomial potentials, which diagrammatic expansion is recognized as an Ising system on a random two-dimensional lattice.

For instance, consider the symmetric case ($U = V$) with a potential of degree 3: it seems we have 4 independent coupling constants (λ and 3 coefficients in the potential); indeed, because of the invariance under the change of variables $X(Y) \to \alpha X(Y) + \beta$, we can assume that $U(x) = \frac{x^2}{2} + g\frac{x^3}{3}$ and we really have two independent coupling constants: λ and g. We also have two possible critical behaviours: if the temperature of the magnetic lattice is well adjusted, we allow for long-range correlation between spins; and if the cosmological constant is well tuned, we will observe larger and larger lattices. The finest critical behaviour will be reached if g is set to some (critical) value g_c and if we let $\lambda \to \lambda_c$ so that spin blocks increase in size with the lattice.

This is the very general situation: in any case, the finest critical behaviour will be obtained for definite values of the potentials when we let the cosmological constant λ approach some critical value (the matter is set to criticality, and waits for the lattice to increase in size in order to manifest its long-range correlations). This we now proceed to study, with the apparatus of orthogonal polynomials.

Quantum Field Theory and String Theory, Edited by
L. Baulieu *et al.*, Plenum Press, New York, 1995

2 Orthogonal Polynomials

We want to compute

$$Z = \int dX\, dY\, e^{-\frac{N}{\lambda} Tr\left[U(X)+V(Y)-XY\right]} \tag{2.2}$$

with

$$U(x) = \sum_{1...p} g_n \frac{x^n}{n} \; ; V(y) = \sum_{1...q} h_n \frac{y^n}{n} \tag{2.3}$$

In terms of angular and radial (that is eigenvalues) variables

$$X = \Omega_1 \begin{pmatrix} x_1 & & 0 \\ & \ddots & \\ 0 & & x_N \end{pmatrix} \Omega_1^{-1} \; , \; Y = \Omega_2 \begin{pmatrix} y_1 & & 0 \\ & \ddots & \\ 0 & & y_N \end{pmatrix} \Omega_2^{-1} \tag{2.4}$$

we obtain (after integrating over angles)

$$Z = \int dx_{1...N}\, dy_{1...N}\, \Delta(x)\Delta(y) \exp\left[-\frac{N}{\lambda} \sum_{i=1}^{N} U(x_i) + V(y_i) - x_i y_i \right] \tag{2.5}$$

where

$$\Delta(x) = \begin{vmatrix} 1 & 1 & \cdots & 1 \\ x_1 & x_2 & \cdots & x_N \\ x_1^2 & x_2^2 & & x_N^2 \\ \vdots & \vdots & \ddots & \vdots \\ x_1^{N-1} & x_2^{N-1} & \cdots & x_N^{N-1} \end{vmatrix} \tag{2.6}$$

We introduce the measure $d\mu(x,y) = dx\, dy\, e^{-\frac{N}{\lambda}\left(U(x)+V(y)-xy\right)}$ and the corresponding scalar product

$$\langle g|f \rangle = \int d\mu(x,y) g(y) f(x) \tag{2.7}$$

Finally, we consider orthonormal polynomials $\langle m|n \rangle = \delta_{m,n}$ with $\langle n|$ and $|n\rangle$ of degree n (bras and kets coincide in the symmetric case $U = V$). Van der Monde determinants are now easily expressed (up to a normalization factor corresponding to the highest degree coefficients of the orthonormal polynomials):

$$\begin{aligned} \Delta(x) &\sim \det_{1\leq i,j\leq N}\left(|i-1\rangle_{(x_j)}\right) \\ \Delta(y) &\sim \det_{1\leq i,j\leq N}\left(\langle i-1|_{(y_j)}\right) \end{aligned} \tag{2.8}$$

3 Position and Momentum Operators

We introduce $\widehat{X}$ and $\hat{P}$:

$$\begin{aligned} \left(\widehat{X}|n\rangle\right)(x) &= x\,|n\rangle(x) \\ \left(\hat{P}|n\rangle\right)(x) &= \frac{\lambda}{N}\frac{d}{dx}|n\rangle \end{aligned} \tag{3.9}$$

which satisfy $[P,X] = \frac{\lambda}{N}$; and similar operators Y, Q with a simple action on bras. The identities:

$$\begin{aligned} \cdot P &= U'(X) - Y \\ Q &= V'(Y) - X \end{aligned} \tag{3.10}$$

are easily checked if we use integration by parts to compute the corresponding matrix elements. As U' and V' have degrees $p-1$ and $q-1$ respectively, it follows that (the matrix) X has one diagonal below the main diagonal and $q-1$ above (and elements on the main diagonal), while P has $(p-1)(q-1)$ diagonals above the main diagonal (and we have dual results for Y, Q). We shall use the following notation for their matrix elements:

$$
\begin{aligned}
X|n\rangle &= \sum_{k=-1}^{q-1} X_k(n)|n-k\rangle \\
P|n\rangle &= \sum_{k=1}^{(p-1)(q-1)} P_k(n)|n-k\rangle \\
\langle n|Y &= \sum_{k=-1}^{p-1} Y_k(n)\langle n-k| \\
\langle n|Q &= \sum_{k=1}^{(p-1)(q-1)} Q_k(n)\langle n-k|
\end{aligned}
\tag{3.11}
$$

In terms of $\widehat{n}$ and $\widehat{\omega} = \widehat{\frac{d}{dn}}$ we have (e.g.) $X = \sum_k e^{k\widehat{\omega}} X_k(\widehat{n})$, where the order matters up to $1/N$ terms only.

4 Classical Operators and Observables

We will only consider the planar limit $N \to \infty$: with $t = \frac{\lambda}{N}n$, the canonical commutation relations become

$$
[\omega, t] = [P, X] = \frac{\lambda}{N}
\tag{4.12}
$$

and we are studying the classical limit (it is also convenient to think of $\frac{N}{\lambda}$ as an inverted temperature - see the partition function (1.1) : we are in the zero temperature limit). $\widehat{X}$ becomes a classical function:

$$
X(\omega, t) = \sum_k e^{k\omega} X_k(Nt/\lambda)
\tag{4.13}
$$

and the classical equations of motion are:

$$
\begin{aligned}
P(\omega, t) &= U'[X(\omega, t)] - Y(-\omega, t) \\
Q(\omega, t) &= V'[Y(\omega, t)] - X(-\omega, t)
\end{aligned}
\tag{4.14}
$$

with

$$
\begin{aligned}
X &= \sqrt{R}e^{-\omega} + \text{polynomial}(e^{\omega}) \\
P &= \frac{t}{\sqrt{R}}e^{\omega} + e^{2\omega}\text{polynomial}(e^{\omega})
\end{aligned}
\tag{4.15}
$$

(to see this, consider the action of X,P on the highest degree monomial in $|n\rangle$).

Before solving (4.14), let us see how we can express geometric observables in terms of these classical functions.

Such an observable is, e.g.: $\langle \frac{Tr}{N}\frac{1}{n}\widehat{X}^n \rangle$, the sum over all surfaces bounded by a loop with length n; or $\langle \frac{Tr}{N}\widehat{X}^n \rangle$, sum over all surfaces bounded by a n-loop with a marked point; or:

$$W(x) = \langle \frac{Tr}{N}\frac{1}{x - \widehat{X}} \rangle \tag{4.16}$$

the one-loop function with a point on the boundary and a boundary cosmological constant equal to x.

The latter is easily written as

$$W(x) = \frac{1}{Z}\int dx_{1...N}\, dy_{1...N}\, e^{\text{action}} \det\!\left(\langle m|_{(y_j)}\right) \frac{1}{N}\sum_{a=1}^{N}\frac{1}{x - x_a}\det\!\left(|n\rangle_{(x_i)}\right) \tag{4.17}$$

where the determinants can be interpreted as the in (out) wavefunctions of N fermions lying in the lowest N levels: $|0\rangle,\ldots,|N-1\rangle$ ($\langle 0|,\ldots,\langle N-1|$), that is: in the zero temperature state. So,

$$W(x) = \frac{1}{N}\sum_{a=0}^{N-1}\langle a|\frac{1}{x - \widehat{X}}|a\rangle = Tr\left(P_{\text{Fermi}}\frac{1}{x - \widehat{X}}\right) \tag{4.18}$$

with P_{Fermi} the projection operator over the N lowest states.

With this expression, it is fairly easy to obtain for the classical function:

$$\partial_\lambda W(x) = \left.\frac{\partial \omega}{\partial X}\right|_\lambda \tag{4.19}$$

ω being a function of X for fixed x through $X(\omega,\lambda) = x$.

Thus, we will be able to investigate the scaling properties of the one-loop function when we know the classical position and momentum.

5   Solution of the Equations of Motion

We introduce for convenience $z = e^\omega$ and we write:

$$P(z,t) = U'[X(z,t)] - Y(1/z,t) \tag{5.20}$$

At the critical point ($t = \lambda_c$) we expect:

$$\begin{aligned}
X - X_c &\sim \omega^q \\
P - P_c &\sim \omega^p \\
Y - Y_c &\sim \omega^p \\
Q - Q_c &\sim \omega^q
\end{aligned} \tag{5.21}$$

(this is the highest order reachable criticality, considering the degrees of $X,\ldots,Q$ in z) which fixes the form of $X(\cdot,\lambda_c),\ldots$:

$$X_c(z) = \frac{(1-z)^q}{z} \;;\; Y_c(z) = \frac{(1-z)^p}{z} \tag{5.22}$$

We now substitute these formulae in (5.20), and, matching negative powers of z, we obtain the critical potentials:

$$U'(\phi) = \frac{1}{2i\pi} \oint \frac{Y_c(1/z)}{X_c(z) - \phi} dX_c(z) \tag{5.23}$$

Then, to extract the behaviour of X for $\lambda \sim \lambda_c$, we write:

$$X = \frac{\sqrt{R}}{z} + a + b\,z + \cdots + d\,z^{q-1} \tag{5.24}$$

and obtain algebraic equations in $a, b, \ldots$ after substitution in (5.20).

Example: symmetric realization of the $(3, 4)$ model

$U = V$ has degree 3; at the critical point:

$$\left| \begin{array}{ccc} X - X_c & \sim & \omega^3 \\ P - P_c & \sim & \omega^4 \end{array} \right. \tag{5.25}$$

which leads to: $\lambda_c = 10; U(\phi) = 3\phi - \frac{3}{2}\phi^2 - \phi^3/3$.

If $X(z,t) = \frac{\sqrt{R}}{z} + a + b\,z + c\,z^2$, we have:

$$\begin{aligned} c &= -R \\ b &= -(3 + 2\,a)\sqrt{R} \\ a &= 2(R - 1) \pm \sqrt{4R^2 - 2R + 7} \end{aligned} \tag{5.26}$$

In the latter equation, we have to choose the minus sign to get the required critical behaviour ($a_c = -3$). And, finally:

$$\lambda = t = 32\,R^3 - 14\,R^2 + 28\,R - 4\,R(4\,R - 1)\sqrt{4R^2 - 2R + 7} \tag{5.27}$$

As for the critical behaviour of X we find:

$$X(\lambda, \omega) \sim (\lambda - \lambda_c)^{1/2} \xi\left(\frac{\omega}{(\lambda - \lambda_c)^{1/6}}\right) \tag{5.28}$$

with ξ the third Chebyshev polynomial: $\xi(\cosh\theta) = \cosh 3\theta$, so that we can write:

$$\mu = (\lambda - \lambda_c)^{1/6} : \left| \begin{array}{ccc} \omega &=& \mu\,\cosh\theta \\ X &\sim& \mu^3\,\cosh 3\theta \end{array} \right. \tag{5.29}$$

6 Scaling Behaviour

The formulae (5.29) are indeed quite general, as we now argue: from

$$[\widehat{\omega}, \widehat{t}] = [\widehat{P}, \widehat{X}] = \frac{\lambda}{N} \tag{6.30}$$

the Poisson bracket $\{P, X\}_{\omega,t}$ is equal to 1.

In the symmetric case, when $U = V$ has degree m (to produce the $(m, m+1)$ model), we expect a critical scaling

$$\begin{aligned} X(\lambda, \omega) &\sim (\lambda - \lambda_c)^a\, \xi\left(\frac{\omega}{(\lambda - \lambda_c)^b}\right) \\ P(\lambda, \omega) &\sim (\lambda - \lambda_c)^c\, \pi\left(\frac{\omega}{(\lambda - \lambda_c)^b}\right) \end{aligned} \tag{6.31}$$

with ξ and π polynomials with degrees m and $m+1$.

Now, to ensure $\{P, X\} = 1$ we shall have: $b = \frac{1}{2m}$, $a = \frac{1}{2}$, $c = \frac{m+1}{2m}$ and

$$m\,\xi\,\pi' - (m+1)\xi'\,\pi = 1 \tag{6.32}$$

And this equation is always satisfied by Chebyshev polynomials:

$$\left| \begin{array}{rcl} \xi(\cosh\theta) &=& \cosh m\theta \\ \pi(\cosh\theta) &=& \cosh(m+1)\theta \end{array} \right. \tag{6.33}$$

So we know the scaling behaviour of X, P and W because $\partial_\lambda W\big|_x = \frac{\partial\omega}{\partial X}\big|_\lambda$ and the former is:

$$\frac{\partial W}{\partial\lambda}\bigg|_\omega + \frac{\partial W}{\partial\omega}\bigg|_\lambda \left(-\frac{\frac{\partial X}{\partial\lambda}\big|_\omega}{\frac{\partial X}{\partial\omega}\big|_\lambda} \right) = \frac{\partial\omega}{\partial X}\bigg|_\lambda \{W, X\}_{\lambda,\omega} \tag{6.34}$$

so that W and X have a Poisson bracket equal to one, and

$$W(x, \lambda) = P(x, \lambda) + \text{function}(x) \tag{6.35}$$

where the unknown function does not contribute to the critical $(\lambda \to \lambda_c)$ behaviour of W.

References

[1] J-M.Daul, V.A.Kazakov, I.K.Kostov, *Rational theories of 2d Gravity from the Two-Matrix Model*, CERN-TH.6834/93, LPTENS 93/7, March 1993.

[2] M.Douglas, Proceedings of the Cargèse Workshop, 1990.

THE SUPER SELF-DUAL MATREOSHKA

Ch. DEVCHAND

Joint Institute for Nuclear Research
Dubna, Russia

and

V. OGIEVETSKY[a]

Physikalisches Institut der Universität Bonn
Bonn, Germany

Abstract: In this talk we review the harmonic space formulation of the twistor transform for the supersymmetric self-dual Yang-Mills equations. The recently established harmonic-twistor correspondence for the N-extended supersymmetric gauge theories is described. It affords an explicit construction of solutions to these equations which displays a remarkable matreoshka-like structure determined by the N=0 core.

1 Introduction

The Yang-Mills self-duality (SDYM) equations are well known Lorentz invariant four dimensional exactly solvable nonlinear systems . Remarkably, these equations afford generalisation to the super self-duality equations for extended super Yang-Mills theories without spoiling their integrability properties . The extended super self-duality equations are therefore further examples of exactly solvable Lorentz invariant four dimensional systems; and the Penrose-Ward twistor transform [1], so succesful for the self-duality equations in complexified four-dimensional space, may be generalised to extended superspaces. The original twistor transform and its supersymmetric generalisations have been found to have a clear and tractable formulation in the language of "harmonic spaces". We therefore call them "harmonic-twistor correspondences". For the N-extended supersymmetric self-duality equations, moreover, this harmonic space formulation [2] of the twistor transform reveals a remarkable "matreoshka"-like structure [3]: Much of the structure of an N-extended self-dual theory is determined by its lower-N sub-theory; and ultimately, by the non-supersymmetric N=0 core. In

---

[a]on leave from JINR, Dubna, Russia

Quantum Field Theory and String Theory, Edited by
L. Baulieu *et al.*, Plenum Press, New York, 1995

particular, given any solution of the N=0 self-duality equations, its most general supersymmetric extension may be recursively constructed. The problem of finding the general local solution of the $N > 0$ super self-duality equations therefore reduces to finding the general solution of the N=0 self-duality equations. The latter completely determines the general N=1 solution, which in turn determines the N=2 solution, and so on. A further consequence of the matreoshka phenomenon is the vanishing of many conserved currents for super self-dual systems, for instance the vanishing of the Yang-Mills stress tensor for N=0 self-dual fields is reflected in the vanishing of the extended supergauge theory supercurrents which contain the stress tensor and its superpartners.

Harmonic (super)spaces contain additional coordinates: harmonics or twistors, which we denote by commuting spinors $u_\alpha^\pm$. The origin of this enlargement is the fact that harmonic spaces are cosets of the (super) Poincare group by some *subgroup* of the rotation group, whereas customary (super)space coordinates parametrise the coset of the (super) Poincare group by the *entire* rotation group. For global considerations harmonics $u_\alpha^\pm$ need to be considered as coordinates on the four-dimensional (super)conformal group factored by its maximal parabolic subgroup [10]. In this talk, however, we limit ourselves to *local* aspects of the self-duality equations.

Originally, harmonic superspaces were introduced [4] as appropriate tools for the construction of unconstrained off-shell $N = 2$ and 3 super Yang-Mills theories; and involved the 'harmonisation' of the internal unitary groups of supersymmetry, with each particular case $(N = 2, 3)$ requiring individual consideration. For the (super) self-duality restrictions, however, one harmonises the rotation group instead. This being N-independent, the harmonisation is universal; and in contrast to the previous aim [4] of constructing off-shell theories, the main aim of the study of the self-dual restrictions [2, 3] is the investigation of the on-shell theory, viz. to solve the (super) self-duality equations of motion.

The self-duality conditions have recently attracted a great deal of renewed interest in view of their reductions to lower-dimensional completely integrable systems [5] and the prospect [19] of unifying lower-dimensional solution methods under the banner of the SDYM twistor transform. The programme has by now advanced rather far, with most known integrable systems having been rederived by the abovementioned reduction. Moreover, there have also appeared papers [7] dealing with reductions of super self-duality equations . Our considerations [3] suggest the interesting possibility that completely integrable supersymmetric systems are merely further layers of the self dual matreoshka.

The main purpose of these lecture notes is to review the harmonic-space formulation of the twistor transform [8, 9, 10, 2, 3]. In section 2 we discuss this formulation for the N=0 case. In section 3 we discuss the super self-duality conditions and in section 4 we review the generalisation of the harmonic-twistor correspondence to N-extended super self-duality equations for all $N > 0$. The latter yields, in particular, a representation of all possible symmetries of these equations, including an important subgroup of diffeomorphisms of the analytic subspace of harmonic superspace. In section 5 we discuss the solution matreoshka: Given an N=0 solution, we show that a purely algorithmic procedure yields solutions of higher N theories.

2 Self-Duality as Harmonic Space Analyticity

The usual self-duality condition for the Yang-Mills field strength

$$F_{\mu\nu} = \frac{1}{2}\epsilon_{\mu\nu\rho\sigma}F_{\rho\sigma} \; , \tag{2.1}$$

basically says that the (0,1) part of the gauge field vanishes. This is better expressed in terms of 2-spinor notation in the form: $f_{\dot\alpha\dot\beta} = 0$ which is equivalent to the statement that the field strengths curvature only contains the (1,0) Lorentz representation, i.e.

$$[\mathcal{D}_{\alpha\dot\alpha} \, , \, \mathcal{D}_{\beta\dot\beta}] = \; \epsilon_{\dot\alpha\dot\beta}f_{\alpha\beta}. \tag{2.2}$$

Now multiplying (2.2) by two commuting spinors $u^{+\dot\alpha}, u^{+\dot\beta}$ mentioned in the Introduction, one can compactly represent it as the vanishing of a curvature

$$[\nabla_\alpha^+, \nabla_\beta^+] = 0 \; , \tag{2.3}$$

where $\nabla_\alpha^+ \equiv u^{+\dot\alpha}\nabla_{\alpha\dot\alpha}$, with linear system

$$\nabla_\alpha^+ \varphi = \; 0 \; . \tag{2.4}$$

This is precisely the Belavin-Zakharov-Ward linear system for SDYM. Now the $u^{+\dot\alpha}$ are actually harmonics [4] on S^2 and it is better to consider these equations in an auxiliary space ('harmonic space') with coordinates $\{x^{\pm\alpha} \equiv x^{\alpha\dot\alpha}u_{\dot\alpha}^{\pm}, u_{\dot\alpha}^{\pm}; u^{+\dot\alpha}u_{\dot\alpha}^- = 1\}$, where the harmonics are defined up to a $U(1)$ phase (see [4, 2, 3]), and gauge covariant derivatives

$$\nabla_\alpha^+ = \partial_\alpha^+ + A_\alpha^+ = \frac{\partial}{\partial x^{-\alpha}} + A_\alpha^+. \tag{2.5}$$

In this space (2.3) is actually not equivalent to the self-duality conditions. We also need

$$[D^{++}, \nabla_\alpha^+] = 0 \; , \tag{2.6}$$

where D^{++} is a harmonic space derivative which acts on negatively-charged harmonic space coordinates to yield their positively-charged counterparts, i.e. $D^{++}u_{\dot\alpha}^- = u_{\dot\alpha}^+$, $D^{++}x^{-\alpha} = x^{+\alpha}$, whereas $D^{++}u_\alpha^+ = D^{++}x^{+\alpha} = 0$. In ordinary x-space, when the harmonics are treated as parameters, the condition (2.6) is actually incorporated in the definition of ∇_α^+ as a *linear* combination of the covariant derivatives. The system (2.3,2.6) is now *equivalent* to SDYM and has been considered by many authors, e.g. [8, 9, 10, 11, 12]; the equivalence holding in spaces of signature (4,0) or (2,2), or in complexified space. In this regard, we should note that for real spaces, our understanding is completely clear for the Euclidean signature. For the (2,2) signature, the situation is richer and more intricate due to the noncompact nature of the rotation group. On the one hand, there appear infinite dimensional representations, and on the other hand, novel subgroups (in particular, the parabolic ones) as well as new cosets (some of them rather intriguing). Our present considerations concern only those signature (2,2) configurations which may be obtained by Wick rotation of (4,0) configurations.

Now, in (2.6) the covariant derivative (2.5) has pure-gauge form

$$\nabla_\alpha^+ = \partial_\alpha^+ + \varphi \partial_\alpha^+ \varphi^{-1}. \tag{2.7}$$

and D^{++} is 'short' i.e. has no connection. This choice of frame is actually inherited from the four-dimensional x-space and is not the most natural one for harmonic space. We may however change coordinates to a basis in which ∇_α^+ is 'short' and D^{++} is 'long' (i.e. acquires a Lie-algebra-valued connection) instead. Namely,

$$\begin{aligned}\nabla_\alpha^+ &= \partial_\alpha^+ \\ \mathcal{D}^{++} &= D^{++} + V^{++},\end{aligned} \tag{2.8}$$

a change of frame tantamount to a gauge transformation by the 'bridge' φ in (2.4). In this basis the SDYM system (2.3,2.6) remarkably takes the form of a Cauchy-Riemann (CR) condition

$$\frac{\partial}{\partial x^{-\alpha}} V^{++} = 0 \tag{2.9}$$

expressing independence of half the x-coordinates. In virtue of passing to this basis the nonlinear SDYM equations (2.1) are in a sense trivialised: Any 'analytic' (i.e. satisfying (2.9)) function $V^{++} = V^{++}(x^{+\alpha}, u^\pm)$ corresponds to some self-dual gauge potential. ¿From any such V^{++}, by solving the *linear* equation

$$D^{++}\varphi = \varphi V^{++} \tag{2.10}$$

for the bridge φ, a self-dual vector potential may be recovered from the harmonic expansion:

$$\varphi \partial_\alpha^+ \varphi^{-1} = u^{+\dot\alpha} A_{\alpha\dot\alpha}; \tag{2.11}$$

the linearity in the harmonics $u^{+\dot\alpha}$ being guaranteed by (2.6). An important comment: It follows from (2.10) that

$$D^{++} \det \varphi = \det \varphi \ \ tr V^{++}.$$

Therefore, for semisimple gauge groups $(tr V^{++} = 0)$ we have

$$D^{++} \det \varphi = 0. \tag{2.12}$$

We may therefore either solve (2.10) for a unimodular bridge, or without worrying about the determinant we may substract traces in (2.11) when calculating the connection.

Solving (2.10) for an *arbitrary* analytic gauge algebra valued function V^{++} yields the *general* self-dual solution. This correspondence between self-dual gauge potentials and holomorphic prepotentials V^{++} is just a transparent formulation of the Penrose-Ward twistor correspondence for SDYM and is a convenient tool for the explicit construction of local solutions of the self-duality equations. For instance the 1-instanton BPST solution

$$A_{\alpha\dot\alpha i}^{j} = \frac{1}{\rho^2 + x^2}\left(\frac{1}{2} x_{\alpha\dot\alpha}\delta_i^j + \epsilon_{i\alpha}x_{\dot\alpha}^j\right), \tag{2.13}$$

corresponds to the analytic function [13, 8, 11]

$$(v^{++j})_i^j = \frac{x^{+j}x_i^+}{\rho^2} \tag{2.14}$$

via the bridge

$$(\varphi_b)_i^j = \left(1 + \frac{x^2}{\rho^2}\right)^{-\frac{1}{2}} \left(\delta_j^i + \frac{x^{+i}x_j^-}{\rho^2}\right). \tag{2.15}$$

Furthermore, in the analytic subspace of harmonic space (with coordinates $\{x^{+\alpha}, u_\alpha^\pm\}$), there exists an especially simple presentation of the infinite-dimensional symmetry group acting on solutions of the self-duality equations. It is the (apparently trivial) transformation $V^{++} \to V^{++'} = g^{++}$, where g^{++} depends in an arbitrary way on V^{++} and its derivatives as well as on the analytic coordinates themselves, modulo gauge transformations $V^{++} \to e^{-\lambda}(V^{++}+D^{++})e^\lambda$, where λ is also an arbitrary analytic function. The situation is the same for any extended supersymmetric gauge theory, as we discuss in section 4.

3 Super Self-Duality

Since extended super Yang-Mills theories are massless theories, the components are classified by helicity and we have the following representation content in theories up to N=3:

$$
\begin{array}{lcccccccc}
helicity: & 1 & \frac{1}{2} & 0 & -\frac{1}{2} & \frac{1}{2} & 0 & -\frac{1}{2} & -1 \\
N = 0 & f_{\alpha\beta} & & & & & & & f_{\dot\alpha\dot\beta} \\
N = 1 & f_{\alpha\beta} & \lambda_\alpha & & & & & \lambda_{\dot\alpha} & f_{\dot\alpha\dot\beta} \\
N = 2 & f_{\alpha\beta} & \lambda_\alpha^i & \overline{W} & & W & & \lambda_{\dot\alpha i} & f_{\dot\alpha\dot\beta} \\
N = 3 & f_{\alpha\beta} & \lambda_\alpha^i & W_i & \chi_{\dot\alpha} & \chi_\alpha & W^i & \lambda_{\dot\alpha i} & f_{\dot\alpha\dot\beta}
\end{array}
\tag{3.16}
$$

In real Minkowski space fields in the left and right triangles are related by CPT conjugation but in complexified space or in a space with signature (4,0) or (2,2), we may set fields in one of the triangles to zero without affecting fields in the other triangle. If we set all the fields in the right (left) triangle to zero, the equations of motion reduce to the super (anti-) self-duality equations. For instance, the equations of motion for the N=3 theory take the form

$$
\begin{aligned}
\epsilon^{\dot\alpha\dot\gamma}\mathcal{D}_{\alpha\dot\gamma}f_{\dot\alpha\dot\beta} + \; \epsilon^{\beta\gamma}\mathcal{D}_{\gamma\dot\beta}f_{\alpha\beta} &= \{\lambda_{\alpha i}, \lambda_{\dot\beta}^i\} + \{\chi_\alpha, \chi_{\dot\beta}\} + [W_i, \mathcal{D}_{\alpha\dot\beta}W^i] + [W^i, \mathcal{D}_{\alpha\dot\beta}W_i] \\
\epsilon^{\dot\gamma\dot\alpha}\mathcal{D}_{\alpha\dot\gamma}\lambda_{\dot\alpha i} &= -\epsilon_{ijk}[\lambda_\alpha^j, W^k] + [\chi_\alpha, W_i] \\
\epsilon^{\gamma\beta}\mathcal{D}_{\gamma\dot\beta}\lambda_\beta^i &= -\epsilon^{ijk}[\lambda_{\dot\beta j}, W_k] + [\chi_{\dot\beta}, W^i] \\
\epsilon^{\dot\gamma\dot\alpha}\mathcal{D}_{\alpha\dot\gamma}\chi_{\dot\alpha} &= -[\lambda_\alpha^k, W_k] \\
\epsilon^{\gamma\beta}\mathcal{D}_{\gamma\dot\beta}\chi_\beta &= -[\lambda_{\dot\beta k}, W^k] \\
\mathcal{D}_{\alpha\dot\beta}\mathcal{D}^{\alpha\dot\beta}W_i &= -2[[W^j, W_i], W_j] + [[W^j, W_j], W_i] + \tfrac{1}{2}\epsilon_{ijk}\{\lambda^{\alpha j}, \lambda_\alpha^k\} + \{\lambda_i^{\dot\alpha}, \chi_{\dot\alpha}\} \\
\mathcal{D}_{\alpha\dot\beta}\mathcal{D}^{\alpha\dot\beta}W^i &= -2[[W_j, W^i], W^j] + [[W_j, W^j], W^i] + \tfrac{1}{2}\epsilon^{ijk}\{\lambda_j^{\dot\alpha}, \lambda_{\dot\alpha k}\} + \{\lambda^{\alpha i}, \chi_\alpha\}
\end{aligned}
\tag{3.17}
$$

On setting the fields in the right-hand triangle to zero, we obtain

$$
\begin{aligned}
\epsilon^{\beta\gamma}\mathcal{D}_{\gamma\dot\beta}f_{\alpha\beta} &= 0 \\
\epsilon^{\gamma\beta}\mathcal{D}_{\gamma\dot\beta}\lambda_\beta^i &= 0 \\
\epsilon^{\dot\gamma\dot\alpha}\mathcal{D}_{\alpha\dot\gamma}\chi_{\dot\alpha} &= -[\lambda_\alpha^k, W_k] \\
\mathcal{D}_{\alpha\dot\beta}\mathcal{D}^{\alpha\dot\beta}W_i &= \tfrac{1}{2}\epsilon_{ijk}\{\lambda^{\alpha j}, \lambda_\alpha^k\}.
\end{aligned}
\tag{3.18}
$$

We see that the spin 1 source current actually factorises into parts from the two triangles, so it manifestly vanishes for super self-dual solutions. The first equation in (3.18) is just the Bianchi identity for self-dual field-strengths. So apart from the self-duality condition (2.1), we have one equation for zero-modes of the covariant Dirac operator in the background of a self-dual vector potential (having (2.1) as integrability condition) and two further non-linear equations. However, as we shall describe in the following sections, any given self-dual vector potential actually *determines* the general (local) solution of the rest of the equations. This is the most striking consequence of the matreoshka phenomenon: the N=0 core determining the properties of the higher-N theories. Another consequence is is that many conserved currents identically vanish in the super self-dual sector. For instance, since self-duality always implies the *source-free* second order Yang-Mills equations, the spin 1 source current vanishes for the entire matreoshka. Moreover, the usual Yang-Mills stress tensor clearly vanishes for self-dual fields:

$$T_{\alpha\dot\alpha,\beta\dot\beta} \equiv f_{\dot\alpha\dot\beta}f_{\alpha\beta} = 0 \; ;$$

and as a consequence of this, once one has put on further layers of the matreoshka, the supercurrents generating supersymmetry transformations, which contain the stress tensor as well as its superpartners also identically vanish for super self-dual fields. In fact, just as the stress tensor factorises into parts from the two triangles in (3.16) , all the supercurrents also factorise in this way. This is best seen in superspace language. The full (non-self-dual) super Yang-Mills theories are conventionally described using super field-strengths defined by the following curvature constraints

$$
\begin{aligned}
N = 1: \quad & [\overline{\mathcal{D}}_{\dot\alpha} \, , \, \mathcal{D}_{\alpha\dot\beta}] = \epsilon_{\dot\alpha\dot\beta} w_\alpha \\
& [\mathcal{D}_\beta \, , \, \mathcal{D}_{\alpha\dot\beta}] = \epsilon_{\beta\alpha} \overline{w}_{\dot\beta} \\[4pt]
N = 2: \quad & \{\mathcal{D}_{\alpha i} \, , \, \mathcal{D}_{\beta j}\} = \epsilon_{ij}\epsilon_{\alpha\beta} W \\
& \{\overline{\mathcal{D}}^i_{\dot\alpha} \, , \, \overline{\mathcal{D}}^j_{\dot\beta}\} = \epsilon^{ij}\epsilon_{\dot\alpha\dot\beta}\overline{W} \\[4pt]
N = 3: \quad & \{\mathcal{D}_{\alpha i} \, , \, \mathcal{D}_{\beta j}\} = \epsilon_{ijk}\epsilon_{\alpha\beta} W^k \\
& \{\overline{\mathcal{D}}^i_{\dot\alpha} \, , \, \overline{\mathcal{D}}^j_{\dot\beta}\} = \epsilon_{\dot\alpha\dot\beta}\epsilon^{ijk}\overline{W}_k.
\end{aligned}
\tag{3.19}
$$

In terms of these superfields the supercurrents take the form

$$
\begin{aligned}
N = 1: \quad & V_{\alpha\dot\alpha} = w_\alpha\overline{w}_{\dot\alpha} \\
N - 2: \quad & V = W\overline{W} \\
N = 3: \quad & V^i_j = W^i\overline{W}_j - \tfrac{1}{3}\delta^i_j W^k\overline{W}_k
\end{aligned}
\tag{3.20}
$$

and the super self-duality equations (eqs.(3.18) and their lower-N truncations) take the compact forms

$$
\begin{aligned}
\overline{w}_{\dot\alpha} &= 0 \\
W &= 0 \\
W^k &= 0
\end{aligned}
\tag{3.21}
$$

which manifestly demonstrate the vanishing of the supercurrents (3.20).

92

4 Super Self-Duality as Harmonic Space Analyticity

In N-independent form, (3.21) can be conveniently written as the following restrictions of the conventional representation-defining constraints for super Yang-Mills [15]:

$$
\begin{aligned}
\{\overline{\mathcal{D}}_{\dot\alpha}^{i}, \overline{\mathcal{D}}_{\dot\beta}^{j}\} + \{\overline{\mathcal{D}}_{\dot\alpha}^{j}, \overline{\mathcal{D}}_{\dot\beta}^{i}\} &= 0 \\
\{\mathcal{D}_{\alpha i}, \mathcal{D}_{\beta j}\} &= 0 = [\mathcal{D}_{\alpha i}, \nabla_{\alpha\beta}] \\
\{\mathcal{D}_{\alpha j}, \overline{\mathcal{D}}_{\dot\beta}^{i}\} &= 2\delta_j^i \nabla_{\alpha\dot\beta} \, .
\end{aligned}
\tag{4.22}
$$

In harmonic superspaces with coordinates

$$
\{x^{\pm\alpha} \equiv u_{\dot\beta}^{\pm} x^{\alpha\dot\beta}, \ \overline{\vartheta}_i^{\pm} \equiv u_{\dot\alpha}^{\pm} \overline{\vartheta}_i^{\dot\alpha}, \ \vartheta^{\alpha i}, \ u_{\dot\alpha}^{\pm}\},
$$

these take the form

$$
\begin{aligned}
\{\mathcal{D}_{\alpha i}, \mathcal{D}_{\beta j}\} &= 0 = \{\overline{\mathcal{D}}^{+i}, \overline{\mathcal{D}}^{+j}\} \\
[\nabla_\alpha^+, \nabla_\beta^+] &= 0 = [\overline{\mathcal{D}}^{+i}, \nabla_\alpha^+] \\
\{\mathcal{D}_{\alpha j}, \overline{\mathcal{D}}^{+i}\} &= 2\delta_j^i \nabla_\alpha^+ \\
[\mathcal{D}_{\alpha i}, \nabla_\beta^+] &= 0,
\end{aligned}
\tag{4.23}
$$

where the gauge covariant derivatives are given by

$$
\begin{aligned}
\mathcal{D}_{\alpha i} &= D_{\alpha i} + A_{\alpha i} \\
\overline{\mathcal{D}}^{+i} &= \overline{D}^{+i} + \overline{A}^{+i} \\
\nabla_\alpha^+ &= \partial_\alpha^+ + A_\alpha^+ ,
\end{aligned}
\tag{4.24}
$$

and satisfy the equations

$$
[D^{++}, \mathcal{D}_{\alpha i}] = [D^{++}, \overline{\mathcal{D}}^{+i}] = [D^{++}, \nabla_\alpha^+] = 0 \, .
\tag{4.25}
$$

The equations (4.23,4.25) are *equivalent* to (4.22) and (4.23) are consistency conditions for the following system of linear equations

$$
\begin{aligned}
\mathcal{D}_{\alpha i}\varphi &= 0 \\
\overline{\mathcal{D}}^{+i}\varphi &= 0 \\
\nabla_\alpha^+\varphi &= 0,
\end{aligned}
\tag{4.26}
$$

This system is extremely redundant, φ allowing the following transformation under the gauge group

$$
\varphi \to e^{-\tau(x^{\alpha\dot\alpha}, \overline{\vartheta}_i^{\dot\alpha}, \vartheta^{\alpha i})} \varphi e^{\lambda(x^{+\alpha}, \overline{\vartheta}_i^+, u_{\dot\alpha}^{\pm})} \, ,
\tag{4.27}
$$

where τ and λ are arbitrary functions of the variables shown, without affecting the constraints (4.23). These constraints therefore allow an economic choice of chiral-analytic basis in which the bridge ϕ and the prepotential V^{++} depend only on *positively* $U(1)$-*charged, barred* Grassmann variables, viz. $\overline{\vartheta}_i^+$, being independent of $\vartheta^{i\alpha}$ and $\overline{\vartheta}_i^-$. In this basis, φ too is independent of $\vartheta^{i\alpha}$ and $\overline{\vartheta}_i^-$; its non-analyticity manifesting itself in its dependence on $x^{-\alpha}$. Moreover, consistently with the commutation relations (4.23), the covariant spinor derivatives take the form $\mathcal{D}_{\alpha i} = \frac{\partial}{\partial \vartheta^{\alpha i}}, \overline{\mathcal{D}}^i = 2\vartheta^{\alpha i}\nabla_\alpha^+$. The super self-duality conditions (4.23,4.25) are therefore equivalent to the *same* system of equations as the N=0 SDYM equations, viz. (2.4,2.6), except that now φ and A_α^+ are superfields depending on $\{x^{\pm\alpha}, \overline{\vartheta}_i^+, u_{\dot\alpha}^\pm\}$ [3]. As for the N=0 case, we may express this system in the form of analyticity conditions for the harmonic space connection superfield V^{++}:

$$\frac{\partial}{\partial x^{-\alpha}}V^{++}(x^{+\alpha}, \overline{\vartheta}^{+i}, u_{\dot\alpha}^\pm) = 0. \tag{4.28}$$

The super SDYM systems are thus equivalent to the CR-like conditions (4.28); and fields solving for instance (3.18) may be obtained by inserting solutions φ of the equation

$$D^{++}\varphi(x^{\pm\alpha}, \overline{\vartheta}_i^+, u_{\dot\alpha}^\pm) = \varphi(x^{\pm\alpha}, \overline{\vartheta}_i^+, u_{\dot\alpha}^\pm)V^{++}(x^{+\alpha}, \overline{\vartheta}_i^+, u_{\dot\alpha}^\pm) \tag{4.29}$$

into the expression

$$\varphi\partial_\alpha^+\varphi^{-1} = u^{+\alpha}A_{\alpha\dot\alpha}(x^{\alpha\dot\alpha}, \overline{\vartheta}_i^{\dot\alpha}), \tag{4.30}$$

(the left side being guaranteed to be linear in u^+), and expanding the superfield vector potential on the right thus:

$$A_{\alpha\dot\beta}(x, \overline{\vartheta}) = A_{\alpha\dot\beta}(x) + \overline{\vartheta}_{\dot\beta i}\lambda_\alpha^i(x) + \epsilon^{ijk}\overline{\vartheta}_{\dot\alpha j}\overline{\vartheta}_i^{\dot\alpha}\nabla_{\alpha\dot\beta}W_k(x) + \epsilon^{ijk}\overline{\vartheta}_{\dot\alpha i}\overline{\vartheta}_j^{\dot\alpha}\overline{\vartheta}_k^{\dot\gamma}\nabla_{\alpha\dot\gamma}\chi_{\dot\beta} , \tag{4.31}$$

to obtain the component multiplet satisfying (3.18). In fact as we have already mentioned, any N=0 solution completely and recursively determines its higher-N extensions. We shall describe this solution matreoshka in the next section.

The most general infinite-dimensional group of transformations of super-self-dual solutions acquires a transparent form in the analytic harmonic superspace with coordinates $\{x^{+\alpha}, \overline{\vartheta}^{+i}, u_{\dot\alpha}^\pm\}$. As for the $N = 0$ case (see the comment at the end of sec.2) it is given by the transformation

$$V^{++} \to V^{++'} = g^{++}(V^{++}, x^{+\alpha}, \overline{\vartheta}^{+i}, u_{\dot\alpha}^\pm), \tag{4.32}$$

where g^{++} is an arbitrary doubly $U(1)$-charged analytic algebra-valued functional, modulo gauge transformations $V^{++} \to e^{-\lambda}(V^{++} + D^{++})e^\lambda$, where λ is also an arbitrary analytic function. This group has an interesting subgroup of transformations

$$V^{++} \to V^{++'} = V^{++}(x^{+'}, \overline{\vartheta}^{+'}, u'), \tag{4.33}$$

induced by diffeomorphisms of the analytic harmonic superspace

$$x^{+\alpha'} = x^{+\alpha'}(x^+, \overline{\vartheta}^+, u), \overline{\vartheta}^{+i'} = \overline{\vartheta}^{+i'}(x^+, \overline{\vartheta}^+, u), u' = u'(x^+, \overline{\vartheta}^+, u). \tag{4.34}$$

It would be of value to know how this group is realised in ordinary superspace and how it contains the Bäcklund transformations of [14], which correspond to a class of transformations (4.32) with $g^{++} = g^{++}(V^{++})$, a functional of V^{++} only.

As we have seen, the equation (4.28) encodes all the super SDYM systems, independently of the extension N. The action for SDYM suggested by [16] is therefore immediately generalisable to arbitrary N thus:

$$S = \int d^4x\, d\overline{\vartheta}_1^+ \ldots d\overline{\vartheta}_N^+\, d^2u\ \ tr\ (\partial^{+\alpha}\zeta_\alpha^{-3-N}\varphi^{-1}D^{++}\varphi), \tag{4.35}$$

which on varying the auxilliary field ζ_α^{-3-N} yields the CR condition (4.28). Although this Lagrange multiplier appears to be dynamical, it does not represent any additional physical degrees of freedom because of the following argument due to [16]. On varying φ, we obtain

$$\varphi^{-1}D^{++}[\varphi\partial^{+\alpha}\zeta_\alpha^{-3-N}\varphi^{-1}] = 0$$

which is actually tantamount to

$$\partial^{+\alpha}\zeta_\alpha^{-3-N} = 0.$$

All local solutions of this equations have the form $\zeta_\alpha^{-3-N} = \partial_\alpha^+ y^{-4-N}$ with arbitrary y^{-4-N}. However ζ_α^{-3-N} occurs in the action via $\partial^{+\alpha}\zeta_\alpha^{-3-N}$, so it is defined only modulo the addition of $\partial^{+\beta}y_{[\alpha\beta]}^{-4-N}$. This arbitrariness in ζ precisely balances its degree of freedom, so the action (4.35) describes no unwanted propagating modes. The action for the N=1 theory presented in [2] is just (4.35) in a different coordinate frame.

5   The Solution Matreoshka

We now discuss the solution of (4.29). Our main result is that given a solution of the N=0 equation (2.10), which we rewrite as

$$D^{++}\varphi_b(x^{\pm\alpha}, u_\alpha^\pm) = \varphi_b(x^{\pm\alpha}, u_\alpha^\pm)v^{++}(x^{+\alpha}, u_\alpha^\pm), \tag{5.36}$$

the solution of the supersymmetric system can be completely determined. Let us consider an N=1 bridge φ in the form

$$\varphi = e^{\bar\vartheta^+\psi^-(x^{\pm\alpha}, u_\alpha^\pm)}\varphi_b(x^{\pm\alpha}, u_\alpha^\pm), \tag{5.37}$$

where φ_b is some (presumed to be known) solution of (5.36), and

$$V^{++}(x^{+\alpha}, \bar\vartheta^+, u_\alpha^\pm) = v^{++}(x^{+\alpha}, u_\alpha^\pm) + \bar\vartheta^+ v^+(x^{+\alpha}, u_\alpha^\pm), \tag{5.38}$$

some arbitrary analytic superfield. In virtue of (4.29) the unknown function ψ^- satisfies

$$D^{++}\psi^- = \varphi_b v^+ \varphi_b^{-1}, \tag{5.39}$$

a first-order equation in which the right-hand side is some known function. It is therefore manifestly integrable, determining the N=1 bridge φ from which the superfield vector potential may be obtained:

$$A_\alpha^+ = -\,\partial_\alpha^+\varphi_b\varphi_b^{-1} - \bar\vartheta^+\nabla_\alpha^+\psi^- \ . \tag{5.40}$$

The coefficient of $\bar\vartheta^+$ in the above superfield vector potential is precisely the spinor field λ_α satisfying the Dirac equation in the background of component vector potential $A_\alpha^+ = -\,\partial_\alpha^+\varphi_b\varphi_b^{-1}$. This N=1 bridge may now be dressed up to an N=2 bridge:

$$\begin{aligned}
\varphi &= e^{\bar\vartheta^{2+}\psi_2^-(\bar\vartheta^{1+}, x^{\pm\alpha})}e^{\bar\vartheta^{1+}\psi_1^-(x^{\pm\alpha})}\varphi_b \\
&= (1 + \bar\vartheta^{2+}\psi_2^-(x^{\pm\alpha}) + \bar\vartheta^{2+}\bar\vartheta^{1+}\psi_{21}^{--}(x^{\pm\alpha}))(1 + \bar\vartheta^{1+}\psi_1^-(x^{\pm\alpha}))\varphi_b(x^{\pm\alpha})
\end{aligned} \tag{5.41}$$

and the N=2 analytic prepotential may be expanded thus

$$V^{++}(x^{+\alpha}, \bar\vartheta^{+i}) = \left(v^{++}(x^{+\alpha}) + \bar\vartheta^{+1}v_1^+(x^{+\alpha})\right) + \bar\vartheta^{2+}\left(v_2^+(x^{+\alpha}) + \bar\vartheta^{1+}v_{21}(x^{+\alpha})\right). \tag{5.42}$$

Once again, in virtue of (4.29) the unknown functions in this ansatz for φ satisfy first-order equations which afford explicit integration; and the N=2 super self-dual multiplet may be explicitly constructed. Now given an N=2 solution we can promote it to an N=3 solution using the matreoshkan ansatz

$$\varphi = e^{\bar{\vartheta}^{3+}\psi_3^-(\bar{\vartheta}^{2+},\bar{\vartheta}^{1+},x^{\pm\alpha})}e^{\bar{\vartheta}^{2+}\psi_2^-(\bar{\vartheta}^{1+},x^{\pm\alpha})}e^{\bar{\vartheta}^{1+}\psi_1^-(x^{\pm\alpha})}\varphi_b, \tag{5.43}$$

where ψ_3^- and ψ_2^- are N=2 and N=1 superfields respectively; this form clearly breaking the internal SU(3) invariance, just as the N=2 ansatz (5.41) breaks the internal SU(2) invariance. Expanding the superfield ψ_2^- as in (5.41) and ψ_3^- as follows:

$$\psi_3^-(\bar{\vartheta}_2^+,\bar{\vartheta}_1^+,x^{\pm\alpha}) = \psi_3^-(x^{\pm\alpha}) + \bar{\vartheta}^{2+}\psi_{32}^{--}(x^{\pm\alpha}) + \bar{\vartheta}^{1+}\psi_{31}^{--}(x^{\pm\alpha}) + \bar{\vartheta}^{2+}\bar{\vartheta}^{1+}\psi_{321}^{---}(x^{\pm\alpha}),$$

and the N=3 analytic superfield thus:

$$\begin{aligned}
V^{++}(x^{+\alpha},\bar{\vartheta}^{+i}) = & \left(v^{++}(x^{+\alpha}) + \bar{\vartheta}^{+1}v_1^+(x^{+\alpha}) + \bar{\vartheta}^{2+}\left(v_2^+(x^{+\alpha}) + \bar{\vartheta}^{1+}v_{21}(x^{+\alpha})\right)\right) \\
& + \bar{\vartheta}^{3+}\left(v_3^+(x^{+\alpha}) + \bar{\vartheta}^{1+}v_{31}(x^{+\alpha}) + \bar{\vartheta}^{2+}\left(v_{32}(x^{+\alpha}) + \bar{\vartheta}^{1+}v_{321}(x^{+\alpha})\right)\right),
\end{aligned} \tag{5.44}$$

again yields a system of first-order equations for the unknown functions in (5.43), thus allowing the explicit construction of the N=3 self-dual multiplet. This matreoshka structure in which successively higher N-superfields are parametrised as N=1 superfields with (N-1)-superfield 'components' is very reminiscent of the Cayley-Dixon procedure of describing division algebras: a complex number as a complex combination of two reals, a quaternion as a complex combination of two complex numbers; and an octonion as a complex combination of two quaternions.

As an example let us take the φ_b for the BPST instanton (2.15) and the simplest v^+ linear in x^+ and having a constant spinorial parameter ζ_i of dimension $[cm]^{-\frac{3}{2}}$:

$$(v^+)_i^j = x^{+j}\zeta_i + x_i^+\zeta^j . \tag{5.45}$$

This yields

$$\psi_i^{-j} = \left(1 + \frac{x^2}{\rho^2}\right)^{-1}\left(x^{-j}\zeta_i + \left(1 + \frac{x^2}{\rho^2}\right)x_i^-\zeta^j - \frac{1}{\rho^2}x_i^- x^{+j}x_l^-\zeta^l\right), \tag{5.46}$$

from which the vector potential may now be found to be

$$A_{\alpha\dot{\alpha}i}^j = \frac{1}{\rho^2 + x^2}(\frac{1}{2}x_{\alpha\dot{\alpha}}\delta_i^j + \epsilon_{i\alpha}x_{\dot{\alpha}}^j) + \bar{\vartheta}_{\dot{\alpha}}\frac{\rho^4}{(\rho^2 + x^2)^2}\left(\epsilon_{i\alpha}\zeta^j + \delta_\alpha^j\zeta_i\right). \tag{5.47}$$

In fact this is solution is related to the N=0 one we started with by a supertranslation with parameter $\rho^2\zeta^\alpha$:

$$x^{+\alpha} \rightarrow x^{+\alpha} + \bar{\vartheta}^+\rho^2\zeta^\alpha.$$

Similarly, using another v^+ linear in x^+, but of the form

$$(v^+)_i^j = x^{+p}c_{pik}\epsilon^{kj}, \tag{5.48}$$

where c_{pik} is a totally symmetric tensor parameter having, like the parameter ζ of the previous example, dimension $[cm]^{-\frac{3}{2}}$. This yields

$$A^j_{\alpha\dot\alpha i} = \frac{1}{\rho^2 + x^2}(\frac{1}{2}x_{\alpha\dot\alpha}\delta^j_i + \epsilon_{i\alpha}x^j_{\dot\alpha}) + \bar\vartheta_{\dot\alpha}c_{\alpha in}\epsilon^{nj}\left(1 + \frac{x^2}{\rho^2}\right), \tag{5.49}$$

a potential not related to the N=0 one by any symmetry transformation. This simple solution, however, does not vanish asymptotically.

Now choosing a v^+ quadratic in x^+ with constant spinorial parameter $\bar\eta^{\dot\alpha}$ of dimension $[cm]^{-\frac{5}{2}}$:

$$(v^+)^j_i = x^{+j}x^+_i u^-_{\dot\alpha}\bar\eta^{\dot\alpha},$$

yields the self-dual vector potential

$$A^j_{\alpha\dot\alpha i} = \frac{1}{\rho^2 + x^2}(\frac{1}{2}x_{\alpha\dot\alpha}\delta^j_i + \epsilon_{i\alpha}x^j_{\dot\alpha}) + \bar\vartheta_{\dot\alpha}\frac{\rho^4}{(\rho^2 + x^2)^2}(\epsilon_{i\alpha}x^j_{\dot\beta} - \delta^j_\alpha x_{i\dot\beta})\bar\eta^{\dot\beta}, \tag{5.50}$$

related to the N=0 one by a superconformal transformation with parameter $\rho^2\bar\eta_{\dot\alpha}$

$$x^{+\alpha} \to x^{+\alpha}(1 - \rho^2\bar\eta_{\dot\alpha}u^{-\dot\alpha}\bar\vartheta^+) \tag{5.51}$$

and is precisely the solution discussed by [17].

These are just some particular examples of our solution generating technique [3]; our method, however, describes *all* local solutions of the super self-duality equations.

6 Conclusion

To conclude we mention some prospects of this approach to self-duality . The vanishing supergauge supercurrents are just the non-gravitational sources for the spin 2 field in supergravity theories. This indicates that the situation in self-dual supergravity is very similar and that our matreoshka is part of a much larger, albeit more intricate, supergravity matreoshka. This gives rise to the prospect of obtaining hyper-kähler manifolds with additional spinorial structure.

Going in the other direction, recent interest in self-duality has concentrated around the Ward conjecture [5] that all lower dimensional solvable systems are reductions of SDYM; and our solution matreoshka promises to yield new (supersymmetric) solvable systems, together with their solutions, by truncation of the analytic data. This would yield a unification of the various existing methods of solving two dimensional systems as different manifestations of the harmonic-twistor correspondence for SDYM.

As we have seen, the spin 1 source currents of all super self-dual theories vanish because they factorise into parts from the two triangles in (3.16). It turns out that we can solve the full (non-self-dual) super Yang-Mills equations, in other words restore these source currents, by intermingling self-dual and anti-self-dual holomorphic data [4]; and this works *exactly* for the N=3 case. Work on the explicit construction of non-self-dual N=3 solutions is in progress.

One of us (V O) would like to gratefully acknowledge receipt of a Humboldt Forschungspreis enabling the performance of this work at Bonn University and to thank the Humboldt Stiftung for financial support to attend the Cargese meeting.

References

[1] R.S. Ward and R.O. Wells, *Twistor geometry and field theory*, Camb. Univ. Press, Cambridge, 1990.

[2] C. Devchand and V. Ogievetsky, Phys. Lett. B297 (1992) 93.

[3] C. Devchand and V. Ogievetsky, hep-th/9306163, Nucl.Phys.B (to appear).

[4] A. Galperin, E. Ivanov, S. Kalitzin, V. Ogievetsky, E. Sokatchev, Class. Quant. Grav. 1 (1984) 469, 2 (1985) 255.

[5] R.S. Ward, Phil.Trans.Roy.Soc. A315 (1985) 451.

[6] L. Mason and G. Sparling, J. Geom. Phys. 8 (1992) 243.

[7] S.J. Gates and H. Nishino, Phys.Lett. B299 (1993) 255, H. Nishino, Maryland prepr. UMDEPP 93 - 144, 145.

[8] S. Kalitzin and E. Sokatchev, Class. Quant. Grav. 4 (1987) L173.

[9] A. Galperin, E. Ivanov, V. Ogievetsky, E. Sokatchev, Ann. Phys. (N.Y.) 185 (1988) 1.

[10] M. Evans, F. Gürsey, V. Ogievetsky, Phys.Rev. D47 (1993) 3496.

[11] O. Ogievetsky, in *Group Theoretical Methods in Physics*, Ed. H.-D. Doebner et al, Springer Lect. Notes in Physics 313 (1988) 548.

[12] N. Markus, Y. Oz, S. Yankielovicz, Nucl. Phys. B379 (1992) 121.

[13] A. Galperin, E. Ivanov, V. Ogievetsky, E. Sokatchev, in *Quantum Field Theory and Quantum Statistics*, vol.2, 233 (Adam Hilger, Bristol, 1987); JINR preprint E2-85-363 (1985).

[14] C. Devchand and A. Leznov, hep-th/9301098, Commun. Math. Phys. (to appear).

[15] A. Semikhatov, Phys.Lett. 120B (1983) 171; I. Volovich, Phys.Lett. 123B (1983) 329.

[16] S. Kalitzin and E. Sokatchev, Phys.Lett. 257B (1991) 151.

[17] V. Novikov, M. Shifman, A. Vainshtein, M. Voloshin, V. Zakharov, Nucl. Phys. B229 (1983) 394.

[18] E. Witten, Phys. Lett. 77B (1978) 394.

THE PHENOMENOLOGY OF STRINGS AND CLUSTERS IN THE 3-d ISING MODEL

Vladimir S. DOTSENKO[*], Marco PICCO and Paul WINDEY

LPTHE[†]
Université Pierre et Marie Curie
Bte 126, 4 Place Jussieu
75252 Paris CEDEX 05, FRANCE

Geoffrey HARRIS and Enzo MARINARI[‡]

Physics Department and NPAC
Syracuse University
Syracuse, NY 13244, USA

Emil MARTINEC

Enrico Fermi Institute and Department of Physics
University of Chicago
Chicago, IL 60637, USA

Abstract: We examine the geometrical and topological properties of surfaces surrounding clusters in the 3-d Ising model. For geometrical clusters at the percolation temperature and Fortuin–Kasteleyn clusters at T_c, the number of surfaces of genus g and area A behaves as $A^{x(g)}e^{-\mu(g)A}$, with x approximately linear in g and μ constant. We observe that cross–sections of spin domain boundaries at T_c decompose into a distribution $N(l)$ of loops of length l that scales as $l^{-\tau}$ with $\tau \sim 2.2$. We address the prospects for a string–theoretic description of cluster boundaries.

---

[*]   Also at the Landau Institute for Theoretical Physics, Moscow

[†]   Laboratoire associé No. 280 au CNRS.

[‡]   Also at Dipartimento di Fisica and INFN, Università di Roma Tor Vergata, Viale della Ricerca Scientifica, 00133 Roma, Italy.

Quantum Field Theory and String Theory, Edited by
L. Baulieu *et al.*, Plenum Press, New York, 1995

1 Introduction

One of the major successes of 20th century physics has been the expression of the critical behavior of a variety of theories of nature in terms of sums over decorated, fluctuating paths. It has thus also been hoped that higher dimensional analogues, theories of fluctuating membranes, also play a fundamental role in characterizing the physics of critical phenomena. In particular, significant effort has been invested in recasting one of the simpler models of phase transitions, the 3–d Ising model , as a theory of strings. These attempts have been stymied by the difficulty in taking the continuum limit of formal sums over lattice surfaces.

In fact, sums over lattice surfaces, built from e.g. plaquettes or polygons, generically fail to lead to a well-defined continuum theory of surfaces. An exception to this rule occurs when the surface discretizations are embedded in $d \leq 1$. In this case, one can exactly solve a large class of toy lattice models which lead to sensible continuum 'bosonic' string theories (at least perturbatively) [2]. Numerically, it is observed that the $d > 1$ versions of these lattice models suffer a 'fingering instability'; the embedded surfaces, for instance are composed of spikes with thickness of the order of the cutoff. It is suspected that the polygonal discretization of the worldsheet (for large volumes) is configured in a polymer–like structure, so that these theories cannot be realized as sums over surfaces in the continuum limit. This instability is anticipated theoretically, since the mass–squared of the dressed identity operator of the bosonic string becomes negative above $d = 1$, presumably generating a uncontrolled cascade of states that tear the worldsheet apart.

In the continuum limit, we know how to evade these problems in special cases through the implementation of supersymmetry and the GSO projection. This additional structure, however, leads to fundamental difficulties in discretizing these theories. In principle, one might hope to somehow guess an appropriate continuum string theory and then show that it embodies the critical behavior of a lattice theory, such as the 3–d Ising model. The prospects for success through such an approach seem rather poor at this time.

Given this state of affairs, we have turned to a more phenomenological approach, in which we attempt to generate 'physical' random surfaces in a particular model and then examine their topological and geometrical properties. We thus have chosen to look at the structure of domain boundaries in the 3–d Ising model. The phenomenology of these self–avoiding cluster boundaries is interesting in its own right, since it describes a large universality class of behavior that is expressed frequently and quite precisely by nature. We also might hope that our observations may be useful in gauging the prospects of success of a string–theoretic description. The Ising model has been employed previously as a means to generate random lattice surfaces [a]; see for instance, the work of David [4], Huse and Leibler [5], Karowski and Thun and Schrader [6]. In a sense, this work extends these studies by looking for new features of the geometry of these lattice surfaces; we

---

[a]Through the use the phrase 'lattice surface' rather than 'surface', we indicate that these objects should *not* be necessarily inferred to be real surfaces in the continuum limit.

also consider boundaries of Fortuin–Kasteleyn clusters as well as 'geometrical' spin domains. Much of our analysis consists of a measurement of the distribution of surfaces as a function of their area A and genus g, $N_g(A)$ [b]. We shall determine the functional form of $N_g(A)$. We also perform block spin measurements of the genus, to determine if a condensation of handles is present on cluster boundaries at all scales. These cluster boundaries are strongly coupled and thus it appears cannot be directly characterized by perturbative string theory. We see that, however, boundaries of spin domains at the Curie temperature are not just strongly–coupled versions of the branched polymer–like objects that attempts to build 'bosonic' random surfaces typically generate. They instead exhibit a richer fractal structure, albeit one not characteristic of surfaces. We show that they obey a new scaling law that describes the distribution $N(l)$ of lengths l of loops that compose cross–sections of cluster boundaries.

2 Ising Clusters and Surfaces

We shall begin by summarizing the basic physical properties of the cluster boundaries that we have analyzed. To a first approximation, a 2–dimensional membrane of area A and curvature matrix K will exact an energy cost [5, 8]

$$H = \mu A + \lambda \int (\mathrm{Tr}K)^2 + \kappa \int \mathrm{Det}K; \tag{2.1}$$

μ is the bare surface tension, λ is referred to as the bending rigidity and κ couples to the Euler characteristic of the surface. In the regime which characterizes random surfaces, the surface tension must be sufficiently small to allow significant thermal fluctuations. Note that the above action does not capture the entire dynamics; it is essential also to keep in mind that Ising cluster boundaries are naturally self–avoiding. We first consider surfaces in the dual lattice that bound 'geometrical clusters' formed from sets of adjacent identical spins. In this case, the Ising dynamics generates an energy penalty proportional to the boundary area; λ_{bare} and $\kappa_{bare} = 0$. The bare surface tension is tuned by the Ising temperature. To put this model in perspective, we note that for real vesicles, for instance, the couplings λ and κ can be quite large; λ ranging from about kT to $100kT$ have been measured [8]. The bending rigidity may be irrelevant in the continuum limit, however. The string coupling [c] is equal to $\exp(-\kappa)$. Through blocking spins, we make an estimate of the renormalization group behavior of κ. Unless κ effectively becomes large in the infrared, the cluster boundaries will fail to admit a surface description in the continuum limit.

The geometrical clusters and their boundaries are *not* present at all scales at the Curie temperature. Instead, for temperatures somewhat below T_c and all temperatures above T_c two huge geometrical clusters comprise a finite fraction of the entire lattice volume. These clusters percolate, that is, they wrap around the entire lattice (we shall

---

[b]The mean genus per Ising configuration is measured in references [6]. A determination of genus as a function of area in an Ising system with anti–periodic boundary conditions have also appeared recently [7].

[c]We ignore distinctions between intrinsic and extrinsic metrics.

consider periodic boundary conditions). Otherwise, the lattice only contains very small clusters that are the size of a few lattice spacings; there are no intermediate size clusters. We can understand this behavior by considering the $T \to \infty$ behavior of these clusters. The spins are distributed randomly with spin up with probability 50%; the problem of constructing clusters from these spins then reduces to pure site percolation with $p = 1/2$. Pure site (or bond) percolation describes the properties of clusters built by identifying adjacent colored bonds (sites), which are colored randomly with probability p. Above a critical value $p = p_c$, the largest of these clusters percolates through the lattice[9]. For the cubic lattice, it is known that an infinite cluster will be generated (in the thermodynamic limit) at $p_c \sim .311$. Thus, the fact that the geometrical clusters have percolated at the Curie point is essentially a consequence of the connectivity of 3-d lattices.

At very low temperatures, however, there are few reversed (minority) spins in the Ising model; these form a few small clusters. As the density of minority spins increases, the clusters become bigger until the largest cluster percolates at some temperature $T_p < T_c$. It has been suggested (see [10] and [5]) that since this minority spin percolation appears to be due to an increase in the concentration of minority spins and not to any long–distance Ising dynamics, that this transition is in the same universality class as pure (bond or site) percolation. We emphasize that the scaling of minority clusters should not correspond to any non–analyticity in the thermodynamic behavior of the Ising model; it should essentially be a 'geometric effect'.

There is another type of cluster, introduced by Fortuin and Kasteleyn [11, 12], that does proliferate over all length scales at the Curie point. These FK clusters are constructed by connecting adjacent identical spins with a temperature dependent probability $p = 1 - \exp(-2\beta)$. They arise naturally in the reformulation of the Ising model as a percolating bond/spin model [13]. For the Ising partition function can be recast as a sum over occupied and unoccupied bonds with partition function

$$Z = \sum_{bonds} p^b (1 - p)^{(N_b - b)} 2^{N_c} \tag{2.2}$$

where $p = 1 - \exp(-2\beta)$, N_b denotes the number of bonds in the entire lattice in which b bonds are occupied and N_c equals the number of clusters that these occupied bonds form. When the factor 2^{N_c} is replaced by q^{N_c}, then (2.2) is the partition function for the q-state Potts model. If we assign a spin to each bond so that all bonds in the same cluster have the same spin, then the factor of q^{N_c} just comes from a sum over spin states. The above partition function can then be viewed as a sum over FK clusters. Using this construction, one can show that the spin-spin correlator in the original Ising model is equal to the pair connectedness function of FK clusters,

$$\langle \sigma(x)\sigma(y) \rangle = \langle \delta_{C_x, C_y} \rangle, \tag{2.3}$$

which equals the probability that points x and y belong to the same FK cluster [14]. It then follows that for $T \geq T_c$, the mean volume of the FK clusters is proportional to the susceptibility of the Ising model, so that indeed FK clusters only just start to percolate at the Curie point. Additionally, the relation (2.3) also implies that the spatial extent of the FK clusters is proportional to the correlation length of the Ising model. Furthermore, scaling arguments [15] demonstrate that at T_c, the volume distribution of FK clusters obeys

$$N(V_{cl}) \simeq V_{cl}^{-\tau}, \quad \tau = 2 + \frac{1}{\delta}, \tag{2.4}$$

where δ denotes the magnetic exponent of the Ising model ($M \simeq B^{1/\delta}$). Thus we see that FK clusters, unlike the geometrical clusters previously discussed, directly encode the critical properties of the Ising model. Indeed, we are necessarily led to study FK clusters in order to measure scaling laws that characterize cluster boundaries of the scale of the Ising correlation length, i.e. boundaries that scale at the Curie point. On the other hand, geometrical cluster boundaries contribute an energy penalty proportional to their individual area; the lattice surface dynamics of FK cluster boundaries, however, cannot be likewise described by a similar physical rule.

In 2–dimensions both the FK clusters and the geometrical clusters percolate at the Curie temperature. The critical properties of these clusters differ, however, since the scaling of geometrical clusters is partially determined by the 'percolative' properties of two–dimensional lattices. These effects are in some sense removed through the FK construction.

3   The Simulation

We now proceed to outline the techniques used in our Monte Carlo simulations. We analyzed data from lattices of size ranging from $L = 32$ to $L = 150$, using about six months of time on RISC workstations. Spin updates were implemented through the efficient Swendsen–Wang algorithm [16]: FK clusters for each lattice configuration are first constructed, then the spins composing each cluster are (all) assigned a new random spin value. We determined our statistical uncertainties via the jackknife technique and extracted exponents through linear least–squared fits. Statistical errors for these exponents were also obtained by using jackknife when fitting. Generally, systematic corrections to scaling and finite-size effects are much larger than our statistical errors.

The main technical difficulty that we encountered was the measurement of the Euler characteristic, equal to $V - E + F$ for a dual surface with V vertices, E edges and F faces. On the simple cubic lattice, the construction of the dual surface is ambiguous for configurations in which 4 plaquettes intersect along the same link, for instance. We found that we could define a consistent set of rules, which we shall discuss in [17], that resolved these ambiguities. These rules are certainly not unique; one would hope that their implementation essentially serves as a regularization that does not affect long–distance scaling laws.

In two dimensions, one can avoid ambiguous intersections on the dual lattice by considering Ising spins on the triangular lattice. Its dual (the honeycomb lattice) is trivalent and thus the Ising spin domains will not be enclosed by self–intersecting paths. This fortuitous situation generalizes to three–dimensions for the Ising model on a body centered cubic (BCC) lattice in which the vertices at the center of each cube are also connected to those in the centers of neighboring cubes. More explicitly, we coupled with equal strength both the 6 nearest and 8 next-nearest Ising spins so that only three plaquettes of the dual lattice meet along a dual link. Since surfaces built dual to this lattice are also naturally self-avoiding, computing the genus is trivial. A depiction of the Wigner–Seitz cell of this lattice (composed of plaquettes in the dual lattice) appears in figure 1.

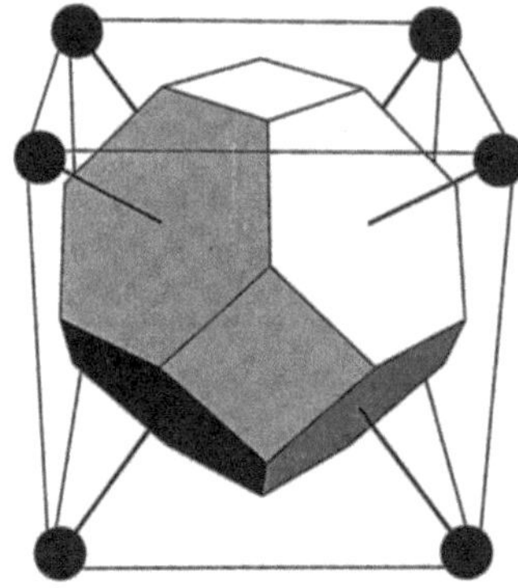

Fig.1. The Wigner–Seitz cell of the BCC lattice with next–nearest neighbor interactions.

Estimating the appropriate critical temperatures also required considerable effort. To find the percolation temperature β_p we used the method discussed by Kirkpatrick [18] in which one measures the fraction f of configurations containing clusters that span the lattice. One plots f versus β for different lattice sizes L; β_p corresponds to the intersection of these curves for different L. On the BCC lattice, we checked this by also determining the temperature at which the mean cluster size scales as a power law in L. From this analysis, we obtained $\beta_p = .0959$ on the BCC lattice and $\beta_p = .232$ on the SC lattice. The value of β_c on the SC lattice has been previously determined to be about .221651 [19]; we also found that $\beta_c \sim .0858$ on the BCC lattice.

4 Results

We now present data from our simulations on both the simple cubic and BCC lattices. We have examined boundaries of FK clusters at T_c, surfaces bounding minority spin domains at T_p and geometrical clusters at T_c. In [17], a more comprehensive discussion of our data will appear, including also results from simulations of the 2–d Ising model and pure bond percolation in 3 dimensions. A more concise summary of some of these results has been presented in [20].

1 Cluster Properties

We begin by discussing a few of the properties of the clusters. Most of the new material, pertaining to the topology of the cluster boundaries, appears in the following sub–sections.

For FK clusters at T_c and minority clusters at T_p, we verified the scaling given in (2.4). This is shown, for example, in figure 2 for FK clusters on an $L = 64$ SC lattice.

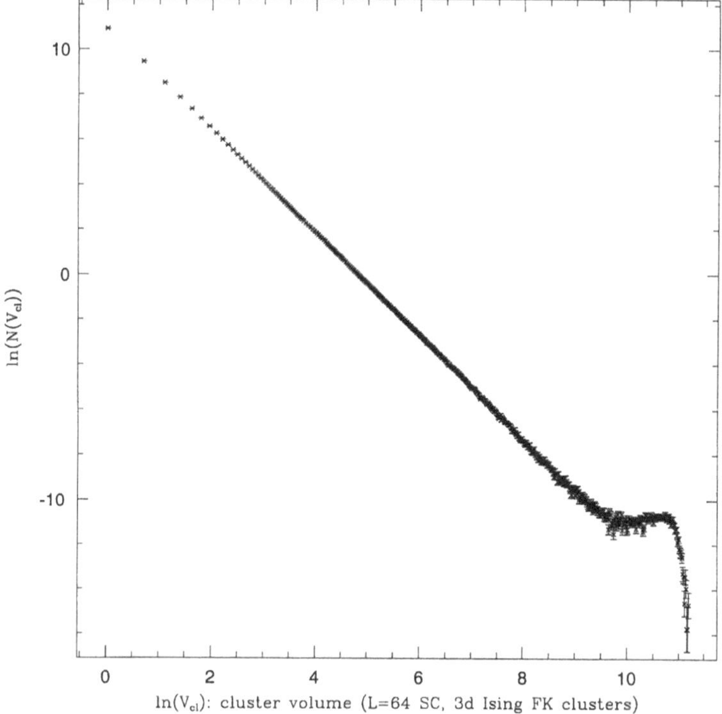

Fig.2. $\ln N(V_{cl})$ vs. $\ln V_{cl}$ for FK clusters on a $N = 64$ SC lattice.

In principle, by carefully determining $\tau = 2+1/\delta$, one might hope to provide concrete numerical evidence for the very reasonable hypothesis that the transition at T_p is in the universality class of pure percolation. In practice, this is quite difficult. The value of the magnetic exponent δ in the 3–d Ising model (as determined through renormalization group methods, for instance) yields the prediction $\tau_{FK} = 2.207(1)$ [21]. The value of τ for pure percolation that one would infer from recent series expansions is $\tau = 2.189(5)$ [9], which is not so different from the FK value. In fact, the power law fits to $N(V_{cl})$ are not very precise, due to large finite volume effects and corrections to scaling. The values we extract from these plots are $\tau_{FK} = 2.25(10)$ (this has been measured by Wang [15]) and $\tau_{geo} = 2.10(5)$ on the largest ($L = 64$ and 100) lattices that we considered. This is a rather poor way to measure these exponents; much more accurate estimates can be obtained through finite–size scaling fits of the mean cluster size as a function of lattice size L. The mean cluster size scales as $L^{\gamma/\nu}$; standard scaling relations and (2.4) give $\tau = (3 + \gamma/\nu d)/(1 + \gamma/\nu d)$ ($d = 3$). Using this technique , we measured $\tau_{FK} = 2.207(3)$ on the SC lattice and $\tau_{geo} = 2.202(3)$ on the BCC lattice. The error on τ_{geo} is in fact probably several times larger than quoted above, due to uncertainties in locating the critical temperature. This measurement of τ_{FK} agrees perfectly with previous values; the measurement of τ_{geo} is not accurate enough to distinguish likely pure percolation behavior from that of percolation of FK clusters.

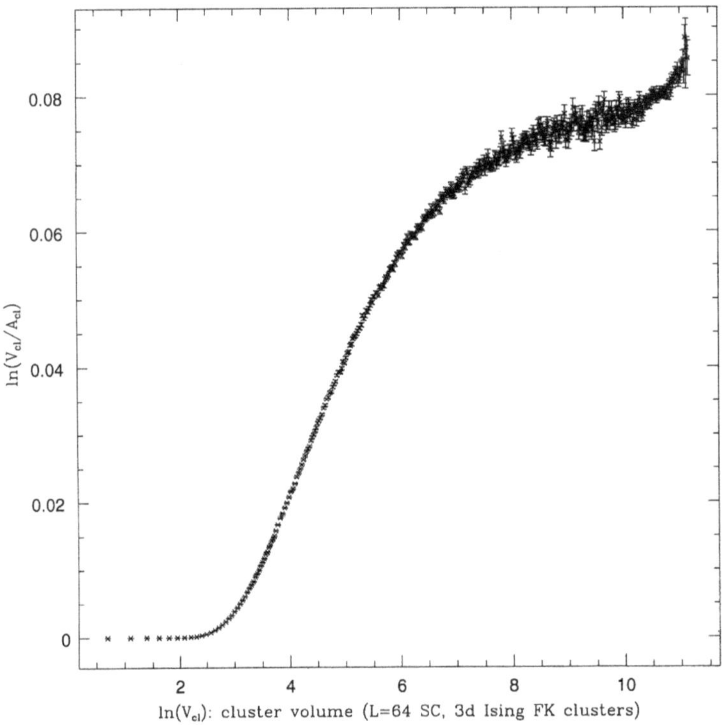

Fig.3. $\ln(V_{cl}/A_{cl})$ vs. $\ln V_{cl}$ for FK clusters on the $L = 64$ SC lattice.

We measured the number of sites on the boundary of each cluster, A_{cl}. A plot of $\ln(V_{cl}/A_{cl})$ vs $\ln(V_{cl})$ for FK clusters on an $L = 64$ BCC lattice appears in figure 3. We see that for very small volumes, the lattice regularization constrains V_{cl} to equal A_{cl} and for intermediate volumes, there is a small deviation from linear scaling (as some interior sites begin to appear). The plateau that appears around $V_{cl} = 3000$ indicates the onset of scaling regime where $A_{cl} \propto V_{cl}$. The growth just at the end of the plot is due to the largest cluster, which wraps around the lattice and merges with itself to form extra interior points. This plateau indicates that the lattice surfaces are not smooth and may be configured as polymer–like networks.

This behavior is not surprising. The observed proportionality of V_{cl} and A_{cl} is well-known in the context of pure percolation in 2 and 3 dimensions [9]. Bonds (or sites) are deleted with a fixed probability in percolation. This implies that holes should be distributed homogeneously with finite measure on percolation clusters; i.e. the boundary length should be proportional to the enclosed volume. Note that FK clusters are constructed by performing percolation on geometrical clusters, so this argument should definitely apply in the FK case. We also found $A_{cl} \propto V_{cl}$ for geometrical clusters at T_p; this observation is consistent with the intuition that the T_p transition is that of pure percolation.

2 Genus Distribution

We now turn to an analysis of the distribution of handles on cluster boundaries. If these boundaries form tangled networks, then the following essentially characterizes the statistics of closed loops in these networks. In the simplest scenario, one might assume that the handles are uncorrelated. It would then follow that $N_g(A)$ asymptotically obeys the Poisson distribution $N_g(A) = \kappa_g(\mu A)^g e^{-\mu A}$, with $\kappa_g \propto 1/g!$. The probability per plaquette of growing a handle is then μ.

We first present a sample of fits to $N_g(A)$ for FK cluster boundaries on the BCC lattice for $L = 64$. In figure 4, we present our data for genus 2 along with a best fit to the functional form

$$N_g(A) = C_g A^{x(g)} e^{-\mu(g)A} \ . \tag{4.5}$$

$N_2(A)$ is peaked near $A = 250$, and the fit is perfect apart from the very small area region, where we expect corrections to scaling to be large. Likewise, the power law plus exponential fit is superb for genus 5 as indicated in figure 5. We find that this functional form fits our data very well for $g \geq 2$ up to about $g = 20$ where our statistics become poor. If we assume that the ansatz (4.5) holds, then it follows that

$$\mu = \mu_{eff} \equiv \frac{\langle A \rangle}{(\langle A^2 \rangle - \langle A \rangle^2)} \tag{4.6}$$

and

$$x(g) = x_{eff} \equiv \frac{\langle A \rangle^2}{(\langle A^2 \rangle - \langle A \rangle^2)} - 1 \ . \tag{4.7}$$

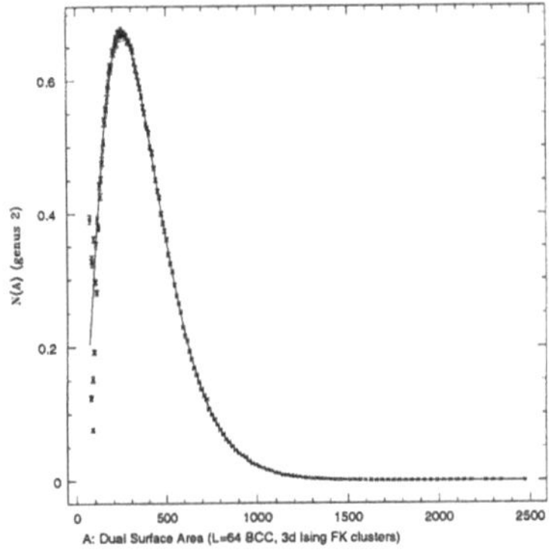 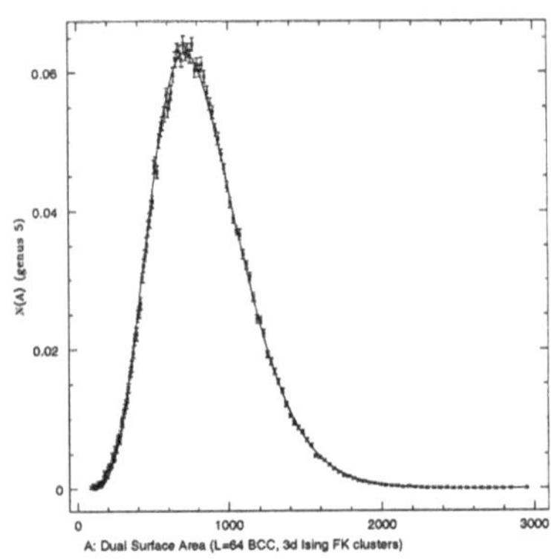

Fig.4. The number of genus 2 surfaces at T_c as a function of dual surface area A for FK clusters on the $L = 64$ BCC lattice, with a best fit to the functional form given in equation 4.5.

Fig.5. As in the previous figure, but for genus 5.

We measured these moments and found that indeed μ_{eff} and x_{eff} agreed very well with the values extracted directly from fits to (4.5). The value $\mu_{eff}^{-1} = 114 \pm 3$ as depicted in figure 6 is proportional to the average surface area (in lattice units) per handle and is independent of genus for $g > 2$. In figure 7 we show the genus dependence of the exponent x, extracted both from moments of the area distribution and from the direct fits. After a transient region for small genus $(g = 0 - 4)$ we find almost linear behavior in the region $g = 5 - 15$ with a slope of 1.25 ± 0.1.

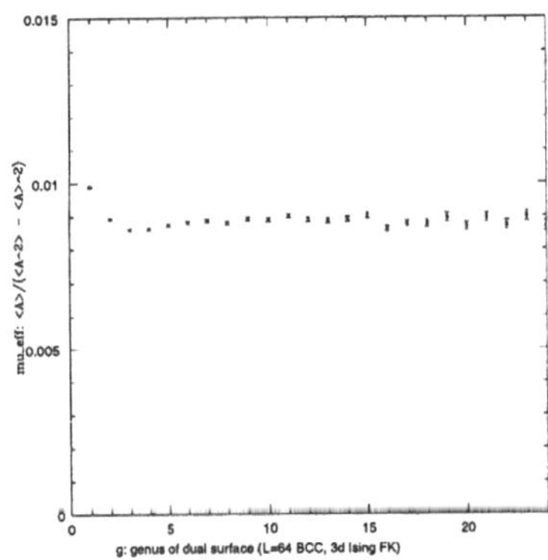 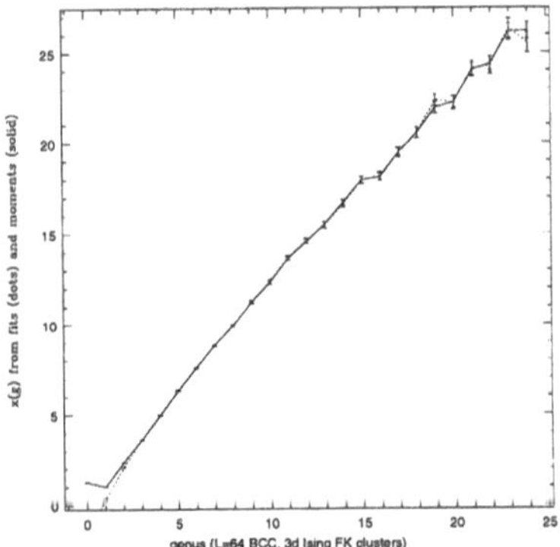

Fig.6. The dependence of μ (extracted from the moments of the area distribution) on genus for FK clusters on the $L = 64$ BCC lattice at T_c.

Fig.7. The dependence of x (extracted from direct fits to (4.5) and moments) on genus for FK clusters on the $L = 64$ BCC lattice at T_c.

The results for FK clusters on the SC lattice are quite similar, though not as clean. We first show the behavior of $N_1(A)$ for $L = 64$ in figure 8. Clearly, here the fit does not work, though one does expect large deviations from asymptotic scaling for

108

surfaces in the range depicted. The fit for genus 5 (figure 9), however, is quite good, although small systematic discrepancies are still notable. Perhaps the regularization needed to define genus in the SC case is partially responsible for these deviations. Again, for the SC lattice, we find that μ appears to be independent of g, though we observe a very small systematic drift. The plot of $x(g)$ vs. g exhibits more curvature than in the BCC case, but the slope in the genus $5 - 15$ region again is about 1.25.

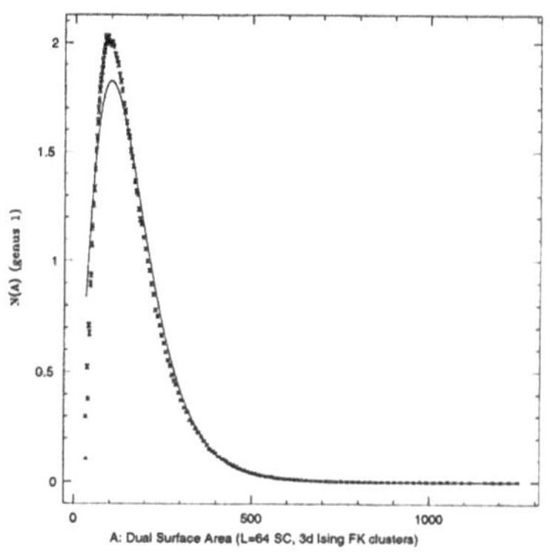

Fig.8. The number of genus 1 surfaces at T_c as a function of dual surface area A for FK clusters on the $L = 64$ SC lattice, with a best fit to the functional form (4.5).

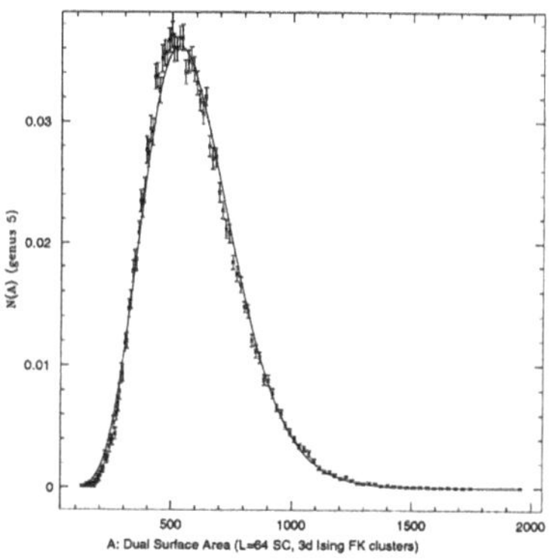

Fig.9. As in the previous figure, but for genus 5.

The deviation of 1.25 from 1 at first glance suggests the presence of significant deviations from Poisson distributed behavior. This may be due, however, to systematic deviations from continuum behavior due to lattice artifacts. The magnitude of these systematic errors is illustrated by the measurement of the dependence of the mean area on genus. If we assume the ansatz (4.5) then we would predict that the mean dual surface area A should increase linearly with genus, obeying

$$\langle A(g) \rangle = \frac{x(g) + 1}{\mu(g)}. \tag{4.8}$$

For FK clusters on the BCC lattice, however, we see by fitting $\ln\langle A \rangle$ to $\ln g$ in the small genus regime that $\langle A \rangle$ is not precisely linear in g; in fact it scales roughly as $g^{.85}$. Note that such a scaling law could not hold asymptotically for large lattices and large areas, since it would imply that surfaces would have more handles than plaquettes. Indeed this effective exponent slowly increases with genus (to roughly .90 at $g = 50$). Thus we observe systematic deviations (of order 15%) of genus dependent exponents from their asymptotic values. This also indicates that the apparent linearity that we observed in $x(g)$ is somewhat deceiving; presumably deviations from linearity would be more apparent if our statistics were better and we could directly fit somewhat higher values of g. The slope of $x(g)$ should decrease with greater g, so that the above estimate of the slope (1.25) may be too large.

The genus behavior of geometrical clusters at T_p is qualitatively quite similar to that of the FK case just discussed. We show fits to $N_2(A)$ and $N_5(A)$ on an $L = 60$ lattice in figures 10 and 11.

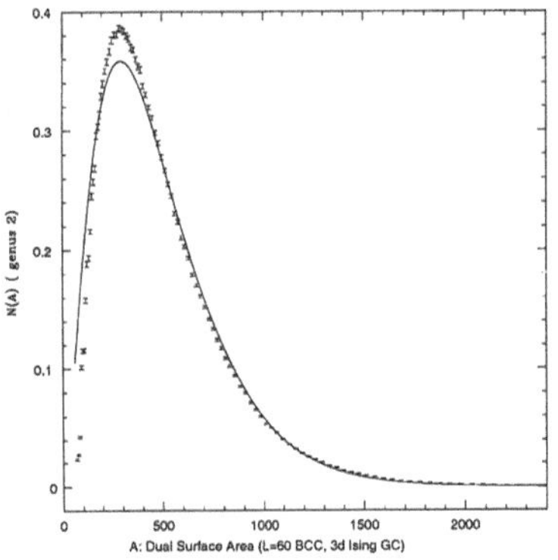

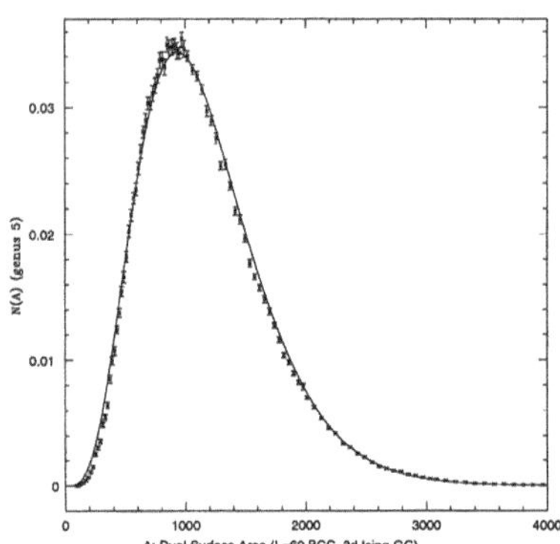

Fig.10. The number of genus 2 surfaces at T_p as a function of dual surface area A bounding minority (geometrical) clusters on the $L = 60$ BCC lattice.

Fig.11. As in the previous figure, but for genus 5.

There are large deviations in the fit for genus 2; for genus 4 and larger, however, the fits are nearly perfect. Again, μ is approximately independent of g, though (from figure 12) we observe transient behavior that is very significant up to genus 10. Again, x is approximately linear in g (as shown in figure 13), with a slope considerably lower than in the FK case; $dx/dg \sim 0.7 \pm 0.1$ in the range $g = 3 - 40$ for Ising minority spin percolation.

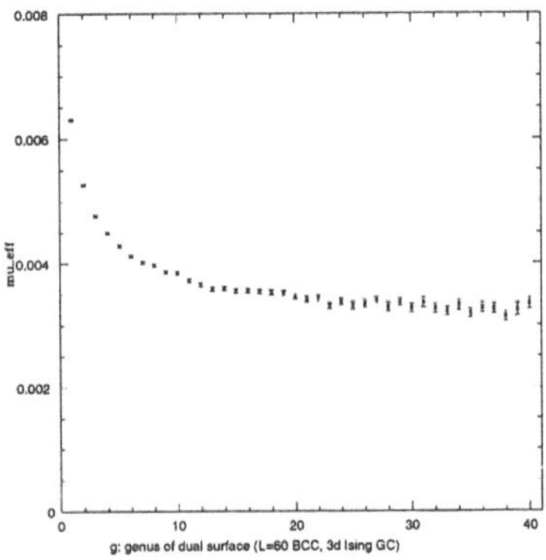

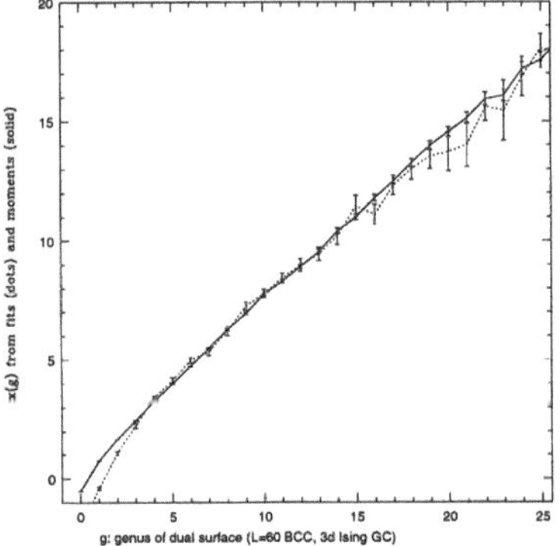

Fig.12. The dependence of μ (extracted from moments) on genus for surfaces bounding minority (geometrical) clusters on the $L = 60$ BCC lattice at T_p.

Fig.13. The dependence of x (extracted from fits and moments) on genus for surfaces bounding minority (geometrical) clusters on the $L = 60$ BCC lattice at T_p.

The same caution as before applies to these slope values; systematic errors could still be quite large, so the actual value of .7 for the slope is not so trustworthy. In this case, potential deviations from asymptotic scaling reveal themselves most clearly through the transient behavior of μ.

From this analysis, we can conclude that the genus distribution of FK cluster boundaries at T_c and geometrical cluster boundaries at T_p is described by the functional form (4.5), with $x(g)$ approximately linear in g and μ constant. The lattices considered, however, are too small to characterize the behavior of x more precisely.

3 Loop Scaling and Blocked Spins

In this section, we will solely be concerned with the structure of boundaries of geometrical clusters as T is increased beyond T_p, particularly to $T = T_c$. Recall that for $T > T_p$, two percolated clusters of opposite sign will span the lattice. For T not so close to T_c, we expect that the characteristics of the Ising interaction will not influence the large–scale structure of these percolating clusters. The percolating clusters (assuming the transition at T_p is indeed in the universality class of pure percolation) should then be described by the 'links, nodes and blobs' picture developed for the infinite clusters of pure percolation in dimensions below $d_c = 6$ [9, 22]. . In this description, the links form the thin backbones of the cluster; they are connected together at the nodes which occur roughly every percolation correlation length ξ. Most of the volume of the cluster consists of dangling ends emanating from the backbones. The backbones do not consist merely of one segment; they contain multiply-connected paths (which close to form the handles that we measure) that form blobs with diameter up to size ξ.

A cross section of the boundaries of these networks of tangled thin tubes would presumably be composed of a set of small lattice–sized loops. To check this, we examined the phase boundaries between up and down spins on planar slices of both the SC and BCC lattices. In figure 14, we show a log–log plot of $N(l)$, the number of loops of length l, versus l taken at the percolation temperature $\beta_p = .232$ on the SC lattice. The curve exhibits a sharp drop–off, indicating indeed that these slices contain only small loops. As we dial the temperature up towards T_c, we find that larger loops begin to appear in the slices. In fact, at T_c, we find loops at all scales; $N(l) \sim l^{-\tau'}$! This scaling is depicted in the log–log plot in figure 15. All of the largest loops must bound the two percolating clusters, since there are no intermediate size geometrical clusters at T_c. The loops themselves have a non–trivial fractal structure; we determined that the number of sites enclosed within a loop of length l scales as $A(l) \sim l^{\delta'}$.

From these measurements, we estimated that $\tau' = 2.06(3)$ and $\delta' = 1.20(1)$. These values are probably not very accurate, however. As in the determination of τ from the behavior of $N(V_{cl})$, corrections to scaling and finite–size effects are a source of large

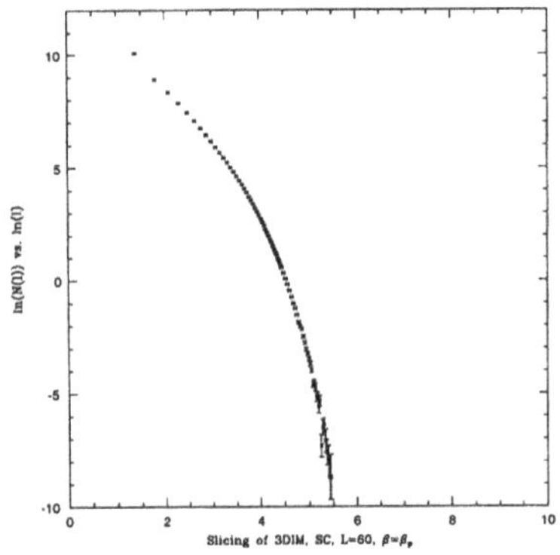
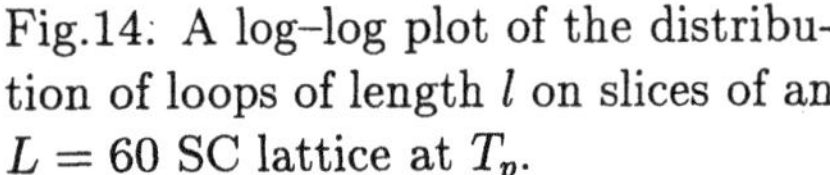
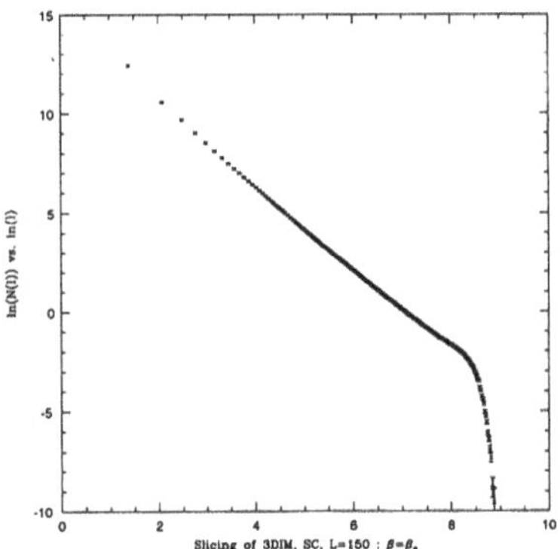

Fig.14: A log–log plot of the distribution of loops of length l on slices of an $L = 60$ SC lattice at T_p.

Fig.15. A log–log plot of the distribution of loops of length l on slices of an $L = 150$ SC lattice at T_c.

systematic errors. These systematic effects were only of order $1 - 2\%$ for δ; thus we suspect that our estimate of δ' is considerably better than that of τ'. Since the boundaries of domains self–intersect on slices of the cubic lattice, we had to pick a prescription (effectively another short–distance regularization) to define loops. Additionally, the enclosed area is not well–defined for loops that wind around the (periodic) lattice. We thus chose to exclude loops with non–zero winding number from consideration. These exponents should obey the relation $\tau' = 1 + \delta'$, which can be derived through scaling arguments [d]. This relation also holds for the corresponding indices that describe the distribution of self–avoiding loops that bound clusters in the 2–d Ising model at the Curie temperature. In that case, $\tau' \sim 2.45$. Finally, we see that the scaling behavior of loops on slices slowly disappears as we continue to increase the Ising temperature. At $\beta = .18$ on $L = 150$ SC lattices, we observed that very large loops were again exponentially suppressed in the distribution $N(l)$.

Should we surprised by the presence of this 'loop scaling' at T_c? The following argument, due to Antonio Coniglio, indicates that this result is at least plausible [24]. First, note that in the $T \to \infty$ limit, the distribution of loops and geometrical clusters is that of pure site percolation with $p = .5$. On the square lattice, $p_c \sim .59$ so that if only half the sites contain identical spins, then the distribution of loops and clusters should be governed by a finite correlation length. Now consider turning on the Ising couplings in the x and y directions. As the spins become correlated, the critical concentration [e] needed for percolation should decrease. At the Curie temperature for the 2–d Ising model ($T_c^{d=2}$) this critical concentration decreases to .5 and geometrical clusters and

---

[d]We thank Bertrand Duplantier [23] for providing us with a derivation of this relation.

[e]Note that we can adjust the relative concentration of up and down spins by also adding a magnetic field.

their boundaries percolate. In two dimensions, this critical concentration cannot be less than .5, since generically two percolating clusters cannot span a single lattice [25]. Imagine next turning on the Ising coupling in the z direction while tuning the x and y couplings to remain at criticality. If the critical concentration remains .5 as the system reaches the 3–d Curie temperature, then one would find a scaling distribution of clusters and boundaries on 2–d slices. On the other hand, we cannot rule out the possibility that the critical concentration again increases above .5; then we would never expect to find scaling of loops on slices of the 3–d Ising model.

We also observed scaling behavior of loops on the BCC lattice. In particular, only small loops were found at T_p while scaling of $N(l)$ with the values $\tau' = 2.23(1)$ and $\delta' = 1.23(1)$ occured at T_c. The uncertainty in the value of T_c probably leads to a significant systematic error in the estimate of these exponents. They do obey the anticipated relation $\tau' = 1 + \delta'$; δ' is not particularly far from the estimate extracted from the SC data. Note that on slices of the BCC lattice, which are triangular, there is no longer any ambiguity in the definition of loops. In this case, we find that $N(l)$ apparently satisfies a power–law distribution, with a temperature–dependent exponent, for all $T > T_c$! This observation can be fully understood theoretically, since the percolation threshold on triangulated lattices equals .5. Therefore, we definitely expect to observe loop scaling at $T = \infty$ with scaling exponents characteristic of 2–d percolation ($\tau' \sim 2.05$ and $\delta' = 1$). Since lowering the temperature increases correlations between spins, we expect to find percolated clusters on slices for all T. For $T < T_c$, however, minority spins cannot percolate on 2–d slices because, as stated above, only one infinite cluster can span a lattice. Thus the minority spins and the loops that enclose them must percolate at T_c on 2–d slices of the 3–d Ising model on the BCC lattice. If we assume that this phenomenon is independent of the particular lattice type, then it follows that loop scaling should always occur at T_c. A similar situation occurs for the 2–d Ising model on the triangular lattice: the distribution $N(l)$ again scales as a power law for all $T > T_c$ because $p_c = 1/2$ on triangulated lattices.

It also seems reasonable that the presence of loop scaling may be related to the vanishing of the surface tension of the Ising model at T_c. The vanishing of the surface tension ensures that the free energy of a system with anti–periodic boundary conditions along one plane (essentially due the insertion of a large loop along the boundary) equals the free energy of a system with periodic boundary conditions.

We now comment on the significance of this scaling. As we noted in the previous two sub–sections, the geometrical cluster boundaries do not in the least resemble surfaces (in the continuum limit) at T_p. The presence of large loops at T_c might indicate that the boundaries grow large long handles. A visual examination of successive slices qualitatively indicates that this is not so. Large loops seemingly always vanish after several consecutive slices. Indeed, it is difficult to envision a smooth surface that decomposes into a scaling distribution of loops along arbitrary slices.

It should also be noted that the exponent τ' is probably not directly related to the magnetic or thermal exponents of the 3–d Ising model. More generally, it may not be associated with the behavior of correlation functions of local operators in a unitary

Table 1. The mean genus per lattice site at T_c for blockings ($L = 8, 16, 32$ and 64) of an $L = 128$ lattice.

lattice	128	64	32	16	8
BCC	.049 (3)	.039 (3)	.037 (3)	.039 (3)	.044 (3)
SC	.021 (2)	.020 (2)	.018 (2)	.015 (2)	.012 (1)

quantum field theory. This is true also for loops bounding clusters in the 2–d Ising model. For in all of these cases, the scaling of geometrical clusters is determined by the geometric effects associated with percolation as well as the long–range correlations due to Ising criticality. Still, this scaling law describes physics that in principle is observable, perhaps by counting domains in sections of crystals that lie in the universality class of 3–d Ising. It would thus be quite interesting to construct a theoretical scheme to compute (approximately) the value of τ'. These loops are significantly 'rougher' than the corresponding boundaries in the 2–d Ising model, since the exponent δ' is lower here. They gain more kinetic energy because they are given an extra dimension in which to vibrate; perhaps this is responsible for their increased roughness.

Ideally, we would like view these loops as string states that evolve in Euclidean time (perpendicular to the slices). Their dynamics is described by the transfer matrix determined from Boltzmann factors associated with their creation, destruction, merging and splitting. We have thus found that the ground state wave functional (string field) of this transfer matrix is peaked around configurations that describe a scaling distribution of loops. These loops seemingly bear little relation to free strings, though, because they interact strongly by splitting and joining every few lattice spacings [f]. One might hope that some sort of perturbative string description could still be viable if the strength of this interaction were just a short–distance artifact; i.e. if the string coupling diminished towards zero in the infrared. To gauge whether this is likely, we blocked spins in our simulations to measure the renormalization group flow of the operator that couples to the total Euler characteristic summed over all cluster boundaries. In particular, during simulations on L=128 SC and BCC lattices, we blocked spins, using the majority rule and letting our random number generator decide ties. At each blocking level, we reconstructed clusters and boundaries and then measured the genus summed over surfaces. We present the results of this analysis in table 1; data was taken at $\beta_c =$ .221651 on the SC lattice and $\beta_c = .0858$ on the BCC lattice.

The results are not so conclusive. In particular, since we lack a very precise determination of the Curie temperature on the BCC lattice, it is likely that by the final blocking the couplings have flowed significantly into either the high or low–temperature regimes. Thus, one should probably not take the increase in genus density in the final two blockings on the BCC lattice seriously. This effect is not a problem on the SC lattice, where we fortunately know the critical temperature (based on previous Monte Carlo Renormalization Group measurements) to very high accuracy. On the other hand, we suspect that the small L blocked values on the SC lattice may be unreliable, due

---

[f] In practice, this makes an analysis of the transfer matrix a formidable task.

to ambiguity in the definition of genus. We can at least infer that the genus density decreases a bit during the first few blockings, indicating that the coupling $\exp(-\kappa)$ does at least slowly diminish at the beginning of the RG flow. There is no clear indication, however, that the flow continues on to the weak string coupling regime. One might also object to our choice of blocking scheme. Indeed, perhaps it might be more appropriate to somehow block the cluster boundaries themselves rather than the spins. In practice this would probably be technically difficult.

5 Assessment

The prospects for passing from the Curie point to the regime in which surfaces are weakly coupled are addressed in the work of Huse and Leibler [5]. They qualitatively map out the phase diagram of a model of self–avoiding surfaces with action (2.1). The large κ (large coupling to total Euler characteristic) regime of their model lies in a droplet crystal phase, where the large percolated surface has shattered into a lattice of small disconnected spheres. Such a configuration maximizes the Euler density; it clearly does not correspond to a theory of surfaces. By estimating the free energy difference between phases, they argue that the transition to this droplet crystal is first order. Given this picture, there seems to be little evidence for the existence of a fixed point describing a weakly coupled theory of surfaces near the Curie point of the Ising model. Nevertheless, we cannot definitely exclude the possibility that there is still some path which we have not considered to a weak–coupling theory.

In conclusion, it appears that evidence of a continuum theory of surfaces has eluded us in our investigation of Ising cluster boundaries. We have found, however, that these cluster boundaries do exhibit an intriguing fractal structure that does not typically appear in models of lattice surfaces.

Acknowledgments

We would like to thank Stephen Shenker for essential discussions which led to our investigations. We also greatly benefited from discussions with Mark Bowick, Francois David, Bertrand Duplantier, John Marko and Jim Sethna. Furthermore, we are indebted to the organizers of this workshop for their efforts and hospitality. We are also grateful to NPAC for their crucial support. This work was supported in part by the Dept. of Energy grants DEFG02-90ER-40560, DEFG02-85ER-40231, the Mathematical Disciplines Institute of the Univ. of Chicago, funds from Syracuse Univ., by the Centre National de la Recherche Scientifique, by INFN and the EC Science grant SC1*0394.

References

[1] E. Fradkin, M. Srednicki and L. Susskind, Phys. Rev. **D21**, (1980) 2885; C. Itzykson, Nucl. Phys. **B210** (1982) 477; A. Casher, D. Fœrster and P. Windey, Nucl. Phys. **B251** (1985) 29; Vl. Dotsenko and A. Polyakov, *in* Advanced Studies in Pure Math. **15** (1987).

[2] E. Brézin and V.A. Kazakov, Phys. Lett. **236B** (1990) 144; M.R. Douglas and S.H. Shenker, Nucl. Phys. **B335** (1990) 635; D. J. Gross and A. A. Migdal, Phys. Rev. Lett. **64** (1990) 127.

[3] G. Parisi, *in Proceedings of the Third Workshop on Current Problems in High Energy Particle Physics*, John Hopkins Conference, Florence 1979; G. Parisi, J.-M. Drouffe and N. Sourlas, Nucl. Phys. **B161** (1979) 397; B. Durhuus, J. Frohlich and T. Jonsson, Nucl. Phys. **B240** (1984) 453; J. Ambjorn, B. Durhuus, J. Frohlich and P. Orland, Nucl. Phys. **B270** (1986) 457; M. E. Cates, Europhys. Lett. **8** (1988) 719.

[4] F. David, Europhys. Lett. **9** (1989) 575.

[5] D. Huse and S. Leibler, J. de Physique **49** (1988) 605.

[6] M. Karowski and H.J. Thun, Phys. Rev. Lett. **54** (1985) 2556; R. Schrader, J. Stat. Phys. **40** (1985) 533.

[7] M. Caselle, F. Gliozzi and S. Vinti, Turin Univ. preprint DFTT–12–93.

[8] F. David, Jerusalem Gravity (1990) 80.

[9] D. Stauffer and A. Aharony, *Introduction to Percolation Theory,* (Taylor and Francis, London, U.K. 1992).

[10] J. Cambier and M. Nauenberg, Phys. Rev. **B34** (1986) 8071.

[11] C.M. Fortuin and P.W. Kasteleyn, Physica **57** (1972) 536.

[12] A. Coniglio and W. Klein, J. Phys. **A13** (1980) 2775.

[13] R.G. Edwards and A.D. Sokal, Phys. Rev. **D38** (1988) 2009.

[14] C.-K. Hu, Phys. Rev. **B29** (1984) 5103.

[15] J.-S. Wang, Physica **A161** (1989) 149.

[16] R.H. Swendsen and J.-S. Wang, Phys. Rev. Lett. **58** (1987) 86.

[17] V. Dotsenko, G. Harris, E. Marinari, E. Martinec, M. Picco and P. Windey, in preparation.

[18] S. Kirkpatrick, in "Ill-Condensed Matter", Les Houches Proceedings, Vol.31, ed. R. Balian, R. Maynard and G. Toulouse (North Holland, Amsterdam 1983) 372.

[19] M. Hasenbusch and K. Pinn, Munster Univ. preprint MS–TIP–92–24.

[20] V. Dotsenko, G. Harris, E. Marinari, E. Martinec, M. Picco and P. Windey, Phys. Rev. Lett. **71** (1993) 811.

[21] C. Itzykson and J-M. Drouffe, *Statistical Field Theory*, Cambridge University Press, Cambridge (1989).

[22] P.G. DeGennes, La Recherche **7** (1976) 919.

[23] Bertrand Duplantier, private communication.

[24] Antonio Coniglio, private communication.

[25] A. Coniglio, C. Nappi, F. Peruggi and L. Russo, J. Phys. **A10** (1977) 205.

CONFORMAL FIELD THEORY TECHNIQUES
IN LARGE N YANG-MILLS THEORY

Michael R. DOUGLAS

Dept. of Physics and Astronomy
Rutgers University
USA

Following some motivating comments on large N two-dimensional Yang-Mills theory, we discuss techniques for large N group representation theory, using quantum mechanics on the group manifold $U(N)$, its equivalence to a quasirelativistic two-dimensional free fermion theory, and bosonization. As applications, we compute the free energy for two-dimensional Yang-Mills theory on the torus to $O(1/N^2)$, and an interesting approximation to the leading answer for the sphere. We discuss the question of whether the free energy for the torus has $R \to 1/R$ invariance.

1 Introduction

The first part of this article is an introduction to what might be called "large N representation theory," Lie group representation theory with the focus on the limit $N \to \infty$ of $SU(N)$ and the other classical groups. This has many applications in physics and mathematics, and good mathematical introductions exist, which tie it to its applications in group theory, soliton theory, combinatorics, and so forth. (See [1], 5.4 for a treatment very much like the one here; see also [2]) Now although it might seem that this theory would be invaluable for studying the large N limit of models with $SU(N)$ symmetry, and some examples in the physics literature are in [3], it does not get as much use as one might expect. Whether this is simply because the language is taking time to standardize, or because the physics really is too diverse to capture in one formalism, I leave for the reader to judge. However a model which seems made to order as an application is large N two-dimensional Yang-Mills theory (YM$_2$ in the following).

I would like to make some comments about the potential physical relevance of this model, which is certainly a long way from realistic four-dimensional models. The starting point is the old idea that QCD could be reformulated as a string theory, which for many reasons can only be a free string in the large N limit. [4] Although our understanding of such things is primitive, it seems clear that a "QCD string" if it exists is a very different type of string than those studied as theories of quantum gravity, and that a low dimensional solvable model might provide a context in which we could discover and understand such a different type of string theory. This line of thought has led to a revival of the study of two-dimensional Yang-Mills theory and QCD with the objective of developing a string representation which would generalize to higher dimensions. The earliest work in this direction concentrated on reproducing Wilson loop expectation values; even in two dimensions these are non-trivial for self-intersecting loops and evidence was found that their structure could be described by a sum of surfaces each with weight exp $-$area, modified by local factors associated with features on the surface such as branch points. [5] More recently the partition function on a closed Riemann surface has been studied and a complete set of rules derived which reproduces it as a sum over surfaces, again with additional features. [6] The idea is not restricted to two dimensions and a D-dimensional lattice formulation exists. [7]

Besides explicit string constructions of this sort, much can be learned from comparing precise results from a field theory and candidate equivalent string theories. Of course if we can get exact results for a theory we do not really need a string formulation but even crude results in higher dimensions are valuable to show whether a string formulation could work at all. The most important issue for the constructions of [6, 7] is that they are essentially strong coupling expansions. They improve on Wilson's original expansion by replacing the expansion parameter $1/g^2$ by exp $-g^2$ (g is a dimensionless bare lattice coupling) but still face the essential problem of the strong coupling expansion for gauge theory – the continuum limit requires the limit of weak bare coupling. As it turns out there is an interesting two-dimensional case which illustrates the problems, namely YM_2 with the sphere as target space.

Assuming the usefulness of these string representations, the next step would be to take a world-sheet continuum limit and hope that this still reproduces the Yang-Mills continuum limit. The results of [6] are similar to topological field theory results, and [8, 9] pointed out that for torus target space there is a natural candidate for comparison, the topological sigma model with torus target space (possibly with coupling to gravity), and that this worked for the torus world-sheet. One consequence of this on arbitrary genus world-sheet would be $R \to 1/R$ duality invariance, and we will look for this at genus two in the YM_2 results.

This relation to topological theory seems very special to $D = 2$, and in higher dimensions we expect a string with less trivial dynamics. However the generic string theory seems to have a trivial world-sheet continuum limit; this is the famous "$c > 1$" or "branched polymer" problem which has been argued to be inevitable in a theory defined as a sum over world-sheets with positive weights. [10] A sensible QCD string must escape this problem, and the additional world-sheet features have to play an essential role in this, leading to the question of which of the many features are essential and which are irrelevant. One would like to understand this before trying to reproduce the string theory with a continuum world-sheet action. It seems to me that this can only be properly understood in $D > 2$; nevertheless the $D = 2$ results do suggest that some features are more important than others.

2   Quantum Mechanics and Group Representations

The prototypical system we study is quantum mechanics on the group manifold $U(N)$. This allows us to quickly classify representations and derive the Weyl character formula. Physically this system is already interesting, since it describes the (global) degrees of freedom of two-dimensional Yang-Mills theory. We go on to discuss calculations of the YM_2 free energy on a Riemann surface.

The natural Hamiltonian is

$$H = \text{tr} \ \left(U \frac{\partial}{\partial U} \right)^2 \equiv \sum_a E^a E^a. \tag{2.1}$$

Here $E^a = \text{tr} \ t^a U d/dU$ generates left rotations of U and represents the Lie algebra $u(N)$. Thus acting on a wave function which could be any matrix element of an irreducible representation R, $\psi(U) = D^{(R)}_{ij}(U)$, $H = C_2(R)$, the second Casimir (normalized so that $C_2(\square) = N$). In fact it is the unique invariant and purely second order linear differential operator on the group manifold, the Laplacian.

To classify representations we should find their characters $\chi_R(U) = \text{tr} \ D^{(R)}(U)$. These will be wave functions invariant under $\psi(U) \to \psi(gUg^{-1})$, so we should make the change of variables $U_{ij} = g_{ik} z_k g^{-1}_{kj}$. (This is familiar from the quantum mechanics of a hermitian matrix [11] and for the group manifold case is much older, going back to Harish-Chandra. [12])

The invariant volume element in these variables is

$$\sqrt{h} = |\Delta(z)|^2 = \tilde{\Delta}(z)^2 \tag{2.2}$$

where $\Delta(z) = \prod_{i<j}(z_i - z_j)$ and $\tilde{\Delta}(z) = \prod_{i<j} \sin \frac{\theta_i - \theta_j}{2} = \Delta(z)/\prod_i z_i^{(N-1)/2}$. The "radial" components of the metric are simply $h_{ij} = \delta_{ij}$. Thus on wave functions independent of g

$$H = -\sum_i \frac{1}{\tilde{\Delta}^2} \frac{d}{d\theta_i} \tilde{\Delta}^2 \frac{d}{d\theta_i}. \tag{2.3}$$

We can rewrite this as

$$H = -\sum_i \left[\frac{1}{\tilde{\Delta}} \frac{d^2}{d\theta_i^2} \tilde{\Delta} - \frac{1}{\tilde{\Delta}} \left(\frac{d^2 \tilde{\Delta}}{d\theta_i^2} \right) \right]. \tag{2.4}$$

For hermitian matrix quantum mechanics Δ was a Vandermonde and the second term, thanks to a non-trivial identity, gave zero. Here, after a similar calculation, the second term is found to equal $-N(N^2 - 1)/12$. [13]

Thus, after redefining the wave functions by $\psi \to \tilde{\Delta}\psi$, we have a theory of N free fermions on the circle. The boundary conditions are also determined by this redefinition; they become periodic (antiperiodic, respectively) for N odd (even). An orthonormal basis for wave functions is Slater determinants

$$\psi_{\vec{n}} = \det_{i,j} z_i^{n_j} \tag{2.5}$$

with energy $E = \sum_i n_i^2 - N(N^2 - 1)/12$. The ground state has fermions distributed symmetrically about $n = 0$, and energy zero, so the Fermi level $n_F = (N - 1)/2$.

Going back to the original wave functions, we have rederived the Weyl character formula (for $U(N)$ actually due to Schur):

$$\chi_{\vec{n}}(\vec{z}) = \frac{\det_{1 \leq i,j \leq N} z_i^{n_j}}{\det_{1 \leq i,j \leq N} z_i^{j-1-n_F}}.$$

(2.6)

In terms of roots and weights, the indices n_i with $n_1 > n_2 > \ldots > n_N$ are the components of the highest weight vector shifted by half the sum of the positive roots (usually denoted $\mu + \rho$) where the basis of the Cartan subalgebra is just $(H_i)_{jk} = \delta_{ij}\delta_{jk}$. In the language of Young tableaux, if h_i is the number of boxes in the i'th row, $n_i = (N-1)/2 + 1 - i + h_i$.

The $U(1)$ charge is $Q = \sum_i n_i$. We can change this by a multiple of N by shifting all the fermions $n_i \rightarrow n_i + a$, but $Q \bmod N$ is correlated with the conjugacy class of the $SU(N)$ representation (in other words the action of the center) reflecting the identification $U(N) \cong SU(N) \times U(1)/\mathbb{Z}_N$.

Interesting observables in this quantum mechanics, invariant under the adjoint action, are the invariant "position" operators

$$W_n = \operatorname{tr} U^n = \sum_i z_i^n$$

(2.7)

and "generalized Hamiltonians"

$$H_m = (-i)^m \sum_i \frac{\partial^m}{\partial \theta_i^m}.$$

(2.8)

For $m > 2$ these are *not* the higher Casimirs $\operatorname{tr} E^m$ but are polynomial in them (see [14, p. 163] for an explicit expression). We will not discuss $m > 2$ further here.

We next go to a second quantized formalism with operators B_{-n}^+ and B_n creating and destroying the fermion mode z^n, and $\psi(\theta) = \sum_n e^{in\theta} B_n$. Then $H = \int d\theta \partial \psi^+ \partial \psi - E_0$. The operators W_n and H_m will become fermion bilinears.

The first simplification of the large N limit now appears. If we never consider operators W_n with $n \sim N$, then fermions near the positive and negative Fermi surfaces completely decouple. We can then speak of a quasi-relativistic Fermi system, with complex chiral left- and right-moving fermions. We should also speak of U raising the left-moving (upper) fermions while lowering the right-movers, and U^{-1} doing the opposite. This suggests that we refer to representations contained in tensor products of $O(N^0)$ fundamentals as "chiral," and their complex conjugates as "anti-chiral." The full representation theory is a product of chiral and anti-chiral sectors. So, let $b_n^+ = B_{-n_F-\epsilon+n}^+$, $b_n = B_{n_F+\epsilon+n}$, $\bar{b}_n^+ = B_{n_F+\epsilon-n}^+$, $\bar{b}_n = B_{-n_F-\epsilon-n}$, where $\epsilon = \frac{1}{2}$ is an choice of definition, introduced to give antiperiodic ($n \in \mathbb{Z} + \frac{1}{2}$) moding for all N. The local operators $\psi(z) = \sum_{n \in \mathbb{Z}+\frac{1}{2}} z^{-n} b_n$, $\psi^+(z)$, $\overline{\psi}(\bar{z}) = \sum_{n \in \mathbb{Z}+\frac{1}{2}} \bar{z}^{-n} \bar{b}_n$, and $\overline{\psi}^+(\bar{z})$ now satisfy standard 2d field theory commutation relations. The standard Fock vacuum ($b_n|0> = b_n^+|0> = 0$ for $n > 0$) corresponds to the identity representation, and higher representations can be built by acting with the bilinears $b^+_{-n}b_{-m}$ and $\overline{b}^+_{-n}\overline{b}_{-m}$. Of these, clearly the simplest are the W_n's which become [a]

$$W_n = \operatorname{tr} U^n \tag{2.9}$$

$$= \oint dz\, z^{-1-n}\psi^+(z)\psi(z) + \oint d\overline{z}\, \overline{z}^{-1+n}\overline{\psi}^+(z)\overline{\psi}(z) \tag{2.10}$$

$$= \sum_m b^+_{n-m}b_m + \overline{b}^+_{m-n}\overline{b}_{-m}. \tag{2.11}$$

We recognize the operators here as the left- and right-moving conformal field theory $U(1)$ currents, and the construction of the W_n's as bosonization:

$$W_n \equiv \alpha_{-n} + \overline{\alpha}_n = \int d\theta e^{in\theta}\partial_\tau \phi(z = e^{n(\tau+i\theta)}, \overline{z} = e^{n(\tau-i\theta)}) \tag{2.12}$$

defining the standard free boson oscillator expansion with

$$\begin{aligned}
\partial_z \phi(z) &= i\sum_{m\in\mathbb{Z}} \alpha_m z^{m-1} \\
[\alpha_m, \alpha_n] &= [\overline{\alpha}_m, \overline{\alpha}_n] = m\delta_{m+n,0} \\
[\alpha_m, \overline{\alpha}_n] &= 0.
\end{aligned} \tag{2.13}$$

Notice that the W_n commute as operators, as they should.

The charges α_0 and $\overline{\alpha}_0$ count fermion numbers. Their sum is constant in our application. There is a normal ordering ambiguity in the definition (2.11) which we use to define it to be zero. As for the difference $\alpha_0 - \overline{\alpha}_0$, clearly it can be changed by operators like

$$\sum_n B^+_{-n_F-n+1}B_{-n_F+n} = \sum_{n\in\mathbb{Z}+\frac{1}{2}} b^+_{-n}\overline{b}_{-n} = \oint \frac{dz}{z}\psi^+(z)\overline{\psi}(z^*). \tag{2.14}$$

In the bosonic language it is winding number;

$$w = \alpha_0 - \overline{\alpha}_0 = -i\oint dz\partial\phi + i\oint d\overline{z}\overline{\partial}\phi = -\frac{1}{2\pi}(\phi(2\pi) - \phi(0)). \tag{2.15}$$

A better way to change the winding number is to turn it on continuously from zero; taking this back to the Fermi picture we are continuously changing the fermion boundary conditions, or equivalently multiplying the wave function by

$$\psi(\vec{z}) \rightarrow \left(\prod_{i=1}^N z_i\right)^s \psi(\vec{z}). \tag{2.16}$$

Taking s from zero to one gives a new state with the same $SU(N)$ quantum numbers but $U(1)$ charge increased by N. We conclude that the original quantum mechanics on $U(N)$ is equivalent to a free bosonic field theory whose zero modes are treated rather asymmetrically: we sum over integer winding numbers, but not over momenta.

We should keep in mind that although the formalism so far suggests a close relationship with two-dimensional conformal field theory, there is no a priori guarantee that the

---

[a]Our CFT conventions are generally as in [15], except that our boson ϕ is $\sqrt{2}$ times theirs; in other words $S_{\text{free}} = \int d^2x(\partial\phi)^2/2\pi$. Our contour integrals always have an implicit $1/2\pi i$, and $\overline{z}$ integrals go counterclockwise.

Hamiltonian or observables in a specific problem will be local in the two dimensions. Of course the positions θ_i parameterize the maximal torus of our original group, so local time evolution on the group manifold will reduce to local two dimensional evolution. On the other hand group multiplication is an example of a natural operation with no locality properties. So not all problems will have simple conformal field theory translations.

We could summarize by saying that the Bose-Fermi correspondence is the large N limit of the Frobenius relation between characters and symmetric polynomials. A character $\chi_{\vec{n}}$ corresponds to a fermion Fock basis state in a simple way; if one takes a "chiral" representation R, with Young tableau with h_i boxes in the i'th row, $1 \leq i \leq r$, this corresponds to the state

$$|\vec{h}> = \chi_{\vec{h}}(U)|0> = b^+_{\epsilon-h_1} b^+_{\epsilon-h_2} \ldots b^+_{\epsilon-h_r} b_{-\epsilon} b_{-\epsilon-1} \ldots b_{-\epsilon-r+1}|0>. \tag{2.17}$$

An alternate basis for class functions is

$$\prod_i (\text{tr } U^i)^{\sigma_i} \tag{2.18}$$

(in terms of the z_i these functions give a basis for the symmetric polynomials) which will correspond to a state built with bosonic operators

$$|\sigma> = \prod_i W_i^{\sigma_i}|0>. \tag{2.19}$$

Orthonormality of the characters gives us

$$\chi_{\vec{n}}(U)|0> = |\vec{n}> = \sum_\sigma <\vec{n}|\sigma> W_i^{\sigma_i}|0>. \tag{2.20}$$

Another application of this is the integration of class functions, which is just expectation values of products of operators. For example,

$$\begin{aligned}
\int dU (TrU^{-2})(TrU)^2 &= <0|(\alpha_2 + \overline{\alpha}_{-2})(\alpha_{-1} + \overline{\alpha}_1)^2|0> \tag{2.21} \\
&= 0
\end{aligned}$$

to all orders in $1/N$. (and for finite $N > 2$, by going back to the non-relativistic fermions.)

The above was all for $U(N)$; the $U(1)$ generator is $Q = H_1 = \int d\theta \psi^+ \partial \psi = \sum_n n B^+_{-n} B_n$ which in the large N limit becomes

$$\begin{aligned}
Q &= \sum_n n b^+_{-n} b_n - \sum_n n \overline{b}^+_{-n} \overline{b}_n \\
&= L_0 - \overline{L}_0 \tag{2.22}
\end{aligned}$$

(but see below.) Constraining this to zero (keeping integer winding numbers) gives representations of the quotient $SU(N)/\mathbb{Z}_N$. If we are interested in $SU(N)$ we have two options. We can choose a representation of $U(N) \cong SU(N) \times U(1)/\mathbb{Z}_N$ for each representation of $SU(N)$, and subtract the $U(1)$ part of the second Casimir, $H_{U(1)} = Q^2/N$ from the Hamiltonian. Or, we can extend our sum over states to get $SU(N) \times$

$U(1)$. A $U(1)$ character of charge 1 is $\chi_1 = \prod_i z_i^{1/N}$, so the appropriate modification is simply to sum over winding numbers k/N, or equivalently fermion sectors with twisted boundary conditions $\psi(e^{2\pi i}z) = e^{2\pi i(1/2+k/N)}\psi(z)$ and $\overline{\psi}(e^{-2\pi i}\overline{z}) = e^{2\pi i(1/2+k/N)}\overline{\psi}(\overline{z})$.

Our quantum mechanical Hamiltonian is the second Casimir, which is not the relativistic Hamiltonian $L_0 + \overline{L}_0$. In terms of relativistic fermions it is

$$
\begin{aligned}
H_{U(N)} &= \sum_{n\in\mathbb{Z}+\frac{1}{2}} (n_F + \epsilon - n)^2(b^+_{-n}b_n + \overline{b}^+_n\overline{b}_{-n}) - E_0 \\
&= NL_0 + N\overline{L}_0 + \sum_{n\in\mathbb{Z}+\frac{1}{2}} n^2 : b^+_{-n}b_n + \overline{b}^+_{-n}\overline{b}_n : \\
&= NL_0 + N\overline{L}_0 + \oint dz\, z^2 : \partial\psi^+\partial\psi : + \oint d\overline{z}\, \overline{z}^2 : \overline{\partial\psi}^+\overline{\partial\psi} : \qquad (2.23)
\end{aligned}
$$

(we know that the vacuum energy in this ground state is zero).

The bosonization of this Hamiltonian is very well known in the context of matrix models, as it is just the Das-Jevicki-Sakita Hamiltonian governing the dynamics of the eigenvalue density in hermitian matrix quantum mechanics. We are retracing the steps of Gross and Klebanov and of Wadia and Sengupta [16] to arrive at it. The difference here is that there is no potential, and the fermions live on a circle. We have a left and right moving decomposition of the boson $\phi(z, \overline{z}) = \phi_L(z) + \phi_R(\overline{z})$, and the standard formulas : $e^{i\phi_L(z)} := \psi(z)$, : $e^{-i\phi_L(z)} := \psi^+(z)$, etc... Substituting into (2.23), we can define the second derivative term by point-splitting the two operators and taking the limit. The result must be a sum of operators of charge zero and dimension $(3,0)$ and $(0,3)$. In fact the operators $\partial^3\phi$ or $\partial\phi\partial^2\phi$ would be unimportant here because they are total derivatives (and there are enough derivatives to kill the winding mode), so the only possibility is (the coefficients are easily checked on low lying states)

$$
\begin{aligned}
H &= -\frac{N}{2}\oint dz\, z : (\partial\phi)^2 : -\frac{N}{2}\oint d\overline{z}\, \overline{z} : (\overline{\partial}\phi)^2 : \qquad (2.24) \\
&\quad +\frac{i}{3}\oint dz\, z^2 : (\partial\phi)^3 : +\frac{i}{3}\oint d\overline{z}\, \overline{z}^2 : (\overline{\partial}\phi)^3 : \\
&\equiv NL_0 + N\overline{L}_0 + H_I.
\end{aligned}
$$

The cubic interaction term in this Hamiltonian is quite natural, as we could see by considering the action of our original (2.1) on states (2.19) – it would contain terms preserving the "string number" (number of traces), as well as terms joining or splitting strings in higher order in $1/N$. [b] H_I is conserved under free time evolution (as are all the H_m's).

Actually there is a slight awkwardness in the bosonic formalism at this point: with our present definitions, (2.22) is not quite correct. The contribution of w to the $U(1)$ charge is Nw and is correctly reproduced by (2.22) only if we take the momentum $p = N$. Although this might sound like a more natural choice, it obscures the large N limit: we will constantly need to expand $\partial\phi = N/2z + O(1)$ to calculate. Rather we take instead $p = 0$ and

$$
Q = Nw + L_0 - \overline{L}_0. \qquad (2.25)
$$

---

[b]See [18] for a complete elaboration of this.

This point is important only if we are interested in the operator Q; in particular (2.24) is correct with $p = 0$.

All this could be done for a general group manifold. Computing singlet wave functions again leads to the Weyl character formula. For the groups $Sp(2N)$ the maximal torus can be taken to be diagonal matrices $\text{diag}(z_i, z_i^{-1})$ and the Weyl group includes both permutations and the reflections $z_i \to z_i^{-1}$. Although we will not try to develop it here, in the large N limit this should produce a free fermion theory on a surface with boundary. For $SO(N)$ at finite N we would need an additional global degree of freedom to incorporate the spinor representations; however these have $C_2 \sim N^2$ so would drop out of our large N considerations.

3 YM$_2$ on the Cylinder and Torus

This is really the same quantum mechanics on a group manifold under a different name. Let us do canonical quantization with our space being a circle of radius 1; time evolution will generate a cylinder of area $A = 2\pi t$. The Hamiltonian is $g^2 \int dx\,\text{tr}\, E^2$, with $E(x)^a = -i\partial/\partial A^a(x)$, and we must impose Gauss' law $D_x E = 0$, which is solved by gauge invariant wave functions, i.e. satisfying $\psi[g^{-1}(x)(\partial_x + A(x))g(x)] = \psi[A(x)]$. A wave function is determined by its value on configurations of constant $A(x)$, and gauge orbits are in one-to-one correspondence with values of the holonomy $U = P \exp i \int_0^{2\pi} A(x)dx$ modulo the adjoint action $U \to g^{-1}Ug$, completing the reduction to the singlet sector of quantum mechanics. The standard large N limit is taken with gauge coupling $g^2 \sim 1/N$ and in two dimensions we can set $g^2 = 1/N$, defining our unit of length. Then time evolution is generated by an $O(N^0)$ free Hamiltonian with an $O(1/N)$ interaction term. [17] The ground state energy E_0 is freely adjustable, say by adding $E_0 \int d^2x\sqrt{g}$ to our original Lagrangian.

The simplest physical quantity is the partition function on the torus,

$$\text{Tr } e^{-2\pi t H} = Z_{1\to 1} + O(N^{-2}). \tag{3.1}$$

The leading term is $O(N^0)$ and the notation "$1 \to 1$" indicates that the string interpretation of this [6] is a sum of (disconnected) maps from genus one world-sheets (at N^0) to a genus one target space.

Since the gauge invariant states correspond directly to our conformal field theory states, and the interaction is subleading, $Z_{1\to 1}$ is almost the standard torus partition function of free $c = 1$ conformal field theory. The "almost" is there because the total charge of our Fermi theory or momentum zero mode $\alpha_0 + \overline{\alpha}_0$ of our rose theory is conserved; thus we have a partition function with this constraint. This is particularly easy to implement in the Bose description; clearly

$$Z_{1\to 1} = q^{E_0} \prod_{n \geq 1} \frac{1}{(1 - q^n)^2} \sum_{w \in \mathbb{Z}} q^{w^2}. \tag{3.2}$$

where $q \equiv e^{-2\pi t}$. The sum over winding modes is a rather uninteresting side effect of the $U(1)$ factor. The nicest way to eliminate it is to decouple the $SU(N)$ and $U(1)$ in the way described above, by summing over winding numbers $w = k/N$. In the large N limit we clearly want to interpret such a sum as an integral; it is Gaussian, giving

$$Z'_{1 \to 1} = \frac{N^2}{\sqrt{2\pi t}} q^{E_0} \prod_{n \geq 1} \frac{1}{(1 - q^n)^2}. \tag{3.3}$$

Another way of saying this is, we have an $SU(N) \times U(1)$ gauge theory with the same coupling constant in both sectors. Using the coupling constant which gives a nice large N limit, g^2/N, gives the extreme weak coupling limit in the $U(1)$ sector. In this limit we cannot see the compact nature of the group $U(1)$.

Subleading corrections to this will have a string interpretation in terms of maps from higher genus world-sheets. We can write an all-orders expression quite explicitly from the free fermion formalism: for $U(N)$,

$$\begin{aligned}
Z_{\text{all} \to 1} &= \sum_R e^{-AC_2(R)/N} \\
&= q^{E_0} \oint \frac{dz}{z} \left[\prod_{m \geq 1} (1 + z q^{m-1/2+(m-1/2)^2/N}) \prod_{n \geq 1} (1 + z^{-1} q^{n-1/2-(n-1/2)^2/N}) \right]^2 .
\end{aligned} \tag{3.4}$$

The contour integral is there to implement the constraint of zero total charge. It complicates the interpretation so again it is useful to do a bosonic calculation. From both the CFT and string points of view, the free energy is a sum over connected diagrams. Expanding $\exp -\frac{A}{N} H_I$ we have the series

$$F_1(A) = F_{1 \to 1} + \sum_{g \geq 2} N^{2-2g} F_{g \to 1} \tag{3.5}$$

with

$$F_{g \to 1}(A) = \frac{1}{(2g-2)!} < (\frac{iA}{3} \oint dz \, z^2 \; : \partial \phi(z)^3 : + \text{c.c.})^{2g-2} >_c \tag{3.6}$$

This is a connected correlation function on the torus (an annulus with z and qz identified). The integrals are taken over contours of constant $|z|$, and since H_I is conserved, we can take $|z|$ to be slightly different for each contour, avoiding any possible singularities. The Green's function (defined by the original oscillator expansion) will be the usual one [15, p. 571] if we use the same prescription as in (3.3) of integrating the boson winding mode:

$$\begin{aligned}
G(\nu_1, \nu_2) &= < \partial_\nu \phi(\nu_1) \partial_\nu \phi(\nu_2) > \\
&= \partial_1^2 \log \theta_1(\nu_1 - \nu_2 | \tau) + \frac{\pi}{t} \\
&= -\wp(\nu_1 - \nu_2 | \tau) - \frac{\pi^2}{3} E_2(\tau) + \frac{\pi}{t}.
\end{aligned} \tag{3.7}$$

$z = e^{2\pi i \nu}$, $\tau = it$ and the Weierstrass function and Eisenstein series are defined in [19]. The $< \bar{\partial} \phi \bar{\partial} \phi >$ propagator is the same with $\nu \to \bar{\nu}$. Also

$$< \partial \phi \bar{\partial} \phi > = -\frac{\pi}{t}. \tag{3.8}$$

If we want a group other than $SU(N) \times U(1)$, the zero mode contribution π/t will be modified.

The first correction will be at $1/N^2$ from two insertions of our interaction Hamiltonian. In changing variables from z to ν we should remember that the normal ordering of (2.24) was defined with respect to the z coordinate. Taking this into account however gives a contribution proportional to the momentum, in other words zero. The sum of terms involving contractions of a pair of operators from the same appearance of H_I (in [6], contributions which can be disconnected by cutting a "tube") vanish (for $\mathrm{Re}\,\tau = 0$). So,

$$F_{2 \to 1} = \frac{2A^2}{3(2\pi)^6} \int_0^1 d\nu \, G(\nu, 0)^3. \tag{3.9}$$

The contour integrals are in the appendix, giving

$$\begin{aligned}
F_{2 \to 1} &= \frac{A^2}{2^5 \cdot 3^4 \cdot 5}(10E_2^3 - 6E_2 E_4 - 4E_6) + \frac{A}{2^4 \cdot 9}(E_4 - E_2^2) \tag{3.10}\\
&= A^2(8q^2 + 64q^3 + \ldots) + A(2q + 12q^2 + \ldots).
\end{aligned}$$

In the interpretation of [6] this is the generating function counting maps without folds from a genus two surface to a torus. These can have two branch points (the A^2 term) or a "handle" (the $O(A)$ term). Since the $U(1)$ piece does not contribute at subleading orders in $1/N$, this is exactly the $SU(N)$ result. One can also see this by expanding $\partial \phi = w/z + \ldots$ and integrating out w, which produces the correction to H_I appropriate for $SU(N)$.

As for $F_{1 \to 1}$, the most striking thing about this answer is how close it is to being a modular form (here of weight 6) in the variable τ. The Eisenstein series E_k for $k \geq 4$ are forms of weight k, and while E_2 is not a form it has a very simple anomaly in its transformation law:

$$\begin{aligned}
E_k(-1/\tau) &= \tau^k E_k(\tau), & k \geq 4 \tag{3.11}\\
E_2(-1/\tau) &= \tau^2(E_2(\tau) + 12/2\pi i \tau). \tag{3.12}
\end{aligned}$$

There is no analog of this at finite N; it is a non-trivial consequence of the quasi-relativistic nature of the degrees of freedom in the large N limit. Indeed, in terms of the original (unrescaled) couplings this is the transformation $g^2 A/N \to N/g^2 A$, very different from familiar strong-weak coupling duality. It is a general property of the torus answers, from (3.6) we see that $F_{g \to 1}/A^{2(g-1)}$ will "almost" be a modular form of weight $6(g-1)$.

It is very tempting to look for a string theory interpretation of this.[c] It is a target space duality invariance, like the $R \to 1/R$ of the free compact boson CFT. In fact if the world-sheet embedding was described by a free (complex) boson we would expect precisely this symmetry.[d] Furthermore, if we found that the (world-sheet) genus g free energy was precisely a modular form with weight proportional to $g - 1$, we could define

---

[c]The following points were developed in discussions with D. Gross and C. Vafa.

[d]See [8] for a more precise version of this.

a combined transformation on area and string coupling which left the total free energy invariant, just as was the case for the compactified $c = 1$ fundamental string. [20]

We should ask whether by changing definitions or modifying YM_2 slightly we could get a truly modular covariant answer. From (3.3), it seems most promising to consider the $SU(N) \times U(1)$ case, though we have no deep understanding of why this choice of group should be better than $SU(N)$, say. Adding a sum over momenta to complement the sum over integer windings appropriate for $U(N)$ does not seem promising, because the Hamiltonian has a term p^3, which would be unbounded below.

We would then like to make two changes: first, extend τ to a complex parameter; second, extend the contour integrals in (3.6) to integrals $\int d^2\nu$. Ways to accomplish the first have been proposed by several physicists. Since the area controls $L_0 + \overline{L}_0$, we need to combine it with a parameter which controls $L_0 - \overline{L}_0$. Now in $D = 2$ there is a theta term for $U(1)$ but not for $SU(N)$. For the $U(N)$ theory we could just add the $U(1)$ theta term, which would weigh the contribution of charge q in the sum over representations by $e^{iq\theta}$. Modding out by $\mathbb{Z}_N$ correlates this with the conjugacy class in $SU(N)$, which in the large N limit is just $L_0 - \overline{L}_0$. However this does not work for $SU(N) \times U(1)$, where there is no correlation. Instead we could consider twisted boundary conditions for the gauge field on the original torus. A twist by a group element C identifies holonomy U at time t with holonomy CU at time 0; gauge invariance $\psi(U) = \psi(gUg^{-1})$ requires that C be in the center, so there are N possible twists $\theta = 0, 1/N, \ldots$ which act as $\exp 2\pi i\theta(L_0 - \overline{L}_0)$. In the large N limit we can consider $\tau = it + \theta$ to be a continuous variable.

The other modification seems necessary if we want a modular covariant answer, but so far we have no real justification for doing this from the YM_2 point of view. Naively we would say that since H_I is conserved we have $A \int d\nu H_I = \int d^2\nu H_I$ but since (3.10) is not modular covariant there must be a subtlety. It is that we ignored short distance singularities. Integrating a meromorphic function $\int d^2\nu f(\nu)$ will not produce divergences but we might get a finite piece which fixes things up. In fact, for $f(\nu)$ with a singularity only at zero, we can write

$$\frac{1}{2i} \int d\nu f(\nu) \wedge d(\overline{\nu} - \nu) = \operatorname{Im} \tau \int_0^1 d\nu f(\nu) + \pi \operatorname{Res} \nu f(\nu)|_{\nu=0} \qquad (3.13)$$

and with this term

$$F_{2\to1}^{inv} = \frac{A^2}{2^5 \cdot 3^4 \cdot 5} \operatorname{Re} \left(10\widehat{E}_2^3 - 6\widehat{E}_2 E_4 - 4E_6\right) \qquad (3.14)$$

where $\widehat{E}_2(\tau) = E_2(\tau) - 3/\pi\operatorname{Im} \tau$ is modular covariant. The diagrams mentioned just above (3.9) now vanish. This is different from (3.10) at $O(A)$ and now there are also terms at $O(A^0)$ and $O(A^{-1})$ whose string interpretation is unclear. Furthermore, while the "chiral" and "antichiral" parts are separately modular covariant, the combination appearing here is a bit strange.

We could certainly imagine other modifications of (3.10) to get modular covariance, but this modification does generalize to all genus and seems relatively natural. A modular covariant answer is a prerequisite for comparison with topological string theory.

There is a striking similarity between the form of the collective field theory (2.24) and the Kodaira-Spencer field theory developed in [21], describing topological string theory on a Calabi-Yau target space, which may be an important clue to the continuum string interpretation. [22]

4 YM$_2$ on Other Riemann Surfaces

Before discussing the formalism let us review some known results for the large N partition function on other topologies. The qualitative structure is quite clear from the expression as a sum over representations,

$$\exp F_{\text{all}\to\text{G}} = \sum_R (\dim R)^{2-2G} \exp -\frac{A}{N} C_2(R), \tag{4.1}$$

and the leading order behavior $\dim R \sim N^{n+\overline{n}}$ for a representation made from n and $\overline{n}$ (anti)fundamentals.

For $G > 1$, at order $1/N^{2s}$ only the finite set of representations with $n + \overline{n} \le s$ can contribute. Although there are interesting things to say about a string theory which reproduces these answers, since no observable exhibits a sum over an infinite number of degrees of freedom, there is no evidence for any non-topological field theory description. In the present framework we will see that although formally we can build higher genus surfaces by sewing cylinders, the "vertex" which accomplishes this is highly non-local.

For $G = 0$, there is an $O(N^2)$ contribution to F. This is in many ways the most interesting case, because the leading behavior of the free energy for higher dimensional Yang-Mills theory is $O(N^2)$. It is quite possible that the qualitative behavior of the two-dimensional problem is of direct relevance for higher dimensions, because any higher dimensional space (and certainly a lattice as well) will contain embedded two-spheres.

Computing an $O(N^2)$ free energy in large N is quite different from the problems we treated above, because we can hope to find a saddle point which dominates the path integral. Of course this is usually the reason we think that a large N limit will simplify a problem; however it brings with it a complication: there can be more than one saddle point, and a phase transition where their actions are equal.

In terms of our present formalism we would expect to describe the saddle point in terms of an expectation value for the boson ϕ, determined by minimizing the Jevicki-Sakita action (2.24). Although this makes sense it is much easier to describe the saddle point in terms of the conjugate variables n_i, since these are time-independent. A new feature of the problem is that since these are discrete, there is an upper bound on their density, and an associated phase transition when the bound is saturated.[23]

To return to our Hamiltonian formalism, the problem in this section is to find the wave functions for the disk and the three-holed sphere. We might expect to be able to represent them as simple conformal field theory states, as is done in string field theory. Let us start with the disk, or its zero-area limit, in other words the wave function $\psi = \delta(U)$. Clearly acting on this state $\text{Tr } U^n = N \ \forall n \ne 0$, or

$$(\alpha_n + \overline{\alpha}_{-n} - N)|D_0> = 0. \tag{4.2}$$

These constraints are easily solved:

$$|D_0> = \exp\ -\sum_{n\geq 1} \frac{1}{n}(\alpha_{-n}\overline{\alpha}_{-n} - N\alpha_{-n} - N\overline{\alpha}_{-n})\ |0>. \tag{4.3}$$

This state can be used to calculate the dimension of a representation:

$$\dim R_{\vec{n}} = <D_0|\vec{n}>. \tag{4.4}$$

(try for example the characters $\frac{1}{2}((\mathrm{Tr}\ U)^2 + \mathrm{Tr}\ U^2)$.)

Another characterization of this state is through boundary conditions of the fields – we put a boundary $\tau = 0$ with $\partial\phi/\partial\tau = N\delta(\theta)$. This is equivalent to taking Neumann boundary conditions, and inserting the operator $:\exp iN\phi(0):$. In the Fermi language this is a rather singular state, which is defined by the boundary condition that all the fermions (eigenvalues) are at $z = 1$. In the non-relativistic formalism, we can produce it by taking the o.p.e. of N fermions, resulting in the state $\prod_{i=0}^{N-1}\partial^i\psi\ |0>$. Taking the inner product of (for example) a character with this state will reproduce the calculation of the dimension of a representation by taking the limit of all $z_i \to 1$ in the Weyl character formula using l'Hôpital's rule.

The YM$_2$ sphere free energy is then

$$
\begin{aligned}
\exp F_{\mathrm{all}\to 0} &= <D_0|e^{-AH}|D_0> \\
&= <: e^{-iN\phi(\tau)}:\ e^{\frac{t}{N}H_I}\ : e^{iN\phi(0)}:>_{\mathrm{cyl}}
\end{aligned}
\tag{4.5}
$$

a correlation function on a cylinder of length τ. We see that we cannot expand $\exp\frac{t}{N}H_I$; the $1/N$ is compensated in correlation functions by the explicit N in the vertex operators. On the other hand the problem is classical; we can rescale $\phi \to N\phi$ and get an overall N^2 in the exponential.

The qualitative behavior of the dominant classical solution is clear. We have a fluid with a conserved particle number; the boundary condition is that at time $t = 0$ it is concentrated at $\theta = 0$; as it evolves it spreads out but must recontract from $t = \tau/2$ on to meet the boundary condition at $t = \tau$. This is not the easiest way to solve the problem, which is treated more simply in [23], but it does make the nature of the large N transition clear: at a critical τ_c, the two edges of the particle distribution meet at $t = \tau_c/2$ and $\theta = \pi$. For $\tau \leq \tau_c$ they do not know they live on a circle; beyond this value it is crucial, and the classical solution is a non-analytic function of τ at this point. (See also [24].)

The expansion of [6] reproduces the $\tau \geq \tau_c$ behavior but has no apparent relation with the $\tau < \tau_c$ behavior. Putting back the g^2 dependence shows that it is valid at strong coupling. Now the main reason for the transition was simply the compactness of the gauge group, just as for the Gross-Witten transition [25], which suggests that higher dimensional theories will have the same problem. This is not just a property of an approximation but of the true continuum result, and the dynamical picture above suggests the strange possibility that in higher dimensions we could have non-analyticity in the local observables at large N, for example in the dependence of a Wilson loop expectation value on its length. (This is not the case in $D = 2$, see [26]).

To try to understand the effects of the various world-sheet features of [6], we could ask how modifying them or removing them affects the answers. Comparison with the derivation there shows that the Ω-points of [6] are represented here by the states $|D_0>$ while the "movable branch points" (those which come with a weight proportional to the area of the world-sheet) correspond to insertions of the interaction H_I (as we saw above for the torus target space). On the sphere we get an interesting result even after dropping the movable branch points. [e] Since we could accomplish this by multiplying H_I by a new coupling constant and taking it to zero, the resulting theory is in a sense continuously connected with the standard YM_2. In fact this modification corresponds to a local modification of the original YM_2 action. [27] Thus it is not inconceivable that the corresponding modification of a $D > 2$-dimensional string theory would be in the same universality class as YM_D. We might take this as a working hypothesis in interpreting string rules like those of [6, 7]: features we cannot change by a local modification of the original Yang-Mills action (like the Ω points) are more likely to be relevant for physics than features (like the movable branch points) that can be so changed.

What makes it all the more interesting is that in $D = 2$ this modification pushes the transition to zero coupling! For simplicity drop the winding modes of the boson, so this result exactly enumerates covers of the sphere branched only at the Ω points. Correlation functions on the cylinder with Neumann boundary conditions are equal to those of a chiral boson on the doubled surface, here a torus of modulus 2τ. Then

$$
\begin{aligned}
\exp F^{\text{modified}}_{\text{all}\to 0} &= <: e^{-iN\phi(\tau)} : \, : e^{iN\phi(0)} :>_{\text{cyl}} \\
&= \frac{1}{\eta(2\tau)} \exp N^2 \left[< \phi(\tau)\phi(0) > - \lim_{z\to 0}(< \phi(z)\phi(0) > + \log z) \right] \\
&= \frac{1}{\eta(2\tau)} \exp -N^2 \log \frac{\theta_1(\tau|2\tau)}{\theta_1'(0|2\tau)} \\
&= \prod_{m\geq 1} \frac{1}{(1-q^{2m})} \left(\frac{1-q^{2m}}{1-q^{2m-1}} \right)^{2N^2} .
\end{aligned}
\tag{4.6}
$$

This is an analytic function for all real $\tau > 0$, with an expansion in $\exp -\tau$ whose terms have a string representation, and whose $\tau \to 0$ behavior is remarkably similar to the "correct" (heat kernel) weak coupling behavior $F \sim -\frac{1}{2}N^2 \log \tau$. Whether this is of any relevance to $D > 2$ is unclear, and the action which defines this theory is quite peculiar, but this suggests that the heat kernel action might not be the last chapter in the story.

We briefly return to $G > 1$. We can build higher dimensional surfaces by sewing, and the appropriate "pants" vertex is well known in the character language:

$$
< V_3| \, |\psi_1 > \, |\psi_2 > \, |\psi_3 > = \int dU dV dW \, \psi_1(UVU^{-1}W)\psi_2(V^{-1})\psi_3(W^{-1})
\tag{4.7}
$$

$$
|V_3 > = \sum_R \frac{1}{\dim R} |\chi_R >_1 \, |\chi_R >_2 \, |\chi_R >_3 .
\tag{4.8}
$$

Since this expression involves group multiplication, a Ward identity analogous to (4.2) would be highly non-local in our two-dimensional auxiliary space. The result is clearly

---

[e]This modification has also been considered by D. Gross.

non-local at each order in $1/N$ as well; it is a sum over finitely many states which separately have no local definition. All this and consideration of what sort of vertex could satisfy the equation $< D_0|V_3 >= 1$ leads to the conclusion that there is not likely to be an expression for the vertex much simpler than (4.8).

A final comment, which we will not apply here (but see [28]). Clearly there are many other group theoretic ideas we could try to fit into this formalism. One interesting one is multiplication of characters, which allows computing tensor product decompositions. Evidently it is much easier to multiply symmetric functions expressed in the basis of power sums (2.19), so this will be a simpler operation in the bosonic language. Since we are multiplying wave functions at the same point on the group manifold, we expect this to be a local operation in our two dimensional CFT language. In fact we can write multiplication of n wave functions as a vertex $|V_n >$, determined by the condition that

$$< \psi_1| < \psi_2| \ldots \mathrm{Tr}\, U_i^k |V_n >= \int dU \,\mathrm{Tr}\, U^k \prod_i \psi_i(U) \tag{4.9}$$

is the same no matter which wave function we multiply by $\mathrm{Tr}\, U^k$. In conformal field theory terms this means that we have a boundary on which n cylinders meet, and the local boundary condition

$$\partial_\tau \phi_i(z) = \partial_\tau \phi_j(z) \qquad \forall i, j. \tag{4.10}$$

This condition only couples mode n to $-n$ and is easy to translate into an oscillator expression for the vertex

$$|V_n >= \exp \sum_{i \neq j} \sum_{n \geq 1} \frac{1}{n} \alpha_{-n}^{(i)} \overline{\alpha}_{-n}^{(j)} \,\, |0 > . \tag{4.11}$$

In a certain sense this is a trivial result; however it would be quite amusing to translate it into the fermionic language, because it would amount to a new proof of the Littlewood-Richardson rule for tensor product decompositions, at least in the large N limit.

Some contour integrals used in section 3: [19]

$$\int_0^1 d\nu\; \partial_\nu^2 \log \theta_1(\nu) = 0$$

$$\int_0^1 d\nu\; \wp(\nu)^2 = \frac{\pi^4}{9} E_4$$

$$\int_0^1 d\nu\; \wp(\nu)^3 = \frac{4\pi^6}{5 \cdot 27} E_6 - \frac{\pi^6}{15} E_4 E_2$$

Acknowledgments

I thank T. Banks, G. Moore, D. Gross, V. Kazakov, J. Minahan, J. Polchinski, R. Rudd, A. Strominger, and C. Vafa for enjoyable discussions, and DOE grant DE-FG05-90ER40559, NSF grants PHY-9157016 and PHY89-04035, and the Sloan Foundation for their support.

References

[1] M. Jimbo and T. Miwa, in Integrable Systems in Statistical Mechanics, eds. G. M. D'Ariano, A. Montorsi and M. G. Rasetti, 1985; World Scientific, Singapore.

[2] Loop Groups, A. Pressley and G. Segal, Oxford, 1986; most relevant to the present discussion are chapters 10 and 14.3.
Bombay lectures on highest weight representations of infinite dimensional Lie algebras, V. G. Kac and A. K. Raina, World Scientific, 1987.

[3] M. Stone, Phys. Rev. B42 (1990) 8399.
A. Jevicki, Nucl. Phys. B376 (1992) 75.
R. Dijkgraaf, G. Moore and R. Plesser, Nucl. Phys. B394 (1993) 356.

[4] A good review is by J. Polchinski, presented at the Symposium on Black Holes, Wormholes, Membranes and Superstrings, Houston, TX, Jan 16-18, 1992, hep-th/9210045.

[5] V. A. Kazakov and I. K. Kostov, Nucl. Phys. B176 (1980) 199.

[6] D. J. Gross and W. Taylor, Nucl. Phys. B400 (1993) 181 and Nucl. Phys. B403 (1993) 395.

[7] I. K. Kostov, SACLAY-SPHT-93-050, June 1993. hep-th/9306110.

[8] M. Bershadsky, S. Cecotti, H. Ooguri, and C. Vafa, Nucl. Phys. B405 (1993) 279.

[9] R. Dijkgraaf and R. Rudd, unpublished.

[10] See J. Ambjorn et. al., Nucl. Phys. B270 (1986) 457 and references there.

[11] E. Brézin, C. Itzykson, G. Parisi and J.-B. Zuber, Comm. Math. Phys. 59 (1978) 35.

[12] Harish-Chandra, Amer. J. Math. 79 (1957) 87-120; for a nice pedagogical treatment which continues with non-compact groups see Non-Abelian Harmonic Analysis, R. Howe and E. C. Tan, Springer-Verlag, 1992.

[13] J. S. Dowker, J. Phys. A (1970) 451.

[14] Compact Lie Groups and their Representations, D. P. Zelobenko, AMS translations vol. 40, AMS, 1973.

[15] Statistical Field Theory, C. Itzykson and J.-M. Drouffe, Cambridge, 1989.

[16] D. J. Gross and I. Klebanov, Nucl. Phys. B352 (1990) 671;
A. M. Sengupta and S. R. Wadia, Int. J. Mod. Phys. A6 (1991) 1961.

[17] S. R. Wadia, Phys. Lett. 93B (1980) 403.

[18] J. A. Minahan and A. P. Polychronakos, Phys. Lett. B312 (1993) 155.

[19] *Introduction to Elliptic Curves and Modular Forms,* N. Koblitz, Springer-Verlag, 1984.
Table of Integrals, Series and Products, I. S. Gradshteyn and I. M. Ryzhik, Academic Press, 1980.

[20] D. J. Gross and I. Klebanov, Nucl. Phys. B344 (1990) 475.

[21] M. Bershadsky, S. Cecotti, H. Ooguri and C. Vafa, HUTP-93-A025, Sep. 1993. hep-th/9309140.

[22] C. Vafa, private communication.

[23] M. R. Douglas and V. A. Kazakov, to appear in Phys. Lett. B; see also Kazakov's lecture in this volume.

[24] J. A. Minahan and A. P. Polychronakos, CERN-TH-7016-93, Sept. 1993. hep-th/9309119.
M. Caselle, A. D'Adda, L. Magnea, S. Panzeri, DFTT-50-93, Sept. 1993. hep-th/9309107.

[25] D. J. Gross and E. Witten, Phys. Rev. D21 (1980) 446.

[26] J.-M. Daul and V. A. Kazakov, LPTENS-93-37, Oct. 1993. hep-th/9310165.
D. V. Boulatov, NBI-HE-93-57, Oct. 1993. hep-th/9310041.

[27] For a discussion of such modifications, see W.-D. Zhao, PUPT-1390, Mar. 1993, or upcoming work by M. R. Douglas, K. Li, and M. Staudacher.

[28] M. R. Douglas, RU-93-13, NSF-ITP-93-39, Mar. 1993. hep-th/9303159.

INTRODUCTION TO DIFFERENTIAL
W-GEOMETRY

Jean-Loup GERVAIS

Laboratoire de Physique Théorique de
l'École Normale Supérieure
24 rue Lhomond, 75231 Paris CEDEX 05, France

Abstract: Ideas recently put forward by Y. Matsuo and the author are summarized on
the example of the simplest (W_3) generalization of two-dimensional gravity.

1   Introduction

In many ways, W-algebras are natural generalizations of the Virasoro algebra. They
were first introduced as consistent operator-algebras involving operators of spins higher
than two[1]. Moreover, the Virasoro algebra is intrinsically related with the Liouville
theory which is the Toda theory associated with the Lie algebra A_1, and this relationship
extends to W-algebras which are in correspondence with the family of conformal Toda
systems associated with arbitrary simple Lie algebras[2]. Another point is that the deep
connection between Virasoro algebra and KdV hierarchy has a natural extension[3] to
W-algebras and higher KdV (KP) hierarchies[4]. On the other hand, W symmetries
exhibit strikingly novel features. First, they are basically non-linear algebras. Since the
transformation laws of primary fields contain higher derivatives, product of primaries are
not primaries at the classical level. Naive tensor-products of commuting representations
do not form representations. A related novel feature is that W-algebras generalize the
diffeomorphisms of the circle by including derivatives of degree higher than one. Going
beyond linear approximation (tangent space) is a highly non-trivial step. Taking
higher order derivatives changes the shape of the world-sheet in the target-space, thus
W-geometry should be related to the extrinsic geometry of the embedding. Finally,
Virasoro algebras are notoriously related to Riemann surfaces. The W-generalization
of the latter notion is a fascinating problem.

In a series of recent papers, we have developed a geometrical framework for the class
of Conformal field Theory CFT mentioned above, where these features emerge from the

Quantum Field Theory and String Theory, Edited by
L. Baulieu *et al.*, Plenum Press, New York, 1995

standard Riemannian geometry of particular manifolds which we called W surfaces[5]–[7]. These references cover quite a lot of material, and the present lecture will go in the opposite way. Leaving the description of the general scheme to refs[5]–[7], we shall, instead, illustrate the ideas by two explicit examples: the 2D gravity case (section 2), and its simplest generalization to the W_3 gravity (section 3). In both cases, we discuss the two current approaches, namely, the conformal one where W-gravity is identified with a conformal Toda (or Liouville theory for 2D gravity) theory, and the light-cone approach.

Before starting, however, let us recall how our general scheme goes. One basic point is to make use of the fact that one deals with integrable models, but geometrically there are two aspects. The first uses extrinsic geometry. In ref.[6], we showed that the A_n-W–geometry corresponds to the embedding of holomorphic two-dimensional surfaces in the complex projective space CP^n. These (W) surfaces are specified by embedding equations of the form $X^A = f^A(z)$, $\overline{X}^A = \overline{f}^A(\overline{z})$, where z, and $\overline{z}$ are the two surface-parameters. The fact that they are functions of a single variable is equivalent to the Toda field-equations, so that this describes W gravity in the conformal gauge. These functions have a natural extension to CP^n using the higher variables z^k, $\overline{z}^k$ of the Toda hierarchy of integrable flows, and this provides a local parametrization of CP^n. The original variables z and $\overline{z}$ are identified with z^1, $\overline{z}^1$, respectively. For the embedding functions the extension is such that they become functions of half of the variables noted $f^A([z]) = f^A(z^0, \cdots z^n)$, and $\overline{f}^A([\overline{z}]) = \overline{f}^A(\overline{z}^0, \cdots \overline{z}^n)$ such that

$$\frac{\partial f^A([z])}{\partial z^k} = \frac{\partial^k f^A([z])}{(\partial z)^k}, \qquad \frac{\partial \overline{f}^A([\overline{z}])}{\partial \overline{z}^k} = \frac{\partial^k \overline{f}^A([\overline{z}])}{(\partial \overline{z})^k} \tag{1.1}$$

One main virtue of the coordinates z^k, $\overline{z}^k$ is that, due to the last equations, higher derivatives in z and $\overline{z}$ are changed to first-order ones, and this is how our geometrical scheme gets rid of the troublesome higher derivatives of the usual approaches.

The second aspect[5][6] only makes use of intrinsic geometries, but introduces a family of associated surfaces in the standard Grassmannians associated with CP^n. This is useful to discuss global aspects by using the fact developed in ref.[6], that W surfaces are instantons of non-linear σ models. We shall not dwell into this aspect in these lectures.

So far this is only for the conformal gauge. Concerning the light-cone approach, our recent insight[7] is that, for any Kähler manifold, there exist changes of coordinates such that metric tensors of the light-cone type come out. This allows us to relate conformal and light-cone descriptions by diffeomorphisms.

Finally, let us stress that we shall remain entirely at the level of classical field theory, thus describing only the $C \to \infty$ limit of the problem. The quantum approach to Toda theories is making steady progress[2]–[20], [18]–[28] but its connection with the present geometrical scheme remains as a fascinating problem for the future.

2 Two-dimensional Gravity

In this part we discuss the case of two-dimensional gravity in some details, using methods that will be later on generalized to W gravity. With general world-sheet parameters ξ, the Weyl anomaly takes the form

$$S = \frac{1}{\gamma} \int d_2\xi (\frac{1}{2}\mathcal{R}\frac{1}{\Delta}\mathcal{R} + \mu\sqrt{-\mathcal{G}}) \tag{2.1}$$

In this expression, Δ is the Laplacian with the 2D-gravity metric $\mathcal{G}_{\alpha,\beta}$, γ is the coupling constant, $\mathcal{R}$ the scalar curvature, and μ the cosmological constant.

1 Conformal gauge

First, choose conformal coordinates z and $\bar{z}$, so that the arc length takes the form[a]

$$ds^2 = 2e^{2\Phi}dzd\bar{z}, \tag{2.2}$$

the Weyl anomaly becomes the Liouville action

$$S = \frac{1}{\gamma} \int d_2z (-\frac{1}{2}\partial\Phi\bar{\partial}\Phi + \mu e^{2\Phi}). \tag{2.3}$$

In general ∂ and $\bar{\partial}$ denote $\partial/\partial z$, and $\partial/\partial\bar{z}$ respectively. At the level of the present discussion we may always set $\mu = 1$ by a shift of ϕ, and we shall do so from now on. The general solution of the Liouville equation is given by the holomorphic decomposition

$$e^{-\Phi} = \sum_{j=1}^{2} \chi_j(z)\bar{\chi}_j(\bar{z}). \tag{2.4}$$

The functions $\chi_j(z)$, and $\bar{\chi}_j(\bar{z})$ are pairs of solutions of the differential equations

$$-\frac{d^2\chi_j(z)}{(dz)^2} + T(z)\chi_j(z) = 0, \quad -\frac{d^2\bar{\chi}_j(\bar{z})}{(d\bar{z})^2} + \overline{T}(\bar{z})\bar{\chi}_j(\bar{z}) = 0, \tag{2.5}$$

where T and $\overline{T}$ are the two non-vanishing components of the stress-energy tensor. They are normalized so that

$$\chi_1(z)\frac{d\chi_2(z)}{dz} - \chi_2(z)\frac{d\chi_1(z)}{dz} = \bar{\chi}_1(\bar{z})\frac{d\bar{\chi}_2(\bar{z})}{d\bar{z}} - \chi_2(\bar{z})\frac{d\chi_1(\bar{z})}{d\bar{z}} = 1. \tag{2.6}$$

At this point, and in the following, we need a simple mathematical lemma concerning differential equations, which we state once for all next.

Lemma 1 *Automatic differential equation.*

Consider N functions f^A of a single variable x, whose Wronskian does not vanish. They satisfy a differential equation of the form

$$f^{(N)A}(x) = \sum_{\ell=1}^{N} \kappa_\ell(x)f^{(N-\ell)A}(x). \tag{2.7}$$

---

[a]This may not be possible globally, but we shall only consider the local aspects

Proof: Recall the definition of the Wronskian:

$$\mathrm{Wr}(f^1, \cdots, f^N) \equiv \begin{vmatrix} f^1 & \cdots & f^N \\ f^{(1)\,1} & \cdots & f^{(1)\,N} \\ \vdots & \cdots & \vdots \\ f^{(N-1)\,1} & \cdots & f^{(N-1)\,N} \end{vmatrix}. \tag{2.8}$$

In this formula, and hereafter, upper indices in between parentheses denote the order of derivatives. Clearly, any of the functions f^A may be trivially written as a linear combination of the type $\sum_B c^A_B f^B$. It is then clear that

$$\begin{vmatrix} f^1 & \cdots & f^N & f \\ f^{(1)\,1} & \cdots & f^{(1)\,N} & f^{(1)} \\ \vdots & \cdots & \vdots & \\ f^{(N)\,1} & \cdots & f^{(N)\,N} & f^{(N)} \end{vmatrix} = 0.$$

Expanding this determinant with respect to the last column immediately gives Eq.2.7. **Q.E.D.** Moreover, one immediately sees that

$$\kappa_1 = \frac{d \ln[\mathrm{Wr}(f^1, \cdots, f^n)]}{dx} \tag{2.9}$$

Returning to our main line, we see that, in the differential equations Eq.2.5 the first order term vanishes, so that the Wronskians $\mathrm{Wr}(\chi)$ and $\mathrm{Wr}(\overline{\chi})$ are constant, and may be chosen equal to one (see Eq.2.6). For our geometrical description this is not appropriate, however. The basic reason is as follows. Under conformal transformations $\delta z = \epsilon(z)$, the χ_j fields[b] transform as primary fields of weight $-1/2$ (that is such that $\chi_j(z)\,(dz)^{-1/2}$ is invariant). This is consistent, since it follows that the Wronskian $\mathrm{Wr}(\chi)$ transforms with weight zero, so that condition Eq.2.6 is conformally invariant. In our geometrical description, these functions will become geometrical objects whose form should not change under conformal transformation, so that they should transform with weight zero. We may change the conformal weights by using a projective description. For this we define

$$f^A = \sqrt{w(z)}\,\chi_{A+1}, \quad \overline{f}^A = \sqrt{\overline{w}(\overline{z})}\,\overline{\chi}_{A+1}, \quad A = 0,\,1, \tag{2.10}$$

so that

$$\mathrm{Wr}(f^0, f^1) = w(z), \quad \mathrm{Wr}(\overline{f}^0, \overline{f}^1) = \overline{w}(z) \tag{2.11}$$

where w and $\overline{w}$ are arbitrary functions of a single variable. Now the f's satisfy a more general differential equation of the type Eq.2.7, where $\kappa_1 \neq 0$. Substituting into Eq.2.4, we derive the arc length

$$ds^2 = dz d\overline{z}\,\frac{\mathrm{Wr}(f^0, f^1)\mathrm{Wr}(\overline{f}^0, \overline{f}^1)}{(f^0 \overline{f}^0 + f^1 \overline{f}^1)^2}, \tag{2.12}$$

---

[b]In the conformal gauge χ and $\overline{\chi}$ are on the same footing. When we discuss properties of chiral component we some times talk about the χ fields as an example. Clearly, the $\overline{\chi}$ fields are analogous.

and, re-arranging the terms, we may write

$$ds^2 = dz d\bar{z} \left(\frac{-(f^{(1)0}\bar{f}^{(1)0} + f^{(1)1}\bar{f}^{(1)1})(f^0\bar{f}^0 + f^1\bar{f}^1)}{(f^0\bar{f}^0 + f^1\bar{f}^1)^2} + \right.$$

$$\left. \frac{(f^{(1)0}\bar{f}^0 + f^{(1)1}\bar{f}^1)(f^0\bar{f}^{(1)0} + f^1\bar{f}^{(1)1})}{(f^0\bar{f}^0 + f^1\bar{f}^1)^2} \right). \tag{2.13}$$

As is well-known $\mathrm{Wr}(\rho f^0, \rho f^1) = \rho^2 \mathrm{Wr}(f^0, f^1)$, for arbitrary function ρ, so that[c] Eq.2.12 is invariant under $f^A \to \rho f^A$, and $\bar{f}^A \to \bar{\rho} \bar{f}^A$. We arrive at a projective structure, following our general scheme[6] where A_n-W geometries are described in CP^n (we will have $n = 1$ in the present case). Let us recall some useful definitions at this point.

Definition 1 *Kähler manifolds*

A Kähler manifold of real dimension $2n$ is complex manifold with a special class of coordinates X^A, $X^{\bar{A}}$, $1 \le A, \bar{A} \le n$, such that the only components of the metric are $G_{A\bar{B}} = G_{\bar{B}A}$, and

$$G_{A\bar{B}} = \partial_A \partial_{\bar{B}} K, \tag{2.14}$$

where K is the Kähler potential.

In general we denote the differential operators $\partial/\partial X^A$, and $\partial/\partial X^{\bar{B}}$ by ∂_A, and $\partial_{\bar{B}}$.

Definition 2 *Complex projective space CP^n*

The complex projective space CP^n is defined from the trivial complex space C^{n+1} with coordinates X^A, $\overline{X}^{\bar{A}}$, $A, \bar{A} = 0, \cdots n$, by identifying any two points $X, \overline{X}$, and Y, $\overline{Y}$ related by the scale transformation

$$X^A = Y^A \rho, \quad \text{and} \quad \overline{X}^{\bar{A}} = \overline{Y}^{\bar{A}} \bar{\rho}. \tag{2.15}$$

This is a simplest non-trivial example of a Kähler manifold. The metric on this space is given by the Fubini-Study equation

$$G_{A\bar{A}} = \left(\delta_{A\bar{A}} \left(\sum_{B=0}^{n} X^B \overline{X}^B \right) - X^{\bar{A}} \overline{X}^A \right) \Big/ (\sum_{B=0}^{n} X^B \overline{X}^B)^2, \tag{2.16}$$

whose Kähler potential is given by

$$K = \ln \sum_{A=0}^{n} X^A \overline{X}^{\bar{A}}. \tag{2.17}$$

Eq.2.16 is invariant under the scale transformation Eq.2.15. Using this freedom, it is customary to parametrize CP^n, by letting one component (say, X^0) equal 1.

Closing the parenthesis, we return to Eq.2.13. Consider the space CP^1. The equations

$$X^A = e^{\zeta} f^A(z), \quad \overline{X}^A = e^{\bar{\zeta}} \bar{f}^A(\bar{z}), \quad A = 0, 1, \tag{2.18}$$

---

[c]this is obviously true by construction.

define a holomorphic change of coordinates in C^2 from X^A to ζ, z. It is easy to see that Eq.2.13 is equivalent to

$$G_{z\bar{z}} \equiv e^{2\Phi} = f^{(1)A}\overline{f}^{(1)\overline{B}}G_{A\overline{B}}\Big|_{X^A=f^A,\overline{X}^{\overline{A}}=\overline{f}^{\overline{A}}} \tag{2.19}$$

By construction, Eq.2.13 is invariant under the rescaling $f \to \rho(z)f$, $\overline{f} \to \overline{\rho}(\overline{z})\overline{f}$. Thus we may let $\zeta = \overline{\zeta} = 0$, and the geometrical meaning of Eqs.2.18, Eq.2.19 is that $z\ \overline{z}$ are parameters of CP^1, such that the Liouville exponential $G_{z,\bar{z}}$ is equivalent to the metric tensor of Fubini-Study. Thus any solution of Liouville equation defines a local holomorphic parametrization of CP^1. Note that, since the change of coordinate is holomorphic, the Kähler condition is preserved. Indeed, one has

$$G_{z\bar{z}} = \partial\overline{\partial}\mathcal{K}, \quad \mathcal{K} = \ln(f^0\overline{f}^0 + f^1\overline{f}^1). \tag{2.20}$$

2 Light-cone gauge

1 The Weyl anomaly

First, rederive the ideas of ref.[8] is the present context. At the level of Eq.2.1, one changes coordinates from $z\ \overline{z}$ to $u\ \overline{u}$, such that

$$e^{2\Phi}dzd\overline{z} = dud\overline{u} + hdu^2 \tag{2.21}$$

The Weyl anomaly Eq.2.1 becomes

$$S = \frac{1}{2\gamma}\int d_2u\left((\overline{D}^2 h)\frac{1}{D\overline{D} - \overline{D}h\overline{D}}(\overline{D}^2 h) + 2\right), \tag{2.22}$$

where D and $\overline{D}$ stand for $\partial/\partial u$, and $\partial/\partial\overline{u}$, respectively. Here the second term is a constant and can be forgotten. Using the obvious relation

$$\frac{1}{D\overline{D} - \overline{D}h\overline{D}} = \frac{1}{D - h\overline{D}}\frac{1}{\overline{D}}$$

we may write after partially integrating

$$S = \frac{1}{\pi}\int d_2u\, h\, \mathcal{T}, \quad \mathcal{T} \equiv \frac{\pi}{2\gamma}\overline{D}^2\frac{1}{D - h\overline{D}}\overline{D}h \tag{2.23}$$

Starting from this expression, we compute

$$\left(D - h\overline{D} - 2(\overline{D}h)\right)\mathcal{T} = \frac{\pi}{2\gamma}\overline{D}^2\frac{D - h\overline{D}}{D - h\overline{D}}\overline{D}h$$

so that we get

$$\left(D - h\overline{D} - 2(\overline{D}h)\right)\mathcal{T} = \frac{\pi}{2\gamma}\overline{D}^3 h \tag{2.24}$$

Thus the deformed conservation law are satisfied. Minimizing S, under the form Eq.2.23, with respect to h, one finds

$$0 = \frac{\pi}{2\gamma}\overline{D}^2\frac{1}{D - h\overline{D}}\overline{D}h = \mathcal{T}$$

so that Polyakov's anomaly equation comes out:

$$\overline{D}^3 h = 0 \tag{2.25}$$

2 Liouville dynamics

Next, following ref.[7], we connect the results just recalled with the conformal-gauge Liouville appproach recalled above. Since the Liouville equations also follow from minimizing S, they are equivalent to Polyakov's anomaly equations. This may be seen as follows. Making use of Eq.2.21, one sees that

$$h = -\frac{\partial \overline{u}}{\partial z}, \quad e^{2\Phi} = \frac{\partial \overline{u}}{\partial \overline{z}},$$

$$(2.26)$$

The last equation is immediately solved, thanks to Eq.2.20, and we find that the change of variables form conformal to light-cone explicitly reads

$$u = z, \quad \overline{u} = \partial K = \frac{f^{(1)0}\overline{f}^0 + f^{(1)1}\overline{f}^1}{f^0\overline{f}^0 + f^1\overline{f}^1}$$

$$(2.27)$$

Computing h in terms of f^1, and $\overline{f}^1$ one finds.

$$h = -\frac{f^{(2)0}\overline{f}^0 + f^{(2)1}\overline{f}^1}{f^0\overline{f}^0 + f^1\overline{f}^1} + \left[\frac{f^{(1)0}\overline{f}^0 + f^{(1)1}\overline{f}^1}{f^0\overline{f}^0 + f^1\overline{f}^1}\right]^2$$

The first term is retransformed by using the lemma which gives $f^{(2)A} = \kappa_1 f^{(1)A} + \kappa_0 f^A$. One gets

$$h = -\kappa_0 - \kappa_1 \overline{u} + \overline{u}^2$$

$$(2.28)$$

For fixed $z = u$, it is quadratic in $\overline{u}$, so that Eq.2.25 is indeed verified. In terms of the $sl(2, R)$ light-cone currents we have $J^+ = 1$, $J^0 = \kappa_1/2$, and $J^- = -\kappa_0$.

3 W_3 gravity

Unfortunately no formula similar to the Weyl-anomaly term Eq.2.1 is known at present. Our starting point will be the generalization of the Liouville equation known as the A_2 Toda equation, where the W_3 algebra is realized by Poisson bracket, and which should thus describe W_3 gravity in the conformal gauge.

1 Conformal gauge

1 W surface

In the A_2 Toda theory there are two fields ϕ_1, ϕ_2. The field equations read

$$\partial\overline{\partial}\phi_1 = -e^{2\phi_1 - \phi_2}, \quad \partial\overline{\partial}\phi_2 = -e^{2\phi_2 - \phi_1}$$

$$(3.1)$$

with solution

$$e^{-\phi_1} = \sum_{j=1}^{3}\chi_j(z)\overline{\chi}_j(\overline{z}), \quad e^{-\phi_2} = \sum_{i<j=1}^{3}\chi_{ij}(z)\overline{\chi}_{ij}(\overline{z})$$

$$(3.2)$$

There are now three pairs of arbitary functions of a single variable χ_j, $\overline{\chi}_j$, the functions χ_{ij}, and $\overline{\chi}_{ij}$ are defined by

$$\chi_{ij} = \mathrm{Wr}(\chi_i, \chi_j), \quad \overline{\chi}_{ij} = \mathrm{Wr}(\overline{\chi}_i, \overline{\chi}_j).$$

$$(3.3)$$

The Toda equations 3.1 are satisfied if

$$\mathrm{Wr}(\chi_1, \chi_2, \chi_2) = 1, \quad \mathrm{Wr}(\overline{\chi}_1, \overline{\chi}_2, \overline{\chi}_2) = 1, \tag{3.4}$$

One sees that the generalization of the Liouville case is very direct. The lemma, combined with the last Wronskian condition shows that the χ_j satisfy a differential equation of the form

$$\chi_j^{(3)} = T\chi_j^{(1)} + W\chi_j \tag{3.5}$$

The potentials are the generators of the W transformations: T is the stress-energy tensor, and W the W_3 charge. Again we relax the condition 3.4 by letting

$$f^A = (w(z))^{1/3}\,\chi_{A+1}, \quad \overline{f}^A = (\overline{w}(\overline{z}))^{1/3}\,\overline{\chi}_{A+1}, \quad A = 0,\,1,\,2 \tag{3.6}$$

so that

$$\mathrm{Wr}(f^0, f^1, f^2) = w(z), \quad \mathrm{Wr}(\overline{f}^0, \overline{f}^1, \overline{f}^2) = \overline{w}(z) \tag{3.7}$$

$$
\begin{aligned}
e^{-\phi_1} &= \frac{\sum_{A=0}^{2} f^A(z)\overline{f}^A(\overline{z})}{\left[\mathrm{Wr}(f^0, f^1, f^2)\mathrm{Wr}(\overline{f}^0, \overline{f}^1, \overline{f}^2)\right]^{1/3}} \\[2mm]
e^{-\phi_2} &= \frac{\sum_{A<B=0}^{2} \mathrm{Wr}(f^A, f^B)\mathrm{Wr}(\overline{f}^A, \overline{f}^B)}{\left[\mathrm{Wr}(f^0, f^1, f^2)\mathrm{Wr}(\overline{f}^0, \overline{f}^1, \overline{f}^2)\right]^{2/3}}
\end{aligned}
\tag{3.8}
$$

Our task is now to give a meaning for these formulae in Riemannian geometry. There are two new features as compared with the Liouville case. First we now have three pairs of functions, but we still only have two variables z and $\overline{z}$. Second, derivatives of second degree appear so that one does not have a direct tangent-space interpretation. These two problems are solved simultaneously according to our general scheme. First, we introduce CP^2 as target space where the equations

$$X^A = f^A(z), \quad \overline{X}^A = \overline{f}^A(\overline{z}), \quad A = 0,\,1,\,2, \tag{3.9}$$

define a holomorphic (W) surface. Second, we will also consider the extrinsic geometry of these surfaces where higher derivatives naturally appear when one defines normals. It is convenient to define

$$\tilde{G}_{a\overline{b}} = \sum_{A,B} f^{(a)A}(z)\overline{f}^{(b)B}(\overline{z})G_{A\overline{B}}\Big|_{X^A = f^A,\,\overline{X}^A = \overline{f}^A} = \tilde{G}_{\overline{b}a} \tag{3.10}$$

Consider first the intrinsic geometry. It is straightforward to see that, according to Eq.3.8, the arc-length on the surface is given by

$$ds^2 \equiv \tilde{G}_{11}dzd\overline{z} = e^{2\phi_1-\phi_2}dzd\overline{z} \tag{3.11}$$

so that this particular exponential of the Toda fields gives the induced metric on the W surface. The complete geometrical interpretation may be obtained following the usual method of differential geometry, namely by considering normals to the surface. There are two of them, noted e_1, and $e_{\bar{1}}$. Their components may be written compactly by the generalized Frenet-Serret formulae

$$e_1^A = \frac{1}{\sqrt{\widetilde{G}_{1\bar{1}}\Delta_2}} \begin{vmatrix} \widetilde{G}_{1\bar{1}} & \widetilde{G}_{2\bar{1}} \\ f^{(1)A} & f^{(2)A} \end{vmatrix}, \quad e_{\bar{1}}^A = 0, \tag{3.12}$$

$$e_1^{\bar{A}} = 0, \quad e_{\bar{1}}^{\bar{A}} = \frac{1}{\sqrt{\widetilde{G}_{1\bar{1}}\Delta_2}} \begin{vmatrix} \widetilde{G}_{1\bar{1}} & \widetilde{G}_{1\bar{2}} \\ \overline{f}^{(1)\bar{A}} & \overline{f}^{(2)\bar{A}} \end{vmatrix}. \tag{3.13}$$

where

$$\Delta_2 \equiv \begin{vmatrix} \widetilde{G}_{1\bar{1}} & \widetilde{G}_{1\bar{2}} \\ \widetilde{G}_{2\bar{1}} & \widetilde{G}_{2\bar{2}} \end{vmatrix} \tag{3.14}$$

It is easy to see that, if we denote by $(\mathbf{X}, \mathbf{Y})$ the inner product $G_{A\bar{B}}(X^A Y^{\bar{B}} + Y^A X^{\bar{B}})$ in CP^2, we have $(e_1, f^{(1)}) = 0$, $(e_1, \overline{f}^{(1)}) = 0$, $(\bar{e}_1, f^{(1)}) = 0$, $(\bar{e}_1, \overline{f}^{(1)}) = 0$, $(e_1, e_1) = (e_{\bar{1}}, e_{\bar{1}}) = 0$, $(e_1, e_{\bar{1}}) = 1$. So that the last equations define orthonormalized normals. With the present notations, the second fundamental form Q is obtained by writing the equivalent expressions

$$f^{(2)A} - f^{(1)A}\frac{\widetilde{G}_{2\bar{1}}}{\widetilde{G}_{1\bar{1}}} = e_1^A Q, \quad \overline{f}^{(2)\bar{A}} - \overline{f}^{(1)\bar{A}}\frac{\widetilde{G}_{2\bar{1}}}{\widetilde{G}_{1\bar{1}}} = e_{\bar{1}}^{\bar{A}} Q \tag{3.15}$$

with $Q = \sqrt{\Delta_2/\widetilde{G}_{1\bar{1}}}$. By explicit computation^d one finds that $\Delta_2 = \exp(3\phi_1)$, so that

$$Q = e^{(\phi_1 + \phi_2)/2} = \sqrt{\frac{\mathrm{Wr}(f^0, f^1, f^2)\mathrm{Wr}(\overline{f}^0, \overline{f}^1, \overline{f}^2)}{\sum_{A=0}^2 f^A(z)\overline{f}^A(\bar{z}) \sum_{C<B=0}^2 \mathrm{Wr}(f^C, f^B)\mathrm{Wr}(\overline{f}^C, \overline{f}^B)}}. \tag{3.16}$$

2   Associated holomorphic change of coordinates

Although the geometrical meaning of the Toda fields is now clear, the description is not covariant under change of parametrization of CP^2, since it involves second derivatives. This problem is solved by using the fact that the Toda field equations are a subsystem of the so-called Toda hierarchy of integrable flows that makes use of 6 real parameters. These parameters provide a parametrization of C^3 where the formulae we summarized may be re-expressed in terms of first order derivatives only, so that their covariance properties become clear. This part is best handled by means of the fermionic method and Hirota equations[4] [5] [6]. We shall nevertheless remain at the same elementary simple-minded level for illustration. In terms of f and $\overline{f}$, the Toda hierarchy equations become extremely simple (see Eq.1.1). The variables are^e z^0, $z^1 \equiv z$, z^2 denoted collectively as $[z]$; and $\bar{z}^0$, $\bar{z}^1 \equiv \bar{z}$, $\bar{z}^2$ denoted collectively as $[\bar{z}]$. The functions $f^A(z)$ and $\overline{f}^{\bar{A}}(\bar{z})$ are extended as holomorphic or antiholomorphic functions $f^A([z])$ and $\overline{f}^{\bar{A}}([\bar{z}])$ solutions of the equations

$$\begin{aligned} \partial_0 f^A([z]) &= f^A([z]), & \partial_2 f^A([z]) &= (\partial_1)^2 f^A([z]) \\ \overline{\partial}_0 \overline{f}^{\bar{A}}([\bar{z}]) &= \overline{f}^{\bar{A}}([\bar{z}]), & \overline{\partial}_2 \overline{f}^{\bar{A}}([\bar{z}]) &= (\overline{\partial}_1)^2 \overline{f}^{\bar{A}}([z]) \end{aligned} \tag{3.17}$$

---

^{d}this is done more powerfully using the fermionic method as explained in ref.[6]

^{e}Upper indices are always covariant indices. When we want to indicate powers such as z square, we put parentheses, and write $(z)^2$.

with the initial conditions

$$f^A([z]) = f^A(z), \quad \overline{f}^{\overline{A}}(\overline{z}) = \overline{f}^{\overline{A}}([\overline{z}]), \quad \text{for } z^0 = z^2 = \overline{z}^0 = \overline{z}^2 = 0 \qquad (3.18)$$

Of course the differential equations in z^0 and $\overline{z}^0$ are trivial and give simple factors $\exp(z^0)$ or $\exp(\overline{z}^0)$. Since we work homogeneously we may forget these variables completely and keep z^0 and $\overline{z}^0$ equal to zero. The point of the differential equations in z^2 is to turn second derivatives in z into a first order derivatives. Thus the matrix of inner product of derivatives Eq.3.10 is extended to CP^2 as

$$G([z], [\overline{z}])_{a\overline{b}} = \sum_{A,\overline{B}} \partial_a f^A([z]) \partial_{\overline{b}} \overline{f}^{\overline{B}}([\overline{z}]) G_{A\overline{B}} \Big|_{X^A = f^A([z]), \overline{X}^{\overline{A}} = \overline{f}^{\overline{A}}([\overline{z}])} \qquad (3.19)$$

where now **only first order derivatives appear**. As a matter of fact, this last equation is but the Riemannian metric obtained by making the holomorphic change of variables $X^A = f^A(z, z^2)$, $\overline{X}^{\overline{A}} = \overline{f}^{\overline{A}}(\overline{z}, \overline{z}^2)$. Thus again, we associate, with any solution of the Toda equation, a conformal parametrization of the complex projective space. Note again that, since the change is conformal, the Kähler condition is preserved. The metric $G([z], [\overline{z}])_{a\overline{b}}$ may be written as

$$G([z], [\overline{z}])_{a\overline{b}} = \partial_a \partial_{\overline{b}} \ln \left[\sum_A f^A([z]) \overline{f}^A([\overline{z}]) \right] \qquad (3.20)$$

Our next topic is the W transformations of f. We shall use the parametrization such that $X^0 = \overline{X}^0 = 1$, so that $f^0([z]) = \overline{f}^0([\overline{z}]) = 1$. Due to these conditions, and according to the lemma, the functions $f^A(z)$ are solution of a differential equation of the form

$$f^{(3)A}(z) = \lambda(z) f^{(2)A}(z) + \mu(z) f^{(1)A}(z) \qquad (3.21)$$

there is no term without derivative since $f^0 = 1$ is a solution. The most general infinitesimal transformation that leaves the condition $f^0 = 1$ invariant is

$$\delta f^A(z) = (\alpha(z)\partial + \beta(z)\partial^2) f^A(z), \qquad (3.22)$$

Higher derivatives may be re-expressed in terms of the first two by means of Eq.3.21. Consider, now the extension to CP^2. To simplify the notation we will let $t = z^2$, so that, according to Eqs.3.17,

$$\partial_t f^A(z, t) = (\partial)^2 f^A(z, t). \qquad (3.23)$$

This is used in practice as follows. Consider

$$\delta_n f(z) = \epsilon(z)(\partial)^n f(z)$$

Clearly one has

$$f^A(z, t) = e^{t(\partial)^2} f^A(z, 0) \qquad (3.24)$$

One writes

$$\delta_n f(z, t) = e^{t(\partial)^2} \delta f(z) = \epsilon(z + 2t\partial) e^{t(\partial)^2} \partial^n f(z) = \epsilon(z + 2t\partial)(\partial)^n f(z, t)$$

where $\epsilon(z + 2t\partial)$ is defined by

$$\epsilon(z + 2t\partial) = \sum_r \frac{\epsilon_r}{r!}(z + 2t\partial)^r, \quad \epsilon_r = (\partial)^r \epsilon(z)|_{z=0} \tag{3.25}$$

In general, one sees that an arbitrary W transformation leads to

$$\delta f(z,t) = \alpha(z + 2t\partial)\partial f(z,t) + \beta(z + 2t\partial)\partial_t f(z,t) \tag{3.26}$$

Eq.3.21 is straightforwardly extended to CP^2 where it takes the form

$$\partial_z \partial_t f^A([z]) = L([z])\partial_t f^A([z]) + M([z])\partial_z f^A([z]). \tag{3.27}$$

Using this relation, and Eq.3.23, we may re-express all higher derivatives in Eq.3.26 in terms of the first-order ones, obtaining a transformation of form

$$\delta f(z,t) = A(z,t)\partial f(z,t) + B(z,t)\partial_t f(z,t) \tag{3.28}$$

which correspond to the change of parametrization

$$\delta z = A(z,t), \quad \delta t = B(z,t), \text{ with } \delta f(z,t) = f(z + \delta z, t + \delta t) - f(z,t). \tag{3.29}$$

Thus W transformations are extended as particular diffeomorphisms of CP^2.

2 Light-cone description

Finally, we connect the above discussion with the light-cone approach to W_3 gravity, which so far was developed independently of the conformal gauge[16],[17]. The basic point, displayed in ref.[7], is that, for any Kähler manifold, there exist parametrizations such that the metric takes a block-form identical to the light-cone metric Eq.2.21 introduced by Polyakov for two-dimensional gravity.

1 Light-cone parametrization for Kähler manifolds

Consider an arbitrary Kähler manifold, introduced by Definition 1. There exists another class of prefered coordinates denoted U^A, and $U^{\overline{A}}$, such that the new metric tensor denoted by the letter H takes a form which is the generalization of the right-hand side of Eq.2.21. The change of coordinates is of the form

$$U^{\overline{A}} = U^{\overline{A}}(X^1, \cdots, X^n; X^{\overline{1}} \cdots, X^{\overline{n}}), \quad U^A = X^A. \tag{3.30}$$

The functions $U^{\overline{A}}(X; \overline{X})$ will be determined next, in such a way that the light-cone metric takes the form

$$H_{\overline{A}\,\overline{B}} = 0, \quad H_{AB} = 2h_{AB}$$

$$H_{A\overline{B}} = H_{\overline{B}A} = \delta_{A,\overline{B}}, \tag{3.31}$$

where h_{AB} will be related to the Kähler potential. Before going on, let us remark that the determinant of H is equal to -1, and that its inverse is given by

$$H^{AB} = 0, \quad H^{\overline{A}\overline{B}} = -2h_{AB}$$

$$H^{A\overline{B}} = H^{\overline{B}A} = \delta_{A,\overline{B}}. \tag{3.32}$$

On the contrary, no general form may be given for the determinant of G or its inverse. This is a very nice point of the light-cone parametrization[f]. Going back to our main line, one finds, by standard computations, that conditions Eqs.3.31 are fulfilled if one has

$$G_{A\overline{B}} = H_{A\overline{C}}\partial_{\overline{B}}U^{\overline{C}} \tag{3.33}$$

$$2h_{AB} = -H_{B\overline{C}}\partial_A U^{\overline{C}} - H_{A\overline{C}}\partial_B U^{\overline{C}} \tag{3.34}$$

These equations are easily solved by using the Kähler potential, obtaining

$$U^{\overline{A}} = H^{\overline{A}C}\partial_C K, \tag{3.35}$$

$$h_{AB} = -\partial_A\partial_B K. \tag{3.36}$$

Thus we reach the important conclusion that, for any Kählerian manifold there is a choice of coordinates such that the metric tensor takes the block-form

$$H = \begin{pmatrix} 2h & 1 \\ 1 & 0 \end{pmatrix} \tag{3.37}$$

which is the same as the one introduced by Polyakov to describe 2D gravity in the light-cone gauge.

2 Light-cone coordinates for W_3 gravity

We finally apply this change of coordinates to CP^2, calling $\overline{u}^1$, and $\overline{u}^2$ the two light-cone coordinates. Combining Eq.3.21 with Eq.3.35, 3.36, one sees that one has

$$\overline{u}^\ell = \frac{\sum_{A=0}^2 \partial_\ell f^A([z])\overline{f}^A([\overline{z}])}{\sum_{A=0}^2 f^A([z])\overline{f}^A([\overline{z}])} \tag{3.38}$$

$$h_{\ell m} = -\frac{\sum_{A=0}^2 \partial_\ell\partial_m f^A([z])\overline{f}^A([\overline{z}])}{\sum_{A=0}^2 f^A([z])\overline{f}^A([\overline{z}])} + \overline{u}^\ell\overline{u}^m$$

Making use of Eq.3.27, this gives

$$h_{11} = -\overline{u}^2 + (\overline{u}^1)^2, \quad h_{12} = -L\overline{u}^2 - M\overline{u}^1 + \overline{u}^1\overline{u}^2$$

$$h_{22} = -u^2(\partial L + L^2 + M) - u^1(\partial M + LM) + (u^2)^2 \tag{3.39}$$

so that h is quadratic in the $\overline{u}$'s. On the other hand, Polyakov's equation for light-cone W_3 gravity takes the form $(\overline{D})^3 h = 0$, and $(\overline{D})^5 A = 0$. This contradiction may be resolved as follows: **The W surface where light-cone physics takes place is not at $z^{(2)} = \overline{z}^{(2)} = 0$, contrary to conformal physics.** Indeed, let us assume that the light-cone W surface has equations

$$z^{(2)} = 0, \quad \overline{u}^2 = \alpha(z)(\overline{u}^1)^2 + \beta(z)\overline{u}^1 + \gamma(z) \tag{3.40}$$

---

[f]inverting the metric tensor is often a pain in the neck.

Then, on this surface, $h_{\ell m}$ is a polynomial in $\overline{U}^1$ of degree $\ell + m$. Thus it seems natural to let

$$h_{11} = h, \quad h_{22} = A \tag{3.41}$$

obtaining

$$(\overline{D})^3 h = 0, \quad (\overline{D})^5 A = 0, \quad \text{with } \overline{D} = \overline{D}_1 \tag{3.42}$$

Clearly, α, β, and γ are arbitrary. They may be precisely changed by an $sl(3)$ transformation. Thus, one is led to conjecture that **the geometrical origin of the $sl(3)$ invariance is a change of the light-cone W surface so that Eq.3.39 keeps its form.**

4 Outlook

These lecture notes left out many aspects, some of which have been developed elsewhere. In particular, quantization has led to many interesting progress. For the Liouville case, it was realized[1, 2] that there exists a quantum-group structure where the mathematical parameter of the quantum-group deformation coincides with Planck's constant. In this theory, the non-commutativity which is inherent in the quantum group – since the co-product in non-symmetric – is brought about by the actual quantization of the system. This appearance of quantum group seems to be very natural geometrically, since one deals with a gravity theory, where the space-time metric is quantized, which seems tantamount to quantizing the 2D space-time itself. Thus an object like a quantum plane should appear, and quantum groups are natural transformations of such "quantum manifolds". For recent progress in this direction, see refs. [18, 19, 28]. The quantum group picture was extended to the quantum A_n Toda theories in ref.[11]. The classical geometry setting summarized in these notes indicates that at the quantum level one should be dealing with quantum complex projective spaces. This remains to be clarified.

Classically, what we describe is also incomplete. In the light-cone connection, the ansatz Eq.3.41 remains somewhat ad hoc, and its Kac-Moody structure should be clarified. Since we use the Toda equation throughout, we stayed on the mass shell. Thus we only get Polyakov equations (Eqs.2.25, 3.42), and not the full anomaly equations that characterize light-cone W gravities. Besides the particular change of gauge we have discussed, the most basic problem is to formulate W gravity without gauge fixing, and to couple it to some appropriate kind of matter from first principle.

References

[1] A.B.Zamolodchikov, *Theor. Math. Phys.* **65** (1985) 1205; V.A. Fateev and A.B.Zamolodchikov, *Nucl. Phys.* **B280 [FS18]** (1987) 644.

[2] A. Bilal and J.-L. Gervais, *Phys. Lett.* **B206** (1988) 412; *Nucl. Phys.* **B314** (1989) 646; *Nucl. Phys.* **B318** (1989) 579.

[3] J.-L. Gervais, *Phys. Lett.* **B160** (1985) 277; K. Yamagishi, *Phys. Lett* **B205** (1988) 406; I. Bakas, *Phys. Lett.* **B213** (1988) 313.

[4] M. Sato, *RIMS Kokyuroku* **439** (1981) 30; E. Date, M. Jimbo, M.Kashiwara and T.Miwa, "Transformation Groups for Soliton Equations" in *Proc. of RIMS Symposium on Non-linear Integrable Systems-Classical Theory and Quantum Theory* (Kyoto, Japan, May 1981) (World Scientific Publication Co. Singapore, 1983); G.Segal and G.Wilson, *Pub. Math. IHES* **61** (1985) 5.

[5] J.-L. Gervais, Y. Matsuo: *Phys. Lett.* **274B** (1992) 309.

[6] J.-L. Gervais, Y. Matsuo: Comm. in Math. Phys. **152** (1993) 317.

[7] J.-L. Gervais, Y. Matsuo: *Phys. Lett.* **312B** (1993) 285.

[8] A. Polyakov: Mod. Phys. Lett. A **2, 11** (1987) 893.

[9] J.-L. Gervais, Comm. in Math. Phys. **130**, 257, (1990) .

[10] J.-L. Gervais, Phys. Lett. **B243**, 85, (1990) .

[11] E. Cremmer, J.-L. Gervais, Comm. in Math. Phys. **134**, 619, (1990).

[12] O. Babelon, Phys. Lett. **B215**, (1988) 523.

[13] J.-L. Gervais, Comm. in Math. Phys. **138** 138, 301, (1991) .

[14] J.-L. Gervais, "Quantum group derivation of 2D gravity-matter coupling" Invited talk at the Stony Brook meeting *String and Symmetry 1991*, LPTENS preprint 91/22, Nucl. Phys. **B391**, 287, (1993).

[15] J.-L. Gervais, J. Schnittger, Phys. Lett. **B315** , 258, (1993).

[16] Y. Matsuo: Phys. Lett. **B227** (1989) 209.

[17] see, for instance, the related reviews in "String and symmetries 91", *in* Proceeding of the Stony Brook Meeting, World Scientific ed.

[18] E.Cremmer, J.-L. Gervais. J.-F. Roussel, "The genus-zero bootstrap of chiral vertex operators in Liouville theory", preprint LPTENS-93/29, Nucl. Phys. to be published.

[19] E.Cremmer, J.-L. Gervais. J.-F. Roussel, "The quantum group structure of 2D gravity and minimal models II: The genus zero chiral bootstrap" preprint LPTENS-93/02, hep-th 9302035, Comm. Math. Phys. to be published.

[20] J.-L. Gervais, J. Schnittger, "The many faces of the quantum Liouville exponentials" preprint LPTENS-93/30, hep-th 9308134, Nucl. Phys. to be published.

TOPOLOGICAL STRINGS
AND QCD in TWO DIMENSIONS

PETR HOŘAVA

Enrico Fermi Institute
University of Chicago
5640 S. Ellis Ave.
Chicago IL 60637, USA

Abstract. I present a new class of topological string theories, and discuss them in two dimensions as candidates for the string description of large-N QCD. The starting point is a new class of topological sigma models, whose path integral is localized to the moduli space of harmonic maps from the worldsheet to the target. The Lagrangian is of fourth order in worldsheet derivatives. After gauging worldsheet diffeomorphisms in this "harmonic topological sigma model," we obtain a topological string theory dominated by minimal-area maps. The bosonic part of this "topological rigid string" Lagrangian coincides with the Lagrangian proposed by Polyakov for the QCD string in higher dimensions.

1. Introduction

Strings undoubtedly represent one of the most interesting and universal patterns in Nature, emerging at vastly different scales and in different physical phenomena. The Regge phenomenology of hadrons and the structure of dual models were first indications that the adequate degrees of freedom for the theory of strong interactions at low energies are qualitatively those of a string theory. These ideas received a more specific support soon after QCD had been identified as the proper theory for strong interactions. Two basic techniques that reveal some form of string structure in QCD are the expansion in the number of colors, and the strong coupling expansion. Yet, even though the success of QCD in the ultraviolet has been enormous, any progress in the identification of the "stringy" fixed point that governs the theory in the infrared has been very limited (for recent reviews, see [1]).

In the meantime, strings have been recognized as playing an important, although still mostly enigmatic, rôle in quantization of gravity. Ever since, string theory of

Quantum Field Theory and String Theory, Edited by
L. Baulieu *et al.*, Plenum Press, New York, 1995

quantum gravity has been changing profoundly the basic concepts of space and time in quantum theory, and promises to do so even more after it is understood better as a second-quantized theory. Among the fascinating tentative results in this direction are the apparent lack of degrees of freedom in string theory at short distances, an extremely soft behavior of strings in super-Planckian collisions, and the existence of a minimal length/enlarged uncertainty principle, to mention just a few. The idea of a possible topological phase of quantum gravity (and string theory) at short distances also arised in this context. Without any doubt, many aspects of string theory have been elucidated in the context of quantum gravity.

At a more mundane level, however, one may return back to old unsolved problems, and start wondering whether the enormous progress accomplished in string theory during last decade could possibly lead to some new insight, and help in the quest for the string theory of QCD in the infrared. This program has been raised recently by Gross [2] and Gross and Taylor [3], in the context of the exactly solvable, pure-glue QCD in two dimensions. The idea is that the theory, despite having no local degrees of freedom, is rich enough in the content that it leads to a non-trivial string theory. It is exactly solvable on any compact spacetime manifold, thus supplying us with lots of information about the possible character of the string theory. The authors of [3] have been able to interpret the results of the $1/N$ expansion for 2D QCD in terms of simple rules for counting maps between surfaces (very similar results were obtained for Wilson loops on the plane some time ago by Kazakov and Kostov, see [4]). The next step is to try and find the worldsheet Lagrangian for the string theory, with the hope that this two dimensional QCD string Lagrangian could grasp some important aspects of the "stringy" fixed point that supposedly describes the real case of our interest, namely four dimensional QCD in the infrared. This represents an extremely interesting program for string theorists.

A successful identification of the worldsheet theory for two dimensional QCD could also have implications for string theory itself. Assuming that the worldsheet theory for the 2D QCD string is found, we can follow standard rules and construct its corresponding string field theory. This theory could be sufficiently complicated, but in this case – unlike in the "fundamental" string theory of quantum gravity – we do know the correct spacetime description of the theory, this being given by the QCD Lagrangian itself! Thus, the 2D QCD string theory could even teach us something about the conundrums of string field theory for quantum gravity.

In this talk I present a new class of topological string theories, and discuss them in two dimensions as candidates for the worldsheet description of the QCD string theory. The presentation is based on unpublished results that will appear elsewhere [5].

In section 2 I review very briefly some aspects of quantum Yang-Mills theory in two dimensions, its relation to topological Yang-Mills theory, and the interpretation of the $1/N$ expansion of the $SU(N)$ theory in terms of maps between surfaces as obtained by the authors of [2-4].

In section 3 , a new class of topological sigma models is defined, as the "worldsheet matter" for the topological string theory that we are about to construct. The path integral of these topological sigma models is localized at the moduli space of harmonic maps from the worldsheet to the target (with respect to a fixed metric on both). To distinguish the theory from the traditional topological sigma model dominated by holomorphic maps, I refer to it as the "harmonic topological sigma model" henceforth. The bosonic part of the Lagrangian is of fourth order in worldsheet derivatives, and is reminiscent of the rigid string Lagrangian proposed by Polyakov [6] some time ago as a candidate for the QCD string in higher dimensions.

In section 4 we gauge worldsheet diffeomorphisms in the harmonic topological sigma model. Instead of introducing a dynamical metric on the worldsheet, we will replace the fixed metric by the induced metric. This completes the analogy with the rigid string, and I refer to the theory as the "topological rigid string theory." The instanton moduli space that dominates the path integral is the space of maps harmonic in their own induced metric, i.e. the moduli space of minimal-area maps.

In section 5 I discuss possible variations on the theme of topological rigid strings in dimensions higher than two, in particular in four dimensions.

2. Yang-Mills Theory and Large-N QCD in Two Dimensions

The recent revival of interest in two dimensional Yang-Mills theory has led to very interesting results. The quantum theory is exactly solvable on any Riemann surface, a fact that can be explained from different points of view. The original one, due to Migdal and recently Rusakov [7], uses a RG-invariant lattice formulation. Another point of view, developed by Witten [8], explains the exact solvability in the continuum theory in terms of localization of the path integral in equivariant cohomology/topological Yang-Mills theory to a moduli space of classical configurations. Since we are interested in the continuum limit of the worldsheet theory for the QCD string, the results of Witten seem to be particularly relevant. In [8] it is shown that after some fermion fields are added to the Yang-Mills theory, the "physical" Yang-Mills Lagrangian can be treated as a BRST invariant observable in the underlying topological Yang-Mills theory. This correspondence then allows one to calculate the partition function of the "physical" theory as a correlation function in the underlying topological theory. In this picture, two dimensional Yang-Mills theory is so special because there is an underlying topological theory that actually governs the calculation of the partition function in the "physical" theory.

The partition function of the Yang-Mills theory on an arbitrary two dimensional surface M of genus G is given by

$$\mathcal{Z} \equiv \int \mathcal{D}A_\mu \, \mathrm{e}^{(1/\tilde{g}^2) \int_M \, \mathrm{d}\mu \, \mathrm{Tr} \, F_{\mu\nu} F^{\mu\nu}} = \sum_R (\dim R)^{2-2G} \mathrm{e}^{-\tilde{g}^2 A C_2(R)/2}, \qquad (0.1)$$

where $\mathrm{d}\mu$ is the Riemannian volume element and A the area of a fixed metric $G_{\mu\nu}$ on M, the sum runs over all irreducible representations of the gauge group, and $C_2(R)$ is the second Casimir of R. For SU(N) we can take the large-N limit that keeps $\lambda \equiv \tilde{g}^2 N$

fixed, and expand the exact result (0.1) in the powers of $1/N$. In [2,3], Gross and Taylor obtained the following expression for the $1/N$ expansion of (0.1):

$$\mathcal{Z} = \sum_{g \geq 0} \frac{1}{N^{2g-2}} \sum_{n,\tilde{n}} e^{-(n+\tilde{n})\lambda A/2} \sum_{i=0}^{2g-2-(n+\tilde{n})(2G-2)} (\lambda A)^i \, \omega_{g,G}^{n,\tilde{n},i}, \qquad (0.2)$$

where $\omega_{g,G}^{n,\tilde{n},i}$ are calcullable coefficients independent of A. The expansion is reminiscent of partition functions of string theory, with $1/N$ being the string coupling constant, λ the string tension, g the worldsheet genus, and n ($\tilde{n}$) the number of orientation-preserving (-reversing) sheets of the map from the worldsheet to the target.

The authors of [2,3] interpreted the coefficients of this expansion in terms of rules for counting specific maps between surfaces. Maps that are allowed to contribute to the sum are branched covers of the target, with collapsed handles and infinitesimal tubes that connect various sheets of the cover (see [2,3] for details). The string partition function can then be rewritten as an integral over the moduli space $\mathcal{M}$ of such coverings of M,

$$\mathcal{Z} = \int_{\mathcal{M}} d\nu \, W(\nu), \qquad (0.3)$$

with the location of the branch points, collapsed handles and connecting tubes serving as coordinates on $\mathcal{M}$. The density $W(\nu)$ to be integrated is given essentially by the symmetry factor of the cover (see [2,3]).

The structure of the $1/N$ expansion and the character of maps that contribute to the rules of [2-4] give us some indications about what we can expect from the QCD string theory. There are no terms in (0.2) with both n and $\tilde{n}$ equal to zero, which indicates that folds of the maps are suppressed and the QCD string is sensitive to its "extrinsic curvature" in the target. This point of view is also supported by the fact that worldsheets of genus zero do not contribute to $\mathcal{Z}$, and by the form of the terms exponential in the area. The fact that both orientation-preserving and orientation-reversing covers contribute indicates that the corresponding string theory is non-chiral.

Notice also that the string theoretical expression (0.2) for the QCD partition function allows us to take the limit $\lambda \to 0$. Although this limit is rather singular from the spacetime point of view, the large-N expansion serves as a nice regulator. Indeed, when we take the $\lambda \to 0$ limit for fixed n and $\tilde{n}$ in (0.2), we obtain finite results. From the point of view of string theory, this corresponds to the zero tension limit, and the only structure carried by the allowed maps are the so-called Ω-points [3]. It seems very reasonable to expect that in the zero tension limit, the string theory of 2D QCD is topological.

3. Harmonic Topological Sigma Models

In the previous section we have seen that the coefficients of the large-N expansion of the QCD partition function on an arbitrary spacetime surface M has the structure of a sum over classes of maps from the worldsheet to the spacetime. The coefficients of this sum are exponentials of the area of the worldsheet that covers the spacetime times finite polynomials in the area. One of the first observations made by the authors

154

of [3] is that the contribution to the leading term in this polynomial is essentially one from each homotopy sector that is allowed to contribute by Kneser's formula. As a first step towards the worldsheet Lagrangian for QCD string theory in two dimensions, let us first try to construct a string theory whose partition function is essentially equal to one in each allowed homotopy sector of maps.

The field configurations of the theory are given solely by the maps from a worldsheet Σ (of genus g) to the spacetime M (of genus G),

$$\Phi : \Sigma \to M. \tag{0.4}$$

We will be using coordinates X^μ on the target, and σ^a on the worldsheet. Before specializing to two target dimensions, we will work with targets of arbitrary real dimension D.

The theory that we are about to construct is a topological sigma model, with the topological gauge symmetry given by all possible deformations of Φ. We do not choose any other fixed geometrical structure either on the worldsheet or in the target except the target metric $G_{\mu\nu}$. In particular, we do not assume a complex structure on M. (In fact, the theory will make perfect sense even for targets on which no complex structure exists.) The gauge-invariant Lagrangian is identically equal to zero, and we use the standard BRST technology to get a gauge-fixed version of the theory. First, we introduce the BRST charge that maps the fields to their corresponding ghosts,

$$[Q, X^\mu] = \psi^\mu, \qquad \{Q, \psi^\mu\} = 0. \tag{0.5}$$

Then we choose a gauge fixing condition. Whereas the standard choice in the traditional topological sigma model is the holomorphicity of Φ, here we will proceed in a different way. Inspired by the results of the $1/N$ expanison of 2D QCD as reviewed in section 2, we are looking for a non-chiral topological sigma model, and we need a gauge fixing condition that prefers neither holomorphic nor anti-holomorphic maps.

To obtain a gauge fixing condition that satisfies these requirements, we pick an auxiliary fixed metric h_{ab} on the worldsheet, define the Laplacian on the maps from the worldsheet to the target in the standard way:

$$\Delta X^\mu \equiv h^{ab}\nabla_a\partial_b X^\mu \equiv h^{ab}\left(\partial_a\partial_b X^\mu + \Gamma^\mu_{\sigma\rho}\,\partial_a X^\sigma\partial_b X^\rho - \Gamma^c_{ab}\,\partial_c X^\mu\right), \tag{0.6}$$

and take the harmonicity condition

$$\Delta X^\mu = 0 \tag{0.7}$$

as the gauge fixing condition.[a]

To implement the gauge fixing, we introduce the antighost/auxiliary BRST multiplet,

$$\{Q, \chi^\mu\} = B^\mu, \qquad [Q, B^\mu] = 0. \tag{0.8}$$

---

[a]It is amusing to note here that the theory of harmonic maps between surfaces (see [9] for details) produces one of the shortest and most elegant analytic proofs of the (purely topological) Kneser formula that seems to be so important for the 2D QCD string.

The tensorial properties of the multiplet are determined by the properties of the gauge fixing condition (0.7), which in our case is a section of $\Phi^{-1}(TM)$. Note that both the ghosts and the antighosts are (fermionic) sections of the same bundle over the worldsheet, an important fact that makes the theory very different from the traditional topological sigma model.

The choice of the gauge fixed Lagrangian that ensures its general covariance and yields the required gauge fixing condition is

$$\mathcal{L} = \int_{\Sigma} \mathrm{d}^2\sigma \, \sqrt{h} \, \{Q, G_{\mu\nu}\chi^{\mu} \left(-\frac{1}{2}B^{\nu} + \Delta X^{\nu} + \frac{1}{2}\Gamma^{\nu}_{\sigma\rho}\chi^{\sigma}\psi^{\rho} \right)\}. \tag{0.9}$$

After performing the BRST commutator and integrating out the auxiliaries, we end up with the following form of the gauge fixed Lagrangian:

$$\mathcal{L} = \int_{\Sigma} \mathrm{d}^2\sigma \, \sqrt{h} \left(\frac{1}{2}G_{\mu\nu}\Delta X^{\mu}\Delta X^{\nu} - G_{\mu\nu}\chi^{\mu}\Delta\psi^{\nu} + R_{\mu\sigma\rho\nu}h^{ab}\partial_a X^{\sigma}\partial_b X^{\rho}\chi^{\mu}\psi^{\nu} \right. \tag{0.10}$$
$$\left. -\frac{1}{2}R_{\mu\sigma\rho\nu}\chi^{\mu}\psi^{\sigma}\chi^{\rho}\psi^{\nu} \right),$$

which is invariant under the on-shell BRST transformation given by

$$[Q, X^{\mu}] = \psi^{\mu}, \quad \{Q, \psi^{\mu}\} = 0, \quad \{Q, \chi^{\mu}\} = \Delta X^{\mu} + \Gamma^{\mu}_{\sigma\rho}\chi^{\sigma}\psi^{\rho}. \tag{0.11}$$

In (0.10) we have used the following definition of the covariant Laplacian on the ghosts,

$$\Delta\psi^{\mu} = h^{ab}\nabla_a\nabla_b\psi^{\mu} \equiv h^{ab}\left(\partial_a\partial_b\psi^{\mu} + 2\Gamma^{\mu}_{\sigma\rho}\partial_a X^{\sigma}\partial_b\psi^{\rho} - \Gamma^{c}_{ab}\partial_c\psi^{\mu} \right) \tag{0.12}$$
$$+ \Gamma^{\mu}_{\sigma\rho}\Delta X^{\sigma}\psi^{\rho} + h^{ab}\partial_a X^{\sigma}\partial_b X^{\rho}\psi^{\nu}\left(R^{\mu}{}_{\rho\sigma\nu} + \partial_{\nu}\Gamma^{\mu}_{\sigma\rho} \right);$$

in this notation, the BRST transformation of ΔX^{μ} is equal to $\Delta\psi^{\mu}$ plus additional terms,

$$[Q, \Delta X^{\mu}] = \Delta\psi^{\mu} - R^{\mu}{}_{\sigma\rho\lambda}h^{ab}\partial_a X^{\sigma}\partial_b X^{\rho}\psi^{\lambda} - \Gamma^{\mu}_{\sigma\rho}\Delta X^{\sigma}\psi^{\rho}. \tag{0.13}$$

This formula will be useful below.

The Lagrangian (0.10) is an exact BRST commutator, and the path integral of the theory is by standard arguments localized at the moduli space of instantons, which in our case are harmonic maps from Σ to M. To distinguish the theory from the traditional topological sigma model, I refer to it as the "harmonic topological sigma model."

The Lagrangian of the harmonic topological sigma model resembles closely the Lagrangian of topological mechanics. First of all, since both the ghost field ψ^{μ} and the antighost χ^{μ} are sections of the same bundle, their zero modes (in a fixed harmonic background X_0^{μ}) satisfy identical equations,

$$\Delta\psi^{\mu} - R^{\mu}{}_{\sigma\rho\nu}h^{ab}\partial_a X_0^{\sigma}\partial_b X_0^{\rho}\psi^{\nu} = 0 \tag{0.14}$$

(and similarly for χ^{μ}), and the ghost number anomaly in the path integral is identically zero.

The absence of quantum anomaly in the ghost number conservation has important implications for the structure of correlation functions. Topological sigma models have

156

typically two classes of observables. The de Rham cohomology ring of the target produces one of them (with the degree of the cohomology class being the ghost number of the observable). If the first homotopy group is non-trivial, there is another class of observables, corresponding to the vacua of the winding sectors (cf. [10]). Assuming that there are no point-like observables of negative ghost numbers, every correlation function that contains at least one observable of non-zero ghost number is zero by the ghost number conservation in the quantum theory.

Given a fixed harmonic map $\Phi_0 : \Sigma \to M$, we can analyze the local structure of the moduli space of harmonic maps around Φ_0 as follows. Consider a small deformation δX^μ of X^μ. The deformed map is harmonic (to lowest order) if δX^μ satisfies the linearized harmonicity condition,

$$\Delta \delta X^\mu - R^\mu{}_{\sigma\rho\nu} h^{ab} \partial_a X_0^\sigma \partial_b X^\rho \, \delta X^\nu = 0. \tag{0.15}$$

This equation, also known in the theory of harmonic maps as the Jacobi equation, is identical to Eqn. (0.14) for the zero modes of ψ^μ. The (integrable) zero modes of ψ^μ are thus tangent to the moduli space of harmonic maps, and the number of linear independent integrable solutions to the Jacobi equation measures the dimension of the moduli space.

Except for the higher order form of the kinetic terms, the only apparent difference between the Lagrangian of the harmonic sigma model and the Lagrangian of topological mechanics is the existence of the curvature-dependent two-fermi term in (0.10). One might naïvely expect that this term will join the curvature-dependent four-fermi term in saturating the fermionic integral over the zero modes. The BRST transformation of ΔX^μ, Eqn. (0.13), indicates that this is not the case: In the path integral, the curvature-dependent two-fermi term gets absorbed into the kinetic term of the fermi fields, and leads to the determinant cancelation in the (exact) Gaussian approximation. Hence, the four-fermi term remains the only one that can be used to saturate the ghost/antighost zero modes in the path integral.

The structure of the partition function of the harmonic sigma model is thus very similar to that of topological mechanics with the same target manifold: In the homotopically trivial sector, harmonic maps are constant maps, and the moduli space is isomorphic to the target. The partition function is equal to the Euler character of the target manifold. Non-trivial homotopy sectors provide interesting stringy corrections to this result, but what is always computed is the Euler character of the moduli space of harmonic maps in the corresponding homotopy class.

The fact that the path integral gives the Euler character of the moduli space of harmonic maps can be proven rigorously, following Atiyah and Jeffrey [11]. The path integral of a given topological field theory computes a regularized Euler number of a specific vector bundle $\mathcal{V}$ over the infinite dimensional manifold $\mathcal{A}$ of all field configuations. In our case, $\mathcal{A}$ contains all maps from Σ to M, and $\mathcal{V}$ is the bundle whose typical section is the left hand side of the gauge fixing condition, i.e. $\mathcal{V}$ is the tangent bundle to $\mathcal{A}$. The regularized Euler number of this infinite dimensional bundle is defined as the Euler number of a restriction of $\mathcal{V}$ to the zero locus of ΔX^μ, and coincides with the Euler character of the moduli space of harmonic maps.

Harmonic topological sigma models in two dimensions

Let us now consider, as an example, the harmonic topological sigma model with a two dimensional target M of genus G. We know that the path integral of the theory is localized at the moduli space of the harmonic maps from Σ to M, and we would like to check explicitly that it computes the Euler character of the moduli space. In the case of maps between surfaces, harmonic maps have been analyzed thoroughly in the mathematical literature (see [9] and references therein), and many important results have been obtained. Consider for instance an arbitrary homotopy class of maps between surfaces with both g and G greater than one, and with non-zero degree. Then there is a theorem [9] claiming that in such a homotopy class, there is exactly one harmonic map! The moduli space thus consists of only one point, there are no zero modes of the fermi fields, and the path integral in this homotopy class gives exactly one. Analogously, in the homotopically trivial sector, the only harmonic maps are the constant maps, and the moduli space coincides with M. There are two zero modes for both ψ^μ and χ^μ, corresponding to translations of the image of Φ in M. These zero modes bring down the curvature in the path integral, which is then equal to the Euler character of the target.

The path integral is similarly simple in the general case with any two dimensional target (see [5]). The moduli space of harmonic maps between surfaces with fixed metrics in a given homotopy class is always non-empty if $G \geq 1$ [9]; the only case when there are no harmonic maps in some homotopy classes corresponds to G equal to zero, i.e. to the spherical target.

Up to this point we have studied the topological sigma model as a matter system on the worldsheet. To get a string theory instead, we have to take care of worldsheet diffeomorphisms. The minimal way of doing so is to take the partition function of the harmonic topological sigma model, and sum over all possible homotopy classes. The sum is naïvely infinite, and we have to factorize by the group of global diffeomorphisms of the worldsheet. The resulting theory is finite, but it is still far from describing the QCD string in two dimensions.

4. Topological Rigid Strings

In the standard way of gauging worldhseet diffeomorphisms in a topological matter system, one introduces a topological gravity multiplet and couples it to the matter system. Here we will proceed in a different way, motivated mainly by the observation that QCD string theory is probably very different from traditional string theory. In particular, the QCD string can be sensitive to the extrinsic geometry, and the worldsheet quantization of the string theory could use a spacetime cutoff instead of the worldsheet cutoff, as e.g. in the rigid string theory. The naïve dimension of X^μ in such a theory is one instead of zero, and the quantum properties of the string are governed by a fixed point that is substantially different from the one usually employed in traditional critical string theory. Unfortunately, such a string theory is very hard to quantize explicitly, and little is known about its properties.

Here we will see how to avoid these shortcomings in a topological version of such a string theory. Indeed, we have noticed above that the bosonic part of the Lagrangian for the harmonic topological sigma model is reminiscent of the rigid string theory studied by Polyakov and others some time ago. The basic discrepancy is that we have used a fixed fiducial metric on the worldsheet (which does not violate the naïve background independence of the theory by virtue of the topological symmetry), whereas the rigid string theory uses the induced metric. We will use this observation here to gauge worldsheet diffeomorphisms as follows. The BRST multiplet combines the topological symmetry of the harmonic topological sigma model with the worldsheet diffeomorphism symmetry,

$$
\begin{aligned}
[Q, X^\mu] &= \psi^\mu + c^a \partial_a X^\mu, \\
\{Q, \psi^\mu\} &= c^a \partial_a \psi^\mu + \phi^a \partial_a X^\mu, \\
\{Q, c^a\} &= c^b \partial_b c^a - \phi^a, \\
[Q, \phi^a] &= c^b \partial_b \phi^a - \phi^b \partial_b c^a.
\end{aligned}
\tag{0.16}
$$

Here c^a is the diffeomorphism ghost, which can be omitted from the (equivariant) theory, under the assumption that we deal exclusively with diffeomorphism invariant objects on which the reduced BRST charge is still nilpotent. We have also introduced the bosonic ghost-for-ghost ϕ^a of ghost number two, which takes care of the overcounting of worldsheet diffeomorphisms in the BRST symmetry.

The gauge fixing condition in the bosonic sector remains formally the same as in the harmonic topological sigma model,

$$
\sqrt{h}\, \Delta X^\mu = 0,
\tag{0.17}
$$

but now we have not introduced any fiducial metric on the worldsheet. Instead, we identify the metric h_{ab} with the induced metric,

$$
h_{ab} = \partial_a X^\mu \partial_b X^\nu G_{\mu\nu}.
\tag{0.18}
$$

Consequently, the Laplacian that enters the gauge fixing condition (0.17) is now given by

$$
\Delta X^\mu \equiv h^{ab} \left(\delta^\mu_\nu - h^{cd} G_{\nu\lambda} \partial_c X^\mu \partial_d X^\lambda \right) \left(\partial_a \partial_b X^\nu + \Gamma^\nu_{\sigma\rho} \partial_a X^\sigma \partial_b X^\rho \right).
\tag{0.19}
$$

One may wonder whether the naïve counting of gauge fixing conditions agrees with the number expected in a diffeomorphism invariant theory. While in the harmonic topological sigma model in D target dimensions the harmonicity condition gives D gauge fixing conditions (as it should), here the requirement that h_{ab} be the induced metric effectively reduces the number of independent components of ΔX^μ by two. Indeed, assuming that the induced metric is smooth and non-degenerate, ΔX^μ coincides with the trace of the second fundamental form of the map $\Phi : \Sigma \to M$, and it is easy to show that ΔX^μ of (0.19) is normal to the image $\Phi(\Sigma)$ of Σ in M,

$$
G_{\mu\nu} \Delta X^\mu \partial_a X^\nu = 0.
\tag{0.20}
$$

In D target dimensions, the number of independent components in (0.17) is thus $D-2$, which leaves exactly two unfixed coordinates for worldsheet diffeomorphism symmetry. In two dimensions, we seem to have no condition at all. What makes the theory nontrivial, however, is the fact that the requirements imposed on the induced metric that lead to (0.20) are generically impossible to satisfy everywhere on the worldsheet. In particular, the gauge fixing condition forbids the existence of folds on the map from Σ to M.

The gauge fixing procedure can be easily completed at higher ghost numbers, and we end up with a topological string Lagrangian of the following form:

$$\mathcal{L} = \int_\Sigma \mathrm{d}^2\sigma \, \sqrt{h} \, \left(G_{\mu\nu} \Delta X^\mu \Delta X^\nu + \text{ghost terms} \right). \tag{0.21}$$

Unlike in the harmonic topological sigma model, h_{ab} in (0.21) is now the induced metric. The exact form of the ghost terms is really not so important for our purposes here, and can be found elsewhere [5]. Indeed, since the theory has been constructed as a topological string theory, we do not need to know the exact form of the Lagrangian in order to compute the partition function. In topological theories, we know a priori what the path integral is computing, and we can frequently do the calculation without invoking the path integral.

In the case at hand, the Lagrangian (0.21) defines a a topological version of the rigid string theory. The ghost-independent part of the Lagrangian coincides with the rigid string theory Lagrangian of [6], and it is natural to refer to the theory as the "topological rigid string." Its path integral is localized at the moduli space of solutions to the gauge fixing solution $\sqrt{h}\Delta X^\mu = 0$. This means that the maps dominating the path integral are harmonic in their own induced metric, i.e. the path integral is localized at the moduli space of minimal-area maps, counted up to worldsheet diffeomorphisms!

In the topological sigma model of section 3 , we knew a priori that the path integral computed the Euler number of the moduli space of harmonic maps, and we confirmed this fact by a direct calculation for the simplest case of two dimensional targets. In the topological rigid string theory, similarly, we know that the path integral gives an (equivariant) Euler number of the moduli space of minimal maps. The moduli spaces of minimal maps can be conveniently parametrized if we adopt the spacetime point of view. Indeed, the minimal-area condition forbids folds of the maps, and only allows branch points (and possibly collapsed handles). The location of branch points in the spacetime then serves as a natural set of coordinates on the moduli space of minimal-area maps. The branch points are indistinguishable, as a result of the worlsheet diffeomorphism symmetry. Consequently, the moduli spaces that dominate the path integral of the topological rigid string are quite similar to the spaces of maps that contribute to the $1/N$ expansion of 2D QCD. More details about this correspondence can be found in [5].

What we have considered thus far is a topological string theory with zero tension. Its partition function is naturally independent of the area of the target. To obtain a string theory with non-zero tension, we would like to add to the Lagrangian a term that behaves as

$$\delta\mathcal{L} = \lambda \int_\Sigma \mathrm{d}^2\sigma \, \sqrt{h} \, . \tag{0.22}$$

160

When evaluated at the maps that dominate the path integral of the topological rigid string theory, i.e. at the minimal-area maps, this term gives

$$\delta \mathcal{L}(\Phi) = \lambda(n + \tilde{n})A, \tag{0.23}$$

where $n, \tilde{n}$ are determined by the homotopy class of the minimal map Φ and represent the number of orientation-preserving and orientation-reversing sheets of the worldsheet over the spacetime, and A is the area of the target. If we add (0.22) to the Lagrangian of the topological rigid string and then evaluate the deformed Lagrangian at the moduli space, we recover exactly the same exponential dependence on the area as observed in 2D QCD.

Of course, this cannot be the whole story. The naïve area term (0.22) does not respect the topological BRST symmetry of the rigid string theory and cannot be added to the Lagrangian without spoiling its BRST invariance. We can, however, make the induced area BRST-invariant by adding ghost-dependent terms to (0.22):

$$\delta \mathcal{L}' = \lambda \int_{\Sigma} \mathrm{d}^2 \sigma \left(\sqrt{h} + \text{ghost terms} \right), \qquad [Q, \delta \mathcal{L}'] = 0. \tag{0.24}$$

This term can be added to the topological rigid string Lagrangian. In the partition function, the ghost corrections to the area term do not change the form of the terms that are exponential in the area. They do affect, however, the integral over the moduli space of minimal-area maps. Indeed, in the path integral of the deformed theory, the ghost corrections to (0.22) will enter the integral over the fermionic zero modes, thus changing the integration over the bosonic zero modes. In the partition function, this will produce a polynomial dependence on the target area in each homotopy sector of maps from the worldsheet to the target.

This procedure is reminiscent of the mechanism that connects the "physical" Yang-Mills theory in the target to a topological version thereof (section 2). It would be very interesting to see whether a similar mechanism can be involved in the worldsheet theory of the 2D QCD string as well.

5. Higher Dimensions

In previous paragraphs, we have presented some evidence in favor of the rigid string picture of QCD, in the exactly solvable case of the two dimensional, pure-glue theory on compact surfaces. We have observed some striking similarities between a topological rigid string theory and the results of the large-N expansion in QCD as obtained in [2-4]. Here I comment on possible extensions of the results to higher dimensions.

Although both the harmonic topological sigma model and the topological rigid string theory can be in principle constructed in arbitrary real dimension, the four-dimensional case is unique. Let us first rewrite the rigid string Lagrangian

$$\mathcal{L} = \int_{\Sigma} \mathrm{d}^2 \sigma \sqrt{h} \, G_{\mu\nu} \Delta X^\mu \Delta X^\nu \tag{0.25}$$

in an equivalent form,

$$\mathcal{L} = \int_{\Sigma} \mathrm{d}^2 \sigma \sqrt{h} \, h^{ab} \nabla_a t^{\mu\nu} \nabla_b t^{\sigma\rho} G_{\mu\sigma} G_{\nu\rho}, \tag{0.26}$$

with $t^{\mu\nu}$ defined by

$$t^{\mu\nu} = \frac{\epsilon^{ab}}{\sqrt{h}} \partial_a X^\mu \partial_b X^\nu. \tag{0.27}$$

In four dimensions, the rigid string theory described by (0.26) is known to have instantons, with the instanton number being the self-intersection number $s(\Phi)$ of the worldsheet in the target. It is easy to show that

$$s(\Phi) = \int_\Sigma \mathrm{d}^2\sigma \; \epsilon^{ab} \nabla_a t^{\mu\nu} \nabla_b t^{\sigma\rho} \epsilon_{\mu\nu\sigma\rho}. \tag{0.28}$$

In the string picture of 4D QCD, $s(\Phi)$ is believed to correspond to the θ angle of the spacetime Yang-Mills theory [12]. It is possible to write down a Lagrangian for a topological rigid string theory in four dimensions, such that its path integral is localized at the moduli space of instantons with the instanton number given by (0.28). One can naturally ask whether such a topological rigid string theory would correspond to topological Yang-Mills theory in four dimensions [13], i.e. to the $SU(N)$ Donaldson theory in $1/N$ expansion. This question should be much easier to answer that its full-fledged physical counterpart.

Acknowledgement

It is a pleasure to thank the organizers for creating such a stimulating environment in Cargèse and for the opportunity to present these results. During the course of the work I have benefitted from discussions with and comments from P. Freund, D. Gross, D. Kutasov, E. Martinec and E. Witten. The research has been supported in part by NSF Grant PHY90-00386 and DOE Grant DEFG02-90ER40560.

References

1. D.J. Gross, "Some New/Old Approaches to QCD," Berkeley/Princeton preprint LBL-33232=PUPT-1355 (November 1992)
   J. Polchinski, "Strings and QCD?," Austin preprint UTTG-16-92 (June 1992)

2. D.J. Gross, "Two Dimensional QCD as a String Theory," *Nucl. Phys.* **B400** (1993) 161

3. D.J. Gross and W. Taylor, "Two Dimensional QCD is a String Theory," *Nucl. Phys.* **B400** (1993) 181; "Twists and Wilson Loops in the String Theory of Two Dimensional QCD," *Nucl. Phys.* **B403** (1993) 395

4. V.A. Kazakov and I.K. Kostov, "Non-Linear Strings in Two Dimensional U(∞) Gauge Theory," *Nucl. Phys.* **B176** (1980) 199
   V.A. Kazakov, "Wilson Loop Average for an Arbitrary Contour in Two-Dimensional U(N) Gauge Theory," *Nucl. Phys.* **B179** (1981) 283
   for recent reviews, see:

V. Kazakov, "A String Project in Multicolour QCD," LPTENS preprint (1993),

I.K. Kostov, "$U(N)$ Gauge Theory and Lattice Strings," Saclay preprint T93/079 (August 1993),

and this volume

5. P. Hořava, "Topological Rigid String Theory and Two Dimensional QCD," Chicago preprint, to appear (November 1993)

6. A.M. Polyakov, "Fine Structure of Strings," *Nucl. Phys.* **B268** (1986) 406; "*Gauge Fields and Strings*," section 10.4 (Harwood Publishers, 1987);

"A Few Projects in String Theory," Princeton preprint PUPT-1394 (April 1993)

7. A.A. Migdal, "Recursion Equations in Gauge Field Theories," *Zh. Eksp. Teor. Fiz.* **69** (1975) 810 [*Sov. Phys. JETP* **42** (1975) 413]

B.Ye. Rusakov, "Loop Averages and Partition Functions in U(N) Gauge Theory on Two-Dimensional Manifolds," *Mod. Phys. Lett.* **A5** (1990) 693; "Large-N Quantum Gauge Theories in Two Dimensions," Tel-Aviv preprint TAUP-2012-92 (December 1992)

8. E. Witten, "On Quantum Gauge Theories in Two Dimensions," *Commun. Math. Phys.* **141** (1991) 153; "Two Dimensional Gauge Theories Revisited," *J. Geom. Phys.* **9** (1992) 303

9. J. Eells and L. Lemaire, "A Report on Harmonic Maps," *Bull. London Math. Soc.* **10** (1980) 1; "Another Report on Harmonic Maps," *Bull. London Math. Soc.* **20** (1988) 385; "On the Construction of Harmonic and Holomorphic Maps between Surfaces," *Math. Ann.* **252** (1980) 27

10. P. Hořava, "Two Dimensional String Theory and the Topological Torus," *Nucl. Phys.* **B386** (1992) 383; "Spacetime Diffeomorphisms and Topological w_∞ Symmetry in Two Dimensional Topological String Theory," Chicago preprint EFI-92-70 (January 1993) hep-th/9302020, to appear in *Nucl. Phys.* **B** (1993)

11. M.F. Atiyah and L. Jeffrey, "Topological Lagrangians and Cohomology," *J. Geom. Phys.* **7** (1990) 119

for a review directed to physicists, see:

M. Blau, "The Mathai-Quillen Formalism and Topological Field Theory," Amsterdam preprint NIKHEF-H/92-07 (March 1992)

12. P.O. Mazur and V.P. Nair, "Strings in QCD and θ Vacua," *Nucl. Phys.* **B284** (1986) 146

13. E. Witten, "Topological Quantum Field Theory," *Commun. Math. Phys.* **117** (1988) 353

CONTINUUM QCD_2
IN TERMS OF DISCRETE RANDOM SURFACES
WITH LOCAL WEIGHTS

Ivan K. KOSTOV

Service de Physique Théorique CE-Saclay, F-91191 Gif-Sur-Yvette, France

The $1/N$ expansion of pure $U(N)$ gauge theory on a two-dimensional manifold $\mathcal{M}$ is reformulated as the topological expansion of a special model of random surfaces defined on a lattice $\mathcal{L}$ covering $\mathcal{M}$. The random surfaces represent branched coverings of $\mathcal{L}$. The Boltzmann weight of each surface is exp[-area] times a product of local factors associated with the branch points. The $1/N$ corrections are produced by surfaces with higher topology as well as by contact interactions due to microscopic tubes, trousers, handles, etc. The continuum limit of this model is the limit of infinitely dense covering lattice $\mathcal{L}$. The construction generalizes trivially to $D > 2$ where it describes the strong coupling phase of the lattice gauge theory. A possible integration measure in the space of continuous surfaces is suggested.

1 Introduction

The conjecture that the large N QCD is equivalent to some kind of free string theory dates from the original paper of 't Hooft [1]. Later on, the study of the loop equations showed that the simplest bosonic string should be abandoned as a possibility because it does not produce the correct contact terms. In the simplest case of two-dimensional space-time the loop equations have been solved [4] and the result for the Wilson loop average interpreted as a sum over minimal surfaces spanning the loop, with Boltzmann weights having both positive and negative signs. (This is the reason why the triviality theorem of Durhuus, Fröhlich and Jonsson [5] is not applicable here.) However, a representation of these weights in terms of a local world-sheet action was not given.

Alternatively, one can try to construct the chromodynamical string starting from the lattice regularization of the gauge theory [2]. The lattice formalism, in which

Quantum Field Theory and String Theory, Edited by
L. Baulieu *et al.*, Plenum Press, New York, 1995

the continuous Euclidean space-time is replaced by a discrete lattice $\mathcal{L}$, provides an unambiguous definition of both gauge theories and random surfaces. In the discretized theory, the gauge field is associated with the links $\ell = < xx' >$ of the lattice and the action is a sum over the elementary cells of the lattice. An example of a simple (but rather pathological) gauge theory which is equivalent to a model of random surfaces is the Weingarten model [6]. It can be described as a $U(N)$ pure lattice gauge theory, with the unitary measure replaced by a Gaussian measure in the space of unrestricted complex $N \times N$ matrices. The idea to represent, by means of an auxiliary Lagrange multiplier field, the $U(N)$ gauge theory in a form similar to the Weingarten model was suggested originally by V. Kazakov [7]. Here and in the subsequent works [8] [9] it was shown that the $1/N$ expansion of the Wilson $U(N)$ lattice gauge theory can be written as the topological expansion of a model of *noninteracting* surfaces on the lattice. Unfortunately this result was restricted to the strong coupling phase of the theory which is separated from the continuum limit by a large-N phase transition [10]. However, in two dimensions, it is possible to avoid this transition by an appropriate choice of the lattice action [11]. Among these actions, a special place is taken by the so called "heat kernel" action which has trivial scaling, that is, reproduces itself when the integration over a link variable is performed [12]. Using the heat kernel action we suggested some time ago a random surface ansatz for the large N limit of the Wilson loops in continuum QCD_2 (Appendix C of ref. [13]).

Very recently, some interesting exact results have been obtained for the partition function of $U(N)$ and $SU(N)$ gauge theories defined on a compact two-dimensional surface of genus G, refs. [14], [15], [16], [17], [18]. An interpretation of these partition functions in terms of surfaces was suggested by Gross and Taylor [15], [16]. The partition function can be written as a sum over oriented coverings of the space-time manifold of genus G and area A, where different sheets of a covering are glued together by means of branch points, microscopic handles and tubes, and what they called "omega points". The "omega points" represent topologically branch points, handles and tubes but have no location on the surface. This representation looks particularly simple on the torus where there are no such singular points at all. On a surface with genus G there are $|2 - 2G|$ "omega points" and it is not clear how they can be incorporated in a local string Lagrangian.

In this lecture we will explain the construction of ref. [13] giving a string representation of the Wilson loops in the large N limit, in such a way that it can be applied also for the partition function Z of the theory defined on an arbitrary $2D$ manifold $\mathcal{M}$, and in all orders in $1/N$.

First we choose a lattice (two-dimensional simplicial complex) $\mathcal{L}$ covering $\mathcal{M}$ and integrate over the gauge field in the 2-cells $c \in \mathcal{L}$. The result is a lattice gauge theory (with heat kernel action) for the residual $U(N)$ variables associated with the links (1-cells) $\ell \in \mathcal{L}$. We assume for simplicity that all 2-cells have the same area λ. Following the same strategy as in [13], we then reformulate the $U(N)$ lattice gauge theory as a theory of $N \times N$ complex matrix field with flat measure and regular interaction and then apply the standard Feynman-'t Hooft diagram technique.

The perturbative $1/N$ expansion of this model will be interpreted in terms of a discretized string theory with interaction constant $1/N$. The leading term is given by the path integral of the free string, that is, by the sum over noninteracting random surfaces on the lattice. The intrinsic geometry of the world sheet of this QCD string allows local singularities which can be interpreted as branch points of the corresponding Riemann surface. The branch points can occur at sites or links or cells of the lattice. The Boltzmann weight of such branched surface is equal to the usual exp [− area] times a product of factors associated with the branch points. These factors appear as the vacuum expectation values of special puncture operators located at the links and cells of the lattice. The world surface is also allowed to make folds but the total contribution of the surfaces with folds vanishes.

The subleading terms in the $1/N$ expansion have double origin. First, there is the contribution of the surfaces with higher topology. Second, the fluctuations of the puncture operators, which are of order $1/N^2$, lead to contact interactions of the surfaces. The vertices of these interactions can be imagined as microscopic tubes, trousers, ets., connecting several punctures.

The continuum limit of this random surface model is the limit of infinitely dense lattice $\mathcal{L}$ covering the manifold $\mathcal{M}$. Of course, for practical calculations it is easier to work with the minimal lattice for given problem, since the choice of the lattice is irrelevant. The connection with the approach of Gross and Taylor [15], [16] can be made by choosing a lattice containing a single cell and $2G$ links, obtained by cutting the manifold $\mathcal{M}$ along the cycles with nontrivial homotopy. In this way the mystery about the "omega points" in the surface interpretation of the free energy is resolved. Indeed, in the continuum limit (dense covering lattice $\mathcal{L}$), an "omega point" takes into account the contribution of a whole class of surfaces distinguished by the position of a *localized* singularity (e.g., branch point or microscopic tube). The sum of the Boltzmann weights of these surfaces is a topological invariant and is not sensitive to the choice of the covering lattice $\mathcal{L}$. This is why the "omega point" has no fixed location in $\mathcal{M}$.

Does this construction give a string theory for QCD_2? The answer depends on what we understand by string theory. Even if the continuum limit of the lattice QCD string is formally obtained as the limit $\lambda \to 0$, we do not know how to write it in the form of a continuum string theory with a local action on the world sheet. The most important step towards the continuum theory, the translation of the ansatz from the language of cellular complexes to the language of differential forms, is still to be done.

Another important unsolved problem is whether it is possible to construct a similar ansatz in $D > 2$ dimensions. Our construction generalizes trivially but it seems to describe only the strong coupling phase of the lattice gauge theory. We argue that the weak coupling phase can be achieved by relaxing the kinematical restrictions obeyed by the surfaces on the lattice, i.e., by enlarging the configuration space of surfaces. Our conjecture is that in the continuum limit these generalized surfaces are characterized by an "intrinsic" tangent plane at each point of the world sheet, representing a new dynamical field and coinciding with the tangent plane induced by the embedding only along the classical trajectories. The path integral for such string is a direct generalization of the phase-space path integral for the relativistic quantum particle.

The lecture has the following plan. In sect. 2 we reformulate the functional integral for the gauge theory as an integral over complex matrix fields with flat measure and regular interaction. This can be achieved by introducing special puncture operators coupled to the moments of the matrix fields and playing the rôle of Lagrange multipliers. In sect. 3 we perform the integration over the matrix fields using the 't Hooft diagrammatic rules and obtain as a result the partition function of a gas of lattice surfaces interacting with the vacuum and among themselves through the puncture operators. In sect. 4 we study the large N limit which is the classical limit for the puncture operators. We find explicitly the weights of the branched points and check the result on some simple examples of Wilson loops. Then we consider the two-loop correlator and the free energy on a compact lattice where the contact interactions should be taken into account. In sect. 5 we discuss what in the random surface ansatz should be modified in order to make it applicable for the Continuum limit of the $D > 2$ gauge theory, and what the random surface itself looks like. In the Appendix we derive the microscopic loop equations for the puncture operators.

2 QCD à la Weingarten

Let us consider the formal path integral of the $U(N)$ gauge theory on the oriented two-dimensional manifold $\mathcal{M}$

$$\mathcal{Z} = \int \prod_{x \in \mathcal{M}} dA_\mu(x) \exp[-\frac{1}{2}N \int \mathrm{tr}F^2 d^2x] \tag{2.1}$$

where the gauge field $A_\mu(x)$ belongs to the Lie-algebra of $U(N)$ and $F = \partial_1 A_2 - \partial_2 A_1 + [A_1, A_2]$ is the field strength. Choose a lattice $\mathcal{L}$ covering the manifold $\mathcal{M}$ and having the structure of a two-dimensional simplicial complex consisting of cells c, links ℓ and sites s. Assume for simplicity that all cells have the same area λ. After performing the integration over the gauge field inside the cells c, the integrand will depend only on the variables

$$U_\ell = \exp[i \int_{x_i}^{x_j} dx_\mu A_\mu(x)] \tag{2.2}$$

associated with the links $\ell = < s_i s_j >$ of the lattice and belonging to the fundamental representation of the group $U(N)$. The result of the integration in each cell c depends only on the holonomy factor associated with its boundary ∂c

$$U_c = \exp[i \oint_{\partial c} dx_\mu A_\mu(x)] = \prod_{\ell \in \partial c} U_\ell \tag{2.3}$$

and equals the heat kernel on the $U(N)$ group manifold.

It is perhaps worth to give an idea of how this can be proved. Let us assume for simplicity that the cell c is convex and choose the axial gauge $A_2 = 0$. Then the interior of the cell can be parametrized as the set of points (x_1, x_2) such that $x_1^d \leq x_1 \leq x_1^u$, $x_2^l(x_1) \leq x_2 \leq x_2^r(x_1)$ The residual gauge freedom can be killed by an an additional gauge condition along the boundary of the cell: $A_1(x_1, x_2^l(x_1)) = 0$. The functional integral over A_1 is Gaussian and is given (up to a multiplicative constant) by the exponential of the classical action

$$\mathcal{Z}_c = \int \prod_{x \in c} dA_2(x) e^{-1/2 \int_c d^2x (\partial_2 A_1)^2} = \text{constant} \times e^{-1/2 \int_c d^2x (\partial_2 A_1^{\text{classical}})^2} \tag{2.4}$$

where the classical solution satisfies the equation $\partial_2^2 A_1^{\text{classical}} = 0$. It is therefore linear in x_2 and depends on an arbitrary function of x_1. The representation

$$A_1^{\text{classical}} = (x_2 - x_2^l(x_1))U^{-1}(x_1)\partial_1 U(x_1) \tag{2.5}$$

where $U(x_1) \in U(N)$ allows to introduce the dependence on the boundary variable (2.3) in a simple way

$$U^{-1}(x_1^u)U(x_1^d) = U_c \tag{2.6}$$

The functional integral over the residual degree of freedom $U(x_1)$ is equal to the partition function of a one-dimensional $U(N)$ chiral field defined on the interval $0 \leq \tau \leq \lambda$, $\tau = \int_{x_1^D}^{x_1}(x_2^r(y) - x_2^l(y))dy$, with boundary condition $U^{-1}(\lambda)U(0) = U_c$

$$e^{S_\lambda(U_c)} = \int_{U^{-1}(\lambda)U(0)=U_c} \prod_{\tau \in [0,\lambda]} [dU(\tau)] e^{-\frac{1}{2}\int_0^\lambda [U^{-1}(\tau)\frac{\partial}{\partial \tau}U(\tau)]^2 d\tau} = \langle I|e^{\frac{\lambda}{2N}[U\partial/\partial U]^2}|U_c\rangle \tag{2.7}$$

where $[dU]$ is the invariant measure on the group $U(N)$. The Laplace operator $\hat{C}_2 = -\text{tr}(U^{-1}\partial/\partial U)^2$ (the quadratic Casimir operator) is diagonalized by the characters $\chi_R(U)$ of the irreducible representations R of the group $U(N)$. If $C_2(R)$ is the corresponding eigenvalue, then the character expansion of the exponential defines the *heat kernel* action

$$e^{S_\lambda(U)} = \sum_R \overline{\chi}_R(I)\chi_R(U)e^{-\frac{\lambda}{2N}C_2(R)} \tag{2.8}$$

After that the partition function (2.1) reduces to

$$Z(\mathcal{L}) = \int_{\ell \in \mathcal{L}}[dU_\ell] \exp[\sum_{c \in \mathcal{L}} S_\lambda(U_c)] \tag{2.9}$$

Now we apply a classical trick used in the mean field analysis of lattice gauge theories [21]. It consists in replacing the invariant measure over the unitary matrices U_ℓ with a homogeneous measure over non restricted $N \times N$ complex matrices V_ℓ. This is done by inserting the Fourier representation of the delta-function $\delta(U_\ell - V_\ell)$ in the integrand for each link

$$[dU] = \prod_{i,j=1}^N dV_i^j d\overline{V}_i^j \int dA_i^j d\overline{A}_i^j \int [dU] \, e^{N\text{tr}[A^\dagger(U-V)+A(U^\dagger - V^\dagger)]} \tag{2.10}$$

The new variables V_ℓ are unrestricted and the unitarity condition $V_\ell^\dagger V_\ell = I$ is imposed only on dynamical level through the complex $N \times N$ matrix Lagrange multiplier A_ℓ.

Now the U_ℓ variables decouple and the integration can be performed independently for each link. The one-link integral

$$e^{F(A)} = \int [dU]e^{N\text{tr}(A^\dagger U + AU^\dagger)} \tag{2.11}$$

has been the object of much attention in the past years. Its large-N limit was calculated explicitly by Brézin and Gross [22] as a function of the eigenvalues of the Hermitian matrix $A^\dagger A$.

After integrating with respect to the U-field we obtain an effective field theory of two coupled $N \times N$ complex matrix fields, A_ℓ and V_ℓ, associated with the oriented links ℓ of the lattice $\mathcal{L}$, where the convention

$$V_{\ell^{-1}} = V_\ell^\dagger, \quad A_{\ell^{-1}} = A_\ell^\dagger \tag{2.12}$$

is assumed. This theory is defined by the partition function

$$\mathcal{Z} = e^{2\mathcal{F}} = \int \prod_\ell \left(dA_\ell dV_\ell e^{-N\mathrm{tr}(A_\ell^\dagger V_\ell + A_\ell V_\ell^\dagger)} \right) \; e^{-\mathcal{E}[A,V]} \tag{2.13}$$

$$\mathcal{E}[A,V] = -\sum_\ell F(A_\ell^\dagger A_\ell) - \sum_c S_\lambda(V_c) \tag{2.14}$$

where V_c denotes the ordered product of link variables along the boundary ∂c of the elementary cell c

$$V_c = \prod_{\ell \in \partial c} V_\ell \tag{2.15}$$

A potentially subtle point is the necessity to define the action for all complex and not only unitary matrices. This has to be done by expanding $\exp S_\lambda(U)$ as a series in U and $U^{-1} = U^\dagger$ and replacing $U \to V$, $U^\dagger \to V^\dagger$.

The measure is now flat and the integrand is regular. Assuming that the action (2.14) is minimized by the trivial field $A = V = 0$, the functional integral can be calculated by expanding the exponential and applying all possible Wick contractions between the A and V fields

$$\langle (A_\ell)^i_j (V_\ell^\dagger)^k_l \rangle = \frac{1}{N}\delta^i_l \delta^k_j, \qquad \langle (A_\ell^\dagger)^i_j (V_\ell)^k_l \rangle = \frac{1}{N}\delta^i_l \delta^k_j \tag{2.16}$$

The potentials F and S are completely determined by two local Ward identities
a) The unitarity condition on the V-field

$$[\mathrm{tr}(\partial/\partial A^\dagger \, \partial/\partial A) - N]e^{F(A)} = 0 \tag{2.17}$$

b) The heat kernel equation

$$\left[2N\frac{\partial}{\partial \lambda} + \mathrm{tr}\left(U\partial/\partial U\right)^2\right]e^{S_\lambda(U)} = 0 \tag{2.18}$$

The second Casimir operator acts in the space of non restricted complex matrices as the projection of the full Laplace operator onto the hypersurface $V^\dagger V = 1$. Therefore the heat kernel equation (2.18) for the V field reads

$$\left[2N\frac{\partial}{\partial \lambda} + \mathrm{tr}\left(V\partial/\partial V - V^\dagger\partial/\partial V^\dagger\right)^2\right]e^{S_\lambda(V)} = 0 \tag{2.19}$$

This equation does not fix the dependence on the radial degrees of freedom. This is done by the additional assumption that $e^{S_\lambda(V)}$ is expandable in the matrix elements of V and $V^\dagger$.

The functions F and S_λ are regular for small fields and can be expanded in the momenta

$$a_n = \frac{\mathrm{tr}}{N}(A^\dagger A)^n, \quad v_n = \frac{\mathrm{tr}}{N}V^n, \ v_{-n} = \frac{\mathrm{tr}}{N}V^{\dagger n}; \qquad n = 1, 2, \ldots \tag{2.20}$$

On the other hand, we are looking for an action which is linear in the traces of the fields. This can be achieved by means of the integral representations

$$e^{F(A)} = \int \prod_{n=1}^{\infty} (da_n da_n^* \ e^{N^2 \frac{a_n^*}{n}[\frac{\mathrm{tr}}{N}(A^\dagger A)^n - a_n]}) \, e^{F[a]} \tag{2.21}$$

and

$$e^{S_\lambda(V)} = \int \prod_{n=1}^{\infty} (dv_n dv_n^* dv_{-n} dv_{-n}^* e^{N^2[\frac{v_n^*}{n}(\frac{\mathrm{tr}}{N}V^n - v_n) + \frac{v_{-n}^*}{n}(\mathrm{tr}V^{\dagger n} - v_{-n})]}) e^{S_\lambda[v]} \tag{2.22}$$

where

$$F[a] = N^2 \sum_{n=1}^{\infty} \sum_{k_1,\ldots,k_n \geq 1} f_{[k_1,\ldots,k_n]} \frac{a_{k_1}\ldots a_{k_n}}{n!} \tag{2.23}$$

and

$$S_\lambda[v] = \sum_{n=1}^{\infty} N^2 \sum_{k_1,\ldots,k_n \neq 0} s_{[k_1\ldots k_n]} \frac{v_{k_1}\ldots v_{k_n}}{n!} \tag{2.24}$$

are the expansions of F and S in terms of the momenta (2.20). The lowest-order coefficients are (see Appendix)

$$f_{[1]} = 1, \ f_{[2]} = -1 \ f_{[1,1]} = 1, \ f_{[3]} = 2, \ f_{[4]} = -5$$

$$s_{[1]} = s_{[-1]} = e^{-\lambda/2}, \ s_{[2]} = s_{[-2]} = (\cosh\frac{\lambda}{N}) - N\sinh\frac{\lambda}{N})e^{-\lambda},$$

$$s_{[1,1]} = s_{[-1,-1]} = [-N\sinh\frac{\lambda}{N} + 2(N\sinh\frac{\lambda}{2N})^2]e^{-\lambda}, s_{[1,-1]} = -e^{-\lambda}$$

$$\tag{2.25}$$

We therefore introduce at each link ℓ and at each cell c a set of auxiliary loop variables

$$a_n(\ell), a_n^*(\ell); \ \ \ell \in \mathcal{L}, \ n = 1, 2, \ldots \tag{2.26}$$

$$v_n(c), v_n^*(c); \ \ \ c \in \mathcal{L}, \ n = \pm 1, \pm 2, \ldots \tag{2.27}$$

and represent the exponent of the action (2.14) as the average with respect of the a^*, v^* fields

$$e^{-\mathcal{E}[A,V]} = \langle e^{-\mathcal{E}[A,V;\ a^*,v^*]} \rangle \tag{2.28}$$

where

$$\mathcal{E}[A, V;\ a^*, v^*] = -N^2\Big[\sum_{\ell \in \mathcal{L}; n > 0} \frac{a_n^*}{n} \frac{\mathrm{tr}}{N}(A_\ell^\dagger A_\ell)^n - \sum_{c \in \mathcal{L}; n > 0} \frac{1}{n}(v_n^* \frac{\mathrm{tr}}{N}V_c^n + v_{-n}^* \frac{\mathrm{tr}}{N}V_c^{\dagger n}) \tag{2.29}$$

and $\langle \ldots \rangle$ is defined by

$$\langle \mathcal{O}[a^*, v^*]\rangle = \int \prod_{\ell \in \mathcal{L}; n > 0} da_n(\ell) da_n^*(\ell) e^{-N^2 \frac{1}{n} a_n^*(\ell) a_n(\ell) + F[a(\ell)]}$$

$$\prod_{c \in \mathcal{L}; n \neq 0} dv_n(c) dv_n^*(c) e^{-N^2 \frac{1}{n} v_n^*(c) v_n(c) + S_\lambda[v(c)]} \ \mathcal{O}[a^*, v^*]$$

$$\tag{2.30}$$

The potentials F and S in the measure (2.30) can be determined from the loop transform of the original constraints (2.17) and (2.19) worked out in the Appendix. The last step is the inversion of the order of the integration with respect to the matrix and loop variables. The integral over the matrix fields can be interpreted as a theory of Weingarten type with fluctuating "coupling constants" described by the action (2.29)

$$e^{\mathcal{F}} = \left\langle \int \prod_\ell \left(dA_\ell^\dagger dA_\ell dV_\ell^\dagger dV_\ell e^{-N\mathrm{tr}(A_\ell^\dagger V_\ell + A_\ell V_\ell^\dagger)} \right) \; e^{-\mathcal{E}[A,V;\, a^*,v^*]} \right\rangle \qquad (2.31)$$

The Feynman diagrams of this theory will be interpreted as random surfaces embedded in the lattice $\mathcal{L}$. The vertices of these diagrams are themselves quantum fields (the a^*, v^* variables).

The effective action of these new gauge-invariant fields is proportional to N^2. Therefore, the limit $N \to \infty$ these fields freeze at their vacuum expectation values and we obtain a theory of noninteracting random surfaces.

3 Branched Random Surfaces

1. Feynman rules, punctures and branch points

By applying the rules of the Feynman-'t Hooft diagram expansion, the partition function of the gauge model 2.14 can be reformulated as the partition function of a gas of lattice surfaces. An important feature of these surfaces is that they can have branch points (that is, local singularities of the curvature) not only at the sites, as it is the case in the original Weingarten model, but also at the links and cells of the lattice. These surfaces are made from surface elements with the topology of a disk, glued together by identifying pairwise pieces of edges with opposite orientations along the links of the lattice (this is the meaning of the contractions (2.16)). There are two types of surface elements associated with the A and V vertices.

a) link-vertices

The vertex $a_n^*(\ell)\mathrm{tr}(A_\ell^\dagger A_\ell)^n$ will be represented graphically by a punctured disk with boundary going along the loop $(\ell\ell^{-1})^n$. The A-matrices are associated with the $2n$ edges and the a-operator - with the puncture. When shrunk to a point, the puncture becomes a branch point of order n located at the link ℓ. This surface element has zero area.

b) plaquette-vertices

The vertex $v_n^*(c)\mathrm{tr}V_c^n$ corresponds to a punctured disk with area $n\lambda$ covering n times the 2-cell c and with boundary going along the loop $(\partial c)^n$. This surface was called in ref. [13] plaquette-vertex of order n or simply n-plaquette. Again the V-matrices are associated with the edges of the boundary and the v^*-operator - with the puncture. A microscopic puncture creates a branch point of order n located at the cell c.

The vertex $v_{-n}^*(c)\mathrm{tr}V^{\dagger n}(\partial c)$ corresponds to the same surface but with the opposite orientation. Note that the v_n^*-operator changes to v_{-n}^* after changing the orientation. This means that the puncture is not just a point, but rather an oriented microscopic loop.

Expanding the exponential in a sum of monomials, we can assign to each monomial a set of multiplaquettes of different orders covering the 2-cells and a set of link elements associated with the 1-cells of the lattice $\mathcal{L}$. After applying many times the contractions (2.16) we obtain a nonzero result only if at each link there are equal numbers of edges belonging to the link and plaquette elements. In this case they form a collection of closed surfaces. The surfaces are built by identifying the edges of the surface elements going along the same link and having opposite orientations. Plaquette elements are glued only to link elements and vice versa.

The weight of each surface is the product of the a^* and v^* operators associated with its elements. As in all matrix field theories, the dependence on the number of colors N is through a factor N^χ where χ is the total Euler characteristic.

The sum over all surface configurations is given by $\exp(\mathcal{F}_0[a^*, v^*])$ where $\mathcal{F}_0[a^*, v^*]$ is the sum over all connected surfaces

$$\mathcal{F}_0[a^*, v^*] = \sum_{S: \partial S = 0} N^{\chi(S)}\, \Omega[a^*, v^*] \tag{3.1}$$

(This trivial exponentiation is the main advantage of the present method compared with the character expansion used in [15], [16].)

In the last formula $\chi(S)$ is the Euler characteristic of the surface, S is its area, and the Ω-factor is the product of all a^*, v^*-operators associated with the elementary disks from which the surface S is composed. If we denote by $\alpha_n(\ell)$ the number of the link-vertices of order n, located at the link ℓ, and by $\beta_n(c)$ - the number of n-plaquettes covering the cell c, then

$$\Omega[a^*, b^*] = \prod_\ell \prod_{n=1}^\infty [a_n^*(\ell)]^{\alpha_n(\ell)} \prod_c \prod_{n \neq 1} [v_n^*(c)]^{\beta_n(c)}, \qquad A(S) = \lambda \sum_{c \in \mathcal{L}} \beta_n(c) \tag{3.2}$$

Now the partition function of the gauge theory is obtained as the expectation value

$$e^\mathcal{F} = \langle e^{\mathcal{F}_0[a^*, v^*]} \rangle \tag{3.3}$$

where the average with respect to the puncture fields a^*, v^* is defined by eq.(2.30).

In this way we succeeded to reformulate the original gauge theory only in terms of the local gauge-invariant puncture operators a_n, a_n^*, v_n, v_n^* whose interaction is of order $1/N^2$. The fields a_n^*, v_n^* are associated with connected surfaces on the lattice $\mathcal{L}$ while the fields a, v are assigned to the vertices (the coefficients in the expansions (2.23) , (2.24)) which describe the contact interactions of surfaces. The vertices produced by the potential of the a_n, v_n fields can be imagined as a microscopic surfaces with several punctures.

One can calculate this partition function by expanding the exponentials in the measure (2.30) and applying all Wick contractions between the a, v and a^*, v^* fields. However, even in the limit $N \to \infty$, the result will not be a sum over noninteracting surfaces. Indeed, the weak interaction ($1/N^2$) is compensated by the N^2 factors in the weights of the vacuum surfaces. Therefore, the large-N free energy will be given by tree-like clusters of spherical surfaces with the form of a cactus which are analogous to the tree diagrams of an ordinary field theory containing tadpoles. This kind of expansion was derived in refs. [7] and [21].

The appearance of tadpoles means that the puncture operators have nonzero vacuum expectation values $\bar{a}_n, \bar{a}_n^*, \bar{v}_n, \bar{v}_n^*$. These are determined by the saddle point equations

$$n\bar{a}_n = \frac{\partial \mathcal{F}_0}{\partial \bar{a}_n^*} = \langle (A^\dagger A)^n \rangle, \quad n\bar{v}_n = \frac{\partial \mathcal{F}_0}{\partial \bar{v}_n^*} = \langle V_c^n \rangle; \quad n\bar{a}_n^* = \frac{\partial F}{\partial \bar{a}_n}, \quad n\bar{v}_n^* = \frac{\partial S_\lambda}{\partial \bar{v}_n} \qquad (3.4)$$

The diagrammatic rules obtained by expanding the fields around their classical values (3.4) do not contain tadpoles. (One can say, alternatively, that the tadpole surfaces are summed up by solving the classical equations of motion (3.4).)

In the "classical" limit $N \to \infty$, the puncture operators $a_n^*(\ell)$ and $v_n^*(c)$ freeze at their classical values and the free energy is given by the sum of noninteracting random surfaces

$$\mathcal{F} = \mathcal{F}_0(\bar{a}^*, \bar{v}^*) + \sum_\ell F[\bar{a}] + \sum_c S_\lambda[v] - \sum_{\ell, n \geq 1} n\bar{a}_n^* \bar{a}_n - \sum_{c, n \neq 0} n\bar{v}_n^* \bar{v}_n \qquad (3.5)$$

2. Irreducible surfaces

In general, the classical fields $\bar{a}^*, \bar{v}^*$ are complicated functions of the coupling constant λ (the area of an elementary cell) and the volume A_{tot} of the lattice. However, it is possible to simplify the whole construction by absorbing the sum over certain class of surfaces in the puncture operators. Namely, the sum over all reducible surfaces containing "baby universes" renormalizing the link vertices can be taken into account by adjusting the classical values $\bar{a}_n^*$ the puncture operators located at links. This is why we will replace the sum $\mathcal{F}_0$ over all connected surfaces by the sum $\mathcal{F}_I$ of irreducible surfaces defined as follows.

Definition:

> A surface S is *reducible* with respect to given link $\ell \in \mathcal{L}$ if it splits into two or more disconnected pieces after being cut along this link. A surface which is not reducible with respect to any $\ell \in \mathcal{L}$ is is called *irreducible*.

If we restrict the sum over surfaces to the irreducible ones, the free energy is obtained by the same formulae (3.4) and (3.5), with $\mathcal{F}_0$ replaced by $\mathcal{F}_I$. In what follows by sum over surfaces we will understand a sum over irreducible surfaces. This allows to simplify drastically the classical equations of motion (3.4) and solve them exactly. It turns out that, with this replacement, the expectation values of the puncture operators a_n^* and v_n^* coincide with the coefficients $f_{[n]}$ and $s_{[n]}$ in the expansions (2.23) and (2.24). The same is true for the nonplanar vertices describing the contact interactions which appear within the $1/N$ expansion. That is, the renormalization of the link and plaquette vertices due to the tadpole surfaces is miraculously compensated by the contribution of the reducible surfaces.

3. Evaluation of the link-vertices

Let us concentrate on the classical limit $N \to \infty$. Instead of solving directly (3.4) we would like to present a short-cut derivation using the hidden unitarity of the V-field.

Consider the Wilson loop average for the closed contour $C = (\ell_1\ell_2...\ell_n)$ on the lattice. It is given, in the limit $N \to \infty$, by a formula similar to (3.1), with the puncture operators replaced by their vacuum expectation values

$$W(C) = \langle \frac{\mathrm{tr}}{N} \prod_{\ell \in C} V(\ell) \rangle = \sum_{S:\partial S = C} \Omega[\bar{a}^*, \bar{v}^*] \tag{3.6}$$

Note that the sum over surfaces involves only these with the topology of a disk.

We will exploit the unitarity condition $VV^\dagger = I$, applied to the loop functional (3.6). It means that the Wilson average $W(C)$ will not change if a backtracking piece $\ell\ell^{-1}$ is added to the contour C

$$W(C\ell\ell^{-1}) = W(C) \tag{3.7}$$

The sum over surfaces spanning the loop $C\ell\ell^{-1}$ can be divided into two pieces

$$W(C\ell\ell^{-1}) = W(C)W(\ell\ell^{-1}) + W_{\mathrm{conn}}(C\ell\ell^{-1}) \tag{3.8}$$

The first term is the sum over all surfaces made of two disconnected parts spanning the loops C and $\ell\ell^{-1}$. The second term contains the rest. The constraint (3.7) is satisfied for all loops if $W(\ell\ell^{-1}) = 1$ and $W_{\mathrm{conn}}(C\ell\ell^{-1}) = 0$. But there is only one irreducible surface spanning the loop $\ell\ell^{-1}$; it contains a single link-vertex $a_1^*(\ell)$ contracting ℓ and ℓ^{-1}. Therefore $\bar{a}_1^* = 1$.

Now consider a surface contributing to the second term W_I. The links ℓ and ℓ^{-1} may be connected to the rest of the surface by means of the same link-vertex or by two different link-vertices. The condition that their total contribution is zero is given by

$$\bar{a}_n^* + \sum_{k=1}^{n-1} \bar{a}_k^* \bar{a}_{n-k}^* = 0, \quad n = 2, 3, ... \tag{3.9}$$

Eq. (3.9) is identical to the loop equation satisfied by the coefficients f_n in the expansion (2.23), which is solved by the Catalan numbers (see the Appendix)

$$\bar{a}_n^* = f_{[n]} = (-)^{(n-1)} \frac{(2n - 2)!}{n!(n - 1)!} \tag{3.10}$$

Since this point is important, let us give another derivation which might be combinatorially more transparent. Consider the sum over surfaces for the trivial Wilson loops

$$W[(\ell\ell^{-1})^n] = \langle \frac{\mathrm{tr}}{N}(V_\ell V_\ell^\dagger)^n \rangle = 1, \quad n = 1, 2, ... \tag{3.11}$$

For each of these Wilson loops the sum over irreducible surfaces contains only finite number of terms, namely, the link-vertices contracting directly the edges of the loop $\ell\ell^{-1}$. For example,

$$W(\ell\ell^{-1}) = \bar{a}_1^*, \quad W[(\ell\ell^{-1})^2] = (\bar{a}_1^*)^2 + \bar{a}_2^*, ... \tag{3.12}$$

Introducing the generating functions

$$w(t) = \sum_{n=0}^{\infty} t^n W[(\ell\ell^{-1})^n] = \frac{1}{1-t}, \quad f(t) = 1 + \sum_{n=1}^{\infty} t^n (\bar{a}_n^*)^n \tag{3.13}$$

we easily find the relation [13]

$$w(t) = f[tw^2(t)] \tag{3.14}$$

which is solved by the generating function for the Catalan numbers

$$f(t) = \frac{1 + \sqrt{1 + 4t}}{2} \tag{3.15}$$

We have found that, after restricting the sum over surfaces to the irreducible ones, the expectation values of the puncture operators associated with the links are the same as the bare ones. That is, the renormalization of the bare link vertices $f_{[n]}$ due to tadpole diagrams is neatly compensated by the contribution of the reducible surfaces.

This is true also for the nonplanar vertices responsible for the contact interactions of surfaces. The equations for the renormalized link vertices obtained from the surface representation of the multiloop correlators $\langle \frac{\text{tr}}{N}(V_\ell V_\ell^\dagger)^{k_1} \frac{\text{tr}}{N}(V_\ell V_\ell^\dagger)^{k_2} ... \rangle = 1$ are equivalent to the loop equations for the bare vertices considered in the Appendix.

4. loop equations

In ref. [13] we proved that the loop equations in the Wilson lattice gauge theory [23] are satisfied by the the sum over surfaces for the Wilson loop average $W(C)$. Let us only write here the general formula which is derived in the same fashion. For any link $\ell \in C$

$$\sum_{n=1}^{\infty} \sum_{c:\partial\ni\ell} \tilde{\beta}_n[(W(C(\partial c)^n) - W(C(\partial c)^{-n})] = \sum_{\ell' \in C} W(C_{\ell\ell'})W(C_{\ell'\ell})[\delta(\ell, \ell') - \delta(\ell^{-1}, \ell')] \tag{3.16}$$

The sum on the l.h.s. goes over the two cells whose boundary contains the link ℓ and the closed loops $C_{\ell\ell'}, C_{\ell'\ell}$ in the contact term are obtained by cutting the links ℓ and ℓ' and reconnecting them in the alternative way.

Eq. (3.16) is the lattice version of the loop equation in the continuum theory derived in [3].

5. The contribution of the surfaces with folds is zero

The weights of the branch points, as they were determined by eq. (3.9), provide a mechanism of suppressing the longitudinal modes of the string or, which is the same, world surfaces having folds. The surfaces with folds have both positive and negative weights and their total contribution to the string path integral is zero.

We have checked this in many particular cases but the general proof is missing. It is perhaps instructive to give one example.

Consider a surface having one fold. It covers three times the interior of a nonselfintersecting loop C and once the rest of the lattice. The change of the orientation of the world surface occurs along the loop. There are two specific points on the loop C at which the curvature has a conical singularity; they can be thought of as the points where the fold is created and annihilated. Each of these points can occur either at a site or at a link (in the last case it coincides with the branch point of order two of a link-vertex). Let us evaluate the total contribution of all irreducible surfaces distinguished by the positions of the two singular points. Denoting by n_0 and $n_1(= n_0)$ the number of sites and links along the loop $\mathcal{C}$, and remembering that a singular point located at a link has to be taken with a weight $f_{[2]} = -1$, we find that the contribution of these surfaces is proportional to

$$[\frac{n_0(n_0 - 1)}{2} + f_{[2]}n_0 n_1 + f_{[2]}^2 \frac{n_1(n_1 - 1)}{2}] - [n_1 + 2f_{[2]}n_1] = 0 \qquad (3.17)$$

The second term on the left hand side contains the contribution of the reducible surfaces which had to be subtracted. A reducible surface arises when the two points are located at the extremities of the same link (n_1 configurations) or when one of the points is a branch point and the other is located at one of the extremities of the same link ($2n_1$ configurations).

6. Evaluation of the plaquette-vertices

Now we are ready to evaluate the expectation values of the fields v_n^*. For this purpose we consider the sum over surfaces (3.6) for the one-cell Wilson loops $W_n = W\{(\partial c)^n\}$; $n = \pm 1, \pm 2, ...$, in the limit of infinite area $A_{\rm tot}$ covered by the lattice $\mathcal{L}$. In this limit the one-cell Wilson loops do not feel the rest of the lattice and coincide with these for the one-plaquette model. (This is a phenomenon specific to two dimensions). These loops can be calculated from the loop equations derived in the Appendix; they coincide with the coefficients s_n in the expansion (2.24).

Let us first observe that the only surfaces that contribute to the Wilson loop are these with minimal area $n\lambda$ covering n times the cell c. All other surfaces have folds (this is true only for an infinite lattice) and their contributions cancel. Therefore the sum (3.6) contains only finite number of terms and can be easily evaluated. Let l be the number of links forming the boundary of the cell c. Then

$$W_n = (\overline{a}_1^*)^{nl}\overline{v}_n^* + nl(\overline{a}_1^*)^{nl-2}(\overline{a}_2^* + (\overline{a}_1^*)^2) \sum_{k=1}^{n-1} \overline{v}_k^*\overline{v}_{n-k}^* + ... \qquad (3.18)$$

One can easily see that, due to the identity (3.9), only the first term survives and therefore

$$\overline{v}_n^* = \overline{v}_{-n}^* = s_{[n]}; \quad n = 1, 2, 3, ... \qquad (3.19)$$

In this way the expectation values of the v^* fields in presence of the vacuum surfaces is not renormalized as well.

A similar statement is true for the nonplanar plaquette vertices: the contact interactions of the puncture operators v_n^* are given by the bare nonplanar link vertices.

7. String representation for the Wilson loops on the infinite plane

Now we are able to formulate the string ansatz for the Wilson loop average as the sum over all irreducible surfaces without folds, spanning the loop $\mathcal{C}$ and having branch points of all orders. It is convenient to extract the trivial Nambu factor exp[-area] from Boltzmann weight of a surface $\mathcal{S}$. The remaining factor $\Omega(\mathcal{S})$ then comes only from the singularities of the world-sheet geometry. Then eq.(3.6) takes the form

$$W(\mathcal{C}) = \sum_{\partial \mathcal{S} = \mathcal{C}} \Omega(\mathcal{S})e^{-\frac{1}{2}\mathrm{Area}(\mathcal{S})} \qquad (3.20)$$

where the factor $\Omega(\mathcal{S})$ is a product of the weights of all branched points of orders 2,3,...
on the world sheet. Let us remind that a branch point of order n means a curvature
$2\pi(n-1)$ concentrated at this point.

A branch point of order n $(n = 1, 2, 3, ...)$ associated with a k-cell $(k = 0, 1, 2$ is
weighed by a factor $\omega_n^{(k)}$ where

$$\omega_n^{(0)} = 1$$
$$\omega_n^{(1)} = f_{[n]} = (-)^{(n-1)}\frac{(2n-2)!}{n!(n-1)!}$$
$$\omega_n^{(2)} = s_{[n]}(s_{[1]})^{-n} = \sum_{m=0}^{n-1} \frac{n!}{(n-m-1)!(m+1)!}\frac{n^{m-1}}{m!}(-\lambda)^m = (1 - \frac{n(n-1)}{2}\lambda + ...)$$

$$(3.21)$$

The coefficient in front of the area in eq.(3.20) comes from the explicit form $s_{[1]} =
e^{-\lambda/2}$ of the simple plaquette weights. The weight of an n-plaquette is $e^{-n\lambda/2}$ times a
polynomial in λ. The exponential contributes to the Nambu factor and the polynomial
counts for the branch point. In the Continuum limit $\lambda \to 0$ one can retain only the
linear term of this polynomial.

8. Vacuum surfaces on the plane

Consider the sum over vacuum surfaces in the large N limit. It follows from 3.5
that the vacuum energy is equal to the sum over surfaces that, after being cut along
all copies of a given link, split only into irreducible pieces. On the infinite plane, the
only surfaces satisfying these conditions are made by identifying the boundaries of two
multiplaquettes and the free energy is given by

$$\mathcal{F} = (\#\text{cells}) \sum_{n=1}^{\infty} \frac{s_n^2}{n} \tag{3.22}$$

In the Continuum limit $\lambda \to 0, A_{\text{tot}} = \lambda\#\text{plaquettes} = $ const, we find, up to an infinite
constant,

$$\mathcal{F} = \frac{A_{\text{tot}}}{\lambda} \sum_{n=1}^{\infty} \frac{1}{n}(1 - n^2\lambda + ...) = \frac{1}{12}A_{\text{tot}} + \text{ infinite constant} \tag{3.23}$$

9. Interactions. Compact lattices

For a compact lattice $\mathcal{L}$ with the topology of a sphere and area A_{tot} the sum over
surfaces becomes much more involved. It will contain a sum over spherical surfaces
representing multiple branched coverings of $\mathcal{L}$. There are infinitely many surfaces contributing to the elementary Wilson loops. For example,

$$W_1 = e^{-\frac{1}{2}\lambda} + e^{-\frac{1}{2}(A_{\text{tot}})} + ... \tag{3.24}$$

Therefore, we have either to modify the weights of the irreducible surfaces, or to use the
same weights as for the infinite lattice but allow contact interactions due to microscopic
tubes, trousers, etc., connecting two, three, etc. surface elements. The amplitudes of
these contact interactions are given by the coefficients in the expansion (2.24).

We are not going to discuss this case in details since there is nothing conceptually
new. As a single example we will calculate the two leading orders of the free energy
due to the surfaces defining simple and double coverings of $\mathcal{L}$.

Let n_i be the number of i-cells of the lattice $\mathcal{L}$ ($i = 0, 1, 2$). The free energy should depend only on the total area $A_{\text{tot}} = n_2\lambda$. The two simple coverings (with the two possible orientations) contribute the term $2e^{-\frac{1}{2}A_{\text{tot}}}$. There are two kinds of double coverings depending on the orientations of the two layers.

Equal orientations

The sum over cylindric surfaces with a cut is given by the same formula as for the two-loop correlator, with the only difference that here the Euler relation gives $n_0 - n_1 + n_2 = 2$. The result is $(1 - \lambda n_2 + \frac{1}{2}\lambda^2)e^{-\lambda A_{\text{tot}}}$. We have to add the term with contact interaction due to the vertex $\omega^{(2)}_{[1,1]} = s_{[1,1]}/(s_{[1]})^2 = (-\lambda + \frac{1}{2}\lambda^2)$ connecting two equally oriented plaquettes: $n_2(1 - \lambda)e^{-A_{\text{tot}}}$.

Opposite orientations

The sum over irreducible surfaces consisting of two sheets with opposite orientations and connected along a tree-like configuration of links, was calculated in Appendix B of [13] and equals $(p - 2)e^{-A_{\text{tot}}}$. We have to add to it the term with contact interaction due to the vertex $f_{[1,-1]} = -1$ connecting two plaquettes with opposite orientations: $-n_2 e^{-A_{\text{tot}}}$.

Collecting all terms, we find the result obtained using the character expansion [18]

$$\mathcal{F} = 2e^{-\frac{1}{2}A_{\text{tot}}} + [-1 - 2A_{\text{tot}} + \frac{1}{2}A_{\text{tot}}^2]e^{-A_{\text{tot}}} + \ldots \tag{3.25}$$

An interesting phenomenon on compact lattices was observed recently by Douglas and Kazakov [18]. The sum over surfaces for the free energy on the sphere is convergent only for sufficiently large total area

$$A_{\text{tot}} > \pi^2 \tag{3.26}$$

The explanation of this singularity is that a t the critical point $A_{\text{tot}} = \pi^2$ the entropy of the branched points overwhelms the energy and the effective string tension vanishes. The phenomenon resembles to what happens with the compactified bosonic string at the Hagedorn radius.

The singularity at $A = \pi^2$ can be observed already in the behaviour of the multiple Wilson loops surrounding a contour of area A. ¿From the explicit expression given in the Appendix (with λ replaced by A) one finds the asymptotics at $n \to \infty$

$$W_n(A) \approx \frac{1}{\sqrt{\pi}}(n\sqrt{A})^{-3/2}\cos(2n\sqrt{A} - 3\pi/4) \tag{3.27}$$

The value of A for which the oscillations disappear is exactly the critical area $A_c = \pi^2$. This is not a surprise since on a compact lattice with area A, the free energy contains a term $\sum_n \frac{1}{n}W_n^2(A)$. (Compare with eq. (3.22).)

Finally, let make a remark concerning the case of a toroidal manifold $\mathcal{M}$. It was nicely demonstrated by Gross and Taylor that the free energy of the gauge theory on the torus can be interpreted as the sum over all connected nonfolding oriented surfaces wrapping the torus, with right combinatorial factors. If we add to this sum the sum over microscopic surfaces (3.22), the result will be exactly the logarithm of the η-function

$$\mathcal{F}_{\text{torus}} = \frac{1}{12}A_{\text{tot}} - \sum_{n=1}^{\infty}\log(1 - e^{-nA_{\text{tot}}}) \tag{3.28}$$

4 Discussion

We have formulated a model of random surfaces (lattice strings) satisfying the back-tracking condition (3.7) and the lattice loop equation (3.16). This model yields the free energy and the loop correlators of the continuum $U(N)$ gauge theory in the limit $N \to \infty$ as well as the $1/N$ corrections.

Nevertheless this is still not a continuum string theory with a local action on the world sheet. The most important step towards the continuum theory, the translation of the random surface ansatz from the language of cellular complexes to the language of differential forms, is still to be done. The main problem to solveis to reformulate the contributions of the puncture operators in terms of local fields defined on the world sheet. In other words, we have to look for a continuum version of the puncture operators (2.26) and (2.27) defined on the 1-cells and 2-cells of the lattice using the language of differential forms.

The weights (3.21) of the branch points consist of a dimensionless "topological" term and a term proportional to the infinitesimal area λ. (The higher powers in λ can be neglected in the continuum limit.) The term proportional to λ is easy to interpret in the Continuum limit: it corresponds to a point-like singularity of the intrinsic curvature weighed by a factor (-1). The factor λ comes from the measure over surfaces, as we have noticed in [13].

The real problem consists in the interpretation of the dimensionless weights. They seem to be associated with puncture operators which are pure derivatives. These are the only singularities in the "topological" limit $\lambda = 0$ of the model. In this limit the vector potential is restricted to be a pure gauge, $A_\mu(x) = U^{-1}(x)\partial_\mu U(x)$. In this "topological" string theory the sum over surfaces bounded by a loop is always 1.

Presumably there can exist only a finite number of local excitations with the lowest dimension. We have to study the classes of universality of local operators on the world sheet to extract what is left from the puncture operators in the Continuum limit.

A Loop Equations for the Puncture Operators

1. Loop equations for the link vertices

The expansion of $F(A)$ in terms of the momenta (2.20) has been found by O'Brien and Zuber [9] in the form (2.23) where the coefficients $f_{[k_1,\dots,k_n]}$ were obtained in the large N limit by a system of algebraic relations suggested originally by Kazakov in [7]. These relations are equivalent to eq. (2.17) written in terms of loop variables

$$\left(\sum_{n\geq 1} n a_{n-1}\partial_n + \frac{1}{N} \sum_{k,n\geq 1} [(n+k+1)a_n a_k \partial_{k+n+1} + \frac{nk}{N^2} a_{n+k-1}\partial_n \partial_k] - 1 \right) e^{F[a]} = 0 \quad (A.1)$$

where we denoted $a_0 = 1, \partial_n = \partial/\partial a_n$. Eq. (A.1) is sufficient to determine the $1/N$ expansion of the link-vertices $f_{[k_1,\dots,k_n]}$. In the large N limit the identity A.1 is equivalent to a system of recursive relations [9]

$$f_{[1]} = 1$$

$$f_{[k_1,...,k_n]} + \sum_{j=2}^{n} k_j f_{[k_1+k_j,k_2,...,\bar{k}_j,...,k_n]} + \sum_{k=1}^{k_1-1} \sum_{\alpha \in P(L)} f_{[k_1-k,L\setminus\alpha]} f_{[k,\alpha]} = 0 \tag{A.2}$$

where $L = \{k_2,...,k_n\}$ and $P(L)$ is the set of all subsets of L, including the empty set, and the bar means omitting the argument below it. The equations for the weights $f_{[k]}$ of the disks [7]

$$f_{[1]} = 1; \quad f_{[n]} + \sum_{k=1}^{n-1} f_{[k]} f_{[n-k]} = 0, \quad n = 1, 2, ... \tag{A.3}$$

are solved by the Catalan numbers

$$f_{[n]} = (-)^{(n-1)} \frac{(2n-2)!}{n!(n-1)!} \tag{A.4}$$

2. loop equations for the plaquette vertices

The coefficients in (2.24) can be found order by order from the character expansion of the heat kernel

$$e^{S_\lambda(V)} = \sum_R \overline{\chi}_R(I) \chi_R(V, V^\dagger) e^{-\lambda(2N)^{-1} C_2(R)}$$
$$= 1 + N(\mathrm{tr}V + \mathrm{tr}V^\dagger)e^{-\lambda/2} + (N^2-1)(\mathrm{tr}V\,\mathrm{tr}V^\dagger - 1)e^{-\lambda} + ... \tag{A.5}$$

$$S_\lambda[v]/N^2 = (v_1 + v_{-1})e^{-\lambda/2} - e^{-\lambda}(v_1 v_{-1} + 1)$$
$$+ \tfrac{1}{2}(v_2 + v_{-2})(\cosh \tfrac{\lambda}{N} - N \sinh \tfrac{\lambda}{N})e^{-\lambda} + \tfrac{1}{2!}(v_1^2 + v_{-1}^2)(-N \sinh \tfrac{\lambda}{N}$$
$$+ 2(N \sinh \tfrac{\lambda}{2N})^2)e^{-\lambda} + ... \tag{A.6}$$

which gives, in the limit $N \to \infty$,

$$s_{[1]} = e^{-\lambda/2},$$
$$s_{[2]} = (\cosh \tfrac{\lambda}{N} - N \sinh \tfrac{\lambda}{N})e^{-\lambda} = (1-\lambda)e^{-\lambda} + \mathcal{O}(\tfrac{1}{N^2})$$
$$s_{1,-1} = -e^{-\lambda}$$
$$s_{[1,1]} = (-N \sinh \tfrac{\lambda}{N} + 2(\sinh \tfrac{\lambda}{2N})^2)e^{-\lambda} = (-\lambda + \tfrac{1}{2}\lambda^2)e^{-\lambda} + \mathcal{O}(\tfrac{1}{N^2}) \tag{A.7}$$

The heat kernel equation (2.19) yields, when applied to both sides of (2.22), the following loop equation (we use the shorthand notation $\partial_n = \partial/\partial v_n$)

$$\left(2\partial/\partial\lambda + \sum_{n\neq 0} |n| v_n \partial_n + \sum_{k,n\neq 0} [|n+k|\theta(nk)v_n v_k \partial_{n+k} + \frac{nk}{N^2} v_{n+k} \partial_n \partial_k]\right) e^{S_\lambda[v]} = 0 \tag{A.8}$$

This identity allows in principle to determine the coefficients $s_{[k_1,...,k_n]}$ in the expansion of S in the loop variables, but it is not triangular and does not produce simple recursion relations. Such recursion relations will be written below for the set of all possible loop correlators

$$W_{k_1,\ldots,k_n} = N^{n-2}\langle \mathrm{tr}U^{k_1}\ldots\mathrm{tr}U^{k_n}\rangle_{\text{connected}} \tag{A.9}$$

The coefficients $s_{[k_1,\ldots,k_n]}$ are related to the elementary Wilson loop correlators (A.9) by a simple transformation, to be defined below.

Let us first remind the expression for the generating function for the $U(N)$ characters with strictly positive chirality (i.e., containing only powers of U and not $U^\dagger$).

$$\sum_{R_+} \chi_R(U)\chi_R(V^{-1}) = \prod_{i,j=1}^{N} \frac{1}{1 - u_i/v_j} = \exp\Big(\sum_{n=1}^{\infty} \frac{1}{n}\mathrm{tr}U^n\mathrm{tr}V^{-n}\Big) \tag{A.10}$$

It is possible to write a similar formula for the nonrestricted sum over the irreducible representations R

$$\delta(U,V) = \sum_{R} \chi_R(U)\chi_R(V^\dagger) = \exp\Big(-\sum_{n=1}^{\infty}\frac{1}{n}(\mathrm{tr}U^n - \mathrm{tr}V^n)(\mathrm{tr}U^{\dagger n} - \mathrm{tr}V^{\dagger n})\Big) \tag{A.11}$$

This formula which makes sense only within the large N expansion and for small arguments U and V, is similar to eq. (22) in the paper by M. Douglas [17]. We checked it in many particular cases but could not find a proof.

Using the above representation of the δ-function, we write the exponent of the lattice action as the integral

$$\begin{aligned}
e^{S_\lambda[V]} &= \int \mathcal{D}U e^{S(U)} \exp\Big(-N\sum_{n\geq 1}\frac{1}{n}(Nv_n - \mathrm{tr}U^n)(Nv_{-n} - \mathrm{tr}U^{-n})\Big)\\
&= \int_{-i\infty}^{i\infty} \prod_{n=1}^{\infty} dy_n dy_{-n} \exp N^2\Big(\sum_{n=1}^{\infty}\frac{1}{n}(y_n y_{-n} - y_n v_{-n} - y_{-n}v_n) + W_\lambda[y]\Big)
\end{aligned} \tag{A.12}$$

where $v_n = \frac{\mathrm{tr}}{N}V^n$, $v_{-n} = \frac{\mathrm{tr}}{N}V^{\dagger n}$ and

$$W_\lambda[y] = \log\int[dU]\exp[S_\lambda(U) + N\sum_{n\neq 0}\frac{y_n}{n}\mathrm{tr}U^n] = \sum_n \frac{1}{n!}\sum_{k_1,\ldots,k_n\neq 0}\frac{y_{k_1}}{k_1}\ldots\frac{y_{k_n}}{k_n}W_{k_1,\ldots,k_n} \tag{A.13}$$

is the generating function for the connected Wilson loop correlators (A.9). This transformation can be written also in the operator form

$$e^{S_\lambda[v]} = \exp\Big(\sum_{n=1}^{\infty}\frac{\partial}{\partial v_n}\frac{\partial}{\partial v_{-n}}\Big) e^{N^2 W_\lambda[v]} \tag{A.14}$$

In the topological limit $\lambda = 0$ all loop correlators become trivial

$$W_{\lambda=0}[y] = \sum_{n=1}^{\infty}\frac{y_n}{n} \tag{A.15}$$

and the transformation (A.14) yields

$$S_0[v] = -\sum_{n=1}^{\infty}(v_n - N)(v_{-n} - N) \tag{A.16}$$

For nonzero λ the relation between the coefficients $s_{[\ldots]}$ in the expansion (2.24) and the Wilson loop correlators is

$$s_{[k_1,\ldots,k_n]} = W_{k_1,\ldots,k_n}, \qquad n \neq 2 \tag{A.17}$$

and, for $n = 2$

$$s_{[k_1,k_2]} = W_{k_1,k_2} - \delta_{k_1,-k_2} \tag{A.18}$$

In particular, $s_n = W_n$, as it has been noticed in [13]. Now we face the problem of evaluating the one-cell Wilson loops.

3. The one-cell Wilson loops

Applying the identity (2.19) and integrating by parts in (A.13), we find the following loop equation

$$\left[2\frac{\partial}{\partial\lambda} + \sum_{n\neq 0}|n|x_n\partial_n + \sum_{n,k\neq 0}\left(nx_nkx_k\partial_{n+k} + \frac{1}{N^2}\theta(kn)(|n|+|k|)x_{n+k}\partial_n\,\partial_k + \right)\right]e^{N^2W[x]} = 0$$

$$\tag{A.19}$$

This equation is a particular case of the Migdal-Makeenko equations derived for a general Wilson loop in [4].

In the limit $N \to \infty$ the loop equation (A.19) yields a system of recursive equations for the connected correlators $W_{k_1\ldots k_n}$

$$(2\frac{\partial}{\partial\lambda}+\sum_{i=1}^{n}k_i)W_{k_1,k_2,\ldots,k_n}+\sum_{i\neq j}^{n}k_ik_jW_{k_i+k_j,k_2,\ldots,\bar{k}_j,\ldots,k_n}+\sum_{i=1}^{n}\sum_{k=1}^{k_i-1}\sum_{\alpha\in P(L)}|k_i|W_{k_i-k,L\setminus\alpha}W_{k,\alpha} = 0$$

$$\tag{A.20}$$

In particular, for the simple Wilson loops $W_n = \langle\frac{\mathrm{tr}}{N}U^n\rangle$, eq. (A.20) reads

$$2\frac{\partial}{\partial\lambda}W_n + n\sum_{k=1}^{n}W_kW_{n-k} = 0 \tag{A.21}$$

This equation has been studied in [12] . The solution of (A.21) is

$$W_n = e^{-n\lambda/2}\sum_{m=0}^{n-1}\frac{(-n\lambda)^m}{(m+1)!} = e^{-n\lambda/2}(1 - \frac{n(n-1)}{2}\lambda + \ldots) \tag{A.22}$$

References

[1] G. 't Hooft, Nucl. Phys. B72 (1974) 461

[2] K. Wilson, Phys. Rev. D10 (1974) 2445; A.M. Polyakov, *unpublished*

[3] Yu. M. Makeenko and A. A. Migdal, Nucl. Phys. B188 (1981) 269, A. A. Migdal, Phys. Rev. 102 (1983) 201

[4] V. Kazakov and I. Kostov, Nucl. Phys. B176 (1980) 199; V. Kazakov, Nucl. Phys. B179 (1981) 283

[5] B.Durhuus, J. Fröhlich and T. Jonsson, Nucl. Phys. B225 (1983) 185; Nucl. Phys. B240 (1984) 453

[6] D. Weingarten, Phys. Lett. 90B (1980)285

[7] V.Kazakov, Phys. Lett. 128B (1983) 316, JETP (Russian edition) 85 (1983) 1887

[8] I. Kostov, Phys. Lett. 138B (1984) 191, 147B (1984) 445

[9] K.H. O'Brien and J.-B. Zuber, Nucl. Phys. B253 (1985) 621, Phys. Lett. 144B (1984) 407

[10] D. Gross and E. Witten, Phys. Rev. D21 (1980) 446; S. Wadia, Chicago preprint EFI 80/15, unpublished

[11] N.S. Manton, Phys. Lett. 96B (1980) 328; P. Menotti and E. Onofri, Nucl. Phys. B190 (1984) 288; C.B. Lang, P. Salomonson and B.S. Skagerstam, Phys. Lett. 107B (1981) 211, Nucl. Phys. B190 (1981) 337

[12] P. Rossi, *Ann. Phys.* 132 (1981) 463

[13] I. Kostov, Nucl. Phys. B265 (1986) 223

[14] B. Rusakov, *Mod. Phys. Lett.* A5 (1990) 693

[15] D. Gross, Princeton preprint PUPT-1356, hep-th/9212149

[16] D. Gross and W. Taylor IV, preprints PUPT-1376 hep-th/9301068 and PUPT-1382 hep-th/9303046

[17] M. Douglas, Preprint RU-93-13 (NSF-ITP-93-39); A. D'Adda, M. Caselle, L. Magnea and S. Panzeri, Preprint hep-th 9304015; J. Minahan and A. Polychronakos, hep-th/9303153

[18] M. Douglas and V. Kazakov, preprint LPTENS-93/20

[19] A. Polyakov, Phys. Lett. B (103) 207

[20] A. Polyakov, preprint PUPT-1394, April 1993

[21] J.-M. Drouffe and J.-B. Zuber, *Phys. Rep.* 102, Nos. 1,2 (1983)1-119, section 4

[22] E. Brézin and D. Gross, Phys. Lett. 97B (1980) 120

[23] A.A. Migdal, *unpublished* (1978); D. Förster, Phys. Lett. 87B (1979) 87; T. Eguchi, Phys. Lett. 87B (1979) 91

STRINGS AND CAUSALITY

Emil MARTINEC

Enrico Fermi Inst. and Dept. of Physics
University of Chicago, Chicago, IL 60637

One of the early concerns in quantum field theory was causality – can one ensure that measurements at spacelike separation do not interfere? General issues of this type were important in the days when little was understood about the underlying dynamics of particle interactions; the hope was to place constraints on the types of allowed dynamics and interactions on the basis of cherished tenets of kinematics. There were in fact two ways to approach the problem. First, one could take the point of view that only the results of scattering experiments had physical content; then causal behavior means that the scattered wave cannot reach the detector before the incident wave strikes the target. We might describe this as global causality, and it is obeyed by tachyon-free string theory S-matrices. However there is a second, local version of causality, which asks whether local measurements commute at spacelike separation. Local causality ensures global causality in ordinary field theory, and typically one can construct global violations from local ones, so it would seem that the two are equivalent. But in quantum field theory the local question is more fundamental, and can be resolved without solving any complicated global problem such as long-distance signal propagation. Local causality rests on the universal properties of field theory, *e.g.* that any manifold locally looks the same, rather than their implementation in any particular spacetime background. For instance we can be sure that an interacting scalar field propagating in a Schwarzschild geometry obeys the axioms of local field theory without having to solve for the full S-matrix of the problem.

The consequences of causality in field theory are dispersion relations arising from the analyticity of correlation functions in the complex plane of the kinematical invariants. These dispersion relations reveal themselves in terms of the analyticity properties of the correlation functions[1]. For instance the N-point correlation function of a scalar field

$$\langle 0|\phi(x_1)\cdots\phi(x_N)|0\rangle = \int ds_{ij}\, G_N(s_{ij})\, \Delta_N(x_i - x_{i+1}; s_{ij}) \qquad (0.1)$$

Quantum Field Theory and String Theory, Edited by
L. Baulieu *et al.*, Plenum Press, New York, 1995

$$\Delta_N = \int dp_i \, \exp[ip_j \cdot (x_j - x_{j+1})] \delta(p_i \cdot p_j - s_{ij}) \prod_k \theta(p_k^0) \qquad (0.2)$$

is analytic in the complex plane of the kinematical invariants s_{ij} (for the two-point function this is the Källen-Lehmann representation of the propagator). The representation splits dynamics contained in the function G_N from the kinematics residing in the basic objects Δ_N. Dispersion relations arise from the deformation of the contour integral over s_{ij} onto cuts.

In quantum particle-field theory, we are used to the idea that dynamics takes place on a given spacetime manifold, with specified causal structure. A generic kinematic structure provides the arena in which dynamics unfolds. Causality of the dynamics is intimately tied to the topology of light cones. One of the great confusions of the present day is how to reconcile fluctuations of the causal structure and causality; how can quantum superposition and causal relationships of events coexist? It is an issue the Hawking paradox frames sharply.

String theory is in an even more primitive state. I think it is fair to say that the issue of what a light cone is in string theory is not yet settled even in the classical theory. The string S-matrix is causal when considered as a function of the center of mass momenta of scattering strings. However the center of mass momentum is conjugate to the center of mass position of the string, which may or may not be causally related to the locations of string interactions which take place locally on the string; it is not a priori obvious how to relate this global causality in the center of mass to some local notion of causality in, say, the loop space that is the arena of string dynamics as we now understand it. There is of course the natural notion that the light cone structure of loop space is induced from the light cone topology of its underlying point manifold. But can there be a different notion of light cone? Indeed, is there any operational sense in which the light cone structure is determined by the point manifold light cones when there are no pointlike objects which could detect this causal structure? For instance, with the infinity of points on the string, there are infinitely many Lorentz invariant quantities one can build; what could be the analog of Δ_N in string field theory?

A number of arguments have been put forward proposing that string theory exhibits a 'generalized uncertainty principle' whereby objects cannot be localized to a region smaller than the string scale $\ell_s = \sqrt{\hbar c \alpha'}$ [2][3][4] [5][6] There are two models for how this principle manifests itself in string theory. The first model is entropic; it is not that there is a limiting size in string theory, but rather that the *probability* of observing pointlike behavior is extremely small (*e.g.* form factors for hard scattering have Gaussian falloff as one begins to probe constituent structure[4], or thermal partition functions exhibit slower growth at high temperature[6]). Such a model would not require a fundamental overhaul of the conceptual foundations of spacetime geometry. The second model is the Heisenberg uncertainty relation – short distance measurements are meaningless and we should reformulate the theory in such a way that it is impossible to ask about them. The irrelevance of spacetime discreteness[5][7] and the smearing introduced by renormalization[3] may point in this direction. If this second model were the proper setting, then one would probably need to abandon the notion of having the string causal structure induced from a point manifold, if not the notion of manifolds and loop space altogether. Obviously the choice of model made by string theory has profound implications for the question of what a horizon (and hence a black hole) is in string theory.

One approach to the causality problem is to work backwards: to take a given string theory and try to deduce the causal structure from a specific set of measurements. This has the unfortunate property that it does not necessarily isolate the kinematics from the dynamics. In addition, the specific measurements we shall choose below are not gauge invariant, and eventually we will have to understand how to separate the causal structure and the gauge dependence. In particle-field theory, the light cone is quite simply found as the boundary of the domain of commutativity of two scalar field operators. We will adopt the same definition for string theory and see where it leads us.

Consider the two-point function in particle field theory

$$\langle 0 | \, [\phi(x), \phi(x')] \, | 0 \rangle = \oint_C \frac{dp^0}{2\pi} \int \frac{d\vec{p}}{(2\pi)^{d-1}} \frac{i \, e^{ip \cdot x}}{p^2 - m^2} \, . \tag{0.3}$$

The contour integral here encircles both the poles of the integrand in the complex p^0 plane. Some elementary algebra yields

$$\langle 0 | \, [\phi(x), \phi(x')] \, | 0 \rangle = \int_0^\infty d\tau \Big(\frac{1}{4\pi\tau}\Big)^{d/2} \Big(\exp i \Big[\frac{(x - x')^2}{4\tau} - m^2\tau\Big] \; - \; h.c. \Big) \, . \tag{0.4}$$

This expression has the following analyticity properties in τ as a function of $(x - x')^2$:

- $(x - x')^2 > 0$: Both terms in (0.4) are real and equal upon continuation to the appropriate imaginary semi-axis; they cancel one another and the field commutator vanishes.

- $(x - x')^2 < 0$: One cannot rotate the contour to the imaginary axis so that the integral converges both at zero and infinity; the two terms don't cancel and the field commutator is nonzero.

Thus the hypersurface $(x - x')^2 = 0$ can be identified as the boundary of causal propagation – the light cone.

The signs of the exponent for $\tau \to 0$ and $\tau \to \infty$ determine the direction of the contour rotation. The sign for $\tau \to \infty$ is controlled by the sign of the particle mass. The sign for $\tau \to 0$ can be read directly from the semiclassical limit of the path integral representation of the two-point function

$$\langle 0 | \, \phi(x)\phi(x') \, | 0 \rangle = \int_{\substack{X(0)=x \\ X(\tau)=x'}} \mathcal{D}X \, \exp \Big[\frac{i}{m\hbar} \int_0^\tau \tfrac{1}{2}\dot{x}^2\Big] \tag{0.5}$$

For short proper time of propagation, the straight-line motion of the particle from initial to final points dominates: $S_{cl} \propto (x - x')^2/\tau$. The sign of the exponent is indeed the sign of $(x - x')^2$.

The calculation in string field theory (as currently understood) is no different. The string field $\Phi(X(\sigma), B(\sigma), C(\sigma))$ creates/destroys entire strings (here $B(\sigma)$, $C(\sigma)$ are Faddeev-Popov ghost coordinates). The two-point function can be written as a path integral[8][9] with the action

$$\langle 0|\, \Phi(X)\Phi(X')\, |0\rangle = \int_{\substack{X_{\text{init}}=X(\sigma) \\ X_{\text{final}}=X'(\sigma)}} \mathcal{D}X \, \exp\left[\tfrac{i}{\hbar\alpha'} \int_0^\tau dt\, d\sigma[-(\partial_t X)^2 + (\partial_\sigma X)^2]\right] \qquad (0.6)$$

In order to simplify the presentation the ghost coordinates have been suppressed here and below. The short-time limit of the action ignores the harmonic forces $(\partial_\sigma X)^2$ since there is no time to react to them. Each point on the string moves ballistically from $X(\sigma)$ to $X'(\sigma)$, so the classical action is

$$S_{\text{cl}} \sim \frac{\int d\sigma (X - X')^2}{\tau} \qquad (0.7)$$

Hence repeating the steps to the point particle field commutator yields $[\Phi, \Phi'] = 0$ if and only if $\int d\sigma (X - X')^2 > 0$. Note that there is an immediate generalization of the causal boundary to an arbitrary curved spacetime

$$\int d\sigma\, \mathcal{I}(X(\sigma), X'(\sigma)) = 0 \qquad (0.8)$$

where $\mathcal{I}(\sigma)$ is the geodesic interval between $X(\sigma)$ and $X'(\sigma)$.

Let me conclude with a few remarks:

- Any given measurement can only reveal a finite amount of information about the relative location of the two strings. So the strings must decide whether they are relatively spacelike or timelike – yes or no – and faced with only these two options, *democratically takes the majority vote* across the string. But note that this answer is quite bizarre since a given measurement can only reflect the average causal relationship of the points on the two strings. Consider the case where $X(\sigma)$ is a pointlike string, $X(\sigma) = x$. Then part of $X'(\sigma)$ can be *outside* the point light cone of x, yet still the measurements interfere; on the other hand, part of $X'(\sigma)$ can be *inside* the point light cone of x and yet the measurements do not interfere! The fact that the commutator calculation can be phrased in terms of semiclassical world sheet physics does mean that no information encoded in the string configuration itself is being propagated outside the point light cone; the Virasoro constraints guarantee that the world sheet causal structure is the one projected from the point spacetime. However the string field is a function on the whole string, not part of it, so it is not clear what the effect is on the propagation of the string field.

- The light cone

$$\int d\sigma (X - X')^2 = 0 \qquad (0.9)$$

  is invariant under conformal transformations of spacetime generated on loop space by the operator $\int d\sigma X(\sigma) \partial/\partial X(\sigma)$ and its cousins, so it does play to some extent the same role that the point light cone does for a point particle.

- Unfortunately the expression (0.9) is rather badly noninvariant under gauge transformations of string theory (reparametrizations of the loops). Thus one might worry that 'spacelike' noncommutation of string fields is only a gauge artifact; however the fact that a light cone calculation yields a similar result [10] suggests that this is not the crux of the problem.

- The semiclassical approach outlined above immediately shows that the evaluation of field commutators in the interacting theory yields results that are strongly dependent on the choice of interaction vertex. For instance, the correlation function

$$\langle 0 | \, \Phi(X_1)[\Phi(X_2), \Phi(X_3)] \, | 0 \rangle \qquad (0.10)$$

  can be evaluated with either the Witten-type[4] or Mandelstam-type[12] overlap. The sign of the exponent in the short-time, semiclassical limit of the path integral will give a vanishing result a) for the Witten vertex if the overlapping 'half-strings' obey condition $\int_a^b (X_2 - X_3)^2 > 0$ (with a the string midpoint and b its endpoint), and b) for the Mandelstam vertex if the segments of strings two and three that overlap (*e.g.* all of string two overlaps with part of string three) are on average causally related[a]. In either case the answer seems not to depend on the parts of the loop arguments outside the segments that overlap, and does depend strongly on the (gauge-dependent) geometry of the overlap. All this points to the necessity of a better understanding of string gauge invariance before causality can be properly addressed.

- We have phrased the problem of causality in the loop representation (position eigenstates $X(\sigma)$). Causality for mass eigenstates involves a convolution

$$\Phi(\{n_\ell\}) = \int \prod_\ell dx_\ell \, H_{n_\ell}(x_\ell) \Phi(\{x_\ell\}) \qquad (0.11)$$

  where $H_{n_\ell}(x_\ell)$ is a Hermite function. This smearing involves $\delta x_\ell \sim 1/\ell$; summing over ℓ, the string wanders logarithmically over all spacetime. The typical string contributing to typical processes is wild; smooth loops have measure zero. But then what is causality? Clearly we need a renormalized notion of the spread of the string and of locality. In the low energy theory, particles are a good approximation to strings; only the center of mass location of the string is effectively measured, and the logarithmically large spread of the string is erased by averaging to leave a finite but nonzero residue of order the string scale. The question is whether in this averaging process one smears out the 'location' of information carried by strings. An effective string size should be set

---

[a]More recently[13], D. Lowe has independently calculated the commutator in the interacting light cone string field theory, with a somewhat different result

by the spacetime energy scale of the process under consideration (as occurs, for example, in [4]). This sort of consideration could result in an effective smearing of the light cone. Similar arguments have been put forward[2] to support the Heisenberg model of the generalized uncertainty principle.

Finally, and perhaps most important, the argument leading to the causality condition assumed that the fluctuation determinants of the string modes do not compete with the exponential of the classical action in determining the convergence of the proper time integral in the path integral representation of the two-point function. Indeed, in particle-field theory the determinant contributes a power law underneath $exp[-S_{cl}]$. However we should know very well that in string theory this is usually not the case due to the exponential growth in the level density of string states; I thank M. Green for pointing this out to me. The effect of including this exponential contribution is to shift the commutator condition by a constant (spacelike) term (see [14] for closely related calculations):

$$\int d\sigma (X - X')^2 = 1 \qquad (0.12)$$

is the causal boundary. Green has suggested that this is another piece of evidence pointing toward the Heisenberg paradigm for the generalized uncertainty relation, since even pointlike states leak information outside the light cone[9]. This spacelike shift appears even in the superstring for certain worldsheet boundary conditions and hence seems unrelated to the presence or absence of a tachyon in the physical spectrum. If these boundary conditions contribute to physical processes, and the spacelike pole is not a gauge artifact[9], then indeed we will have to revise our notions of geometry in string theory.

Acknowledgements: I want to thank Michael Green for a stimulating discussion and an explanation of his work. This work was supported in part by Dept. of Energy grant DEFG02-90ER-40560.

References

[1] O. Kallen, Properties of Vacuum Expectation Values of Field Operators, in *Relations de Dispersion et Particules Élémentaires*, Les Houches 1960.

[2] D. Amati, M. Ciafaloni, and G. Veneziano, Can Spacetime be Probed below the Planck Scale?, Phys. Lett. B216 (1989) 41.

[3] Kenichi Konishi, Giampiero Paffuti, and Paolo Provero, Minimum Physical Length and the Generalized Uncertainty Principle in String Theory, Phys. Lett. B234 (1990) 276.

[4] D. Gross and P. Mende, String Theory Beyond the Planck Scale, Nucl. Phys. B303 (1988) 407.

[5] Igor Klebanov and Leonard Susskind, Continuum Strings from Discrete Field Theories, Nucl. Phys. B309 (1988) 175.

[6] J.J. Atick and E. Witten, The Hagedorn Transition and the Number of Degrees of Freedom of String Theory, Nucl. Phys. B310 (1988) 291-334.

[7] Emil J. Martinec, Criticality, Catastrophes, and Compactifications, In Brink, L. (ed.) et al.: Physics and mathematics of strings (V.G. Knizhnik memorial vol.), 389-433.

[8] F. Lorenzo, J. Mittelbrun, M. Medrano, and G. Sierra, Quantum Mechanical Amplitude for String Propagation, Phys. Lett. B171 (1986) 369; C. Ordonez, M. Rubin, and R. Zucchini, Polyakov Path Integrals with Ghosts: Closed Strings, and One-loop Amplitudes, Phys. Lett. B215 (1988) 103; D. Birmingham and C. Torre, BRST Extension of a String Propagator, Phys. Lett. B205 (1988) 289.

[9] A. Cohen, G. Moore, P. Nelson, and J. Polchinski, An Off-Shell Propagator for String Theory, Nucl. Phys. B267 (1986) 143; An Invariant String Propagator, in *Unified String Theories*, M. Green and D. Gross (eds.), World Scientific.

[10] E. Martinec, The Light Cone in String Theory, Class. Quant. Grav. 10 (1993) L187.

[11] E. Witten, Noncommutative Geometry and String Field Theory, Nucl. Phys. B268 (1986) 253.

[12] S. Mandelstam, Dual-Resonance Models, Phys. Reports 13C (1974) 259.

[13] D. Lowe, String Causality, Santa Barbara preprint UCSBTH-93-34, Sep 1993 (hep-th 9310009).

[14] M.B. Green, Space-Time Duality and Dirichlet String Theory, Phys. Lett. B266 (1991) 325; The Influence of World Sheet Boundaries on Critical Closed String Theory, Phys. Lett. B302 (1993) 29; Point-Like States for Type 2B Superstrings, preprint QMW-91-02 (revised).

LOOP EQUATION AND
AREA LAW IN TURBULENCE

A.A. MIGDAL

Physics Department, Princeton University
Jadwin Hall, Princeton, NJ 08544-1000
E-mail: migdal@math.princeton.edu

Abstract: The incompressible fluid dynamics is reformulated as dynamics of closed loops C in coordinate space. We derive explicit functional equation for the pdf of the circulation $P_C(\Gamma)$ which allows the scaling solutions in inertial range of spatial scales. The pdf decays as exponential of some power of Γ^3/A^2 where A is the minimal area inside the loop.

1 Introduction

Incompressible fluid dynamics underlies the vast majority of natural phenomena. It is described by famous Navier-Stokes equation

$$\dot{v}_\alpha = \nu \partial_\beta^2 v_\alpha - v_\beta \partial_\beta v_\alpha - \partial_\alpha p; \ \partial_\alpha v_\alpha = 0 \tag{1.1}$$

which is nonlinear, and therefore hard to solve. This nonlinearity makes life more interesting, though, as it leads to turbulence. Solving this equation with appropriate initial and boundary conditions we expect to obtain the chaotic behavior of velocity field.

The simplest boundary conditions correspond to infinite space with vanishing velocity at infinity. We are looking for the translation invariant probability distribution for velocity field, with infinite range of the wavelengths. In order to compensate for the energy dissipation, we add the usual random force to the Navier-Stokes equations, with the short wavelength support, corresponding to large scale energy pumping.

One may attempt to describe this probability distribution by the Hopf generating functional (the angular bracket denote time averaging, or ensemble averaging over

Quantum Field Theory and String Theory, Edited by
L. Baulieu *et al.*, Plenum Press, New York, 1995

realizations of the random forces)

$$Z[J] = \left\langle \exp\left(\int d^3r J_\alpha(r) v_\alpha(r) \right) \right\rangle \tag{1.2}$$

which is known to satisfy linear functional differential equation

$$\dot{Z} = H\left[J, \frac{\delta}{\delta J} \right] Z \tag{1.3}$$

similar to the Schrödinger equation for Quantum Field Theory, and equally hard to solve. Nobody managed to go beyond the Taylor expansion in source J, which corresponds to the obvious chain of equations for the equal time correlation functions of velocity field in various points in space. The same equations could be obtained directly from Navier-Stokes equations , so the Hopf equation looks useless.

In this work[a] we argue, that one could significantly simplify the Hopf functional without loosing information about correlation functions. This simplified functional depends upon the set of 3 periodic functions of one variable

$$C : r_\alpha = C_\alpha(\theta); \; 0 < \theta < 2\pi \tag{1.4}$$

which set describes the closed loop in coordinate space. The correlation functions reduce to certain functional derivatives of our loop functional with respect to $C(\theta)$ at vanishing loop $C \to 0$.

The properties of the loop functional at large loop C also have physical significance. Like the Wilson loops in Gauge Theory, they describe the statistics of large scale structures of vorticity field, which is analogous to the gauge field strength.

In Appendix A we recover the expansion in inverse powers of viscosity by direct iterations of the loop equation .

In Appendix B we study the matrix formulation of the Navier-Stokes equation, which may serve as a basis of the random matrix description of turbulence.

In Appendix C we study the reduced dynamics, corresponding to the functional Fourier transform of the loop functional. We argue, that instead of 3D Navier-Stokes equations one can use the 1D equations for the Fourier loop $P_\alpha(\theta, t)$.

In Appendix D we discuss the relation between the initial data for velocity field and the P field, and we find particular realisation for these initial data in terms of the Gaussian random variables.

In Appendix E we introduce the generating functional for the scalar products $P_\alpha(\theta)P_\alpha(\theta')$. The advantage of this functional over the original $\Psi[C]$ functional is the smoother continuum limit.

In Appendix F we discuss the possible numerical implementations of the reduced loop dynamics.

In Appendix G we show uniqueness of the tensor area law within certain class of functionals.

In Appendix H we present the modern view at the old problem of the minimal surface.

In Appendix I we show that the triple Kolmogorov correlation function corresponds to a vanishing correlation of vorticity with two velocity fields.

---

[a]see also [4] where this approach was initiated and [5] where its relation with the generalized Hamiltonian dynamics and the Gibbs-Boltzmann statistics was established

2 The Loop Calculus

We suggest to use in turbulence the following version of the Hopf functional

$$\Psi\left[C\right] = \left\langle \exp\left(\frac{\imath}{\nu} \oint dC_\alpha(\theta)v_\alpha\left(C(\theta)\right)\right)\right\rangle \tag{2.1}$$

which we call the loop functional or the loop field. It is implied that all angular variable θ run from 0 to 2π and that all the functions of this variable are 2π periodic.[b] The viscosity ν was inserted in denominator in exponential, as the only parameter of proper dimension. As we shall see below, it plays the role, similar to the Planck's constant in Quantum mechanics, the turbulence corresponding to the WKB limit $\nu \to 0$. [c]

As for the imaginary unit $\imath$, there are two reasons to insert it in the exponential. First, it makes the motion compact: the phase factor goes around the unit circle, when the velocity field fluctuates. So, at large times one may expect the ergodicity, with well defined average functional bounded by 1 by absolute value. Second, with this factor of $\imath$, the irreversibility of the problem is manifest. The time reversal corresponds to the complex conjugation of Ψ, so that imaginary part of the asymptotic value of Ψ at $t \to \infty$ measures the effects of dissipation.

The loop orientation reversal $C(\theta) \to C(2\pi - \theta)$ also leads to the complex conjugation, so it is equivalent to the time reversal. This symmetry implies, that any correlator of odd/even number of velocities should be integrated odd/even number of times over the loop, and it must enter with an imaginary/real factor. Later, we shall use this property in the area law.

We shall often use the field theory notations for the loop integrals,

$$\Psi\left[C\right] = \left\langle \exp\left(\frac{\imath}{\nu} \oint_C dr_\alpha v_\alpha\right)\right\rangle \tag{2.2}$$

This loop integral can be reduced to the surface integral of vorticity field

$$\omega_{\mu\nu} = \partial_\mu v_\nu - \partial_\nu v_\mu \tag{2.3}$$

by the Stokes theorem

$$\Gamma_C[v] \equiv \oint_C dr_\alpha v_\alpha = \int_S d\sigma_{\mu\nu}\omega_{\mu\nu}; \quad \partial S = C \tag{2.4}$$

This is the well-known velocity circulation, which measures the net strength of the vortex lines, passing through the loop C. Would we fix initial loop C and let it move with the flow, the loop field would be conserved by the Euler equation, so that only the viscosity effects would be responsible for its time evolution. However, this is not what we are trying to do. We take the Euler rather than Lagrange dynamics, so that the loop is fixed in space, and hence Ψ is time dependent already in the Euler equations. The difference between Euler and Navier-Stokes equations is the time irreversibility, which leads to complex average Ψ in Navier-Stokes dynamics.

---

[b]This parametrization of the loop is a matter of convention, as the loop functional is parametric invariant.

[c]One could also insert any numerical parameter in exponential, but this factor could be eliminated by space- and/or time rescaling.

It is implied that this field $\Psi\,[C]$ is invariant under translations of the loop $C(\theta) \to C(\theta)+const$. The asymptotic behavior at large time with proper random forcing reaches certain fixed point, governed by the translation- and scale invariant equations, which we derive in this paper.

The general Hopf functional (1.2) reduces for the loop field for the following imaginary singular source

$$J_\alpha(r) = \frac{\imath}{\nu} \oint_C dr'_\alpha \delta^3 \left(r' - r\right) \tag{2.5}$$

The Ψ functional involves connected correlation functions of the powers of circulation at equal times.

$$\Psi[C] = \exp\left(\sum_{n=2}^\infty \frac{\imath^n}{n!\,\nu^n}\,\langle\langle \Gamma_C^n[v] \rangle\rangle\right) \tag{2.6}$$

This expansion goes in powers of effective Reynolds number, so it diverges in turbulent region. There, the opposite WKB approximation will be used.

Let us come back to the general case of the arbitrary Reynolds number. What could be the use of such restricted Hopf functional? At first glance it seems that we lost most of information, described by the Hopf functional, as the general Hopf source J depends upon 3 variables x, y, z whereas the loop C depends of only one parameter θ. Still, this information can be recovered by taking the loops of the singular shape, such as two infinitesimal loops R_1, R_2, connected by a couple of wires

Fig.1.

The loop field in this case reduces to

$$\Psi\,[C] \to \left\langle \exp\left(\frac{\imath}{2\nu}\Sigma_{\mu\nu}^{R_1}\omega_{\mu\nu}(r_1) + \frac{\imath}{2\nu}\Sigma_{\mu\nu}^{R_2}\omega_{\mu\nu}(r_2)\right)\right\rangle \tag{2.7}$$

where

$$\Sigma_{\mu\nu}^R = \oint_R dr_\nu r_\mu \tag{2.8}$$

is the tensor area inside the loop R. Taking functional derivatives with respect to the shape of R_1 and R_2 prior to shrinking them to points, we can bring down the product of vorticities at r_1 and r_2. Namely, the variations yield

$$\delta\Sigma_{\mu\nu}^R = \oint_R (dr_\nu \delta r_\mu + r_\mu d\delta r_\nu) = \oint_R (dr_\nu \delta r_\mu - dr_\mu \delta r_\nu) \tag{2.9}$$

where integration by parts was used in the second term.

One may introduce the area derivative $\frac{\delta}{\delta\sigma_{\mu\nu}(r)}$, which brings down the vorticity at the given point r at the loop.

$$-\nu^2 \frac{\delta^2 \Psi\,[C]}{\delta\sigma_{\mu\nu}(r_1)\delta\sigma_{\lambda\rho}(r_2)} \to \langle \omega_{\mu\nu}(r_1)\omega_{\lambda\rho}(r_2)\rangle \tag{2.10}$$

The careful definition of these area derivatives are or paramount importance to us. The corresponding loop calculus was developed in[2] in the context of the gauge theory. Here we rephrase and further refine the definitions and relations established in that work.

The basic element of the loop calculus is what we suggest to call the spike derivative, namely the operator which adds the infinitesimal Λ shaped spike to the loop

$$D_\alpha(\theta, \epsilon) = \int_\theta^{\theta+2\epsilon} d\phi \left(1 - \frac{|\theta + \epsilon - \phi|}{\epsilon}\right) \frac{\delta}{\delta C_\alpha(\phi)} \qquad (2.11)$$

The finite spike operator

$$\Lambda(r, \theta, \epsilon) = \exp\left(r_\alpha D_\alpha(\theta, \epsilon)\right) \qquad (2.12)$$

adds the spike of the height r. This is the straight line from $C(\theta)$ to $C(\theta + \epsilon) + r$, followed by another straight line from $C(\theta + \epsilon) + r$ to $C(\theta + 2\epsilon)$.

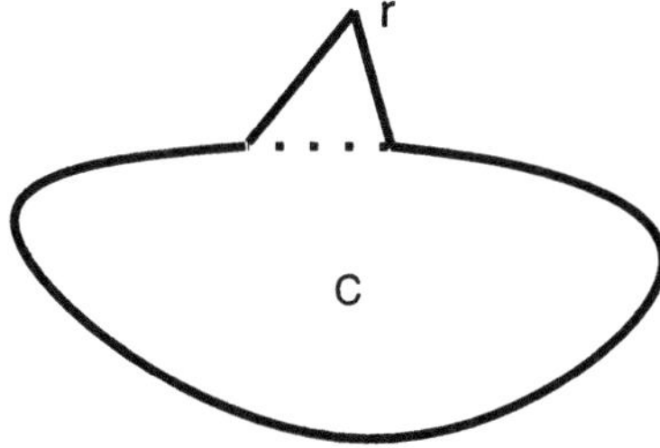

Fig.2.

Note, that the loop remains closed, and the slopes remain finite, only the second derivatives diverge. The continuity and closure of the loop eliminates the potential part of velocity; as we shall see below, this is necessary to obtain the loop equation.

In the limit $\epsilon \to 0$ these spikes are invisible, at least for the smooth vorticity field, as one can see from the Stokes theorem (the area inside the spike goes to zero as ϵ). However, taking certain derivatives prior to the limit $\epsilon \to 0$ we can obtain the finite contribution.

Let us consider the operator

$$\Pi(r, r', \theta, \epsilon) = \Lambda\left(r, \theta, \frac{1}{2}\epsilon\right) \Lambda(r', \theta, \epsilon) \qquad (2.13)$$

By construction it inserts the smaller spike on top of a bigger one, in such a way, that a polygon appears.

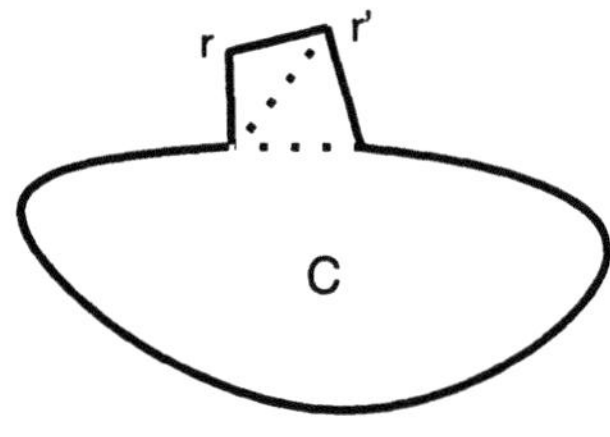

Fig.3.

Taking the derivatives with respect to the vertices of this polygon r, r' , setting $r = r' = 0$ and antisymmetrising, we find the tensor operator

$$\Omega_{\alpha\beta}(\theta, \epsilon) = -\imath\nu D_\alpha\left(\theta, \frac{1}{2}\epsilon\right) D_\beta(\theta, \epsilon) - \{\alpha \leftrightarrow \beta\} \tag{2.14}$$

which brings down the vorticity, when applied to the loop field

$$\Omega_{\alpha\beta}(\theta, \epsilon)\Psi[C] \xrightarrow{\epsilon \to 0} \omega_{\alpha\beta}\left(C(\theta)\right)\Psi[C] \tag{2.15}$$

The quick way to check these formulas is to use formal functional derivatives

$$\frac{\delta\Psi[C]}{\delta C_\alpha(\theta)} = C'_\beta(\theta)\frac{\delta\Psi[C]}{\delta\sigma_{\alpha\beta}(C(\theta))} \tag{2.16}$$

Taking one more functional derivative derivative we find the term with vorticity times first derivative of the δ function, coming from the variation of $C'(\theta)$

$$\frac{\delta^2\Psi[C]}{\delta C_\alpha(\theta)\delta C_\beta(\theta')} = \delta'(\theta - \theta')\frac{\delta\Psi[C]}{\delta\sigma_{\alpha\beta}(C(\theta))} + C'_\gamma(\theta)C'_\lambda(\theta')\frac{\delta^2\Psi[C]}{\delta\sigma_{\alpha\gamma}(C(\theta))\,\delta\sigma_{\beta\lambda}(C(\theta'))} \tag{2.17}$$

This term is the only one, which survives the limit $\epsilon \to 0$ in our relation (2.15).

So, the area derivative can be defined from the antisymmetric tensor part of the second functional derivative as the coefficient in front of $\delta'(\theta - \theta')$. Still, it has all the properties of the first functional derivative, as it can also be defined from the above first variation. The advantage of dealing with spikes is the control over the limit $\epsilon \to 0$, which might be quite singular in applications.

So far we managed to insert the vorticity at the loop C by variations of the loop field. Later we shall need the vorticity off the loop, in arbitrary point in space. This can be achieved by the following combination of the spike operators

$$\Lambda(r, \theta, \epsilon)\,\Pi(r_1, r_2, \theta + \epsilon, \delta);\ \delta \ll \epsilon \tag{2.18}$$

This operator inserts the Π shaped little loop at the top of the bigger spike, in other words, this little loop is translated by a distance r by the big spike.

Taking derivatives, we find the operator of finite translation of the vorticity

$$\Lambda(r, \theta, \epsilon)\,\Omega_{\alpha\beta}(\theta + \epsilon, \delta) \tag{2.19}$$

and the corresponding infinitesimal translation operator

$$D_\mu(\theta,\epsilon)\Omega_{\alpha\beta}(\theta+\epsilon,\delta) \tag{2.20}$$

which inserts $\partial_\mu\omega_{\alpha\beta}\left(C(\theta)\right)$ when applied to the loop field.

Coming back to the correlation function, we are going now to construct the operator, which would insert two vorticities separated by a distance. Let us note that the global Λ spike

$$\Lambda\left(r,0,\pi\right)=\exp\left(r_\alpha\int_0^{2\pi}d\phi\left(1-\frac{|\phi-\pi|}{\pi}\right)\frac{\delta}{\delta C_\alpha(\phi)}\right) \tag{2.21}$$

when applied to a shrunk loop $C(\phi)=0$ does nothing but the backtracking from 0 to r

$$0 \rightleftarrows r$$

Fig.4.

This means that the operator

$$\Omega_{\alpha\beta}(0,\delta)\Omega_{\lambda\rho}(\pi,\delta)\Lambda\left(r,0,\pi\right) \tag{2.22}$$

when applied to the loop field for a shrunk loop yields the vorticity correlation function

$$\Omega_{\alpha\beta}(0,\delta)\Omega_{\lambda\rho}(\pi,\delta)\Lambda\left(r,0,\pi\right)\Psi[0]=\langle\omega_{\alpha\beta}(0)\omega_{\lambda\rho}(r)\rangle \tag{2.23}$$

The higher correlation functions of vorticities could be constructed in a similar fashion, using the spike operators. As for the velocity, one should solve the Poisson equation

$$\partial_\mu^2 v_\alpha(r)=\partial_\beta\omega_{\beta\alpha}(r) \tag{2.24}$$

with the proper boundary conditions , say, $v=0$ at infinity. Formally,

$$v_\alpha(r)=\frac{1}{\partial_\mu^2}\partial_\beta\omega_{\beta\alpha}(r) \tag{2.25}$$

This suggests the following formal definition of the velocity operator

$$V_\alpha(\theta,\epsilon,\delta)=\frac{1}{D_\mu^2(\theta,\epsilon)}D_\beta(\theta,\epsilon)\Omega_{\beta\alpha}(\theta,\delta);\ \delta\ll\epsilon \tag{2.26}$$

$$V_\alpha(\theta,\epsilon,\delta)\Psi[C]\overset{\delta,\epsilon\to0}{\longrightarrow}v_\alpha\left(C(\theta)\right)\Psi[C] \tag{2.27}$$

Another version of this formula is the following integral

$$V_\alpha(\theta,\epsilon,\delta)=\int d^3\rho\frac{\rho_\beta}{4\pi|\rho|^3}\Lambda\left(\rho,\theta,\epsilon\right)\Omega_{\alpha\beta}(\theta+\epsilon,\delta) \tag{2.28}$$

where the Λ operator shifts the Ω by a distance ρ off the original loop at the point $r=C(\theta+\epsilon)$.

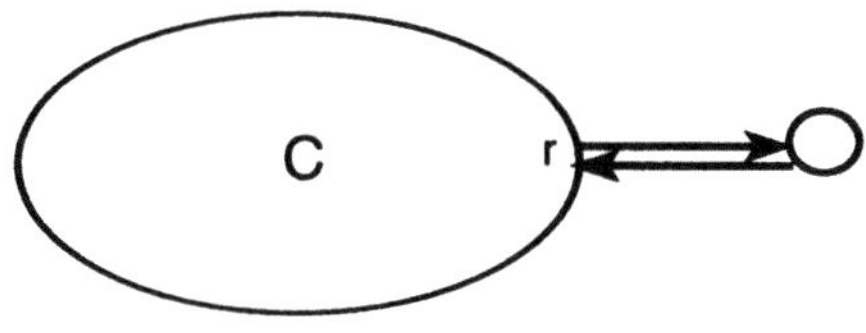

Fig.5.

3 Loop Equation

Let us now derive exact equation for the loop functional. Taking the time derivative of the original definition, and using the Navier-Stokes equation we get in front of exponential

$$\oint_C dr_\alpha \frac{\imath}{\nu} \left(\nu \partial_\beta^2 v_\alpha - v_\beta \partial_\beta v_\alpha - \partial_\alpha p \right) \tag{3.1}$$

The term with the pressure gradient yields zero after integration over the closed loop, and the velocity gradients in the first two terms could be expressed in terms of vorticity up to irrelevant gradient terms, so that we find

$$\oint_C dr_\alpha \frac{\imath}{\nu} \left(\nu \partial_\beta \omega_{\beta\alpha} - v_\beta \omega_{\beta\alpha} \right) \tag{3.2}$$

Replacing the vorticity and velocity by the operators discussed in the previous Section we find the following loop equation (in explicit notations)

$$-\imath \dot{\Psi}[C] = \tag{3.3}$$

$$\oint dC_\alpha(\theta) \left(D_\beta(\theta,\epsilon)\Omega_{\beta\alpha}(\theta,\epsilon) + \frac{1}{\nu} \int d^3\rho \frac{\rho_\gamma}{4\pi|\rho|^3} \Lambda\left(\rho,\theta,\epsilon\right) \Omega_{\gamma\beta}(\theta+\epsilon,\delta)\Omega_{\beta\alpha}(\theta,\delta) \right) \Psi[C]$$

The more compact form of this equation, using the notations of [2], reads

$$\imath\,\nu\dot{\Psi}[C] = \mathcal{H}_C\Psi \tag{3.4}$$

$$\mathcal{H}_C \equiv \nu^2 \oint_C dr_\alpha \left(\imath\,\partial_\beta \frac{\delta}{\delta\sigma_{\beta\alpha}(r)} + \int d^3r' \frac{r'_\gamma - r_\gamma}{4\pi|r-r'|^3} \frac{\delta^2}{\delta\sigma_{\beta\alpha}(r)\delta\sigma_{\beta\gamma}(r')} \right)$$

Now we observe that viscosity ν appears in front of time and spatial derivatives, like the Planck constant $\hbar$ in Quantum mechanics. Our loop Hamiltonian $\mathcal{H}_C$ is not hermitian, due to dissipation. It contains the second loop derivatives, so it represents a (nonlocal!) kinetic term in loop space.

So far, we considered so called decaying turbulence, without external energy source. The energy

$$E = \int d^3r \frac{1}{2} v_\alpha^2 \tag{3.5}$$

would eventually all dissipate, so that the fluid would stop. In this case the loop wave function Ψ would asymptotically approach 1

$$\Psi[C] \xrightarrow{t\to\infty} 1 \tag{3.6}$$

In order to reach the steady state, we add to the right side of the Navier-Stokes equation the usual Gaussian random forces $f_\alpha(r, t)$ with the space dependent correlation function

$$\langle f_\alpha(r, t) f_\beta(r', t') \rangle = \delta_{\alpha\beta} \delta(t - t') F(r - r') \tag{3.7}$$

concentrated at at small wavelengths, i.e. slowly varying with $r - r'$.

Using the identity

$$\langle f_\alpha(r, t) \Phi[v(.)] \rangle = \int d^3r' F(r - r') \frac{\delta \Phi[v(.)]}{\delta v_\alpha(r')} \tag{3.8}$$

which is valid for arbitrary functional Φ we find the following imaginary potential term in the loop Hamiltonian

$$\delta \mathcal{H}_C \equiv \imath\, U[C] = \frac{\imath}{\nu} \oint_C dr_\alpha \oint_C dr'_\alpha F(r - r') \tag{3.9}$$

Note, that orientation reversal together with complex conjugation changes the sign of the loop Hamiltonian, as it should. The potential part involves two loop integrations times imaginary constant. The first term in the kinetic part has one loop integration, one loop derivative times imaginary constant. The second kinetic term has one loop integration, two loop derivatives and real constant. The left side of the loop equation has no loop integrations, no loop derivatives, but has a factor of $\imath$.

The relation between the potential and kinetic parts of the loop Hamiltonian depends of viscosity, or, better to say, it depends upon the Reynolds number, which is the ratio of the typical circulation to viscosity. In the viscous limit, when the Reynolds number is small, the loop wave function is close to 1. The perturbation expansion in $\frac{1}{\nu}$ goes in powers of the potential, in the same way, as in Quantum mechanics. The second (nonlocal) term in kinetic part of the Hamiltonian also serves as a small perturbation (it corresponds to nonlinear term in the Navier-Stokes equation). The first term of this perturbation expansion is just

$$\Psi[C] \to 1 - \int \frac{d^3k}{(2\pi)^3} \frac{\widetilde{F}(k)}{2\nu^3\, k^2} \left| \oint_C dr_\alpha e^{\imath\, kr} \right|^2 \tag{3.10}$$

with $\widetilde{F}(k)$ being the Fourier transform of $F(r)$. This term is real, as it corresponds to the two-velocity correlation. The next term comes from the triple correlation of velocity, and this term is purely imaginary, so that the dissipation shows up.

This expansion can be derived by direct iterations in the loop space as in [2], inverting the operator in the local part of the kinetic term in the Hamiltonian. This expansion is discussed in Appendix A. The results agree with the straightforward iterations of the Navier-Stokes equations in powers of the random force, starting from zero velocity.

So, we have the familiar situation, like in QCD, where the perturbation theory breaks because of the infrared divergences. For arbitrarily small force, in a large system, the region of small k would yield large contribution to the terms of the perturbation expansion. Therefore, one should take the opposite WKB limit $\nu \to 0$.

In this limit, the wave function should behave as the usual WKB wave function, i.e. as an exponential

$$\Psi[C] \rightarrow \exp\left(\frac{\imath \, S[C]}{\nu}\right) \tag{3.11}$$

The effective loop Action $S[C]$ satisfies the loop space Hamilton-Jacobi equation

$$\dot{S}[C] = -\imath U[C] + \oint_C dr_\alpha \int d^3 r' \frac{r'_\gamma - r_\gamma}{4\pi |r - r'|^3} \frac{\delta S}{\delta \sigma_{\beta\alpha}(r)} \frac{\delta S}{\delta \sigma_{\beta\gamma}(r')} \tag{3.12}$$

The imaginary part of $S[C]$ comes from imaginary potential $U[C]$, which distinguishes our theory from the reversible Quantum mechanics. The sign of $\Im S$ must be positive definite, since $|\Psi| < 1$. As for the real part of $S[C]$, it changes the sign under the loop orientation reversal $C(\theta) \rightarrow C(2\pi - \theta)$.

At finite viscosity there would be an additional term

$$-\nu \oint_C dr_\alpha \partial_\beta \frac{\delta S[C]}{\delta \sigma_{\beta\alpha}(r)} - \imath \nu \oint_C dr_\alpha \int d^3 r' \frac{r'_\gamma - r_\gamma}{4\pi |r - r'|^3} \frac{\delta^2 S[C]}{\delta \sigma_{\beta\alpha}(r)\delta \sigma_{\beta\gamma}(r')} \tag{3.13}$$

on the right of (3.12). As for the term

$$-\oint_C dr_\alpha \imath \, (\partial_\beta S[C]) \frac{\delta S[C]}{\delta \sigma_{\beta\alpha}(r)} \tag{3.14}$$

which formally arises in the loop equation, this term vanishes, since $\partial_\beta S[C] = 0$. This operator inserts backtracking at some point at the loop without first applying the loop derivative at this point. As it was discussed in the previous Section, such backtracking does not change the loop functional. This issue was discussed at length in [2], where the Leibnitz rule for the operator $\partial_\alpha \frac{\delta}{\delta \sigma_{\beta\gamma}}$ was established

$$\partial_\alpha \frac{\delta f(g[C])}{\delta \sigma_{\beta\gamma}(r)} = f'(g[C])\partial_\alpha \frac{\delta g[C]}{\delta \sigma_{\beta\gamma}(r)} \tag{3.15}$$

In other words, this operator acts as a first order derivative on the loop functional with finite area derivative (so called Stokes type functional). Then, the above term does not appear.

The Action functional $S[C]$ describes the distribution of the large scale vorticity structures, and hence it should not depend of viscosity. In terms of the above connected correlation functions of the circulation this corresponds to the limit, when the effective Reynolds number $\frac{\Gamma_C |v|}{\nu}$ goes to infinity, but the sum of the divergent series tends to the finite limit. According to the standard picture of turbulence, the large scale vorticity structures depend upon the energy pumping, rather than the energy dissipation.

It is understood that both time t and the loop size[d] $|C|$ should be greater then the viscous scales

$$t \gg t_0 = \nu^{\frac{1}{2}}\mathcal{E}^{-\frac{1}{2}}; \quad |C| \gg r_0 = \nu^{\frac{3}{4}}\mathcal{E}^{-\frac{1}{4}} \tag{3.16}$$

where $\mathcal{E}$ is the energy dissipation rate.

---

[d] As a measure of the loop size one may take the square root of the minimal area inside the loop.

202

It is defined from the energy balance equation

$$0 = \partial_t \left\langle \frac{1}{2} v_\alpha^2 \right\rangle = \nu \left\langle v_\alpha \partial^2 v_\alpha \right\rangle + \left\langle f_\alpha v_\alpha \right\rangle \tag{3.17}$$

which can be transformed to

$$\frac{1}{4} \nu \left\langle \omega_{\alpha\beta}^2 \right\rangle = 3F(0) \tag{3.18}$$

The left side represents the energy, dissipated at small scale due to viscosity, and the right side - the energy pumped in from the large scales due to the random forces. Their common value is $\mathcal{E}$.

We see, that constant $F(r - r')$, i.e., $\tilde{F}(k) \propto \delta(k)$ is sufficient to provide the necessary energy pumping. However, such forcing does not produce vorticity, which we readily see in our equation. The contribution from this constant part to the potential in our loop equation drops out (this is a closed loop integral of total derivative). This is important, because this term would have the wrong order of magnitude in the turbulent limit - it would grow as the Reynolds number.

Dropping this term, we arrive at remarkably simple and universal functional equation

$$\dot{S}[C] = \oint_C dr_\alpha \int d^3r' \frac{r'_\gamma - r_\gamma}{4\pi |r - r'|^3} \frac{\delta S}{\delta \sigma_{\beta\alpha}(r)} \frac{\delta S}{\delta \sigma_{\beta\gamma}(r')} \tag{3.19}$$

The stationary solution of this equation describes the steady distribution of the circulation in the strong turbulence. Note, that the stationary solutions come in pairs $\pm S$. The sign should be chosen so, that $\Im S > 0$, to provide the inequality $|\Psi| < 1$.

4 Scaling Law

The 'Hamilton-Jacobi' equation without the potential term (3.19) allows the family of the scaling solutions

$$S[C] = t^{2\kappa - 1} \phi \left[\frac{C}{t^\kappa} \right] \tag{4.1}$$

with arbitrary index κ. The scaling function satisfies the equation

$$(2\kappa - 1)\phi[C] - \kappa \oint_C dr_\alpha \frac{\delta \phi[C]}{\delta \sigma_{\beta\alpha}(r)} r_\beta = \oint_C dr_\alpha \int d^3r' \frac{r'_\gamma - r_\gamma}{4\pi |r - r'|^3} \frac{\delta \phi[C]}{\delta \sigma_{\beta\alpha}(r)} \frac{\delta \phi[C]}{\delta \sigma_{\beta\gamma}(r')} \tag{4.2}$$

The left side here was computed, using the chain rule differentiation of functional.

Asymptotically, at large time, we expect the fixed point, which is the homogeneous functional

$$S_\infty[C] = |C|^{2 - \frac{1}{\kappa}} f \left[\frac{C}{|C|} \right] \tag{4.3}$$

zeroing the right side of our 'kinetic' functional equation

$$0 = \oint_C dr_\alpha \int d^3r' \frac{r'_\gamma - r_\gamma}{4\pi |r - r'|^3} \frac{\delta S_\infty[C]}{\delta \sigma_{\beta\alpha}(r)} \frac{\delta S_\infty[C]}{\delta \sigma_{\beta\gamma}(r')} \tag{4.4}$$

The Kolmogorov scaling [1] would correspond to

$$\kappa = \frac{3}{2} \tag{4.5}$$

in which case one can express the S functional in terms of $\mathcal{E}$

$$S[C] = \mathcal{E}t^2\phi\left[\frac{C}{\sqrt{\mathcal{E}t^3}}\right] \tag{4.6}$$

One can easily rephrase the Kolmogorov arguments in the loop space. The relation between the energy dissipation rate and the velocity correlator reads

$$\mathcal{E} = \langle v_\alpha(r_0)v_\beta(0)\partial_\beta v_\alpha(0)\rangle \tag{4.7}$$

where the point splitting at the viscous scale r_0 is introduced. Such splitting is necessary to avoid the viscosity effects; without the splitting the average would formally reduce to the total derivative and vanish.

Instead of the point splitting one may introduce the finite loop of the viscous scale $|C| \sim r_0$, and compute this correlator in presence of such loop. This reduces to the WKB estimates

$$\omega_{\alpha\beta}(r) \to \frac{\delta S[C]}{\delta\sigma_{\alpha\beta}(r)}; \; v_\alpha(r) = \int d^3r' \frac{r'_\gamma - r_\gamma}{4\pi|r - r'|^3}\omega_{\alpha\gamma}(r') \tag{4.8}$$

Using the generic scaling law for S we find

$$\omega \sim r_0^{-\frac{1}{\kappa}}; \; v \sim r_0^{1-\frac{1}{\kappa}}; \; \mathcal{E} \sim r_0^{2-\frac{3}{\kappa}} \tag{4.9}$$

We see, that the energy dissipation rate would stay finite in the limit of the vanishing viscous scale only for the Kolmogorov value of the index. This argument looks rather cheap, but I think it is basically correct. The constant value of the energy dissipation rate in the limit of vanishing viscosity arises as the quantum anomaly in the field theory, through the finite limit of the point splitting in the correspondent energy current.[e]

There is another version of this argument, which I like better. The dynamics of Euler fluid in infinite system would not exist, for the non-Kolmogorov scaling. The extra powers of loop size would have to enter with the size L of the whole system, like $\left(\frac{|C|}{L}\right)^\epsilon$. So, in the regime with finite energy pumping rate $\mathcal{E}$ the infinite Euler system can exist only for the Kolmogorov index. This must be the essence of the original Kolmogorov reasoning [1].

The problem is that nobody proved that such limit exists, though. Within the usual framework, based on the velocity correlation functions, one has to prove, that the infrared divergences, caused by the sweep, all cancel for the observables. Within our framework these problems disappear, as we shall see later.

As for the correlation functions in inertial range, unfortunately those cannot be computed in the WKB approximation, since they involve the contour shrinking to a double line, with vanishing area inside. Still, most of the physics can be understood in loop terms, without these correlation functions. The large scale behavior of the loop functional reflects the statistics of the large vorticity structures, encircled by the loop.

---

[e]I am grateful to A. Polyakov and E. Siggia for inspiring comments on this subject.

5 Loop Equation for the Circulation PDF

The loop field could serve as the generating function for the PDF $P_C(\Gamma)$ for the circulation. The Fourier integral

$$P_C(\Gamma) = \int_{-\infty}^{\infty} \frac{dg}{2\pi\nu} \exp\left(\frac{\imath g}{\nu}\left(\oint_C dr_\alpha v_\alpha(r) - \Gamma\right)\right) \tag{5.1}$$

can be analyzed in the same way as the loop field before. The only difference is that the factors of g appear in front of various terms. These factors can be replaced by

$$g \to \imath\nu\frac{\partial}{\partial\Gamma} \tag{5.2}$$

acting on $P_C(\Gamma)$.

As a result we find

$$\frac{\partial}{\partial\Gamma}\dot{P}_C(\Gamma) = -\oint_C dr_\alpha \int d^3r' \frac{r'_\gamma - r_\gamma}{4\pi|r - r'|^3} \frac{\delta^2 P_C(\Gamma)}{\delta\sigma_{\beta\alpha}(r)\delta\sigma_{\beta\gamma}(r')} \tag{5.3}$$
$$+\nu\frac{\partial}{\partial\Gamma}\oint_C dr_\alpha\partial_\beta\frac{\delta P_C(\Gamma)}{\delta\sigma_{\beta\alpha}(r)} - U[C]\frac{\partial^3 P_C(\Gamma)}{\partial\Gamma^3}$$

All the imaginary units disappear, as they should. As for the viscosity and forcing, these terms can be neglected in inertial range in the same way as before. The only new thing is that one has to assume that $\Gamma \gg \nu$ in inertial range in addition to above assumptions about the size of the loop.

In absence of these terms there are no dimensional parameters so that the following scaling laws hold (with the same index κ as before)

$$P_C(\Gamma) = t^{2\kappa-1}F\left[\frac{C}{t^\kappa}, \frac{\Gamma}{t^{2\kappa-1}}\right] \tag{5.4}$$

The factor $t^{2\kappa-1}$ came from the normalization of probability density. Note, that this is more general law than before. Here we do not have to use the WKB approximation for the PDF. In other words, the whole PDF rather than just its decay at large Γ satisfies the scaling law.

The steady distribution would have the form of

$$P_C(\Gamma) \to \frac{1}{\Gamma}\Phi\left[\frac{C}{\Gamma^{\frac{\kappa}{2\kappa-1}}}\right] \tag{5.5}$$

where the scaling functional Φ satisfies the homogeneous equation

$$\oint_C dr_\alpha \int d^3r' \frac{r'_\gamma - r_\gamma}{4\pi|r - r'|^3} \frac{\delta^2\Phi[C]}{\delta\sigma_{\beta\alpha}(r)\delta\sigma_{\beta\gamma}(r')} = 0 \tag{5.6}$$

with the normalization condition

$$1 = \int_{-\infty}^{\infty} \frac{d\Gamma}{\Gamma}\Phi\left[\frac{C}{\Gamma^{\frac{\kappa}{2\kappa-1}}}\right] \tag{5.7}$$

In principle, there could be different scaling functions for positive and negative Γ, rather than just absolute value $|\Gamma|$ prescription. This would correspond to above mentioned violation of the time reversal symmetry. However, as we mentioned above, there is no exact relation which would eliminate the symmetric solution.

The Kolmogorov triple correlation function vanishes for vorticities (see Appendix I), so that there is no restriction on the asymmetric part of the circulation PDF. Nevertheless, the Kolmogorov scaling $\kappa = \frac{3}{2}$ seems to me the most likely possibility, by the reasons discussed in the previous section.

The homogeneous loop equation requires some boundary conditions at large loops, to provide a meaningful solution. The asymptotic decrease of PDF

$$P_C(\Gamma) \sim \exp\left(-Q\left[\frac{C}{\Gamma^{\frac{\kappa}{2\kappa-1}}}\right]\right), Q \to \infty \tag{5.8}$$

would lead to the same WKB equation as before

$$\oint_C dr_\alpha \int d^3 r' \frac{r'_\gamma - r_\gamma}{4\pi|r - r'|^3} \frac{\delta Q[C]}{\delta \sigma_{\beta\alpha}(r)} \frac{\delta Q[C]}{\delta \sigma_{\beta\gamma}(r')} = 0 \tag{5.9}$$

We are studying this equation in the next section.

6   Tensor Area Law

The Wilson loop in QCD decreases as exponential of the minimal area, encircled by the loop, leading to the quark confinement. What is the similar asymptotic law in turbulence? The physical mechanisms leading to the area law in QCD are absent here. Moreover, there is no guarantee, that $\Psi[C]$ always decreases with the size of the loop.

This makes it possible to look for the simple Anzatz, which was not acceptable in QCD, namely

$$S[C] = s\left(\Sigma^C_{\mu\nu}\right) \tag{6.1}$$

where

$$\Sigma^C_{\mu\nu} = \oint_C r_\mu dr_\nu \tag{6.2}$$

is the tensor area encircled by the loop C. The difference between this area and the scalar area is the positivity property. The scalar area vanishes only for the loop which can be contracted to a point by removal of all the backtracking. As for the tensor area, it vanishes, for example, for the 8 shaped loop, with opposite orientation of petals.

Thus, there are some large contours with vanishing tensor area, for which there would be no decrease of the Ψ functional. In QCD the Wilson loops must always decrease at large distances, due to the finite mass gap. Here, the large scale correlations are known to exist, and play the central role in the turbulent flow. So, I see no convincing arguments to reject the tensor area Anzatz.

This Anzatz in QCD not only was unphysical, it failed to reproduce the correct short-distance singularities in the loop equation. In turbulence, there are no such singularities. Instead, there are the large-distance singularities, which all should cancel in the loop equation.

It turns out, that for this Anzatz the (turbulent limit of the) loop equation is satisfied automatically, without any further restrictions. Let us verify this important property. The first area derivative yields

$$\omega_{\mu\nu}^{C}(r) = \frac{\delta S}{\delta \sigma_{\mu\nu}(r)} = 2\frac{\partial s}{\partial \Sigma_{\mu\nu}^{C}} \tag{6.3}$$

The factor of 2 comes from the second term in the variation

$$\frac{\delta \Sigma_{\alpha\beta}^{C}}{\delta \sigma_{\mu\nu}(r)} = \delta_{\alpha\mu}\delta_{\beta\nu} - \delta_{\alpha\nu}\delta_{\beta\mu} \tag{6.4}$$

Note, that the right side does not depend on r. Moreover, you can shift r aside from the base loop C, with proper wires inserted. The area derivative would not change, as the contribution of wires drops.

This implies, that the corresponding vorticity $\omega_{\mu\nu}^{C}(r)$ is space independent, it only depends upon the loop itself. The velocity can be reconstructed from vorticity up to irrelevant constant sterms

$$v_{\beta}^{C}(r) = \frac{1}{2} r_{\alpha} \omega_{\alpha\beta}^{C} \tag{6.5}$$

This can be formally obtained from the above integral representation

$$v_{\beta}^{C}(r) = \int d^{3}r' \frac{r_{\alpha} - r_{\alpha}'}{4\pi|r - r'|^{3}}\omega_{\alpha\beta}^{C} \tag{6.6}$$

as a residue from the infinite sphere $R = |r'| \to \infty$. One may insert the regularizing factor $|r'|^{-\epsilon}$ in ω, compute the convolution integral in Fourier space and check that in the limit $\epsilon \to 0^{+}$ the above linear term arises. So, one can use the above form of the loop equation, with the analytic regularization prescription.

Now, the $v\omega$ term in the loop equation reads

$$\oint_{C} dr_{\gamma}\, v_{\beta}^{C}(r)\, \omega_{\beta\gamma}^{C} \propto \Sigma_{\gamma\alpha}^{C}\, \omega_{\alpha\beta}^{C}\, \omega_{\beta\gamma}^{C} \tag{6.7}$$

This tensor trace vanishes, because the first tensor is antisymmetric, and the product of the last two antisymmetric tensors is symmetric with respect to $\alpha\gamma$.

So, the positive and negative terms cancel each other in our loop equation, like the "income" and "outcome" terms in the usual kinetic equation. We see, that there is an equilibrium in our loop space kinetics.

From the point of view of the notorious infrared divergencies in turbulence, the above calculation explicitly demonstrates how they cancel. By naive dimensional counting these terms were linearly divergent. The space isotropy lowered this to logarithmic divergency in (6.6), which reduced to finite terms at closer inspection. Then, the explicit form of these terms was such, that they all cancelled.

This cancellation originates from the angular momentum conservation in fluid mechanics. The large loop C creates the macroscopic eddy with constant vorticity $\omega_{\alpha\beta}^{C}$ and linear velocity $v^{C}(r) \propto r$. This is a well known static solution of the Navier-Stokes equation. The eddy is conserved due to the angular momentum conservation. The only nontrivial thing is the functional dependence of the eddy vorticity upon the shape and size of the loop C. This is a function of the tensor area $\Sigma^C_{\mu\nu}$, rather than a general functional of the loop.

Combining this Anzatz with the space isotropy and the Kolmogorov scaling law, we arrive at the tensor area law

$$\Psi[C] \propto \exp\left(-B\left(\frac{\mathcal{E}}{\nu^3}\left(\Sigma^C_{\alpha\beta}\right)^2\right)^{\frac{1}{3}}\right) \tag{6.8}$$

The universal constant B here must be real, in virtue of the loop orientation symmetry. When the orientation is reversed $C(\theta) \to C(2\pi - \theta)$, the loop integral changes sign, but its square, which enters here, stays invariant. Therefore, the constant in front must be real. The time reversal tells the same, since *both* viscosity ν and the energy dissipation rate $\mathcal{E}$ are time-odd. Therefore, the ratio $\frac{\mathcal{E}}{\nu^3}$ is time-even, hence it must enter $\Psi[C]$ with the real coefficient. Clearly, this coefficient B must be positive, since $|\Psi[C]| < 1$.

Note, however, that we did not prove this law. The absence of decay for large twisted loops with zero tensor area is suspicious. Also, the physics seems to be different from what we expect in turbulence. The uniform vorticity, even a random one, as in this solution, contrasts the observed intermittent distribution. Besides, there clearly must be corrections to the asymptotic law, whereas the tensor area law is *exact*. This is far too simple. We discussed this unphysical solution mostly as a test of the loop technology.

7 Scalar Area Law

Let us now study the scalar area law, which is a valid Anzatz for the asymptotic decay of the circulation PDF. The set of equations for the minimal surface is summarized in Appendix A. All we need here is the following representation

$$A \to \frac{1}{2L_\Gamma^2} \int\int d\sigma_{\mu\nu}(x)d\sigma_{\mu\nu}(y)\exp\left(-\pi\frac{(x-y)^2}{L_\Gamma^2}\right) \tag{7.1}$$

where $L_\Gamma = |\Gamma|^{\frac{3}{4}}\mathcal{E}^{-\frac{1}{4}}$. The distance $(x-y)^2$ is measured in 3-space and integration goes along the minimal surface. It is implied that its size is much larger than L_Γ.

In this limit the integration over, say, y can be performed along the local tangent plane at x in small vicinity $y - x \sim L_\Gamma$, after which the factors of L_Γ cancel. We are left then with the ordinary scalar area integral

$$A \to \frac{1}{2}\int d\sigma_{\mu\nu}(x)d\sigma_{\mu\nu}(y)\delta^2(x-y) \to \int d^2x\sqrt{g} \tag{7.2}$$

In the previous, regularized form the area represents so called Stokes functional[2], which can be substituted into the loop equation. The area derivative of the area reads

$$\frac{\delta A}{\delta\sigma_{\mu\nu}(x)} = \frac{1}{L_\Gamma^2}\int d\sigma_{\mu\nu}(y)\exp\left(-\pi\frac{(x-y)^2}{L_\Gamma^2}\right) \tag{7.3}$$

In the local limit this reduces to the tangent tensor

$$\frac{\delta A}{\delta\sigma_{\mu\nu}(x)} \to \int d\sigma_{\mu\nu}(y)\delta^2(x-y) = t_{\mu\nu}(x) \tag{7.4}$$

It is implied that the point x approaches the contour from inside the surface, so that the tangent tensor is well defined

$$t_{\mu\nu}(x) = t_\mu n_\nu - t_\nu n_\mu \tag{7.5}$$

Here t_μ is the local tangent vector of the loop, and n_ν is the inside normal to the loop along the surface.

The second area derivative of the regularized area in this limit is just the exponential

$$\frac{\delta^2 A}{\delta\sigma_{\alpha\beta}(x)\delta\sigma_{\gamma\delta}(y)} = \frac{1}{L_\Gamma^2}\exp\left(-\pi\frac{(x-y)^2}{L_\Gamma^2}\right) \tag{7.6}$$

Should we look for the higher terms of the asymptotic expansion at large area we would have to take into account the shape of the minimal surface, but in the thermodynamical limit we could neglect the curvature of the loop and use the planar disk.

Let us use the general WKB form of PDF

$$P_C(\Gamma) = \frac{1}{\Gamma}\exp\left(-Q\left(\frac{A}{t^{2\kappa}},\frac{\Gamma}{t^{2\kappa-1}}\right)\right) \tag{7.7}$$

We shall skip the arguments of effective action Q. We find on the left side of the loop equation

$$\partial_t Q\partial_\Gamma Q - \partial_t\partial_\Gamma Q \tag{7.8}$$

On the right side we find the following integrand

$$\left((\partial_A Q)^2 - \partial_A^2 Q\right)\frac{\delta A}{\delta\sigma_{\alpha\beta}(r)}\frac{\delta A}{\delta\sigma_{\gamma\beta}(r')} - \partial_A Q\frac{\delta^2 A}{\delta\sigma_{\alpha\beta}(r)\delta\sigma_{\gamma\beta}(r')} \tag{7.9}$$

The last term drops after the r' integration in virtue of symmetry. The leading terms in the WKB approximation on both sides are those with the first derivatives. We find

$$\partial_t Q\partial_\Gamma Q = (\partial_A Q)^2 \oint_C dr_\alpha\frac{\delta A}{\delta\sigma_{\alpha\beta}(r)}\int d^3 r'\frac{r_\gamma - r_\gamma'}{4\pi|r-r'|^3}\frac{\delta A}{\delta\sigma_{\gamma\beta}(r')} \tag{7.10}$$

In the last integral we substitute above explicit form of the area derivatives and perform the $d^3 r'$ integration first. In the thermodynamical limit only the small vicinity $r' - y \sim L_\Gamma$ contributes, and we find

$$\int d^3 r'\frac{r_\gamma - r_\gamma'}{4\pi|r-r'|^3}\frac{\delta A}{\delta\sigma_{\gamma\beta}(r')} \to L_\Gamma^2\int d\sigma_{\gamma\beta}(y)\frac{r_\gamma - y_\gamma}{4\pi|r-y|^3} \tag{7.11}$$

This integral logarithmically diverges. We compute it with the logarithmic accuracy with the following result

$$\int d\sigma_{\gamma\beta}(y)\frac{r_\gamma - y_\gamma}{4\pi|r-y|^3} \propto \frac{t_\beta}{\pi}\ln\frac{L_\Gamma^2}{A} \tag{7.12}$$

The meaning of this integral is the average velocity in the WKB approximation. This velocity is tangent to the loop, up to the next correction terms at large area.

Now, the emerging loop integral vanishes due to symmetry

$$\oint_C dr_\alpha t_\beta t_{\alpha\beta} = 0 \tag{7.13}$$

as the line element dr_α is directed along the tangent vector t_α, and the tangent tensor $t_{\alpha\beta}$ is antisymmetric. Similar mechanism was used in the tensor area solution, only there the cancellations emerged at the global level, after the closed loop integration. Here the right side of the loop equation vanishes locally, at every point of the loop. Anyway, we see, that the scalar area indeed represents the steady solution of the loop equation in the leading WKB approximation.

It might be instructive to compare this solution with another known exact solution of the Euler dynamics, namely the Gibbs solution

$$P[v] = \exp\left(-\beta \int d^3 r \frac{1}{2} v_\alpha^2\right) \tag{7.14}$$

For the loop functional it reads

$$\Psi_C(\gamma) = \exp\left(-\frac{\gamma^2}{2\beta} \oint_C dr_\alpha \oint_C dr'_\beta \delta^3(r - r')\right) \tag{7.15}$$

The integral diverges, and it corresponds to the perimeter law

$$\oint_C dr_\alpha \oint_C dr'_\beta \delta^3(r - r') \to r_0^{-2} \oint_C |dr| \tag{7.16}$$

where r_0 is a small distance cutoff. For the PDF it yields

$$P(\Gamma) \propto \exp\left(-\frac{\Gamma^2 \beta r_0^2}{2 \oint_C |dr|}\right) \tag{7.17}$$

When the Gibbs solution is substituted into the loop equation, we observe the same thing. Average velocity is tangent to the loop, which leads to vanishing integrand in the loop equation. The difference is that in our case this is true only asymptotically, there are next corrections.

The shape of the function Q is not fixed by this equation in the leading WKB approximation. In a scale invariant theory it is natural to expect the power law

$$Q\left(\frac{A}{t^{2\kappa}}, \frac{\Gamma}{t^{2\kappa-1}}\right) \to \text{const}\ \left(\Gamma^{2\kappa} A^{1-2\kappa}\right)^\mu \tag{7.18}$$

There is one more arbitrary index μ involved. Even for the Kolmogorov law $\kappa = \frac{3}{2}$ the Γ dependence remains unknown.

8 Discussion

So, we found two asymptotic solutions of the loop equation in the turbulent limit, not counting the Gibbs solution. It remains to be seen, which one (if any) is realized in turbulent flows. The tensor area solution is mathematically cleaner, but its physical meaning contradicts the intermittency paradigm. It corresponds to the uniform vorticity with random magnitude and random direction, rather that the regions of high vorticity interlaced with regions of low vorticity, observed in the turbulent flows.

The recent numerical experiments[6] favor the scalar area rather than the tensor one. Also, the Kolmogorov scaling was observed in these experiments. The Reynolds number was only ~ 100 which was too small to make any conclusions. We have to wait for the experiments (real or numerical) with the Reynolds numbers few orders of magnitude larger.

The scalar area is less trivial than the tensor one. The minimal area as a functional of the loop cannot be represented as any explicit contour integral of the Stokes type, therefore it corresponds to infinite number of higher correlation functions present. Moreover, there could be several minimal surfaces for the same loop, as the equations for the minimal surface are nonlinear. Clearly, the one with the least area should be taken.

The natural generalization of this solution is the string Anzatz where the sum over all surfaces bounded by the loop is taken

$$P_C(\Gamma) = \sum_{S:\partial S = C} \exp\left(-Q\left(\frac{A}{t^{2\kappa}}, \frac{\Gamma}{t^{2\kappa-1}}\right)\right) \tag{8.1}$$

At large loop the minimal Q terms will remain. The extremum condition

$$\delta Q = \frac{\partial Q}{\partial A} \delta A = 0 \tag{8.2}$$

will be satisfied for the minimal surface.

However, the sum over random surfaces is not well defined. The recent studies[3] indicate that the typical closed surfaces degenerate to branched polymers. For the surface bounded by a fixed loop this cannot happen, of course. Still nobody knows how to compute such sums. The loop equation in principle allows to systematically compute the corrections to the area law as the WKB expansion.

The WKB solution is incomplete so far. The leading term in the loop equation is annihilated by arbitrary function of the area (scalar or tensor). The similar ambiguity was present in the Gibbs solution, where arbitrary function of the Hamiltonian satisfied the Liouville equation for the velocity PDF. In that case the ambiguity was removed by extra requirement of thermodynamic limit: only the exponential of the hamiltonian would agree with the factorization of the PDF for two remote parts of the system.

What could be a similar requirement here? The area of the minimal surface represents the effective volume of the system at large loop. The circulation can be written as a surface integral of vorticity, which makes the circulation an extensive variable at this surface. The average vorticity

$$\overline{\omega} = \frac{\Gamma}{A} \tag{8.3}$$

represents an intensive quantity. The thermodynamic limit would then correspond to

$$Q = Aq(\overline{\omega}) \tag{8.4}$$

Comparing this with the previous formula for Q (7.18) we conclude that $\mu = 1$. In this case

$$Q = \mathrm{const}\ A\overline{\omega}^{2\kappa} \tag{8.5}$$

In principle, there could be two different laws for positive and negative Γ, due to violation of the time reversal invariance

$$Q \rightarrow q_{\pm} A |\bar{\omega}|^{2\kappa} \tag{8.6}$$

Another line of argument might start with an assumption of decorrelated average vorticity ω_i at various parts S_i of the area A_0 of the minimal surface. The net circulation, adding up from the large number $n \sim \frac{A}{A_0} \gg 1$ of independent random terms $\omega_i A_0$ would be a gaussian variable as a consequence of the law of large numbers. We would have then

$$Q \sim \frac{\Gamma^2}{n A_0^2 \omega_i^2} = \frac{\Gamma^2}{A A_0 \omega_i^2} \tag{8.7}$$

This would agree with the previous estimate at

$$\mu = \frac{1}{\kappa}, \ A_0 \omega_i^2 \sim A^{1-\frac{1}{\kappa}} \tag{8.8}$$

so that

$$Q \rightarrow \text{const } \Gamma^2 A^{\frac{1}{\kappa}-2} \tag{8.9}$$

The natural assumption here would be that the vorticity variance ω_i does not scale with the area A, so that

$$A_0 \sim A^{1-\frac{1}{\kappa}} \tag{8.10}$$

The Gaussian behavior (with $\kappa = \frac{3}{2}$) was observed in numerical experiments [6], but the Reynolds number was too low to make conclusions at this point. There could be a scaling function

$$Q = q\left(\Gamma^2 A^{\frac{1}{\kappa}-2}\right) \tag{8.11}$$

which starts linearly and then grows as a power, say, $q(x) = (1 + a\,x)^\kappa$. I suggest that this function should be studied in real and numerical experiments. This would teach us something new about turbulence.

Acknowledgments

I am grateful to V. Borue, I. Goldhirsh, A. Majda, D. McLaughlin, A. Polyakov and V. Yakhot for stimulating discussions. These results were also discussed at the summer schools in Cargese and Chernogolovka. I would like to thank the organisers and participants of both of these schools for creating the wondeful atmosphere of an independent scientific thinking.

A Loop Expansion

Let us outline the method of direct iterations of the loop equation. The full description of the method can be found in [2]. The basic idea is to use the following representation of the loop functional

$$\Psi[C] = 1 + \sum_{n=2}^{\infty} \frac{1}{n} \left\{ \oint_C dr_1^{\alpha_1} \cdots \oint_C dr_n^{\alpha_n} \right\}_{\text{cyclic}} W_{\alpha_1 \ldots \alpha_n}^n (r_1, \ldots r_n) \tag{A.1}$$

This representation is valid for every translation invariant functional with finite area derivatives (so called Stokes type functional). The coefficient functions W can be related to these area derivatives. The normalization $\Psi[0] = 1$ for the shrunk loop is implied.

In general case the integration points $r_1, \ldots r_n$ in (A.1) are cyclicly ordered around the loop C. The coefficient functions can be assumed cyclicly symmetric without loss of generality. However, in case of fluid dynamics, we are dealing with so called abelian Stokes functional. These functionals are characterized by completely symmetric coefficient functions, in which case the ordering of points can be removed, at expense of the extra symmetry factor in denominator

$$\Psi[C] = 1 + \sum_{n=2}^{\infty} \frac{1}{n!} \oint_C dr_1^{\alpha_1} \cdots \oint_C dr_n^{\alpha_n} W_{\alpha_1 \ldots \alpha_n}^n (r_1, \ldots r_n) \tag{A.2}$$

The incompressibility conditions

$$\partial_{\alpha_k} W_{\alpha_1 \ldots \alpha_n}^n (r_1, \ldots r_n) = 0 \tag{A.3}$$

does not impose any further restrictions, because of the gauge invariance of the loop functionals. This invariance (nothing to do with the symmetry of dynamical equations!) follows from the fact, that the closed loop integral of any total derivative vanishes. So, the coefficient functions are defined modulo such derivative terms. In effect this means, that one may relax the incompressibility constraints (A.3), without changing the loop functional.

To avoid confusion, let us note, that the physical incompressibility constrains are not neglected. They are, in fact, present in the loop equation, where we used the integral representation for the velocity in terms of vorticity. Still, the longitudinal parts of W drop in the loop integrals.

The loop calculus for the abelian Stokes functional is especially simple. The area derivative corresponds to removal of one loop integration, and differentiation of the corresponding coefficient function

$$\frac{\delta \Psi[C]}{\delta \sigma_{\mu\nu}(r)} = \sum_{n=1}^{\infty} \frac{1}{n!} \oint_C dr_1^{\alpha_1} \cdots \oint_C dr_n^{\alpha_n} \hat{V}_{\mu\nu}^{\alpha} W_{\alpha, \alpha_1 \ldots \alpha_n}^{n+1} (r, r_1, \ldots r_n) \tag{A.4}$$

where

$$\hat{V}_{\mu\nu}^{\alpha} \equiv \partial_\mu \delta_{\nu\alpha} - \partial_\nu \delta_{\mu\alpha} \tag{A.5}$$

In the nonabelian case, there would also be the contact terms, with W at coinciding points, coming from the cyclic ordering [2]. In abelian case these terms are absent, since W is completely symmetric.

As a next step, let us compute the local kinetic term

$$\hat{L}\Psi[C] \equiv \oint_C dr_\nu \partial_\mu \frac{\delta\Psi[C]}{\delta\sigma_{\mu\nu}(r)} \tag{A.6}$$

Using above formula for the loop derivative, we find

$$\hat{L}\Psi[C] = \sum_{n=1}^{\infty} \frac{1}{n!} \oint_C dr^\alpha \oint_C dr_1^{\alpha_1} \dots \oint_C dr_n^{\alpha_n} \partial^2 W^{n+1}_{\alpha,\alpha_1\dots\alpha_n}(r,r_1,\dots r_n) \tag{A.7}$$

The net result is the second derivative of W with respect to one variable. Note, that the second term in $\hat{V}^\alpha_{\mu\nu}$ dropped, as the total derivative in the closed loop integral.

As for the nonlocal kinetic term, it involves the second area derivative off the loop, at the point r', integrated over r' with the corresponding Green's function. Each area derivative involves the same operator $\hat{V}$, acting on the coefficient function. Again, the abelian Stokes functional simplifies the general framework of the loop calculus. The contribution of the wires cancels here, and the ordering does not matter, so that

$$\frac{\delta^2\Psi[C]}{\delta\sigma_{\mu\nu}(r)\delta\sigma_{\mu'\nu'}(r')} = \sum_{n=0}^{\infty} \frac{1}{n!} \oint_C dr_1^{\alpha_1} \dots \oint_C dr_n^{\alpha_n} \hat{V}^\alpha_{\mu\nu} \hat{V}'^{\alpha'}_{\mu'\nu'} W^{n+2}_{\alpha,\alpha',\alpha_1\dot{s}\alpha_n}(r,r',r_1,\dots r_n) \tag{A.8}$$

Using these relations, we can write the steady state loop equation as follows.

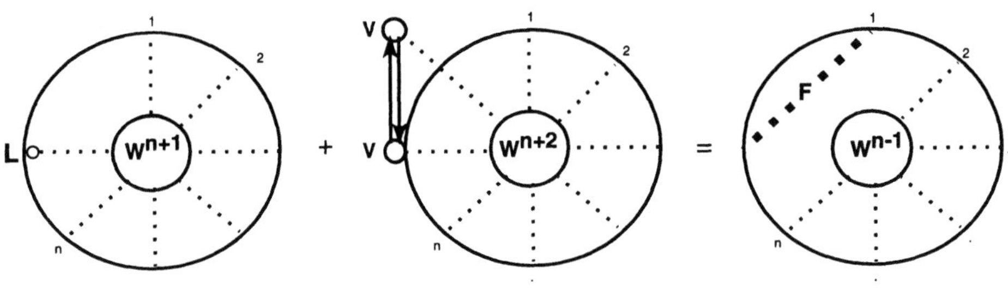

Fig.6.

Here the light dotted lines symbolize the arguments α_k, r_k of W, the big circle denotes the loop C, the tiny circles stand for the loop derivatives, and the pair of lines with the arrow denote the Green's function. The sum over the tensor indexes and the loop integrations over r_k are implied.

The first term is the local kinetic term, the second one is the nonlocal kinetic term, and the right side is the potential term in the loop equation. The heavy dotted line in this term stands for the correlation function F of the random forces. Note that this term is an abelian Stokes functional as well.

The iterations go in the potential term, starting with $\Psi[C] = 1$. In the next approximation, only the two loop correction $W^2_{\alpha_1\alpha_2}(r_1,r_2)$ is present. Comparing the terms, we note, that nonlocal kinetic term reduces to the total derivatives due to the space

214

symmetry (in the usual terms it would be $\langle v\omega \rangle$ at coinciding arguments), so we are left with the local one.

This yields the equation

$$\nu^3 \partial^2 W^2_{\alpha\beta}(r - r') = F(r - r')\delta_{\alpha\beta} \tag{A.9}$$

modulo derivative terms. The solution is trivial in Fourier space

$$W^2_{\alpha\beta}(r - r') = -\int \frac{d^3k}{(2\pi)^3}\exp\left(\imath\, k(r - r')\right)\delta_{\alpha\beta}\frac{\widetilde{F}(k)}{\nu^3 k^2} \tag{A.10}$$

Note, that we did not use the transverse tensor

$$P_{\alpha\beta}(k) = \delta_{\alpha\beta} - \frac{k_\alpha k_\beta}{k^2} \tag{A.11}$$

Though such tensor is present in the physical velocity correlation, here we may use $\delta_{\alpha\beta}$ instead, as the longitudinal terms drop in the loop integral. This is analogous to the Feynman gauge in QED. The correct correlator corresponds to the Landau gauge.

The potential term generates the four point correlation FW^2. which agrees with the disconnected term in the W^4 on the left side

$$W^4_{\alpha_1\alpha_2\alpha_3\alpha_4}\left(r_1, r_2, r_3, r_4\right) \to W^2_{\alpha_1\alpha_2}\left(r_1 - r_2\right)W^2_{\alpha_3\alpha_4}\left(r_3 - r_4\right) + \tag{A.12}$$
$$W^2_{\alpha_1\alpha_3}\left(r_1 - r_3\right)W^2_{\alpha_2\alpha_4}\left(r_2 - r_4\right) + W^2_{\alpha_1\alpha_4}\left(r_1 - r_4\right)W^2_{\alpha_2\alpha_3}\left(r_2 - r_3\right)$$

In the same order of the loop expansion, the three point function will show up. The corresponding terms in kinetic part must cancel among themselves, as the potential term does not contribute. The local kinetic term yields the loop integrals of $\partial^2 W^3$, whereas the nonlocal one yields $\widehat{V}W^2\,\widehat{V'}W^2$, integrated over d^3r' with the Greens's function $\frac{(r-r')}{4\pi|r-r'|^3}$. The equation has the structure

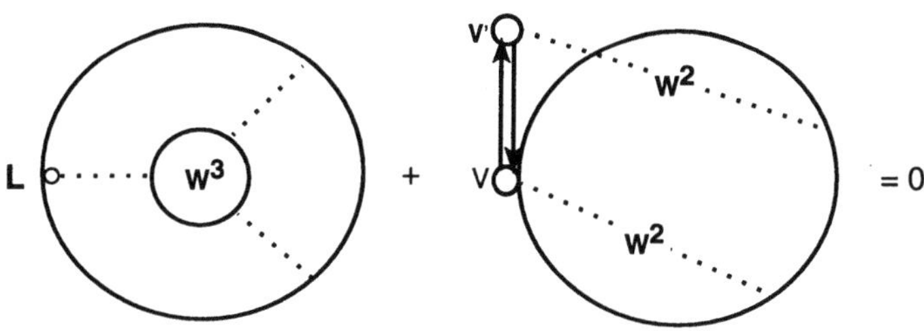

Fig.7.

Now it is clear, that the solution of this equation for W^3 would be the same three point correlator, which one could obtain (much easier!) by direct iterations of the Navier-Stokes equation.

The purpose of this painful exercise was not to give one more method of developing the expansion in powers of the random force. We rather verified that the loop equations are capable of producing the same results, as the ordinary chain of the equations for the correlation functions.

In above arguments, it was important, that the loop functional belonged to the class of the abelian Stokes functionals. Let us check that our tensor area Anzatz

$$\Sigma^C_{\alpha\beta} = \oint_C r_\alpha dr_\beta \tag{A.13}$$

belongs to the same class. Taking the square we find

$$\left(\Sigma^C_{\alpha\beta}\right)^2 = \oint_C dr_\beta \oint_C dr'_\beta r_\alpha r'_\alpha = -\frac{1}{2}\oint_C dr_\beta \oint_C dr'_\beta (r - r')^2 \tag{A.14}$$

where the last transformation follows from the fact, that only the cross term in $(r - r')^2$ yields nonzero after double loop integration.

Any expansion in terms of the square of the tensor area reduces, therefore to the superposition of multiple loop integral of the product of $(r_i - r_j)^2$, which is an example of the abelian Stokes functional. In the limit of large area, this could reduce to the fractional power. An example could be, say

$$\Psi[C] \stackrel{?}{=} \exp\left(B\left(1 - \left(1 + \frac{\mathcal{E}\left(\Sigma^C_{\alpha\beta}\right)^2}{\nu^3}\right)^{\frac{1}{3}}\right)\right) \tag{A.15}$$

One could explicitly verify all the properties of the abelian Stokes functional. This example is not realistic, though, as it does not have the odd terms of expansion. In the real world such terms are present at the viscous scales. According to our solution, this asymmetry disappears in inertial range of loops (which does not apply to velocity correlators at inertial range, as those correspond to shrunk loops).

B Matrix Model

The Navier-Stokes equation represents a very special case of nonlinear PDE. There is a well known galilean invariance

$$v_\alpha(r, t) \rightarrow v_\alpha(r - ut, t) + u_\alpha \tag{B.1}$$

which relates the magnitude of velocity field with the scales of time and space. [f] Let us make this relation more explicit.

First, let us introduce the vorticity field

$$\omega_{\mu\nu} = \partial_\mu v_\nu - \partial_\nu v_\mu \tag{B.2}$$

and rewrite the Navier-Stokes equation as follows

$$\dot{v}_\alpha = \nu\partial_\beta\omega_{\beta\alpha} - v_\beta\omega_{\beta\alpha} - \partial_\alpha w; \quad w = p + \frac{v^2}{2} \tag{B.3}$$

---

[f] At the same time it tells us that the constant part of velocity if frame dependent, so that it better be eliminated, if we would like to have a smooth limit at large times. Most of notorious large scale divergences in turbulence are due to this unphysical constant part.

216

This w is the well known enthalpy density, to be found from the incompressibility condition $divv = 0$, i.e.

$$\partial^2 w = \partial_\alpha v_\beta \omega_{\beta\alpha} \tag{B.4}$$

As a next step, let us introduce "covariant derivative" operator

$$D_\alpha = \nu \partial_\alpha - \frac{1}{2} v_\alpha \tag{B.5}$$

and observe that

$$2\left[D_\alpha D_\beta\right] = \nu \omega_{\beta\alpha} \tag{B.6}$$

$$2D_\beta\left[D_\alpha D_\beta\right] + h.c. = \nu \partial_\beta \omega_{\beta\alpha} - v_\beta \omega_{\beta\alpha} \tag{B.7}$$

where $h.c.$ stands for hermitian conjugate.

These identities allow us to write down the following dynamical equation for the covariant derivative operator

$$\dot{D}_\alpha = D_\beta\left[D_\alpha D_\beta\right] - D_\alpha W + h.c. \tag{B.8}$$

As for the incompressibility condition, it can be written as follows

$$\left[D_\alpha D_\alpha^\dagger\right] = 0 \tag{B.9}$$

The enthalpy operator $W = \frac{w}{\nu}$ is to be determined from this condition , or, equivalently

$$[D_\alpha [D_\alpha W]] = [D_\alpha, D_\beta [D_\alpha D_\beta]] \tag{B.10}$$

We see, that the viscosity disappeared from these equations. This paradox is resolved by extra degeneracy of this dynamics: the antihermitian part of the D operator is conserved. Its value at initial time is proportional to viscosity.

The operator equations are invariant with respect to the time independent unitary transformations

$$D_\alpha \to S^\dagger D_\alpha S; \; S^\dagger S = 1 \tag{B.11}$$

and, in addition, to the time dependent unitary transformations with

$$S(t) = \exp\left(\frac{1}{2\nu} t u_\beta \left(D_\beta - D_\beta^\dagger\right)\right) \tag{B.12}$$

corresponding to the galilean transformations.

One could view the operator D_α as the matrix

$$\langle i|D_\alpha|j\rangle = \int d^3 r \psi_i^\star(r) \nu \partial_\alpha \psi_j(r) - \frac{1}{2}\psi_i^\star(r) v_\alpha(r) \psi_j(r) \tag{B.13}$$

where the functions $\psi_j(r)$ are the Fourier of Tchebyshev functions depending upon the geometry of the problem.

The finite mode approximation would correspond to truncation of this infinite size matrix to finite size N. This is not quite the same as leaving N terms in the mode expansion of velocity field. The number of independent parameters here is $O(N^2)$ rather then $O(N)$. It is not clear whether the unitary symmetry is worth paying such a high price in numerical simulations!

The matrix model of Navier-Stokes equation has some theoretical beauty and raises hopes of simple asymptotic probability distribution. The ensemble of random hermitian matrices was recently applied to the problem of Quantum Gravity [3], which led to a genuine breakthrough in the field.

Unfortunately, the model of several coupled random matrices, which is the case here, is much more complicated then the one matrix model studied in Quantum Gravity. The dynamics of the eigenvalues is coupled to the dynamics of the "angular" variables, i.e. the unitary matrices S in above relations. We could not directly apply the technique of orthogonal polynomials, which was so successful in the one matrix problem.

Another technique, which proved to be successful in QCD and Quantum Gravity is the loop equations. This method, which we are discussing at length in this paper, works in field theory problems with hidden geometric meaning. The turbulence proves to be an ideal case, much simpler then QCD or Quantum Gravity.

C   The Reduced Dynamics

Let us now try to reproduce the dynamics of the loop field by a simpler Ansatz

$$\Psi[C] = \left\langle \exp\left(\frac{\imath}{\nu} \oint dC_\alpha(\theta) P_\alpha(\theta)\right)\right\rangle \tag{C.1}$$

The difference with original definition (2.1) is that our new function $P_\alpha(\theta)$ depends directly on θ rather then through the function $v_\alpha(r)$ taken at $r_\alpha = C_\alpha(\theta)$. This is the $d \to 1$ dimensional reduction we mentioned before. From the point of view of the loop functional there is no need to deal with field $v(r)$, one could take a shortcut.

Clearly, the reduced dynamics must be fitted to the Navier-Stokes dynamics of original field. With the loop calculus, developed above, we have all the necessary tools to build this reduced dynamics.

Let us assume some unknown dynamics for the P field

$$\dot{P}_\alpha(\theta) = F_\alpha\left(\theta, [P]\right) \tag{C.2}$$

and compare the time derivatives of original and reduced Ansatz. We find in (C.1) instead of (3.2)

$$\frac{\imath}{\nu} \oint dC_\alpha(\theta) F_\alpha\left(\theta, [P]\right) \tag{C.3}$$

Now we observe, that P' could be replaced by the functional derivative, acting on the exponential in (C.1) as follows

$$\frac{\delta}{\delta C_\alpha(\theta)} \leftrightarrow -\imath\nu P'_\alpha(\theta) \tag{C.4}$$

This means, that one could take the operators of the Section 2, expressing velocity and vorticity in terms of the spike operator, and replace the functional derivative as above. This yields the following formula for the spike derivative

$$D_\alpha(\theta,\epsilon) = -\imath\nu \int_\theta^{\theta+2\epsilon} d\phi \left(1 - \frac{|\theta+\epsilon-\phi|}{\epsilon}\right) P'_\alpha(\phi) = -\imath\nu \int_{-1}^1 d\mu \, \mathrm{sgn}(\mu) P_\alpha\left(\theta + \epsilon(1+\mu)\right) \tag{C.5}$$

218

This is the weighted discontinuity of the function $P(\theta)$, which in the naive limit $\epsilon \to 0$ would become the true discontinuity. However, the function $P(\theta)$ has in general the stronger singularities, then discontinuity, so that this limit cannot be taken yet.

Anyway, we arrive at the dynamical equation for the P field

$$\dot{P}_\alpha = \nu D_\beta \Omega_{\beta\alpha} - V_\beta \Omega_{\beta\alpha} \qquad (C.6)$$

where the operators V, D, Ω of the Section 2 should be regarded as the ordinary numbers, with definition (C.5) of D in terms of P.

All the functional derivatives are gone! We needed them only to prove equivalence of reduced dynamics to the Navier-Stokes dynamics.

The function $P_\alpha(\theta)$ would become complex now, as the right side of the reduced dynamical equation is complex for real $P_\alpha(\theta)$.

Let us discuss this puzzling issue in more detail. The origin of imaginary units was the factor of $\imath$ in exponential of the definition of the loop field. We had to insert this factor to make the loop field decreasing at large loops as a result of oscillations of the phase factors. Later this factor propagated to the definition of the P field.

Our spike derivative D is purely imaginary for real P, and so is our Ω operator. This makes the velocity operator V real. Therefore the $D\Omega$ term in the reduced equation (C.6) is real for real P whereas the $V\Omega$ term is purely imaginary.

This does not contradict the moments equations, as we saw before. The terms with even/odd number of velocity fields in the loop functional are real/imaginary, but the moments are real, as they should be. The complex dynamics of P simply doubles the number of independent variables.

There is one serious problem, though. Inverting the spike operator D_α we implicitly assumed, that it was antihermitian, and could be regularized by adding infinitesimal negative constant to D_α^2 in denominator. This, indeed, works perturbatively, in each term of expansion in time, or that in size of the loop, as we checked. However, beyond this expansion there would be a problem of singularities, which arise when $D_\alpha^2(\theta)$ vanishes at some θ.

In general, this would occur for complex θ, when the imaginary and real part of $D_\alpha^2(\theta)$ simultaneously vanish. One could introduce the complex variable

$$e^{\imath\theta} = z; \ e^{-\imath\theta} = \frac{1}{z}; \ \oint d\theta = \oint \frac{dz}{\imath z} \qquad (C.7)$$

where the contour of z integration encircles the origin around the unit circle. Later, in course of time evolution, these contours must be deformed, to avoid complex roots of $D_\alpha^2(\theta)$.

D Initial Data

Let us study the relation between the initial data for the original and reduced dynamics. Let us assume, that initial field is distributed according to some translation invariant probability distribution, so that initial value of the loop field does not depend on the constant part of $C(\theta)$.

One can expand translation invariant loop field in functional Fourier transform

$$\Psi[C] = \int DQ \delta^3 \left(\oint d\phi Q(\phi) \right) W[Q] \exp \left(\imath \oint d\theta C_\alpha(\theta) Q_\alpha(\theta) \right) \qquad (D.1)$$

which can be inverted as follows

$$\delta^3 \left(\oint d\phi Q(\phi) \right) W[Q] = \int DC \Psi[C] \exp \left(-\imath \oint d\theta C_\alpha(\theta) Q_\alpha(\theta) \right) \qquad (D.2)$$

Let us take a closer look at these formal transformations. The functional measure for these integrations is defined according to the scalar product

$$(A, B) = \oint \frac{d\theta}{2\pi} A(\theta) B(\theta) \qquad (D.3)$$

which diagonalizes in the Fourier representation

$$A(\theta) = \sum_{-\infty}^{+\infty} A_n e^{\imath n \theta}; \ A_{-n} = A_n^\star \qquad (D.4)$$

$$(A, B) = \sum_{-\infty}^{+\infty} A_n B_{-n} = A_0 B_0 + \sum_1^\infty a_n' b_n' + a_n'' b_n''; \ a_n' = \sqrt{2} \Re A_n, a_n'' = \sqrt{2} \Im A_n \qquad (D.5)$$

The corresponding measure is given by an infinite product of the Euclidean measures for the imaginary and real parts of each Fourier component

$$DQ = d^3 Q_0 \prod_1^\infty d^3 q_n' d^3 q_n'' \qquad (D.6)$$

The orthogonality of Fourier transformation could now be explicitly checked, as

$$\int DC \exp \left(\imath \int d\theta C_\alpha(\theta) \left(A_\alpha(\theta) - B_\alpha(\theta) \right) \right) \qquad (D.7)$$

$$= \int d^3 C_0 \prod_1^\infty d^3 c_n' d^3 c_n'' \exp \left(2\pi \imath \left(C_0 \left(A_0 - B_0 \right) + \sum_1^\infty c_n' \left(a_n' - b_n' \right) + c_n'' \left(a_n'' - b_n'' \right) \right) \right)$$

$$= \delta^3 \left(A_0 - B_0 \right) \prod_1^\infty \delta^3 \left(a_n' - b_n' \right) \delta^3 \left(a_n'' - b_n'' \right)$$

Let us now check the parametric invariance

$$\theta \to f(\theta); \ f(2\pi) - f(0) = 2\pi; \ f'(\theta) > 0 \qquad (D.8)$$

The functions $C(\theta)$ and $P(\theta)$ have zero dimension in a sense, that only their argument transforms

$$C(\theta) \to C\left(f(\theta) \right); \ P(\theta) \to P\left(f(\theta) \right) \qquad (D.9)$$

The functions $Q(\theta)$ and $P'(\theta)$ in above transformation have dimension one

$$P'(\theta) \to f'(\theta) P'\left(f(\theta) \right); \ Q(\theta) \to f'(\theta) Q\left(f(\theta) \right) \qquad (D.10)$$

so that the constraint on Q remains invariant

$$\oint d\theta Q(\theta) = \oint df(\theta) Q\left(f(\theta) \right) \qquad (D.11)$$

The invariance of the measure is easy to check for infinitesimal reparametrization

$$f(\theta) = \theta + \epsilon(\theta); \quad \epsilon(2\pi) = \epsilon(0) \tag{D.12}$$

which changes C and (C, C) as follows

$$\delta C(\theta) = \epsilon(\theta)C'(\theta); \quad \delta(C, C) = \oint \frac{d\theta}{2\pi}\epsilon(\theta)2C_\alpha(\theta)C'_\alpha(\theta) = -\oint \frac{d\theta}{2\pi}\epsilon'(\theta)C_\alpha^2(\theta) \tag{D.13}$$

The corresponding Jacobian reduces to

$$1 - \oint d\theta\epsilon'(\theta) = 1 \tag{D.14}$$

in virtue of periodicity.

This proves the parametric invariance of the functional Fourier transformations. Using these transformations we could find the probability distribution for the initial data of

$$P_\alpha(\theta) = -\nu \int_0^\theta d\phi Q_\alpha(\phi) \tag{D.15}$$

The simplest but still meaningful distribution of initial velocity field is the Gaussian one, with energy concentrated in the macroscopic motions. The corresponding loop field reads

$$\Psi_0[C] = \exp\left(-\frac{1}{2}\oint dC_\alpha(\theta)\oint dC_\alpha(\theta')f\left(C(\theta) - C(\theta')\right)\right) \tag{D.16}$$

where $f(r - r')$ is the velocity correlation function

$$\langle v_\alpha(r)v_\beta(r')\rangle = \left(\delta_{\alpha\beta} - \partial_\alpha\partial_\beta\partial_\mu^{-2}\right)f(r - r') \tag{D.17}$$

The potential part drops out in the closed loop integral.

The correlation function varies at macroscopic scale, which means that we could expand it in Taylor series

$$f(r - r') \to f_0 - f_1(r - r')^2 + \ldots \tag{D.18}$$

The first term f_0 is proportional to initial energy density,

$$\frac{1}{2}\left\langle v_\alpha^2\right\rangle = \frac{d - 1}{2}f_0 \tag{D.19}$$

and the second one is proportional to initial energy dissipation rate

$$\mathcal{E}_0 = -\nu\left\langle v_\alpha\partial_\beta^2 v_\alpha\right\rangle = 2d(d - 1)\nu f_1 \tag{D.20}$$

where $d = 3$ is dimension of space.

The constant term in (D.18) as well as $r^2 + r'^2$ terms drop from the closed loop integral, so we are left with the cross term rr'

$$\Psi_0[C] \to \exp\left(-f_1 \oint dC_\alpha(\theta)\oint dC_\alpha(\theta')C_\beta(\theta)C_\beta(\theta')\right) \tag{D.21}$$

This is almost Gaussian distribution: it reduces to Gaussian one by extra integration

$$\Psi_0[C] \to \text{const} \int d^3\omega \exp\left(-\omega_{\alpha\beta}^2\right)\exp\left(2\imath\sqrt{f_1}\omega_{\mu\nu}\oint dC_\mu(\theta)C_\nu(\theta)\right) \tag{D.22}$$

The integration here goes over all $\frac{d(d-1)}{2} = 3$ independent $\alpha < \beta$ components of the antisymmetric tensor $\omega_{\alpha\beta}$. Note, that this is ordinary integration, not the functional one. The physical meaning of this ω is the random constant vorticity at initial moment.

At fixed ω the Gaussian functional integration over C

$$\int DC \exp\left(\imath \oint d\theta \left(\frac{1}{\nu} C_\beta(\theta) P_\beta'(\theta) + 2\sqrt{f_1}\omega_{\alpha\beta} C_\alpha'(\theta) C_\beta(\theta)\right)\right) \tag{D.23}$$

can be performed explicitly, it reduces to solution of the saddle point equation

$$P_\beta'(\theta) = 4\nu\sqrt{f_1}\omega_{\beta\alpha} C_\alpha'(\theta) \tag{D.24}$$

which is trivial for constant ω

$$C_\alpha(\theta) = \frac{1}{4\nu\sqrt{f_1}}\omega_{\alpha\beta}^{-1} P_\beta(\theta) \tag{D.25}$$

The inverse matrix is not uniquely defined in odd dimensions, since $\mathrm{Det}\,\omega_{\alpha\beta} = 0$. However, the resulting pdf for P is unique. This is the Gaussian probability distribution with the correlator

$$\langle P_\alpha(\theta) P_\beta(\theta')\rangle = 2\imath\nu\sqrt{f_1}\omega_{\alpha\beta}\mathrm{sign}(\theta' - \theta) \tag{D.26}$$

Note, that antisymmetry of ω compensates that of the sign function, so that this correlation function is symmetric, as it should be. However, it is antihermitean, which corresponds to purely imaginary eigenvalues. The corresponding realization of the P functions is complex!

Let us study this phenomenon for the Fourier components. Differentiating the last equation with respect to θ and Fourier transforming we find

$$\langle P_{\alpha,n} P_{\beta,m}\rangle = \frac{4\nu}{m}\delta_{-nm}\sqrt{f_1}\omega_{\alpha\beta} \tag{D.27}$$

This cannot be realized at complex conjugate Fourier components $P_{\alpha,-n} = P_{\alpha,n}^\star$ but in our case $P_{\alpha,-n}$ and $P_{\alpha,n}$ are independent complex random variables, because the function $P_\alpha(\theta)$ is complex. The simplest realization is

$$P_{\alpha,n} = 4\nu\sqrt{f_1}\omega_{\alpha,\beta} C_{\beta,n}; \; P_{\beta,-m} = \frac{1}{2m}\overline{C}_{\beta,m} \tag{D.28}$$

where

$$C_{\beta,n} = a_{\beta,n} + \imath b_{\beta,n} \tag{D.29}$$

with $a_{\beta,n}, b_{\beta,n}$ being independent Gaussian random numbers with unit dispersion.

As for the constant part $P_{\alpha,0}$ of $P_\alpha(\theta)$, it is not defined, but it drops from equations in virtue of translational invariance.

E W-Functional

The difficulties of turbulence are hidden in the loop equation, but they show up, if you try to solve it numerically. The main problem is that one cannot get rid of the cutoffs $\epsilon, \delta \to 0$ in the definitions of the spike derivatives. These cutoffs are designed to pick up the singular contributions in the angular integrals, but with finite number of modes, such as Fourier harmonics there would be no singularities. We did not find any way to truncate degrees of freedom in the P equation, without violating the parametric invariance. It very well may be, that this invariance would be restored in the limit of large number of modes, but it looks that there are too much ambiguity in the finite mode approximation.

After some attempts, we found the simpler version of the loop functional, which can be studied analytically in the turbulent region. This is the generating functional for the scalar products $P_\alpha(\theta_1)P_\alpha(\theta_2)$

$$W[S] = \left\langle \exp\left(- \oint d\theta_1 \oint d\theta_2 S(\theta_1,\theta_2) P_\alpha(\theta_1) P_\alpha(\theta_2) \right) \right\rangle \tag{E.1}$$

where, as before, the averaging goes over initial data for the P field.

The time derivative of this W-functional

$$\dot{W} = -2 \left\langle \oint d\theta_1 \oint d\theta_2 S(\theta_1,\theta_2) P_\alpha(\theta_1) \dot{P}_\alpha(\theta_2) \exp\left(- \oint d\theta_1 \oint d\theta_2 S(\theta_1,\theta_2) P_\alpha(\theta_1) P_\alpha(\theta_2) \right) \right\rangle \tag{E.2}$$

can be expressed in terms of functional derivatives of W by replacing

$$P_\alpha(\phi_1) P_\alpha(\phi_2) \to \frac{\delta}{\delta S(\phi_1, \phi_2)} \tag{E.3}$$

for every scalar product of P fields, which arise after expansion of the spike derivatives (C.5), (2.14), (2.26) in the scalar product

$$P_\alpha(\theta_1)\dot{P}_\alpha(\theta_2) = \nu P_\alpha(\theta_1) D_\beta(\theta_2) \Omega_{\beta\alpha}(\theta_2) - P_\alpha(\theta_1) V_\beta(\theta_2) \Omega_{\beta\alpha}(\theta_2) \tag{E.4}$$

This equation has the structure

$$\dot{W} = \oint d^2\theta S(\theta_1,\theta_2) \left(A_2 \left[\frac{\delta}{\delta S}\right] W + A_3 \left[\frac{\delta}{\delta S}\right] D^{-2}(\theta,\epsilon) W \right) \tag{E.5}$$

where $A_k[X]$ stands for the $k-$ degree homogeneous functional of the function $X(\theta_1,\theta_2)$.

The operator D^{-2} is also the homogeneous functional of the negative degree $k = -1$. It can be written as follows

$$D^{-2}(\theta,\epsilon) W[S] = \int_0^\infty d\tau W[S + \tau U] \tag{E.6}$$

with

$$U(\theta_1,\theta_2) = \epsilon^{-2} \, \text{sgn}(\theta + \epsilon - \theta_1) \, \text{sgn}(\theta + \epsilon - \theta_2) \tag{E.7}$$

F Possible Numerical Implementation

The above general scheme is fairly abstract and complicated. Could it lead to any practical computation method? This would depend upon the success of the discrete approximations of the singular equations of reduced dynamics.

The most obvious approximation would be the truncation of Fourier expansion at some large number N. With Fourier components decreasing only as powers of n this approximation is doubtful. In addition, such truncation violates the parametric invariance which looks dangerous.

It seems safer to approximate $P(\theta)$ by a sum of step functions, so that it is piecewise constant. The parametric transformations vary the lengths of intervals of constant $P(\theta)$, but leave invariant these constant values. The corresponding representation reads

$$P_\alpha(\theta) = \sum_{l=0}^{N} \left(p_\alpha(l+1) - p_\alpha(l)\right) \Theta\left(\theta - \theta_l\right); \; p(N+1) = p(1), \; p(0) = 0 \qquad (\text{F.1})$$

It is implied that $\theta_0 = 0 < \theta_1 < \theta_2 \ldots < \theta_N < 2\pi$. By construction, the function $P(\theta)$ takes value $p(l)$ at the interval $\theta_{l-1} < \theta < \theta_l$.

We could take $\dot{P}(\theta)$ at the middle of this interval as approximation to $\dot{p}(l)$.

$$\dot{p}(l) \approx \dot{P}(\bar{\theta}_l); \; \bar{\theta}_l = \frac{1}{2}\left(\theta_{l-1} + \theta_l\right) \qquad (\text{F.2})$$

As for the time evolution of angles θ_l , one could differentiate (F.1) in time and find

$$\dot{P}_\alpha(\theta) = \sum_{l=0}^{N} \left(\dot{p}_\alpha(l+1) - \dot{p}_\alpha(l)\right) \Theta\left(\theta - \theta_l\right) - \sum_{l=0}^{N} \left(p_\alpha(l+1) - p_\alpha(l)\right) \delta(\theta - \theta_l)\dot{\theta}_l \qquad (\text{F.3})$$

from which one could derive the following approximation

$$\dot{\theta}_l \approx \frac{(p_\alpha(l) - p_\alpha(l+1))}{(p_\mu(l+1) - p_\mu(l))^2} \int_{\bar{\theta}_l}^{\bar{\theta}_{l+1}} d\theta \dot{P}_\alpha(\theta) \qquad (\text{F.4})$$

The extra advantage of this approximation is its simplicity. All the integrals involved in the definition of the spike derivative (C.5) are trivial for the stepwise constant $P(\theta)$. So, this approximation can be in principle implemented at the computer. This formidable task exceeds the scope of the present work, which we view as purely theoretical.

G Uniqueness of the Tensor Area Law

Let us address the issue of the uniqueness of the tensor area solution. Let us take the following Ansatz

$$S[C] = f\left(\oint_C dr_\alpha \oint_C dr'_\alpha W(r - r')\right) \qquad (\text{G.1})$$

When substituted into the static loop equation (with the area derivatives computed in Appendix A), it yields the following equation for the correlation function $W(r)$

$$0 = \oint_C dr_\alpha \oint_C dr'_\beta \oint_C dr''_\gamma U_{\alpha\beta\gamma}(r, r', r'') \qquad (\text{G.2})$$

$$U_{\alpha\beta\gamma}(r, r', r'') = W(r - r')\widehat{V}^\alpha_{\beta\gamma}W(r' - r'') + \text{permutations}$$

$$\widehat{V}^\alpha_{\mu\nu} = \delta_{\alpha\nu}\partial_\mu - \delta_{\alpha\mu}\partial_\nu$$

The derivative f' of the unknown function drops from the static equation.

This equation should hold for arbitrary loop C. Using the Taylor expansion for the Stokes type functional [2], we can argue, that the coefficient function U must vanish up to the total derivatives. An equivalent statement is that the third area derivative of this functional must vanish. Using the loop calculus (see Appendix A) we find the following equation

$$0 = \widehat{V}^{\alpha}_{\mu\nu}\widehat{V'}^{\alpha'}_{\mu'\nu'}\widehat{V''}^{\alpha''}_{\mu''\nu''}U_{\alpha\alpha'\alpha''}(r,r',r'') \tag{G.3}$$

which should hold for arbitrary r,r',r''. This leads to the overcomplete system of equations for $W(r)$ in general case. However, for the special case $W(r) = r^2$ which corresponds to the square of the tensor area

$$\Sigma^2_{\alpha\beta} = -\frac{1}{2}\oint_C dr_\alpha \oint_C dr'_\alpha (r-r')^2 \tag{G.4}$$

the system is satisfied as a consequence of certain symmetry. In this case we find in the loop equation

$$2\oint_C dr_\alpha \oint_C dr'_\beta (r-r')^2 \oint_C dr''_\alpha \left(r'_\beta - r''_\beta\right) \propto \Sigma^C_{\alpha\beta} \oint_C dr_\alpha \oint_C dr'_\beta (r-r')^2 \tag{G.5}$$

The last integral is symmetric with respect to permutations of α, β, whereas the first factor $\Sigma^C_{\alpha\beta}$ is antisymmetric, hence the sum over $\alpha\beta$ yields zero, as we already saw above.

It was assumed in above arguments, that the loop C consist of only one connected part. Let us now consider the more general situation, with arbitrary number n of loops $C_1,\ldots C_n$. The corresponding Ansatz would be

$$S_n[C_1,\ldots C_n] = s_n\left(\Sigma^1,\ldots \Sigma^n\right) \tag{G.6}$$

where Σ^i are tensor areas.

This function should obey the same WKB loop equations in each variable. Introducing the loop vorticities

$$\omega^k_{\mu\nu} = 2\frac{\partial s_n}{\partial \Sigma^k_{\mu\nu}} \tag{G.7}$$

which are constant on each loop, we have to solve the following problem. What are the values of $\omega^k_{\mu\nu}$ such that the single velocity field $v_\alpha(r)$ could produce them?

We do not see any other solutions, but the trivial one, with all equal $\omega^k_{\mu\nu}$ and linear velocity, as before. This would correspond to

$$s_n\left(\Sigma^1,\ldots \Sigma^n\right) = s_1\left(\Sigma\right); \ \Sigma_{\mu\nu} = \sum_{k=1}^n \Sigma^k_{\mu\nu} = \oint_{\uplus C_k} r_\mu dr_\nu \tag{G.8}$$

The loop equation would be satisfied like before, with $C = \uplus C_k$. This corresponds to the additivity of loops

$$S_n[C_1,\ldots C_n] = S_1[\uplus C_k] \tag{G.9}$$

Note, that such additivity is the opposite to the statistical independence, which would imply that

$$S_n[C_1,\ldots C_n] = \sum S_1[C_k] \tag{G.10}$$

The additivity could also be understood as a statement, that any set of n loops is equivalent to a single loop for the Abelian Stokes functional. Just connect these loops by wires, and note that the contribution of wires cancels. So, if the area law holds for *arbitrary* single loop, than it must be additive.

This assumption may not be true, though, as it often happens in the WKB approximation. There is no single asymptotic formula, but rather collections of different WKB regions, with quantum regions in between. In our case, this corresponds to the following situation.

Take the large circular loop, for which the WKB approximation holds, and try to split it into two large circles. You will have to twist the loop like the infinity symbol ∞, in which case it intersects itself. At this point, the WKB approximation might break, as the short distance velocity correlation might be important near the self-intersection point. This may explain the paradox of the vanishing tensor area for the ∞ shaped loop. From the point of view of our area law such loop is not large at all.

H Minimal Surfaces

Let us present here the modern view at the classical theory of the minimal surfaces. The minimal surface can be described by parametric equation

$$S : r_\alpha = X_\alpha \left(\xi_1, \xi_2 \right) \tag{H.1}$$

The function $X_\alpha(\xi)$ should provide the minimum to the area functional

$$A[X] = \int_S \sqrt{d\sigma_{\mu\nu}^2} = \int d^2\xi \sqrt{\mathrm{Det}\ G} \tag{H.2}$$

where

$$G_{ab} = \partial_a X_\mu \partial_b X_\mu, \tag{H.3}$$

is the induced metric. For the general studies it is sometimes convenient to introduce the unit tangent tensor as an independent field and minimize

$$A[X, t, \lambda] = \int d^2\xi \left(e_{ab} \partial_a X_\mu \partial_b X_\nu \, t_{\mu\nu} + \lambda \left(1 - t_{\mu\nu}^2 \right) \right) \tag{H.4}$$

From the classical equations we will find then

$$t_{\mu\nu} = \frac{e_{ab}}{2\lambda} \partial_a X_\mu \partial_b X_\nu \ ; t_{\mu\nu}^2 = 1, \tag{H.5}$$

which shows equivalence to the old definition.

For the actual computation of the minimal area it is convenient to introduce the auxiliary internal metric g_{ab}

$$A[X, g] = \frac{1}{2} \int_S d^2\xi \, \mathrm{tr}\ g^{-1} G \sqrt{\mathrm{Det}\ g}. \tag{H.6}$$

The straightforward minimization with respect to g_{ab} yields

$$g_{ab} \, \mathrm{tr}\ g^{-1} G = 2 G_{ab}, \tag{H.7}$$

226

which has the family of solutions

$$g_{ab} = \lambda G_{ab}. \tag{H.8}$$

The local scale factor λ drops from the area functional, and we recover original definition. So, we could first minimize the quadratic functional (H.6) with respect to $X(\xi)$ (the linear problem), and then minimize with respect to g_{ab} (the nonlinear problem).

The crucial observation is the possibility to choose conformal coordinates, with the diagonal metric tensor

$$g_{ab} = \delta_{ab}\rho, \ g_{ab}^{-1} = \frac{\delta_{ab}}{\rho}, \ \sqrt{\text{Det } g} = \rho; \tag{H.9}$$

after which the local scale factor ρ drops from the integral

$$A[X, \rho] = \frac{1}{2} \int_S d^2\xi \partial_a X_\mu \partial_a X_\mu. \tag{H.10}$$

However, the ρ field is implicitly present in the problem, through the boundary conditions.

Namely, one has to allow an arbitrary parametrization of the boundary curve C. We shall use the upper half plane of ξ for our surface, so the boundary curve corresponds to the real axis $\xi_2 = 0$. The boundary condition will be

$$X_\mu(\xi_1, +0) = C\left(f(\xi_1)\right), \tag{H.11}$$

where the unknown function $f(t)$ is related to the boundary value of ρ by the boundary condition for the metric

$$g_{11} = \rho = G_{11} = (\partial_1 X_\mu)^2 = C_\mu'^2 f'^2 \tag{H.12}$$

As it follows from the initial formulation of the problem, one should now solve the linear problem for the X field, compute the area and minimize it as a functional of $f(.)$. As we shall see below, the minimization condition coincides with the diagonality of the metric at the boundary

$$[\partial_1 X_\mu \partial_2 X_\mu]_{\xi_2=+0} = 0 \tag{H.13}$$

The linear problem is nothing but the Laplace equation $\partial^2 X = 0$ in the upper half plane with the Dirichlet boundary condition (H.11). The solution is well known

$$X_\mu(\xi) = \int_{-\infty}^{+\infty} \frac{dt}{\pi} \frac{C_\mu\left(f(t)\right) \xi_2}{\left(\xi_1 - t\right)^2 + \xi_2^2} \tag{H.14}$$

The area functional can be reduced to the boundary terms in virtue of the Laplace equation

$$A[f] = \frac{1}{2} \int d^2\xi \partial_a \left(X_\mu \partial_a X_\mu\right) = -\frac{1}{2} \int_{-\infty}^{+\infty} d\xi_1 \left[X_\mu \partial_2 X_\mu\right]_{\xi_2=+0} \tag{H.15}$$

Substituting here the solution for X we find

$$A[f] = -\frac{1}{2\pi} \Re \int_{-\infty}^{+\infty} dt \int_{-\infty}^{+\infty} dt' \frac{C_\mu(f(t)) \, C_\mu(f(t'))}{(t - t' - \imath 0)^2} \tag{H.16}$$

This can be rewritten in a nonsingular form

$$A[f] = \frac{1}{4\pi} \int_{-\infty}^{+\infty} dt \int_{-\infty}^{+\infty} dt' \frac{(C_\mu(f(t)) - C_\mu(f(t')))^2}{(t - t')^2} \qquad \text{(H.17)}$$

which is manifestly positive.

Another nice form can be obtained by integration by parts

$$A[f] = \frac{1}{2\pi} \int_{-\infty}^{+\infty} dt f'(t) \int_{-\infty}^{+\infty} dt' \, f'(t') C_\mu'(f(t)) \, C_\mu'(f(t')) \log|t - t'| \qquad \text{(H.18)}$$

This form allows one to switch to the inverse function $\tau(f)$ which is more convenient for optimization

$$A[\tau] = \frac{1}{2\pi} \int_{-\infty}^{+\infty} df \int_{-\infty}^{+\infty} df' C_\mu'(f) \, C_\mu'(f') \log|\tau(f) - \tau(f')| \qquad \text{(H.19)}$$

In the above formulas it was implied that $C(\infty) = 0$. One could switch to more traditional circular parametrization by mapping the upper half plane inside the unit circle

$$\xi_1 + i\xi_2 = i\frac{1 - \omega}{1 + \omega} \; ; \omega = re^{i\alpha} \; ; r \le 1. \qquad \text{(H.20)}$$

The real axis is mapped at the unit circle. Changing variables in above integral we find

$$X_\mu(r, \alpha) = \Re \int_{-\pi}^{\pi} \frac{d\theta}{\pi} C_\mu(\phi(\theta)) \left(\frac{1}{1 - r\exp(i\alpha - i\theta)} - \frac{1}{1 + \exp(-i\theta)} \right) \qquad \text{(H.21)}$$

Here

$$\phi(\theta) = f\left(\tan\frac{\theta}{2}\right). \qquad \text{(H.22)}$$

The last term represents an irrelevant translation of the surface, so it can be dropped. The resulting formula for the area reads

$$A[\phi] = \frac{1}{4\pi} \int_{-\pi}^{\pi} d\theta \int_{-\pi}^{\pi} d\theta' \frac{(C_\mu(\phi(\theta)) - C_\mu(\phi(\theta')))^2}{|e^{i\theta} - e^{i\theta'}|^2} \qquad \text{(H.23)}$$

or, after integration by parts and inverting parametrization

$$A[\theta] = \frac{1}{2\pi} \int_{-\pi}^{\pi} d\phi \int_{-\pi}^{\pi} d\phi' C_\mu'(\phi) \, C_\mu'(\phi') \log\left|\sin\frac{\theta(\phi) - \theta(\phi')}{2}\right| \qquad \text{(H.24)}$$

Let us now minimize the area as a functional of the boundary parametrization $f(t)$ (we shall stick to the upper half plane). The straightforward variation yields

$$0 = \Re \int_{-\infty}^{+\infty} dt' \frac{C_\mu(f(t')) \, C_\mu'(f(t))}{(t - t' + i0)^2} \qquad \text{(H.25)}$$

which duplicates the above diagonality condition (H.13). Note that in virtue of this condition the normal vector $n_\mu(x)$ is directed towards $\partial_2 X_\mu$ at the boundary. Explicit formula reads

$$n_\mu\left(C(f(t))\right) \propto \Re \int_{-\infty}^{+\infty} dt' \, \frac{C_\mu(f(t'))}{(t - t' + \imath 0)^2} \tag{H.26}$$

Let us have a closer look at the remaining nonlinear integral equation (H.25). In terms of inverse parametrization it reads

$$0 = \Re \int_{-\infty}^{+\infty} df \, \frac{C'_\mu(f) C'_\mu(f')}{\tau(f) - \tau(f') + \imath 0} \tag{H.27}$$

Introduce the vector set of analytic functions

$$F_\mu(z) = \int_{-\infty}^{+\infty} \frac{df}{\pi} \, \frac{C'_\mu(f)}{\tau(f) - z} \tag{H.28}$$

which decrease as z^{-2} at infinity. The discontinuity at the real axis

$$\Im F_\mu(\tau + \imath 0) = C'_\mu(f) f'(\tau) \tag{H.29}$$

Which provides the implicit equation for the parametrization $f(\tau)$

$$\int d\tau \Im F_\mu(\tau + \imath 0) = C_\mu(f) \tag{H.30}$$

We see, that the imaginary part points in the tangent direction at the boundary. As for the boundary value of the real part of $F_\mu(\tau)$ it points in the normal direction along the surface

$$\Re F_\mu \propto n_\mu \tag{H.31}$$

Inside the surface there is no direct relation between the derivatives of $X_\mu(\xi)$ and $F_\mu(\xi)$.

The integral equation (H.25) reduces to the trivial boundary condition

$$F_\mu^2(t + \imath 0) = F_\mu^2(t - \imath 0) \tag{H.32}$$

In other words, there should be no discontinuity of F_μ^2 at the real axis. The solution compatible with analyticity in the upper half plane and z^{-2} decrease at infinity is

$$F_\mu^2(z) = (1 + \omega)^4 \, P(\omega); \quad \omega = \frac{\imath - z}{\imath + z} \tag{H.33}$$

where $P(\omega)$ defined by a series, convergent at $|\omega| \le 1$. In particular this could be a polynomial. The coefficients of this series should be found from an algebraic minimization problem, which cannot be pursued forward in general case.

The flat loops are trivial though. In this case the problem reduces to the conformal transformation mapping the loop onto the unit circle. For the unit circle we have simply

$$C_1 + \imath C_2 = \omega; \ F_1 = \imath F_2 = -\frac{(1+\omega)^2}{2}; \ P = 0. \tag{H.34}$$

Small perturbations around the circle or any other flat loop can be treated in a systematic way, by a perturbation theory.

I Kolmogorov Triple Correlation and Time Reversal

Are there any restrictions on the circulation PDF from the known asymmetry of velocity correlations, in particular, the Kolmogorov triple correlation? The answer is that the Kolmogorov correlation does not imply the asymmetry of *vorticity* correlations.

Taking the tensor version of the $\frac{4}{5}$ law in arbitrary dimension d

$$\langle v_\alpha(0) v_\beta(0) v_\gamma(r) \rangle = \frac{\mathcal{E}}{(d-1)(d+2)} \left(\delta_{\alpha\gamma} r_\beta + \delta_{\beta\gamma} r_\alpha - \frac{2}{d} \delta_{\alpha\beta} r_\gamma \right) \tag{I.1}$$

and differentiating, we find that

$$\langle v_\alpha(0) v_\beta(0) \omega_{\gamma\lambda}(r) \rangle = 0 \tag{I.2}$$

So, the odd vorticity correlations could, in fact, be absent, in spite of the asymmetry of the velocity distribution.

References

[1] A.N. Kolmogorov. *The local structure of turbulence in incompressible viscous fluid for very large Reynolds' numbers.* C. R. Acad. Sci. URSS, 30(4):301–305, 1941.

[2] A. Migdal, *Loop equations and $\frac{1}{N}$ expansion*, Physics Reports , 201 , 102,(1983).

[3] D Gross, A Migdal, Physical Review Letters, 717, **64**, (1990), and Nuclear Physics B340, 333,(1990), M. Douglas and S. Shenker, Nuclear Physics B335, 635, (1990), E. Brezin, V. Kazakov, Physical Letters, 144, **236B**, (1990).

[4] A.A. Migdal, *Loop equation in turbulence*, PUPT-1383, March '93, hep-th/9303130.

[5] A.A. Migdal, *Turbulence as statistics of vortex cells*, PUPT-1409, hep-th/9306152.

[6] Makoto Umeki, Tokyo University Preprint , July 93, hep-th/9307144.

THE TWO-DIMENSIONAL STRING AS A TOPOLOGICAL FIELD THEORY[a]

Sunil MUKHI

Tata Institute of Fundamental Research
Homi Bhabha Road, Bombay 400 005, India

Abstract: A certain topological field theory is shown to be equivalent to the compactified $c = 1$ string. This theory is described in both Landau-Ginzburg and Kazama-Suzuki formulations. The genus-g partition function and genus-0 multi-tachyon correlators of the $c = 1$ string are shown to be calculable in this approach. The KPZ formulation of non-critical string theory has a natural relation to this topological model.

1 Introduction

For many years it used to be said that string theory is consistent only in 26 spacetime dimensions. With the understanding that any CFT with total central charge $c = 26$ is a consistent background for bosonic string propagation, this statement had to be modified. It remained true that in some sense the simplest known background for the bosonic string was the one with 26 flat, noncompact spacetime dimensions.

The relevance of the concept of "simplest known background" stems from the fact that we do not yet fully understand background-independent string field theory. In this situation, the theory in the simplest background may reasonably be expected to furnish clues about the nature of the full theory. For example, arbitrary backgrounds in a field theory tend to spontaneously break some or all of the global symmetries. In such backgrounds we would be hard pressed to discover the full symmetry structure of the theory. Thus, we must look for backgrounds which preserve as many symmetries as possible – indeed, this is probably the most reasonable definition of "simplest background".

With this in mind, it becomes clear that the noncompact 26-dimensional background is far from being the simplest one. Indeed, just compactifying the spacetime on a torus of appropriate shape can produce[3] large numbers of Kac-Moody currents, whose integrals are unbroken symmetry generators. But even better is the background with

---

[a]Based on work done in collaboration with C. Vafa[1] and with D. Ghoshal[2]

Quantum Field Theory and String Theory, Edited by
L. Baulieu *et al.*, Plenum Press, New York, 1995

just 2 spacetime dimensions, traditionally described by scalar fields $X(z,\bar{z})$ and $\phi(z,\bar{z})$ on the world-sheet. The unbroken symmetries of this theory correspond to infinitely many holomorphic currents satisfying the wedge subalgebra of W_∞[4, 5, 6]. This is the most symmetric known phase of the bosonic string. More precisely, the maximum number of unbroken symmetries arise when the field $X(z,\bar{z})$ is compactified on a circle of radius $\frac{1}{\sqrt{2}}$, the so-called self-dual radius.

The two-dimensional string background (also known as "$c = 1$ string theory") has the following energy-momentum tensor for its matter sector:

$$T(z) = -\frac{1}{2}\partial X \partial X - \frac{1}{2}\partial\phi\partial\phi + \sqrt{2}\partial^2\phi \tag{1.1}$$

where $X(z,\bar{z})$ and $\phi(z,\bar{z})$ are conventionally known as the "matter" and "Liouville" fields respectively. The presence of a total derivative term in $T(z)$ corresponds to an extra term in the worldsheet action proportional to $\int \sqrt{g}R^{(2)}\phi$. This term renders the functional integral ill-defined, and to stabilise the theory we must add a cosmological term $\mu \int \sqrt{g} \exp(\alpha\phi)$. This means that Liouville momentum is not conserved. Every correlation function carries a power of μ equal to minus the amount by which it violates Liouville momentum conservation, in multiples of $\sqrt{2}$.

In particular, it is easy to see that the partition function of the $c = 1$ string in genus g has the behaviour

$$Z_g(\mu) \sim \mu^{2-2g} \tag{1.2}$$

Indeed, for the two-dimensional string compactified on the self-dual radius, the partition function is known explicitly from matrix models[7] and is given by

$$Z_g^{\text{self}-\text{dual}}(\mu) = \mu^{2-2g}\chi_g, \qquad \chi_g = \frac{B_{2g}}{2g(2g-2)} \tag{1.3}$$

where χ_g is the virtual Euler characteristic of the moduli space of genus-g Riemann surfaces, which is proportional to the Bernoulli number B_{2g}.

It has long been a challenge to reproduce this result from a continuum formulation of two-dimensional string theory. Other matrix-model results for this theory, particularly the correlation functions of "tachyons" in genus 0, have been re-derived in the continuum only after considerable difficulty and after resorting to somewhat arbitrary continuations of the parameters of the theory[8, 9, 10, 11, 12]. One of the crucial difficulties in continuum calculations is that the theory is not perturbative in the cosmological constant μ (as is evident from Eq.1.2). Another problem is that standard CFT techniques produce the amplitudes for local vertex operators inserted at fixed points on the worldsheet, which must then be integrated over the worldsheet to give the physical amplitudes. In a theory coupled to gravity, physical answers for correlations of integrated operators (which are typically quite simple) should not have to depend on first obtaining the local answer and explicitly performing a complicated integral.

In what follows I will show that there is a topological field theory which is equivalent to $c = 1$ string theory at the self-dual radius, and that the manifestly topological formulation permits the computation of the partition function and amplitudes in any genus, and at nonzero cosmological constant, without resorting to perturbative treatments. This means that we have a continuum formulation of this string theory which is apparently as powerful as the very successful matrix-model description. This discovery may lend itself to generalization and teach us something about string theory itself rather than about some specific background, unlike the matrix models which have so far resisted attempts to go beyond the specific backgrounds which they represent.

2 Topological Symmetry of String Backgrounds

Topological symmetry was thought to be a special property of some string backgrounds. More recently, it has become clear that it is present in *any* string background in which there is at least one (possibly anomalous) $U(1)$ current[13]. The basic idea is that in conformal gauge, even the bosonic string possesses four chiral fields of the right spins and statistics to form a twisted (hence topological) $N = 2$ superconformal algebra. In general, this algebra is generated by bosonic fields $T(z)$ and $J(z)$ of spins 2 and 1 respectively, and fermionic fields $G^+(z)$ and $G^-(z)$ which also have spins 2 and 1 respectively. They obey the following singular OPE relations:

$$
\begin{aligned}
T(z)T(w) &\sim \frac{2T(w)}{(z-w)^2} + \frac{\partial T(w)}{(z-w)} \\
T(z)G^\pm(w) &\sim \frac{1}{2}\frac{(3 \mp 1)G^\pm(z)}{(z-w)^2} + \frac{\partial G^\pm(w)}{(z-w)} \\
T(z)J(w) &\sim \frac{c^T/3}{(z-w)^3} + \frac{J(w)}{(z-w)^2} + \frac{\partial J(w)}{(z-w)} \\
J(z)G^\pm(w) &\sim \pm\frac{G^\pm(w)}{(z-w)} \\
J(z)J(w) &\sim \frac{c^T/3}{(z-w)^2} \\
G^\pm(z)G^\pm(w) &\sim 0 \\
G^+(z)G^-(w) &\sim \frac{c/3}{(z-w)^3} + \frac{J(w)}{(z-w)^2} + \frac{T(w)}{(z-w)}
\end{aligned}
\tag{2.1}
$$

For the $c = 1$ string, the bosonic fields $T(z), J(z)$ in the above algebra arise from the full stress-energy tensor (including ghost contribution) and the ghost-number current, while the fermionic fields $G^\pm(z)$ are the BRST current and the antighost. It turns out that the OPEs among these fields are not actually closed, but on modifying some of the fields by total derivative terms, one finds a closed algebra which is precisely the twisted $N = 2$ superconformal algebra above, with some fixed value of the topological central charge c^T.

Explicitly, the generators of the twisted $N = 2$ algebra are as follows:

$$
\begin{aligned}
T(z) &= T(z)^{(\text{matter})} + T(z)^{(\text{ghost})} \\
G^+(z) &= c(z)T(z)^{(\text{matter})} + \frac{1}{2} : c(z)T(z)^{(\text{ghost})} : + x\partial^2 c(z) + y\partial(c(z)\partial\eta(z)) \\
G^-(z) &= b(z) \\
J(z) &= : c(z)b(z) : - y\partial\eta(z)
\end{aligned}
\tag{2.2}
$$

where $\partial\eta(z)$ is a $U(1)$ current formed out of the fields in the matter sector of the background. (Here we will work with the case where the current arises as the derivative of a free scalar field, although the more general case is obvious.) If the η field has a background charge Q_η, then it is easy to check that the constants x and y must take the following values in order to get a closed algebra:

$$
\begin{aligned}
x &= \frac{1}{2}(3 + Q_\eta y) \\
y &= \frac{1}{2}(-Q_\eta + \sqrt{Q_\eta^2 - 8})
\end{aligned}
\tag{2.3}
$$

The topological central charge of this algebra is found to be $c^T = 6x$.

Clearly, the structure described above depends on the choice of both a background and a particular $U(1)$ current of that background. For example, if we choose the $c = 1$ string, the current can be associated to either the Liouville scalar $\phi(z, \bar{z})$ or the matter scalar $X(z, \bar{z})$. Thus, we have:

$$
\begin{array}{ccccccc}
\eta = \phi, & Q_\phi = 2\sqrt{2} & \rightarrow & x = -\frac{1}{2}, & y = -\sqrt{2}, & c^T = -3 \\
\eta = X, & Q_X = 0 & \rightarrow & x = \frac{3}{2}, & y = i\sqrt{2}, & c^T = 9
\end{array}
\tag{2.4}
$$

Another interesting case is the critical bosonic string, in which one can choose the $U(1)$ current to come from any of 26 free scalars, all non-anomalous, and one obtains $c^T = 9$ in each case.

The case of interest here will be the non-critical $c = 1$ background. Thus, in principle both the choices of $U(1)$ current above are available to us. The first choice, in which it is associated to the Liouville field, may seem more natural since this field is generally present in noncritical backgrounds[14]. However, that choice has the following disadvantage. If the action is perturbed by the cosmological operator, as it must be to produce a well-defined functional integral, the Liouville field satisfies

$$
\bar{\partial}\partial\phi + \mu e^\phi = 0
\tag{2.5}
$$

Thus, the current $\partial\phi$ is no longer holomorphic, and we conclude that the η-current in the $N = 2$ algebra cannot be identified with the Liouville field except at zero cosmological constant. No such restriction applies if we choose instead the $c = 1$ matter field (the "time" coordinate of the string). Hence we make that choice, and conclude that $c = 1$ string theory is described by a topological $N = 2$ algebra with central charge $c^T = 9$.

This observation has one more or less immediate consequence. The various states in the BRST cohomology of the $c = 1$ string must be classified by an $N = 2$ topological algebra. It is well-known[15, 16, 17, 5] that there is a plethora of physical states in this theory, including tachyons, discrete states of ghost number 1 (the "remnants" of transverse gravitons and other tensor excitations of the string propagating in two dimensions), and discrete states of ghost number 0 (the so-called "ground ring" elements). It can be argued that the only states among these which are actually primaries of the topological algebra are the tachyons. This applies at any compactification radius for the X coordinate, and both the types of tachyons that generically exist (the special discrete tachyons as well as the intermediate ones) are always primary. The discrete "remnant" states of ghost number 1 are secondaries of the $N = 2$ algebra. On the other hand, the discrete ground ring states, of ghost number 0, are both primary and secondary. These states would normally be termed null, except that (as is familiar[18, 15]) the projection from the chiral algebra to the Fock representation fails to be an isomorphism, due to the presence of background charges. Hence null vectors in the module of the chiral algebra can be non-vanishing Fock-space states.

In our notation, tachyon operators are represented by

$$T_k^{\pm} = c \, \exp\frac{1}{\sqrt{2}}(2 \mp k)\phi \, \exp\frac{i}{\sqrt{2}}kX \tag{2.6}$$

where the label k takes all real values in the noncompact case, and integer values at the self-dual radius. The total conformal dimension of these operators is of course 0, which is a necessary condition for them to be in the BRST cohomology. However, one can also consider more general operators ("off-shell tachyons") described by

$$T_{k_1,k_2}^{\pm} = c \, \exp\frac{1}{\sqrt{2}}(2 \mp k_1)\phi \, \exp\frac{i}{\sqrt{2}}k_2X \tag{2.7}$$

Here, we fix the Liouville momentum k_1 to be a positive integer, while the matter momentum k_2 takes values in $k_1, k_1 - 2, \cdots, -k_1$. These fields have *negative* conformal dimension $k_2^2 - k_1^2$, hence by well-known theorems they cannot be in the BRST cohomology. In fact, they are BRST exact. But some secondaries of the topological algebra above these off-shell tachyons are in the BRST cohomology, and in fact are just the various types of discrete states.

Thus we see that the presence of a twisted $N = 2$ topological algebra in the $c = 1$ string actually explains the variety of physical states which were discovered over the last few years[19, 20, 21]. The tachyons, whose existence was expected and is rather well-understood, are really the only basic states of the theory. The two types of discrete states, both of which seemed a little exotic at first, are in fact nothing but particular secondaries with respect to the topological algebra. Some details concerning this phenomenon may be found in Ref.[1], although the full picture probably contains more that has not yet been worked out.

3 A Coset Model of the $c = 1$ String

Besides organising the various physical states of the $c = 1$ string elegantly, the topological symmetry described above does something more fundamental. As it stands, this symmetry looks accidental, in that there is nothing known about the basics of string theory which would predict it. But since it exists, one could imagine looking for alternative formulations of various string backgrounds in which the topological symmetry is manifest. These formulations might then be exactly solvable in the way that topological theories often are. For the special case of the $c = 1$ background (at the self-dual radius) we will argue that there is a manifestly topological model which fulfils these requirements: it is equivalent to the $c = 1$ string, but is "more solvable", in the sense that it permits the computation of amplitudes which are not calculable (as far as we know) within the conventional KPZ[22] or DDK[23] formulations.

The model for which we are searching cannot be predicted from some general principles, but we do know from the discussion of the previous section that it must possess a topological symmetry with central charge $c^T = 9$. We will look for this model among a general class of $N = 2$ supersymmetric CFT's, the Landau-Ginzburg and Kazama-Suzuki models. First we concentrate on the Kazama-Suzuki (KS) description.

The construction of KS models is based on the following facts. Let us start with an $N = 1$ supersymmetric WZW model, based on a group G, and gauge the adjoint action of a subgroup H. Then, if and only if the coset G/H is a Kähler manifold, the model so obtained has $N = 2$ supersymmetry[24]. Applying this idea to the case where $G = SL(2, R)$ and $H = U(1)$, we find a series of models with central charge $c = 3k/(k - 2)$, where k is the level of the $SL(2, R)$ current algebra. After twisting the model so obtained to make it topological, one finds as usual that the central charge of the theory becomes zero, but there is a topological central charge c^T which has the same value as the central charge before twisting. Accordingly, if we want $c^T = 9$ in this class of models, the unique choice is to take level $k = 3$. Curiously, $SL(2, R)$ at level 3 also occurs in the KPZ description of $c = 1$ string theory, and we will see below that this is not a coincidence – in fact, the construction described here will illuminate a few long-standing mysteries of the KPZ approach.

Let us now look at the KS model at level 3 in some detail. The stress-energy tensor of the supersymmetric G model is, as usual,

$$T^{(G)}(z) =: J^+(z)J^-(z) : + (J^3(z))^2 - \frac{1}{2}(b(z)\partial c(z) + c(z)\partial b(z)) \tag{3.1}$$

where J^+, J^-, J^3 are the spin-1 $SL(2, R)$ currents, b, c are spin-$\frac{1}{2}$ fermions which serve to supersymmetrise the theory, and the coefficient of the first term is unity precisely at $k = 3$.

One can write the stress-energy tensor of the coset model as the difference of the (supersymmetric) G and H stress-energy tensors. However, for our purposes it is more convenient to use an alternative formulation in which to the G stress-energy we *add* a gauge contribution and then pass to a BRST cohomology on this larger space[25]. Thus we have

$$T(z) = T^{(G)}(z) + T^{(\text{gauge})}(z) \tag{3.2}$$

We will write this out explicitly below, but first let us see what the other generators of the $N = 2$ algebra look like. Following Kazama-Suzuki[24], we find (at $k = 3$)

$$\begin{aligned}
G^+(z) &= c(z)J^+(z) \\
G^-(z) &= c(z)J^-(z) \\
J_{N=2}(z) &= 3 : c(z)b(z) : - 2J^3(z)
\end{aligned} \tag{3.3}$$

Here we have labelled the $U(1)$ current of the $N = 2$ algebra explicitly to distinguish it from other $U(1)$ currents in the problem.

So far, this is an untwisted $N = 2$ superconformal algebra, with central charge $3k/(k-2)$ as mentioned earlier. Now we render it topological through the twist

$$T(z) \quad \rightarrow \quad T(z) + \frac{1}{2}\partial J_{N=2}(z) \tag{3.4}$$

As a result, the final stress-energy tensor (to be compared with Eqs.(3.1),(3.2)) is

$$T(z) = \;:J^+(z)J^-(z):\; + (J^3(z))^2 - \partial J^3(z) - 2b(z)\partial c(z) + c(z)\partial b(z) + T^{(\text{gauge})}(z) \tag{3.5}$$

This twist has a rather miraculous consequence. The spins of all the fields in $T^{(G)}$ have changed, and we now have an $SL(2,R)$ current multiplet (J^+, J^3, J^-) of spins $(2,1,0)$ respectively. Moreover, the free fermions (b,c) have changed their spins from $(\frac{1}{2}, \frac{1}{2})$ to $(2, -1)$. The currents are now reminiscent of those in the KPZ description of $c = 1$ string theory, where they described the gravitational or Liouville sector, while the fermions have become identical to the usual ghost system of bosonic string theory! The central charges are found to be $c = 27$ for the twisted currents and of course $c = -26$ for the fermions.

Now we add in the $U(1)$ gauge sector of the theory by making the gauge choice

$$A_z(z) = \partial X(z), \qquad \overline{A}_{\overline{z}}(\overline{z}) = -\overline{\partial X}(\overline{z}) \tag{3.6}$$

and defining the free scalar field $X(z, \overline{z}) = X(z) - \overline{X}(\overline{z})$. Fixing the gauge in this way requires the introduction of a pair of fermionic ghosts (B, C) of spins $(1, 0)$, hence we get

$$T^{(\text{gauge})}(z) = -\frac{1}{2}\partial X(z)\partial X(z) - B(z)\partial C(z) \tag{3.7}$$

along with the $U(1)$ BRST charge

$$Q_{U(1)} = \int c(z)(J^3(z) - :c(z)b(z): - \frac{i}{\sqrt{2}}\partial X(z)) \quad + \text{c.c} \tag{3.8}$$

So far, everything except the current-algebra sector has been reduced to free fields. We now choose to represent the currents also in terms of free fields, using the Wakimoto representation:

$$\begin{aligned}
J^+(z) &= \;:\beta(z)\gamma(z)^2:\; - \sqrt{2}\gamma(z)\partial\phi(z) + 3\partial\gamma(z) \\
J^3(z) &= \;:\beta(z)\gamma(z):\; - \frac{1}{\sqrt{2}}\partial\phi(z) \\
J^-(z) &= \;\beta(z)
\end{aligned} \tag{3.9}$$

where (β, γ) are commuting ghosts and ϕ is a free scalar field. Since we know the spins of the currents (after twisting of course), it is easy to deduce that the ghosts have spins $(0, 1)$ respectively. This means they contribute a total central charge of $+2$. But the total central charge of the twisted current algebra is 27, so we conclude that the field ϕ has a central charge $c_\phi = 25$, which means it must have a background charge $Q_\phi = 2\sqrt{2}$. Thus, ϕ is clearly identical to the Liouville field in $c = 1$ string theory.

To summarise, the full Hilbert space of our topological theory is

$$\mathcal{H}: \qquad \mathcal{H}_\phi \;\oplus\; \mathcal{H}_X \;\oplus\; \mathcal{H}_{b,c} \;\oplus\; \mathcal{H}_{B,C} \;\oplus\; \mathcal{H}_{\beta,\gamma}$$

$$\text{spins}: \qquad (0) \qquad (0) \qquad (2,-1) \qquad (1,0) \qquad (0,1) \tag{3.10}$$

$$\text{central charge}: \quad 25 \qquad 1 \qquad -26 \qquad -2 \qquad 2$$

while the physical Hilbert space is obtained from this by taking the quotient with the two BRST charges

$$\begin{aligned}
G^+ &= \int cJ^+ = \int c(\beta\gamma^2 - \sqrt{2}\gamma\partial\phi + 3\partial\gamma) \\
Q_{U(1)} &= \int C(\beta\gamma - cb - \partial X^-)
\end{aligned} \tag{3.11}$$

(We have defined $X^\pm = \frac{1}{\sqrt{2}}(\phi \mp iX)$.)

Note the remarkable fact that the first three Hilbert spaces above are isomorphic to the Hilbert space of the conventional $c = 1$ string quantised in the DDK formalism. The Liouville field arises from the Wakimoto representation for the $SL(2,R)$ current algebra, the $c = 1$ matter scalar field comes from the $U(1)$ gauge field, and the ghosts are just the fermions of the original supersymmetric WZW model, after twisting. The remaining two Hilbert spaces consist of first-order pairs of the same spins and opposite statistics, so it is quite reasonable to expect that they cancel out in some sense. Thus, we have found a very likely candidate for a manifestly topological description of $c = 1$ string theory.

There are two ways to make this connection more convincing. One is to directly compute the double cohomology of the BRST charges above on the full Hilbert space. This has been described in detail in Ref.[1] and will not be repeated here. The result turns out to be isomorphic to the complete cohomology in the DDK formalism, described in detail in Ref.[5]. Another way is to explore the relation of this model to the KPZ formalism of $c = 1$ string theory.

4 Relation to KPZ

Let us recall the KPZ[22] formulation of non-critical string theory. One starts with an $SL(2,R)$ current algebra at level k, and twists the Sugawara stress-energy tensor by

$$T_{SL(2,R)} \quad \rightarrow \quad T_{SL(2,R)} - \partial J^3(z) \tag{4.1}$$

The total central charge of this twisted system is

$$c = \frac{3k}{k-2} + 6k \tag{4.2}$$

Next, one gauges the parabolic subgroup generated by $J^-(z)$[26]. This introduces a pair of anticommuting ghosts (B,C) of spins $(1,0)$, along with a BRST charge

$$Q_{KPZ} = \int B(z)(J^-(z) - 1) \tag{4.3}$$

Note that in this procedure, the gauge field disappears completely, unlike in the case where we gauge the subgroup generated by J^3. This is due to the fact that the constraint is first-class[27] in the present case.

Now, to the above system (which is collectively supposed to represent the Liouville, or gravitational, degree of freedom on the worldsheet) we couple by hand a $c = 1$ matter field $X(z, \bar{z})$ and a pair of anticommuting ghosts (b, c) of spins $(2, -1)$, and impose the usual string BRST cohomology via

$$Q_{BRST} = \int c(T_{\mathrm{SL(2,R)}} + T_X + \frac{1}{2}T_{\mathrm{ghost}}) \tag{4.4}$$

Summarising, the total Hilbert space is *precisely* the same as in Eq.(3.10) above, but instead of the two BRST charges in Eq.(3.11), we have to impose the ones in Eqs.(4.3) and (4.4).

Now to argue the equivalence of our coset model to the KPZ theory, it only remains to prove that the two double cohomologies on the same Hilbert space are isomorphic. The proof follows most simply from the following result[28]

$$\beta^{-1} Q_{BRST} = G^+ + \{Q_{U(1)}, *\} \tag{4.5}$$

where "$*$" represents some combination of the fields in the theory, whose detailed form is unimportant. Now on the left hand side, if we pass to the cohomology of Q_{KPZ} then we can set $\beta = 1$ (recalling that in the Wakimoto representation, $J^- = \beta$) and we are left with Q_{BRST}. On the right hand side, if we pass to the cohomology of $Q_{U(1)}$ then the second term vanishes. Now the equation just tells us that the operators with respect to which we want to take the next cohomology are equal, hence the double cohomologies are equivalent.

This completes the proof that the topological coset model is equivalent to $c = 1$ string theory. In the next sections we will see why it is a "more solvable" formulation of the theory.

5   Correlation Functions

In order to show that the $k = 3$ coset model is exactly solvable, we need to investigate the available techniques to solve models of this kind. Let us first consider a different class of models: the $N = 2$ supersymmetric minimal models labelled by a positive integer k. After twisting and coupling to gravity, these models are believed to represent special points in the moduli space of the $k + 1$ matrix models[29]. In continuum language, these represent the $(k, 1)$ minimal models of central charge $1 - 6(k - 1)^2/k$ coupled to ordinary gravity, from which the (k, k') models can be obtained by adding marginal perturbations.

For this class of models, the first powerful technique to be discovered was the Landau-Ginzburg description[30] . In this approach, one identifies the above theories with the infrared fixed points of the $N = 2$ Landau-Ginzburg (LG) theory with a single superfield, and superpotential X^{k+2}. In this description, it is known how to

explicitly compute correlation functions (after coupling to gravity) at least in genus
0[31, 32].

The gravitational primaries in the LG theory coupled to gravity are described by
$1, X, X^2, \ldots, X^k$. Let us denote the primary X^r by U_r. Gravitational secondaries are
obtained by multiplying these with the usual fields σ_n of pure topological gravity. The
selection rules for the correlator $\langle \sigma_{n_1}(U_{r_1}) \ldots \sigma_{n_N}(U_{r_N}) \rangle$ are

$$\sum_{i=1}^{N} \left(\frac{r_i}{k+2} - 1 \right) + \sum_{i=1}^{N} n_i = (2g - 2) \frac{k+3}{k+2} \tag{5.1}$$

This picture can be suitably modified for our purposes. For positive k, Landau-
Ginzburg theory flows in the infrared to the Kazama-Suzuki coset $SU(2)_k/U(1)$. Since
for many purposes the $SL(2, R)_k/U(1)$ coset is like the coset of $SU(2)$ at level *minus k*,
the model we have described above should be the infrared fixed point of the LG theory
with $k = -3$, hence with superpotential X^{-1}. This requires some extra work to define
carefully, but we will see below that sphere correlators for $c = 1$ string theory can be
extracted from this formalism by making only some very broad assumptions.

To start with, let us restrict to gravitational primaries and set $k = -3$ in the selection
rule above. The resulting equation is genus-independent, and remarkably simple:

$$\sum_{i=1}^{N} (r_i + 1) = 0 \tag{5.2}$$

This gives us the first clue about the identification of these fields with the physical fields
in the $c = 1$ string (at the self-dual radius). The tachyons in this string theory have
discrete momenta, labelled by integers k_i, satisfying momentum conservation in every
genus:

$$\sum_{i=1}^{N} k_i = 0 \tag{5.3}$$

Thus, we are tempted to make the identification $k_i = r_i + 1$, as a result of which we
would claim that the tachyons in the LG formulation are gravitational primaries:

$$T_k = X^{k-1} \qquad \text{(LG)} \tag{5.4}$$

Note that tachyons with all positive and negative integer momenta are required, so
we must allow fields $U_r = X^r$ for all positive and negative integer values of r. It is less
clear that *all* of these are gravitational primaries, and for us it will be enough to treat
only X^r for positive r as primaries.

In this framework, it was first noticed by Cecotti and Vafa[33] that the correct 4-
point function of tachyons in $c = 1$ string theory can be obtained, using techniques that
were developed to deal with polynomial superpotentials. Indeed, it is now clear that
one can obtain the tachyon N-point function for all N, using Landau-Ginzburg theory
with superpotential X^{-1}. This has been worked out in Ref.[2] using a generalization
(to $k = -3$) of a formalism due to Losev[32] in which contact terms are dealt with
explicitly, and one finds a recursion formula (on the sphere) determining correlators of
N tachyons in terms of those of $N - 1$ tachyons.

Let us illustrate this first for the 4-point function. Losev's formula for this is

$$
\begin{aligned}
\langle X^{r_1} X^{r_2} X^{r_3} X^{r_4} \rangle_V \;=\;& \frac{d}{dt_4} \langle X^{r_1} X^{r_2} X^{r_3} \rangle_{V+t_4 X^{r_4}} \big|_{t_4=0} \\
& + \langle C_V(X^{r_1}, X^{r_4}) X^{r_2} X^{r_3} \rangle_V + \langle X^{r_1} C_V(X^{r_2}, X^{r_4}) X^{r_3} \rangle_V \\
& + \langle X^{r_1} X^{r_2} C_V(X^{r_3}, X^{r_4}) \rangle_V
\end{aligned}
\tag{5.5}
$$

where $C_V(X^i, X^j)$ is a contact term between two operators, for which an expression is known in the case of polynomial LG theory, where it is described by the so-called Saito pairing. For $k = -3$, it turns out[2] that the correct contact term is much simpler, and is given by

$$
\begin{aligned}
C_V(X^i, X^j) \;=\;& (i+j) X^{i+j}, \qquad (i+j) < 0 \\
\;=\;& 0, \qquad (i+j) > 0
\end{aligned}
\tag{5.6}
$$

The only remaining information required is that the three-point function is given by[31]

$$
\langle X^{r_1} X^{r_2} X^{r_3} \rangle_V = res\left(\frac{X^{r_1} X^{r_2} X^{r_3}}{V'} \right)
\tag{5.7}
$$

With all this, the four-point function is easily calculated and one gets

$$
\langle T_{k_1} T_{k_2} T_{k_3} T_{k_4} \rangle_{g=0} = -\frac{1}{2}|k_1 + k_2| - \frac{1}{2}|k_1 + k_3| - \frac{1}{2}|k_2 + k_3| + 1
\tag{5.8}
$$

which is precisely the tachyon 4-point correlator coming from matrix models of $c = 1$ string theory[34] and from perturbative techniques in the DDK formalism[8].

The same technique can now be applied to the recursive formula for N-point functions, due to Losev and Polyubin in the context of polynomial LG theories:

$$
\begin{aligned}
\langle X^{r_1} X^{r_2} \ldots X^{r_N} \rangle_V \;=\;& \frac{d}{dt_N} \langle X^{r_1} X^{r_2} \ldots X^{r_{N-1}} \rangle_{V+t_N X^{r_N}} \big|_{t_N=0} \\
& + \langle C_V(X^{r_1}, X^{r_N}) X^{r_2} \ldots X^{r_{N-1}} \rangle_V + \cdots \\
& + \langle X^{r_1} X^{r_2} \ldots C_V(X^{r_{N-1}}, X^{r_N}) \rangle_V
\end{aligned}
\tag{5.9}
$$

Choosing the kinematic region $k_1 < 0$ and $k_2, k_3, \ldots, k_N > 0$, we find[2]

$$
\langle T_{k_1} T_{k_2} \ldots T_{k_N} \rangle_{g=0} = (k_1 + 1)(k_1 + 2) \cdots (k_1 + N - 3)
\tag{5.10}
$$

which is precisely the matrix-model result!

6   Partition Function

An alternative method which is in principle even more powerful, is to identify these theories with the Kazama-Suzuki (KS) coset models based on $SU(2)_k/U(1)$. It has been shown[35] that algebraic-geometry techniques can be brought to bear on this problem, resulting in expressions for arbitrary correlators in arbitrary genus. These are described as integrals of products of Chern classes of certain bundles over moduli space.

In the KS formulation, the gravitational primaries are $1, g_{11}, g_{11}^2, \ldots, g_{11}^k$ (where g_{ab} is the $SL(2, R)$-valued matrix field of the WZW model). The primary g_{11}^r is equivalent to the Landau-Ginzburg primary X^r so we again assign it the label U_r. Gravitational secondaries, and the selection rules for generic correlators, are the same as in the previous section.

Starting again with genus 0, the 4-point function of gravitational primaries was explicitly computed for $k > 0$ using algebraic-geometry techniques, in Ref.[35]. This computation is performed by choosing a section of the relevant bundle and finding its divisor. The final result is

$$\langle U_{r_1} U_{r_2} U_{r_3} U_{r_4}\rangle_{g=0} = \frac{1}{2}(\min(q_1 + q_2, q_3 + q_4) + \min(q_1 + q_3, q_2 + q_4)$$
$$+\min(q_1 + q_4, q_2 + q_3)) - \frac{k+1}{k+2} \tag{6.1}$$

where $q_i = r_i/(k+2)$ are the $U(1)$ charges of the fields.

This result can now explicitly be continued to the case of interest to us. Inserting $k = -3$ and using the identification in Eq.(5.4), we easily find the expression in Eq.(5.8). Thus the KS formulation also permits the calculation of the tachyon four-point function, although in practice the LG formulation was the simpler one for this case.

Let us finally consider the case of higher genus. Here, two basic results are known. From Ref.[35], we know that the genus-g partition function of the $SU(2)_k/U(1)$ KS model, when continued to $k = -3$, is precisely the (virtual) Euler characteristic $\hat{\chi}_g$ of the moduli space of genus-g Riemann surfaces with no punctures. Now, from the matrix model compactified at the self-dual radius, we have an explicit result for the genus-g partition function (subject to the assumption that the nonsinglet sector can be ignored), and the answer is

$$Z_g = \frac{B_{2g}}{2g(2g-2)} \tag{6.2}$$

where B_{2g} are the Bernoulli numbers. Remarkably, it has been shown[36, 37, 38] that this is just the virtual Euler characteristic of the moduli space of genus-g Riemann surfaces. So, our topological formulation of $c = 1$ string theory is powerful enough to reproduce the genus-g partition function, directly in a continuum approach. It is easy to check that genus-g correlation functions of the cosmological operator ($T_0 = X^{-1} = g_{11}^{-1}$) are also obtained correctly in our approach. Correlators of arbitrary (discrete) tachyons in higher genus have yet to be computed explicitly.

7 Conclusions

We have collected enough evidence to show that not only does our manifestly topological model correctly describe $c = 1$ string theory, but it also allows the explicit computation of correlators which are either impossible, or very difficult to calculate in any other continuum formulation (including the conventional conformal-gauge DDK description). It should be stressed that the correlators described above are at nonzero cosmological constant, and powers of the cosmological constant can easily be inserted in the right places in the above formulae, although we have chosen to omit them for simplicity of presentation. In contrast, the DDK continuum formulation requires a perturbative treatment of the cosmological operator, in addition to certain prescriptions to allow *negative* numbers of insertions of this operator.

The work described here puts $c = 1$ string theory on the same footing as the $c < 1$ theory, for which a manifestly topological description has long been known. However, it leaves some interesting open questions which need to be addressed in the Lagrangian approaches (LG or KS). First of all, the role of gravitational descendants in the topological theory needs to be elucidated. Since such fields are labelled by two integers, it is tempting to conjecture that they are related to the discrete states of the $c = 1$ string. However, this identification has so far not been clarified sufficiently. Well-known properties of $c = 1$ string theory like the presence of a ground ring, and the W_∞ symmetry algebra, appear explicitly in the coset CFT[1], but are less obvious in the Lagrangian formulations. Indeed, the important question is not how the KS coset (described in terms of conformal fields) is related to the DDK $c = 1$ string, but rather, how exactly the Lagrangian formulations of this theory (which are exactly solvable) are related to the CFT formulation.

Acknowledgements

I wish to thank the organisers of the Cargese Workshop for their kind invitation to lecture on this work, and for their generous hospitality at Cargese. I am grateful to Cumrun Vafa and Debashis Ghoshal for collaboration on the papers whose results are described here.

References

[1] S. Mukhi and C. Vafa, Nucl. Phys. B407 (1993) 667.

[2] D. Ghoshal and S. Mukhi, in preparation.

[3] K.S. Narain, Phys. Lett. 169B (1986) 41.

[4] E. Witten, Nucl. Phys. B373 (1992) 187.

[5] E. Witten and B. Zwiebach, Nucl. Phys. B377 (1992) 55.

[6] I. Klebanov and A.M. Polyakov, Mod. Phys. Lett. A6 (1991) 3273.

[7] D. Gross and I. Klebanov, Nucl. Phys. B344 (1990) 475.

[8] P. DiFrancesco and D. Kutasov, Phys. Lett. 261B (1991) 385.

[9] V. Dotsenko, Mod. Phys. Lett. A6 (1991) 3601.

[10] Y. Kitazawa, Phys. Lett. 265B (1991) 262.

[11] D. Ghoshal and S. Mahapatra, Mod. Phys. Lett. A8 (1993) 197.

[12] S. Govindarajan, T. Jayaraman, V. John and P. Majumdar, Phys. Rev. D48 (1993) 839.

[13] B. Gato-Rivera and A. Semikhatov, Phys. Lett. 288B (1992) 38.

[14] M. Bershadsky, W. Lerche, D. Nemeschansky and N. Warner, Nucl. Phys. B401 (1993) 304.

[15] S. Mukherji, S. Mukhi and A. Sen, Phys. Lett. 266B (1991) 337.

[16] P. Bouwknegt, J. McCarthy and K. Pilch, Comm. Math. Phys. 145 (1992) 541.

[17] B. Lian and G. Zuckerman, Phys. Lett. 266B (1991) 21.

[18] M. Kato and S. Matsuda, in "Conformal Field Theory and Solvable Lattice Models", Adv. Studies in Pure Math. 16, Ed. M. Jimbo, T. Miwa and A. Tsuchiya (Kinokuniya, 1988).

[19] D. Gross, I. Klebanov and M. Newman, Nucl. Phys. B350 (1991) 621.

[20] M. Bershadsky and I. Klebanov, Phys. Rev. Lett. 65 (1990) 3088.

[21] A.M. Polyakov, Mod. Phys. Lett. A6 (1991) 635.

[22] V. Knizhnik, A.M. Polyakov and A.B. Zamolodchikov, Mod. Phys. Lett. A3 (1988) 819.

[23] F. David, Mod. Phys. Lett. A3 (1988) 1651;
J. Distler and H. Kawai, Nucl. Phys. B321 (1989) 509.

[24] Y. Kazama and H. Suzuki, Nucl. Phys. B321 (1989) 232.

[25] K. Gawedzki and A. Kupiainen, Nucl. Phys. B320 (1989) 625.

[26] A. Alekseev and S. Shatashvili, Nucl. Phys. B323 (1989) 625;
M. Bershadsky and H. Ooguri, Comm. Math. Phys. 126 (1989) 49.

[27] R. Dijkgraaf, E. Verlinde and H. Verlinde, Nucl. Phys. B371 (1992) 269.

[28] V. Sadov, as quoted in Ref.[1].

[29] K. Li, Nucl. Phys. B354 (1991) 711.

[30] W. Lerche, C. Vafa and N. Warner, Nucl. Phys. B324 (1989) 427.

[31] R. Dijkgraaf, E. Verlinde and H. Verlinde, Nucl. Phys. B352 (1991) 59.

[32] A. Losev, ITEP Preprint PRINT-92-0563 (Jan 1993), hep-th/9211089.

[33] S. Cecotti and C. Vafa, as quoted in Ref.[1].

[34] G. Moore, Nucl. Phys. B368 (1992) 557;
G. Mandal, A. Sengupta and S. Wadia, Mod. Phys. Lett. A6 (1991) 1465;
K. Demeterfi, A. Jevicki and J. Rodrigues, Nucl. Phys. B362 (1991) 173;
J. Polchinski, Nucl. Phys. B362 (1991) 125.

[35] E. Witten, Nucl. Phys. B371 (1992) 191.

[36] J. Harer and D. Zagier, Invent. Math. 185 (1986) 457.

[37] R.C. Penner, Commun. Math. Phys. 113 (1987) 299; J. Diff. Geom. 27 (1988) 35.

[38] J. Distler and C. Vafa, Mod. Phys. Lett. A6 (1991) 259.

LINEAR SYSTEMS
FOR 2D POINCARÉ SUPERGRAVITIES

Hermann NICOLAI

II. Institut für Theoretische Physik
Universität Hamburg
Luruper Chaussee 149
22761 Hamburg, F.R.G.

1 Introduction

This contribution contains a summary of [1], which generalizes the linear systems that were derived already some time ago for the dimensionally reduced field equations of Einstein Yang-Mills theories [2, 3] and their locally supersymmetric extensions [4, 5]. These reductions correspond to solutions of the field equations, which depend on two coordinates only and thus possess at least two commuting Killing vectors. The construction of [1] differs from earlier treatments, which were all based on the (super)conformal gauge, in that it allows for non-trivial topologies of the two dimensional world sheets by taking into account the topological degrees of freedom of the world sheet, i.e. its moduli and supermoduli. These constitute extra physical (but non-propagating) degrees of freedom not present in the corresponding flat space integrable sigma models, and affect the dynamics in a non-trivial fashion. In particular, there is a "back reaction" of the matter fields on the topological degrees of freedom, in contrast to conformal field theories, where the moduli determining the background can be freely chosen. The spectral parameter t entering the linear system is now not only a function of the "dilaton" field as in [3, 4], but also depends on the moduli and super-moduli of the world sheet. It is subject to a pair of differential equations, whose integrability condition yields one of the equations of motion obtained by dimensional reduction of Einstein's equations. Apart from these intriguing new structures, an important motivation for investigating the $2d$ supergravity models is the search for new symmetries generalizing the Geroch group [6] and the "hidden symmetries" of dimensionally reduced supergravities [7, 8, 9]. The results obtained in [1] indicate that, if such extensions of the Geroch group exist, they are likely to involve the topological degrees of freedom. It should be stressed, however, that even for the known classes of solutions, the global structure of the Geroch group is not fully understood (see [3] for a discussion).

Quantum Field Theory and String Theory, Edited by
L. Baulieu *et al.*, Plenum Press, New York, 1995

The $2d$ supergravity models are most conveniently derived from matter coupled supergravity theories in three dimensions, i.e. locally supersmmetric non-linear sigma models as recently formulated in [10]. This procedure has the advantage that in three dimensions, all finite dimensional symmetries are manifest because the matter degrees of freedom are uniformly represented by scalars and spinors rather than tensor fields (as would be the case in dimensions $d > 3$). The models resemble conformal field theories in several respects, but there are also differences. For instance, the equations of motion of the left and right moving degrees of freedom can no longer be disentangled, because there exist genuine solitonic solutions mixing left and right movers such as the gravitational "colliding plane wave" solutions of [11] also considered in [5]. Furthermore, locally supersymmetric theories exist up to $N = 16$ (where N is the number of local supersymmetries), whereas in conformal supergravity, only $N \leq 4$ is possible. The difference is perhaps more easily understood by considering the (super)gravitational fields, which do not carry propagating degrees of freedom. The bosonic ones originate from the dreibein in three dimensions, which by a partial gauge fixing of the Lorentz group $SO(1,2)$ can be cast into the form[a]

$$e_m{}^a = \begin{pmatrix} e_\mu{}^\alpha & \rho A_\mu \\ 0 & \rho \end{pmatrix} \Longrightarrow e_a{}^m = \begin{pmatrix} e_\alpha{}^\mu & -e_\alpha{}^\nu A_\nu \\ 0 & \rho^{-1} \end{pmatrix} \tag{1.1}$$

For the $3d$ gravitino, we have an analogous decomposition in terms of flat indices

$$\psi_a = (\psi_\alpha, \psi_2) \tag{1.2}$$

Dimensional reduction to two dimensions therefore gives rise to a "dilaton" ρ and a Kaluza-Klein vector field A_μ in addition to the zweibein $e_\mu{}^\alpha$, which is the only gravitational degree of freedom in conformal field theory. Similarly, the decomposition (1.2) gives rise to an extra degree of freedom, namely the "dilatino" ψ_2, which may be viewed as the superpartner of ρ. None of these fields possesses propagating degrees of freedom. For the bosonic fields, this can be seen by substituting (1.1) into the $3d$ Einstein action and discarding the dependence on the third (spacelike) coordinate x^2, which yields

$$-\tfrac{1}{4}e^{(3)}R^{(3)} = -\tfrac{1}{4}\rho e R^{(2)} - \tfrac{1}{16}e\rho^3 A_{\mu\nu}A^{\mu\nu} \tag{1.3}$$

with $A_{\mu\nu} := \partial_\mu A_\nu - \partial_\nu A_\mu$. Evidently, the conformal factor does not decouple even in the classical theory as the Euler density $eR^{(2)}$ is multiplied by the dilaton field ρ. Instead, there is now an equation of motion relating the world sheet curvature to matter sources. The field ρ can be identified with a function of the coordinates (for axisymmetric stationary solutions of Einstein's equations, it is usually taken to be a cylindrical coordinate, see [3]). Nevertheless, it modifies the dynamics of the matter fields through its appearance in their equations of motion. It also plays an essential role in establishing one-loop finiteness of the dimensionally reduced models [12]. The vector field A_μ is auxiliary, but offers the possibility of introducing a cosmological

---

[a]Here $m, n, \ldots$ and $a, b, \ldots = 0, 1, 2$, are curved and flat indices in three dimensions, respectively, while the corresponding indices in two dimensions will be denoted by $\mu, \nu, \ldots$ and $\alpha, \beta, \ldots = 0, 1$, respectively. The metric has signature $(+ - -)$.

constant through a non-vanishing expectation value $A_{\mu\nu} \propto \epsilon_{\mu\nu}$. In previous work, this cosmological constant has always been assumed to vanish, and we will also set it equal to zero here. Elimination of the field strength $A_{\mu\nu}$ will then produce only quartic spinor terms, which we ignore, so effectively $A_{\mu\nu} = 0$. However, it must be emphasized that inclusion of the associated field equation into the linear system, which has so far not been accomplished, may provide the crucial missing link in understanding the hidden symmetries that may exist beyond the Geroch group.

As already mentioned, previous studies are based on the special superconformal gauge

$$e_\mu{}^\alpha = \lambda \, \delta_\mu^\alpha \quad , \quad \psi_\alpha = \gamma_\alpha \theta \tag{1.4}$$

which simplifies the equations of motion considerably. This gauge choice is always possible *locally*, but it misses important global aspects because the information about the conformal structure of the world sheet is hidden in the transition functions between local charts in this gauge. Consequently, a change of conformal structure must be accompanied by a corresponding change of atlas if (1.4) is to be maintained. If one wants to vary the conformal structure without having to change the atlas, one must make the dependence on the topological degrees of freedom explicit. In order to do so, one parametrizes the conformal structure over a fixed atlas in terms of Beltrami and super-Beltrami differentials. In the context of conformal field theory and string theory, such a formulation was proposed in [13] and further investigated in [14, 15]; it was also used in studies of higher loop amplitudes in superstring theory [16].

Although all results can be formulated in the Euclidean metric relevant to the study of stationary axisymmetric solutions, we will be working with a Lorentzian worldsheet throughout. A technical reason for this is the occurrence of Majorana Weyl spinors in two dimensions, which are here described as real one component (anticommuting) spinors. As is well known, Majorana Weyl spinors in two dimensions exist only for Lorentzian signature, but not for Euclidean signature. This does not necessarily preclude a Euclidean description, which would require complex spinors. However, by complexifying the spinors, one doubles the number of fermionic degrees of freedom. In a theory with an even number of fermions, this problem can be circumvented by rewriting d real spinors in terms of $\frac{d}{2}$ complex spinors, but some of the previously manifest symmetries would be lost in general. Quite apart from these technical points, however, the study of Lorentzian world sheets is of interest in its own right. These differ from the more familiar Euclidean world sheets (Riemann surfaces) in various respects, one of which is the unavoidability of singularities for higher genus surfaces: a globally Lorentzian surface which is everywhere smooth must have Euler characteristic $\chi = 2 - 2g - n = 0$ [17] (where g is the genus and n the number of punctures). This leaves only the cylinder ($g = 0, n = 2$) and the torus ($g = 1, n = 0$) as everywhere smooth Lorentzian world sheets, so all other worldsheets must have singularities. These observations are also of some physical interest, for instance in two dimensional quantum cosmology (see e.g.[18]), where they imply the existence of catastrophic "naked" singularities for two dimensional observers. Unfortunately, owing to the lack of literature dealing with the geometry of "Lorentzian Riemann surfaces" from either a mathematical or a physical point of view, many elementary questions remain open for the time being. We will proceed nonetheless, assuming that the known results about ordinary Riemann surfaces can be taken over mutatis mutandis.

2 Equations of Motion

We will now list the equations of motion obtained after dimensional reduction to two dimensions, making use of the chiral formalism developed and explained in [1]. For notational simplicity, we will write down the formulas for $N = 16$ supergravity [19] only, the generalization to other N being straightforward (see [10] for a comprehensive discussion of these models). The model is a locally supersymmetric sigma model based on the non-compact coset space $E_{8(+8)}/SO(16)$. The E_8 generators are decomposed into the 120 generators $X^{IJ} = -X^{JI}$ of the $SO(16)$ subgroup and 128 remaining generators Y^A, which transform as the irreducible spinor representation of $SO(16)$. Thus $I, J, ... = 1, ..., 16$ are $SO(16)$ vector indices and $A, B, ... = 1, ..., 128$ are $SO(16)$ spinor indices. The matter fermions $\chi^{\dot{A}}$ transform under the conjugate spinor representation labeled by dotted indices $\dot{A}, \dot{B}, ... = 1, ..., 128$. In addition, the model contains the non-propagating fields ρ and $e_\mu{}^\alpha$ (from the dreibein (1); remember that we put $A_\mu = 0$) and 16 gravitinos $(\psi_\alpha^I, \psi_2^I)$ (with the same decomposition as in (2)), whose vector part is further split into irreducible components according to $\psi_\alpha^I = \tilde{\psi}_\alpha^I + \gamma_\alpha \theta^I$ with $\gamma^\alpha \tilde{\psi}_\alpha^I = 0$. The bosonic sector of the $N = 16$ theory is governed by a non-linear sigma model; thus, the bosonic fields are described by a matrix $\mathcal{V}(x) \in E_8$, which is subject to the transformations

$$\mathcal{V}(x) \longrightarrow g^{-1}\mathcal{V}(x)h(x) \tag{2.1}$$

where g is a rigid E_8 transformation, and $h(x)$ a local $SO(16)$ transformation. From $\mathcal{V}$, one defines the "composite fields" Q_μ^{IJ} and P_μ^A

$$\mathcal{V}^{-1}\partial_\mu \mathcal{V} = \tfrac{1}{2}Q_\mu^{IJ}X^{IJ} + P_\mu^A Y^A \tag{2.2}$$

This definition immediately implies the integrability relations

$$D_\mu P_\nu^A - D_\nu P_\mu^A = 0$$

$$\partial_\mu Q_\nu^{IJ} - \partial_\nu Q_\mu^{IJ} + 2Q_\mu^{K[I}Q_\nu^{J]K} + \tfrac{1}{2}\Gamma_{AB}^{IJ}P_\mu^A P_\nu^B = 0 \tag{2.3}$$

where the $SO(16)$ covariant derivative D_μ is defined by means of the connection Q_μ^{IJ} defined in (2.2).

In [1], an anholonomic chiral basis was used for the derivatives and differentials. For the derivatives, we have

$$\mathcal{D}_{\dot{+}} := \partial_{\dot{+}} - \mu_{\dot{+}}{}^{\dot{-}}\partial_{\dot{-}} \quad , \quad \mathcal{D}_{\dot{-}} := \partial_{\dot{-}} - \mu_{\dot{-}}{}^{\dot{+}}\partial_{\dot{+}} \tag{2.4}$$

where $\mu_{\dot{+}}{}^{\dot{-}}$ and $\mu_{\dot{-}}{}^{\dot{+}}$ are the Beltrami differentials and the dots indicate that the corresponding indices are curved. In terms of these derivatives, left and right moving scalar fields satisfy $\mathcal{D}_- f = 0$ and $\mathcal{D}_+ \overline{f} = 0$, respectively; they are the real analogues of holomorphic and anti-holomorphic functions. The dual basis differentials are

$$\mathcal{D}x^+ := \frac{dx^+ + \mu_-{}^+ dx^-}{1 - \mu_+{}^- \mu_-{}^+} \quad , \quad \mathcal{D}x^- := \frac{dx^- + \mu_+{}^- dx^+}{1 - \mu_+{}^- \mu_-{}^+} \quad , \tag{2.5}$$

In terms of this basis, we define

$$Q_\mu^{IJ} dx^\mu = Q_+^{IJ} \mathcal{D}x^+ + Q_-^{IJ} \mathcal{D}x^- \quad , \quad P_\mu^A dx^\mu = \mathcal{P}_+^A \mathcal{D}x^+ + \mathcal{P}_-^A \mathcal{D}x^- \tag{2.6}$$

The chiral components of other fields are defined similarly (for the fermions, the proper definition is explained in [1]). The above integrability relations then become

$$\mathrm{D}_+ \mathcal{P}_-^A - \mathrm{D}_- \mathcal{P}_+^A = 0$$

$$\mathrm{D}_+ Q_-^{IJ} - \mathrm{D}_- Q_+^{IJ} + 2 Q_+^{K[I} Q_-^{J]K} + \tfrac{1}{2} \Gamma_{AB}^{IJ} \mathcal{P}_+^A \mathcal{P}_-^B = 0 \tag{2.7}$$

where $\mathrm{D}_\pm$ always denotes the fully covariant derivative (see [1] for the precise definition).

We then get the following equations of motion. The Dirac equation is

$$-i\rho^{-1/2} \mathrm{D}_+ \left(\rho^{1/2} \chi_-^{\dot{A}} \right) = -\tfrac{1}{2} i \Gamma_{A\dot{A}}^I \psi_{2-}^I \mathcal{P}_+^A + \tfrac{1}{\sqrt{2}} \Gamma_{A\dot{A}}^I \psi_+^{I-} \mathcal{P}_-^A$$

$$-i\rho^{-1/2} \mathrm{D}_- \left(\rho^{1/2} \chi_+^{\dot{A}} \right) = +\tfrac{1}{2} i \Gamma_{A\dot{A}}^I \psi_{2+}^I \mathcal{P}_-^A + \tfrac{1}{\sqrt{2}} \Gamma_{A\dot{A}}^I \psi_-^{I+} \mathcal{P}_+^A \tag{2.8}$$

For the scalar fields, we find

$$\rho^{-1} \mathrm{D}_- \left(\rho (\mathcal{P}_+^A - i\sqrt{2} \Gamma_{A\dot{A}}^I \chi_+^{\dot{A}} \psi_{2+}^I + 2 \Gamma_{A\dot{A}}^I \chi_-^{\dot{A}} \psi_+^{I-}) \right) +$$

$$+ \rho^{-1} \mathrm{D}_+ \left(\rho (\mathcal{P}_-^A - i\sqrt{2} \Gamma_{A\dot{A}}^I \chi_-^{\dot{A}} \psi_{2-}^I - 2 \Gamma_{A\dot{A}}^I \chi_+^{\dot{A}} \psi_-^{I+}) \right) =$$

$$= \tfrac{1}{8} i \Gamma_{AB}^{IJ} \left(-\sqrt{2} \mathcal{P}_+^B \Gamma_{\dot{A}\dot{B}}^{IJ} \chi_-^{\dot{A}} \chi_-^{\dot{B}} + \sqrt{2} \mathcal{P}_-^B \Gamma_{\dot{A}\dot{B}}^{IJ} \chi_+^{\dot{A}} \chi_+^{\dot{B}} \right)$$

$$+ \Gamma_{AB}^{IJ} \left((\sqrt{2} \psi_{2+}^I \theta_+^J - \psi_{2-}^I \psi_+^{J-}) \mathcal{P}_-^B + (\sqrt{2} \psi_{2-}^I \theta_-^J - \psi_{2+}^I \psi_-^{J+}) \mathcal{P}_+^B \right)$$

$$\tag{2.9}$$

¿From Rarita Schwinger equations in three dimensions, we deduce the following equations,

$$\sqrt{2} \mathrm{D}_+ \theta_-^I - \mathrm{D}_- \psi_+^{I-} = \tfrac{1}{\sqrt{2}} i \Gamma_{A\dot{A}}^I \chi_-^{\dot{A}} \mathcal{P}_+^A$$

$$\sqrt{2} \mathrm{D}_- \theta_+^I - \mathrm{D}_+ \psi_-^{I+} = \tfrac{1}{\sqrt{2}} i \Gamma_{A\dot{A}}^I \chi_+^{\dot{A}} \mathcal{P}_-^A \tag{2.10}$$

and

$$\mathrm{D}_- (\rho \psi_{2+}^I) = \tfrac{1}{\sqrt{2}} i \mathcal{D}_+ \rho \, \psi_-^{I+}$$

$$\mathrm{D}_+ (\rho \psi_{2-}^I) = -\tfrac{1}{\sqrt{2}} i \mathcal{D}_- \rho \, \psi_+^{I-} \tag{2.11}$$

as well as the "super-Virasoro conditions"

$$\mathcal{S}_+^I := \mathrm{D}_+ (\rho \psi_{2+}^I) - i \mathcal{D}_+ \rho \, \theta_+^I + \rho \Gamma_{A\dot{A}}^I \chi_+^{\dot{A}} \mathcal{P}_+^A = 0$$

$$\mathcal{S}_-^I := \mathrm{D}_- (\rho \psi_{2-}^I) + i \mathcal{D}_- \rho \, \theta_-^I - \rho \Gamma_{A\dot{A}}^I \chi_-^{\dot{A}} \mathcal{P}_-^A = 0 \tag{2.12}$$

corresponding to the variation of the traceless gravitino modes $\tilde\psi^I_\pm$.

In the gravitational sector, the $3d$ Einstein equations give rise to several equations after dimensional reduction. For the dilaton field ρ we get

$$\mathrm{D}_+\mathrm{D}_-\rho = -\mathrm{D}_+(\rho\psi^{I+}_-\psi^I_{2+}) - \mathrm{D}_-(\rho\psi^{I-}_+\psi^I_{2-}) \tag{2.13}$$

Thus, ρ would be a free field without the contributions from the super-Beltrami differentials, consistent with the fact that its superpartners $\rho\psi^I_2$ would also be free for vanishing super Beltrami differentials.

For the curvature scalar, a similar calculation leads to

$$\mathcal{R} = \mathcal{P}^A_+\mathcal{P}^A_- + 2i\mathrm{D}_+(\psi^{I+}_-\theta^I_+) - 2i\mathrm{D}_-(\psi^{I-}_+\theta^I_-)$$
$$-\sqrt{2}\rho^{-1}\mathrm{D}_+(\rho\psi^I_{2-}\theta^I_-) - \sqrt{2}\rho^{-1}\mathrm{D}_-(\rho\psi^I_{2+}\theta^I_+) \tag{2.14}$$

The variation of the off diagonal components of the zweibein gives the "Virasoro conditions"

$$\mathcal{T}_{++} := \mathrm{D}_+\mathrm{D}_+\rho + \rho\mathcal{P}^A_+\mathcal{P}^A_+ - i\sqrt{2}\rho\chi^{\dot A}_+\mathrm{D}_+\chi^{\dot A}_+ - 2\Gamma^I_{A\dot A}\rho\psi^{I-}_+\chi^{\dot A}_-\mathcal{P}^A_+$$
$$-i\sqrt{2}\Gamma^I_{A\dot A}\rho\chi^{\dot A}_+\psi^I_{2+}\mathcal{P}^A_+ - 2\sqrt{2}\Gamma^I_{A\dot A}\rho\chi^{\dot A}_+\theta^I_+\mathcal{P}^A_+$$
$$+\mathrm{D}_+\big(\rho(\psi^I_{2-}\psi^{I-}_+ + \sqrt{2}\theta^I_+\psi^I_{2+})\big) = 0$$
$$\mathcal{T}_{--} := \mathrm{D}_-\mathrm{D}_-\rho + \rho\mathcal{P}^A_-\mathcal{P}^A_- + i\sqrt{2}\rho\chi^{\dot A}_-\mathrm{D}_-\chi^{\dot A}_- + 2\Gamma^I_{A\dot A}\rho\psi^{I+}_-\chi^{\dot A}_+\mathcal{P}^A_-$$
$$-i\sqrt{2}\Gamma^I_{A\dot A}\rho\chi^{\dot A}_-\psi^I_{2-}\mathcal{P}^A_- + 2\sqrt{2}\Gamma^I_{A\dot A}\rho\chi^{\dot A}_-\theta^I_-\mathcal{P}^A_-$$
$$+\mathrm{D}_+\big(-\rho(\psi^I_{2+}\psi^{I+}_- + \sqrt{2}\theta^I_-\psi^I_{2-})\big) = 0 \tag{2.15}$$

With $\tilde e := (1 - \mu_+^-\mu_-^+)\det e_\mu{}^\alpha$, we can rewrite the equations containing $\mathrm{D}_\pm\mathrm{D}_\pm\rho$ in terms of the modified conformal factor

$$\lambda \equiv \exp\sigma := \left(\frac{\tilde e}{\mathcal{D}_+\rho\mathcal{D}_-\rho}\right)^{1/2} \tag{2.16}$$

which transforms as a scalar. Modulo fermionic terms from (2.13), we have

$$\mathrm{D}_\pm\mathrm{D}_\pm\rho = -2\mathcal{D}_\pm\sigma\mathcal{D}_\pm\rho \tag{2.17}$$

This result suggests an interpretation of the fields ρ and σ as longitudinal target space degrees of freedom [5].

3 The Linear System

As explained in [3, 5], the construction of the linear system requires the replacement of the matrix $\mathcal{V}(x)$ by another matrix $\hat{\mathcal{V}}$ depending on a spectral parameter t, viz.

$$\mathcal{V}(x) \longrightarrow \hat{\mathcal{V}}(x,t) \tag{3.1}$$

The occurrence of a spectral parameter in linear systems (Lax pairs) for non-linear equations is, of course, a well known phenomenon. However, the linear system constructed here differs from others in that the spectral parameter t not only depends on the dilaton field ρ as in the purely bosonic theories (see [3, 5]), but also on the topological degrees of freedom via the Beltrami and super Beltrami differentials, see (3.5) below. This feature is entirely due to the interaction of the (super)gravitational degrees of freedom with the matter fields, and distinguishes locally supersymmetric integrable systems from flat space models with or without rigid supersymmetry. Moreover, the spectral parameter t, in terms of which the emergence of affine Kac Moody algebras in these models can be directly understood, now becomes a dynamical quantity of its own because the equations determining it themselves obey an integrability constraint that gives rise to one of the equations of motion.

The linear system can be parametrized as follows

$$\hat{\mathcal{V}}^{-1}\mathcal{D}_{+}\hat{\mathcal{V}} = \tfrac{1}{2}\hat{\mathcal{Q}}_{+}^{IJ}X^{IJ} + \hat{\mathcal{P}}_{+}^{A}Y^{A}$$

$$\hat{\mathcal{V}}^{-1}\mathcal{D}_{-}\hat{\mathcal{V}} = \tfrac{1}{2}\hat{\mathcal{Q}}_{-}^{IJ}X^{IJ} + \hat{\mathcal{P}}_{-}^{A}Y^{A} \tag{3.2}$$

where the hatted quantities $\hat{\mathcal{Q}}$ and and $\hat{\mathcal{P}}$ depend on t in contrast to $\mathcal{Q}$ and $\mathcal{P}$, which do not (see (2.6)). They are given by

$$\hat{\mathcal{Q}}_{+}^{IJ} = \mathcal{Q}_{+}^{IJ} - \sqrt{2}\frac{t}{(1+t)^2}\left(i\Gamma_{A\dot{B}}^{IJ}\chi_{+}^{\dot{A}}\chi_{+}^{\dot{B}} + 8\psi_{2+}^{[I}\theta_{+}^{J]}\right)$$

$$-16\sqrt{2}i\frac{t^2}{(1+t)^4}\,\psi_{2+}^{I}\psi_{2+}^{J} + 8\frac{t}{(1-t)^2}\,\psi_{2-}^{[I}\psi_{+}^{J]-}$$

$$\hat{\mathcal{Q}}_{-}^{IJ} = \mathcal{Q}_{-}^{IJ} + \sqrt{2}\frac{t}{(1-t)^2}\left(-i\Gamma_{A\dot{B}}^{IJ}\chi_{-}^{\dot{A}}\chi_{-}^{\dot{B}} + 8\psi_{2-}^{[I}\theta_{-}^{J]}\right)$$

$$+16\sqrt{2}i\frac{t^2}{(1-t)^4}\,\psi_{2-}^{I}\psi_{2-}^{J} - 8\frac{t}{(1+t)^2}\,\psi_{2+}^{[I}\psi_{-}^{J]+}$$

$$\hat{\mathcal{P}}_{+}^{A} = \frac{1-t}{1+t}P_{+}^{A} + 2\sqrt{2}i\frac{t(1-t)}{(1+t)^3}\Gamma_{A\dot{A}}^{I}\chi_{+}^{\dot{A}}\psi_{2+}^{I} - 4\frac{t}{1-t^2}\Gamma_{A\dot{A}}^{I}\chi_{-}^{\dot{A}}\psi_{+}^{I-}$$

$$\hat{\mathcal{P}}_{-}^{A} = \frac{1+t}{1-t}P_{-}^{A} - 2\sqrt{2}i\frac{t(1+t)}{(1-t)^3}\Gamma_{A\dot{A}}^{I}\chi_{-}^{\dot{A}}\psi_{2-}^{I} - 4\frac{t}{1-t^2}\Gamma_{A\dot{A}}^{I}\chi_{+}^{\dot{A}}\psi_{-}^{I+} \tag{3.3}$$

where $[I, J]$ denotes antisymmetrization in the indices I, J with strength one. A somewhat lengthy calculation now establishes that, with the exceptions described below, all equations of motion given in the preceding section as well as the integrability condition (2.3) can be obtained by imposing the generalized integrability constraint

$$\mathrm{D}_{+}(\hat{\mathcal{V}}^{-1}\mathcal{D}_{-}\hat{\mathcal{V}}) - \mathrm{D}_{-}(\hat{\mathcal{V}}^{-1}\mathcal{D}_{+}\hat{\mathcal{V}}) + [\hat{\mathcal{V}}^{-1}\mathcal{D}_{+}\hat{\mathcal{V}}, \hat{\mathcal{V}}^{-1}\mathcal{D}_{-}\hat{\mathcal{V}}] = 0 \tag{3.4}$$

Note that the derivatives to the left are covariant, since otherwise we would have to include a commutator term $\hat{\mathcal{V}}^{-1}[\mathcal{D}_{+},\mathcal{D}_{-}]\hat{\mathcal{V}}$ on the right hand side. In addition, one must make use of the following set of differential equations for the spectral parameter

$$t^{-1}\mathcal{D}_{+}t = \frac{1-t}{1+t}\rho^{-1}\mathcal{D}_{+}\rho - \frac{4t}{1-t^2}\psi_{+}^{I-}\psi_{2-}^{I}$$

$$t^{-1}\mathcal{D}_{-}t = \frac{1+t}{1-t}\rho^{-1}\mathcal{D}_{-}\rho + \frac{4t}{1-t^2}\psi_{-}^{I+}\psi_{2+}^{I} \tag{3.5}$$

Since these are first order equations, their solution $t = t(x, w)$ involves one integration constant w. We stress that the linear system (3.3) gives rise to *all* fermionic field equations, whereas the super Virasoro conditions (2.12) were missed in [4]. The only equations of motion that cannot be recovered from (3.3) are (2.13), (2.14), (2.15) and the Maxwell equation for A_μ, i.e. precisely the equations obtained by dimensional reduction of the $3d$ Einstein equations. Remarkably, however, the equations (3.5) are themselves subject to an integrability constraint that yields one of the missing equations! Namely, for

$$D_-(t^{-1}\mathcal{D}_+ t) - D_+(t^{-1}\mathcal{D}_- t) =$$
$$= -\frac{4t}{1-t^2}\rho^{-1}\Big(D_+ D_- \rho + D_+(\rho\psi_-^{I+}\psi_{2+}^I) + D_-(\rho\psi_+^{I-}\psi_{2-}^I)\Big) \tag{3.6}$$

to vanish we must impose (2.13). To recover the equations of motion (2.14) and (2.15), it has been proposed in [3] to incorporate the conformal factor into the linear system replacing the matrix $\widehat{\mathcal{V}}$ by the pair $(\lambda, \widehat{\mathcal{V}})$; due to the presence of a central charge in the Kac Moody algebra [9], the multiplication of two such pairs involves a non-trivial group two-cocycle. However, this proposal has so far only been shown to work for the bosonic theories in the special gauge (1.4). We have so far not found a way to include the Maxwell equation into the linear system (3.3). Nonetheless, these observations strongly suggest that there exists yet another generalization of (3.3) that also gives rise to the remaining equations of motion and that includes the spectral parameter as one of the dynamical fields. The dependence of t on the topological degrees of freedom has not been considered in earlier work where the relevant field configurations were assumed to be asymptotically flat for the Euclidean reduction and topologically trivial for colliding plane waves. Observe also that the poles at $t = -1$ and $t = +1$ in (3.3) and (3.5) are associated with the positive and negative chirality components of the bosonic and fermionic fields, respectively.

4   Outlook

In [3, 5] it is explained that the space of stationary axisymmetric or colliding plane wave solutions can be identified with the infinite dimensional coset speace

$$\mathcal{M}_{restr} = G^\infty/H^\infty \tag{4.1}$$

where G^∞ is the Kac Moody group corresponding to the group G (with $G = SL(2, \mathbf{R})$ for pure gravity and $G = E_8$ for $N = 16$ supergravity) and depends on the constant spectral parameter w, and H^∞ is its "maximal compact subgroup". The precise definition of H^∞ and the coset space $\mathcal{M}_{restr}$ is, however, somewhat subtle due to the x-dependence of t. E.g. for $G = SL(n, \mathbf{R})$ and $H = SO(n)$, H^∞ is defined to be the set of matrices $h(x, t) \in G$, which is invariant under the Cartan type involution [9, 3]

$$\tau^\infty : h(x, t) \longrightarrow h^T(x, \frac{1}{t}) \tag{4.2}$$

¿From (3.3), one can verify that the involution τ^∞ leaves the expressions $\widehat{\mathcal{V}}^{-1}\mathcal{D}_\pm\widehat{\mathcal{V}}$ invariant, which therefore belong to the Lie algebra of H^∞. The groups G^∞ and H^∞ act on $\widehat{\mathcal{V}}$ according to

$$\widehat{\mathcal{V}}(x,t) \longrightarrow g^{-1}(w)\widehat{\mathcal{V}}(x,t)h(x,t) \tag{4.3}$$

generalizing the action (3.1) of the corresponding finite dimensional groups G and H on $\mathcal{V}(x)$. The elements of the coset space $\mathcal{M}_{restr}$ are then defined to be the equivalence classes of matrices $\widehat{\mathcal{V}}(x,t)$ with respect to the "gauge group" H^∞. In view of the fact that G^∞ "does not know" about x, it is quite remarkable how the x-dependence of the elements of $\mathcal{M}_{restr}$, and thereby of the solutions of the gravitational field equations, emerges from this definition.

To overcome the restriction to topologically trivial solutions and to incorporate configurations involving the topological degrees of freedom, a bigger coset space may be needed. From string theory we know that the configuration space of pure $2d$ gravity is nothing but the moduli space $\mathcal{M}_0$ of the corresponding Riemann surface (this is a finite dimensional space at each genus, but since we are interested in solutions for arbitrary genus, a universal moduli space of the type discussed in [7] would perhaps be more appropriate). Defining the total "moduli space of solutions" as

$$\mathcal{M} := \frac{\text{solutions of field equations}}{\text{gauge transformations}} \tag{4.4}$$

we see that $\mathcal{M}$ must contain both $\mathcal{M}_0$ as well as $\mathcal{M}_{restr}$. Now, owing to the "back reaction" of matter on the geometry discussed previously, it seems very unlikely that $\mathcal{M}$ is the direct product of $\mathcal{M}_0$ and $\mathcal{M}_{restr}$. A most intriguing question is whether $\mathcal{M}$ can be represented as a coset space like $\mathcal{M}_{restr}$ above, but now with bigger groups $G^{\infty\infty} \supset G^\infty$ and $H^{\infty\infty} \supset H^\infty$. It appears likely, however, that this question cannot be settled before yet another extension of the linear system involving the Kaluza Klein vector A_μ and its equation of motion has been found.

In [5], the conserved Kac Moody current was shown to take the form

$$\mathcal{J}^\mu = \epsilon^{\mu\nu}\partial_\nu\left(\frac{\partial\widehat{\mathcal{V}}}{\partial w}\widehat{\mathcal{V}}^{-1}\right) \tag{4.5}$$

The associated conserved charges are given by

$$\int \left(\mathcal{J}_+\mathcal{D}x^+ + \mathcal{J}_-\mathcal{D}x^-\right) \tag{4.6}$$

where the integral is to be performed along a spacelike "hyper-surface" $x^0 = const$. On a topologically non-trivial Lorentzian world sheet, this set may decompose into several disconnected components, and consequently there may be more than one conserved charge at a given instant. The algebraic structure and the interrelation between these charges remain to be understood[b].

Acknowledgements: I would like to thank the organizers for inviting me to this very stimulating workshop in the beautiful surroundings of Cargese.

---

[b]I am grateful to K. Pohlmeyer for a discussion on this point and for alerting me to [22], where this phenomenon has been studied in a somewhat different context.

References

[1] H. Nicolai, *New Linear systems for 2D Poincaré Supergravities*, preprint DESY 93-122 (1993), hep-th 9309052

[2] D. Maison, Phys. Rev. Lett. **41** (1978) 521;
V.A. Belinskii and V.E. Zakharov, Zh. Eksp. Teor. Fiz. **75** (1978) 1955; **77** (1979) 3

[3] P. Breitenlohner and D. Maison, Ann. Inst. Poincaré **46** (1987) 215

[4] H. Nicolai, Phys. Lett. **194B** (1987) 402;
H. Nicolai and N.P. Warner, Commun. Math. Phys. **125** (1989) 384

[5] H. Nicolai, in *Recent Aspects of Quantum Fields*, Proceedings, Schladming 1991, eds. H. Mitter and H. Gausterer, Springer Verlag, 1991

[6] R. Geroch, J. Math. Phys. **12** (1971) 918, **13** (1972) 394

[7] E. Cremmer, S. Ferrara and J. Scherk, Phys.Lett. **B74** (1978) 61;
E. Cremmer and B. Julia, Nucl.Phys. **B159** (1979) 141

[8] B. Julia, in *Superspace and Supergravity*, eds. S.W. Hawking and M. Rocek, Cambridge University Press, 1980

[9] B. Julia, in *Unified Theories and Beyond*, Proc. 5th Johns Hopkins Workshop on Current Problems in Particle Theory, Johns Hopkins University, Baltimore, 1982;
in *Vertex Operators in Mathematics and Physics*, eds. J. Lepowsky, S. Mandelstam and I. Singer, Springer Verlag, Heidelberg, 1984

[10] B. de Wit, H. Nicolai and A. Tollsten, Nucl. Phys. **B392** (1993) 3

[11] K. Khan and R. Penrose, Nature **229** (1970) 286;
V. Ferrari and J. Ibanez, Gen. Rel. Grav. **19** (1987) 405

[12] B. de Wit, M. Grisaru, E. Rabinovici and H. Nicolai, Phys. Lett. **B286** (1992) 78

[13] L. Baulieu and M. Bellon, Phys. Lett. **196B** (1987) 142

[14] R. Dick, *Chiral Fields on Riemann Surfaces and String Vertices*, Thesis (Hamburg University, 1990);
Lett. Math. Phys. **18** (1989) 67; Fortschr. Physik **40** (1992) 519

[15] M. Knecht, S. Lazzarini and R. Stora, Phys. Lett. **262B** (1991) 25; **273B** (1991) 63.

[16] E. Verlinde and H. Verlinde, Phys. Lett. **192B** (1987) 95;
J. Atick, G. Moore and A. Sen, Nucl. Phys. **B307** (1988) 221, **B308** (1988) 1

[17] N. Steenrod, *The Topology of Fiber Bundles*, Princeton University Press, Princeton NJ, 1951

[18] A. R. Cooper, L. Susskind and L. Thorlacius, Nucl. Phys. **B363** (1993) 132

[19] N. Marcus and J. Schwarz, Nucl. Phys. **B228** (1983) 145

[20] J.L. Gervais and B. Sakita, Nucl. Phys. **B34** (1971) 632

[21] D. Friedan and S. Shenker, Nucl. Phys. **B281** (1987) 509

[22] T. Kornhass, *Klassische Erhaltungsgrössen auf verzweigten Stringtra-jektorien der Nambu-Goto Theorie*, Thesis (Freiburg University, 1991)

QUANTIZATION OF MIRROR SYMMETRY

Hirosi OOGURI

Research Institute for Mathematical Sciences
Kyoto University, Kyoto 606-01, Japan

1 Introduction

In the 80's, major progress in the string theory had been made on the world–sheet, and various techniques to study correlation functions of conformal field theories in two–dimensions has been developed. However we are still at a rather preliminary stage in understanding the string theory itself. Especially target–space aspects of the string theory are just beginning to be uncovered. Also, to understand even perturbative aspects of the quantum string theory , we need to develop a practical method to perform integrations over the moduli space of Riemann surface to compute perturbative amplitudes explicitly.

In recent years, some progress has been made in this direction. It is realized that some of the superstring computations can be transformed into solvable finite dimensional problems involving topological field theories [1]. A typical example is a computation of Yukawa couplings [2], [3]. In the standard compactification of the heterotic string theory on a Calabi-Yau manifold M, the world–sheet theory is constructed from a supersymmetric sigma–model on M plus four free bosons and fermions corresponding to the uncompactified dimensions. The sigma–model on M has the $N = 2$ superconformal symmetry generated by the energy–momentum tensor T, a pair of supercurrents $G^{\pm}$ and the $U(1)$ current J. In the minimal scenario, the $E_8 \times E_8$ gauge group is broken on M down to $E_8 \times E_6$ in the low energy effective theory. Each left–handed fermion in the four uncompactified dimensions transforming as $\overline{\mathbf{27}}$ of E_6 corresponds to what is called a (c, c)–primary field with the left and the right $U(1)$ charges (q_L and q_R) equal to 1. A (c, c)–primary field ϕ_i is a primary field of the superconformal field theory , but satisfies an extra-condition that it is annihilated by $G^+_{-1/2,L}$ and $G^+_{-1/2,R}$, i.e.

$$G^+_{-1/2,L}\phi(z,\bar{z}) = \oint_z \frac{dw}{2\pi i} G^+_L(w)\phi(z,\bar{z}) = 0 \tag{1.1}$$

Quantum Field Theory and String Theory, Edited by
L. Baulieu *et al.*, Plenum Press, New York, 1995

$$G^+_{-1/2,R}\phi(z,\bar{z}) = \oint_{\bar{z}} \frac{d\bar{w}}{-2\pi i} G^+_L(\bar{w})\phi(z,\bar{z}) = 0. \tag{1.2}$$

Each (c,c)–primary field ϕ_i is related to an element of the cohomology $H^{1,1}$ of M ($i = 1, ..., \dim H^{1,1}$). Similarly each left–handed fermion in $\mathbf{27}$ corresponds to a (c,a)–primary field $\tilde{\phi}_i$ with the $U(1)$ charges $q_L = 1$, $q_R = -1$ associated to an element of $H^{2,1}(M)$. A (c,a)–primary field is annihilated by $G^-_{-1/2,L}$ and $G^+_{-1/2,R}$,

$$G^-_{-1/2,L}\tilde{\phi}_i = G^+_{-1/2,R}\tilde{\phi}_i = 0. \tag{1.3}$$

The Yukawa coupling C_{ijk} among $\overline{\mathbf{27}}$'s should then be related a three–point amplitude of the (c,c)–primary fields ϕ_i, ϕ_j, ϕ_k on a two–dimensional sphere, and similarly for the Yukawa coupling $\tilde{C}_{ijk}$ among $\mathbf{27}$'s. We should, however, also remember that, in the Yukawa coupling, two of the external states are spacetime fermions. The vertex operators for these spacetime fermions should be in the Ramond sector, and their sigma–model parts are constructed by applying the spectral flow operators on the (c,c) (or (c,a)) primary fields. Thus C_{ijk} (or $\tilde{C}_{ijk}$) is equal to the three–point amplitude of the (c,c) (or (c,a)) primary fields, but with two insertions of the spectral flow operators on the sphere.

The insertion of the two spectral flow operators effectively changes the spins and the weights of operators on the sphere, and this generates what is called the twisting of the $N = 2$ sigma–model into a topological sigma–model [1],[4]. To be precise, the spectral flow operators which appear in C_{ijk} and $\tilde{C}_{ijk}$ are different in their ways in twisting the right–moving sector. Thus the same $N = 2$ sigma–model gives two inequivalent topological models. In the case of C_{ijk} among $\overline{\mathbf{27}}$'s, the spectral flow operator changes the spins of the supercurrents G^+_L and G^+_R from 3/2 to 1 and those for G^-_L and G^-_R from 3/2 to 2. Thus we may regard G^+_L and G^+_R as the BRST currents of the topological sigma–model. This procedure is called the A–twist, and the resulting topological sigma–model is called the A–model [5]. On the other hand, the spectral flow operator for $\tilde{C}_{ijk}$ among $\mathbf{27}$'s transforms G^+_L and G^-_R into the spin–1 BRST currents, and this generates what is called the B–twist of the sigma–model. By definition (1.2), the (c,c)–primary fields commute with the BRST operators of the A–model and are regarded as physical operators in this model. On the other hand, the (c,a)–primary fields are physical operators in the B–model.

In this way, the computations of the Yukawa coupling reduce to the ones in the topological A– and B–models. In the A–model, the BRST symmetries generated by G^+_L and G^+_R imply that the instanton approximation is exact [2], [5]. Thus the Yukawa coupling among $\overline{\mathbf{27}}$ is given by a sum over holomorphic maps from S^2 into M as

$$C_{ijk} = \int_M k_i \wedge k_j \wedge k_k + \sum_s N_s^{(0)} s_i s_j s_k \frac{\exp(-\sum_i s_i t^i)}{1 - \exp(-\sum_i s_i t^i)} \tag{1.4}$$

where k_i, k_j, k_k are elements of $H^{1,1}(M)$ corresponding to the (c,c)–primary fields ϕ_i, ϕ_j, ϕ_k, and $N_s^{(0)}$ is the number of spheres holomorphically embedded in M such that k_i evaluated on them is equal to s_i. The Kähler class k of M can be expressed as a linear combination of the basis of $H^{1,1}(M)$ as $k = \sum_i t^i k_i$, and the expansion coefficient

t^i is called the Kähler moduli of M. The combination $\sum_i s_i t^i$ in the exponent in the above is equal to the sigma–model action evaluated on the instanton. It is known that $N_s^{(0)}$ is generically independent of the complex structure of M, and so is the Yukawa coupling C_{ijk} in the A–model.

On the other hand, in the B–model, the Yukawa coupling $\widetilde{C}_{ijk}$ does not depend on the Kähler structure of M [3], [5]. This is because any deformation of the Kähler structure on M is BRST trivial in the B–model. Thus, in particular, $\widetilde{C}_{ijk}$ is independent of the volume of M. We can then evaluate the Yukawa coupling in the infinite volume limit of M, and the computation reduces to a problem of the classical algebraic geometry. As we mentioned in the above, each (c, a)–primary field corresponds to an element of $H^{2,1}(M)$ which generates an infinitesimal deformation of the complex structure on M. Thus the Yukawa coupling among the (c, a)–primary fields is given by

$$\widetilde{C}_{ijk} = \int_M \Omega \wedge \frac{\partial^3}{\partial y^i \partial y^j \partial y^k} \Omega \tag{1.5}$$

where Ω is the unique holomorphic 3–form on M, and y^i, y^j, y^k are the complex moduli on M. In many examples, it is possible to compute the Yukawa coupling $\widetilde{C}_{ijk}$ among **27** explicitly using the technique of the variation of the Hodge structure.

The Yukawa coupling C_{ijk} among $\overline{\mathbf{27}}$ is much more difficult to evaluate directly. There is a combinatorial method to compute the number $N_s^{(0)}$ for small values of s_i, but we need to know this for all s_i's in order to compute C_{ijk} by using the formula (1.4). Fortunately, in some cases, the computation of C_{ijk} reduces to that of $\widetilde{C}_{ijk}$. It is when the Calabi–Yau manifold M has its mirror partner $\widetilde{M}$. A pair of Calabi–Yau manifold $(M, \widetilde{M})$ is called a mirror pair when the A–model on M is equivalent as a quantum field theory to the B–model on $\widetilde{M}$ (and vice versa). This in particular means that the number of $\overline{\mathbf{27}}$'s for the heterotic string on M is equal to the number of **27**'s for the one on $\widetilde{M}$ and vice versa. This equivalence among the two topological field theories is called the mirror symmetry . The existence of a mirror pair had been conjectured in various places [6],[7],[5], but an explicit construction was first given in [9]. If $(M, \widetilde{M})$ is a mirror pair, the Yukawa coupling among $\overline{\mathbf{27}}$'s on M can be related to the Yukawa coupling among **27**'s on $\widetilde{M}$. In the paper [10], this idea was implemented in the case when M is the quintic 3–fold in $\mathbf{CP}^4$ to compute the Yukawa coupling C_{ijk}. As a by–product, they derived the number $N_s^{(0)}$ of S^2 in M for all values of s by expanding C_{ijk} as in (1.4). Thus when M has a mirror partner $\widetilde{M}$, it is possible to compute the Yukawa couplings among $\overline{\mathbf{27}}$ as well as among **27**'s.

Since the mirror symmetry is an equivalence of quantum field theories, it should hold whether the world–sheet is S^2 or a surface of higher genus. Thus it should also be useful in studying higher loop amplitudes in the string theory . This is what I would like to discuss in the following.

This talk is based on my recent works with Michael Bershadsky, Sergio Cecotti and Cumrun Vafa [11], [12]. I would like to thank them for the collaboration.

2 Quantum Amplitudes of Topological String

It is realized recently that some higher loop amplitudes in the superstring theory can also be related to the topological string computations [12], [13]. At one–loop, a topological string amplitude is given by

$$\mathcal{F}_1 = \frac{1}{2} \int_{\mathcal{M}_1} \frac{d^2\tau}{Im\tau} trace\left[(-1)^{F_L+F_R} F_L F_R q^{L_0} \bar{q}^{\bar{L}_0}\right],\tag{2.1}$$

where the trace is taken over the Ramond Hilbert space of the supersymmetric sigma-model on the Calabi-Yau 3-fold, $\mathcal{M}_1$ is the moduli space of a torus, F_L and F_R are the left and right $U(1)$ charges, and $q = e^{2\pi i\tau}$. This can be related to the one–loop correction to the R^2 term in the low energy effective theory in the type II superstring theory compactified on M [12], [13]. It is also found in [12] that the same quantity $\mathcal{F}_1$ can also be related to the one–loop contribution to the threshold correction to the gauge coupling [14] in the heterotic string theory .

At g–loop $(g > 1)$, a canonical amplitude in the topological string on the Calabi–Yau 3–fold M is

$$\mathcal{F}_g = \int_{\mathcal{M}_g} d^{3g-3}m d^{3g-3}\overline{m} \langle \prod_{a=1}^{3g-3} \int_{\Sigma_g} \mu_a G_L^- \int_{\Sigma_g} \bar{\mu}_a G_R^- \rangle \tag{2.2}$$

where $\mathcal{M}_g$ is the moduli space of the genus–g surface Σ_g, and μ_a is the Beltrami-differential associated to the modulus m_a. This expression is for the A–model. In the B–model, we simply replace G_R^- in the above by G_R^+. After the twisting [1], [4], the supercurrent G^- becomes an operator of dimension 2, thus the integral $\int \mu_a G_L^-$ is well-defined on Σ_g and so is $\int \bar{\mu}_a G_R^-$. The supersymmetric sigma–model can be defined for a Calabi–Yau of any dimensions, but the 3–fold enjoys a special status for the following reason. Because of the chiral anomaly in the $U(1)$ current, $\overline{\partial} J \sim \hat{c} R$, where $\hat{c} = \dim M$ and R is the Ricci curvature on the world–sheet, the fermion number is violated on Σ_g by the amount $(3g - 3) \cdot \dim M$. This can be compensated by the $(3g - 3)$–insertions of G^- in (2.2) only if $\dim M = 3$. Thus $\mathcal{F}_g$ at $g > 1$ is non–vanishing only when M is 3–fold.

It is shown in [12], [13] that $\mathcal{F}_g$ for $g > 1$ computes superpotential terms involving the gravi–photon in the effective theory of the type II superstring. The vertex operator for the gravi–photon belongs to the Ramond–Ramond sector. Thus an insertion of the gravi–photon vertex operator twists the sigma–model, and in some case the sigma-model is transformed into the topological model. This is how the superstring amplitudes are related to the topological string amplitudes $\mathcal{F}_g$. In the following, we develop methods to evaluate $\mathcal{F}_g$ and explore its geometric properties. For more details on the superstring interpretation of $\mathcal{F}_g$, we refer the reader to [12].

3 Holomorphic Anomaly

In the A–model, we saw that the Yukawa coupling C_{ijk} depends holomorphically on the Kähler moduli t^i as in (1.4), but does not depend on the moduli of the complex structure on M. Conversely, in the B–model, the Yukawa coupling $\widetilde{C}_{ijk}$ in the B–model depends holomorphically on the complex moduli y^i (1.5) but does not depend on the Kähler moduli. Some of these properties are inherited by the higher loop amplitudes $\mathcal{F}_g$'s. By using the Ward identity of the $N = 2$ superconformal symmetry, it is shown in [11] and [12] that $\mathcal{F}_g$ in the A–model (B–model) is independent of the complex moduli (Kähler moduli). On the other hand, in these papers, it is also found that their holomorphicities are slightly broken due to contributions from the boundary of the moduli space $\mathcal{M}_g$. Let us briefly explain how this happens.

Let us consider the case of the A–model. For the B–model, we can simply interchange G_R^+ and G_R^-. In the sigma–model, a variation with respect to the modulus parameter t^i is realized by an insertion of the operator

$$\int_{\Sigma_g} G_{-1/2,L}^- G_{-1/2,R}^- \phi_i \tag{3.1}$$

where ϕ_i is a (c, c)–primary field associated to the modulus t^i. Since the sigma–model itself is a unitary theory with the hermiticity condition $(G^-)^\dagger = G^+$, a variation with respect to $\bar{t}^i$ should be generated by an insertion of

$$\int_{\Sigma_g} G_{-1/2,L}^+ G_{-1/2,R}^+ \overline{\phi}_i, \tag{3.2}$$

i.e.

$$\frac{\partial}{\partial \bar{t}^i} \mathcal{F}_g = \int_{\mathcal{M}_g} d^{3g-3}m\, d^{3g-3}\overline{m} \langle \int_{\Sigma_g} G_{-1/2,L}^+ G_{-1/2,R}^+ \overline{\phi}_i, \prod_{a=1}^{3g-3} \int_{\Sigma_g} \mu_a G_L^- \int_{\Sigma_g} \overline{\mu}_a G_R^- \rangle. \tag{3.3}$$

If there were no G^- inserted on Σ_g, we could move $G_{-1/2,L}^+$ and $G_{-1/2,R}^+$ around the surface Σ_g and show that the right–hand side vanishes. However because of $\int_{\Sigma_g} \mu_a G_L^-$ and $\int_{\Sigma_g} \overline{\mu}_a G_R^-$, we must pick up the commutators of G^+ and G^-. They generate insertions of the energy–momentum tensor T, which can be converted into derivatives with respect to the moduli m_a of the Riemann surface. Thus the integrand in the right–hand side becomes a total derivative in the moduli space $\mathcal{M}_g$. We must then evaluate contributions from the boundary of $\mathcal{M}_g$. The non–vanishing contribution to $\overline{\partial}\mathcal{F}_g$ which arises from the boundary of $\mathcal{M}_g$ is called the holomorphic anomaly.

The holomorphic anomaly at $g > 1$ is evaluated in [12] with the result

$$\frac{\partial}{\partial \bar{t}^i} \mathcal{F}_g = \frac{1}{2} \overline{C_{ijk}} e^{2K} G^{j\bar{j}} G^{k\bar{k}} \left(D_j D_k \mathcal{F}_{g-1} + \sum_{r=1}^{g-1} D_j \mathcal{F}_r D_k \mathcal{F}_{g-r} \right) \tag{3.4}$$

where $G_{i\bar{j}}$ is the Zamolodchikov metric on the moduli space of the sigma–model which appears in the operator product expansion

$$\phi_i(z)\overline{\phi}_j(w) \simeq \frac{G_{i\bar{j}}}{|z - w|^2}, \tag{3.5}$$

K is the Kähler potential for the metric, $G_{i\bar{j}} = \partial_i \overline{\partial}_{\bar{j}} K$, and D_i is an appropriate covariant derivative ($\mathcal{F}_g$ is a section of some line–bundle over the moduli space of the sigma–model. See [12] for more detail). In the right–hand side, the repeated indices j, k are summed over the (c, c)–primary states with the $U(1)$ charges $q_L = q_R = 1$. In the case of the B-model, we can simply replace $\overline{C_{ijk}}$ in this formula by $\overline{\tilde{C}_{ijk}}$ and the Zamolodchikov metric $G_{i\bar{j}}$ and the Kähler potential K by the corresponding objects in the B-model. The holomorphic anomaly at $g=1$ is also derived in [11] as

$$\frac{\partial^2}{\partial t^i \partial \overline{t^j}} \mathcal{F}_1 = \frac{1}{2} tr(C_i \overline{C_j}) - \frac{\chi}{24} G_{i\bar{j}}, \tag{3.6}$$

where tr is a trace over the (c,c)-primary states of all $U(1)$ charges q_L, q_R. Especially when M is a 3-fold, this becomes

$$\frac{\partial^2}{\partial t^i \partial \overline{t^j}} \mathcal{F}_1 = \frac{1}{2} C_{ikl} \overline{C_{jkl}} e^{2K} G^{k\bar{k}} G^{l\bar{l}} - (\frac{\chi}{24} - 1) G_{i\bar{j}}. \tag{3.7}$$

Here the sums over the indices j, k are restricted to those with $q_L = q_R = 1$.

Thus, unlike the case of the Yukawa coupling, the higher loop amplitude $\mathcal{F}_g$ is not holomorphic on the moduli space of the sigma–model. Its anti–holomorphic dependence, however, is captured by the anomaly equations (3.4), (3.6). In various cases, these equations can be integrated to derive the moduli dependence of $\mathcal{F}_g$. This will be explained in the next sections.

The similar technique can be used to prove that $\mathcal{F}_g$ in the A-model is independent of the complex moduli. In the case of the A-model, the variation with respect to the complex moduli is realized by an insertion of

$$\int_{\Sigma_g} G^-_{-1/2,L} G^+_{-1/2,R} \phi_i. \tag{3.8}$$

As in the case of the holomorphic anomaly discussed in the above, we can move G^+_R around the surface Σ_g, pick up a commutator with G^-_R and generate an insertion of the energy–momentum tensor which is converted into a derivative with respect to the moduli m of Σ_g. Then the computation again reduces to the boundary of $\mathcal{M}_g$. It turned out that the contribution from the boundary vanishes in this case [12]. Thus $\mathcal{F}_g$ in the A-model is independent of the complex moduli. Similarly one can show that $\mathcal{F}_g$ in the B-model is independent of the Kähler moduli of M.

4 One–Loop

Let us examine the property of $\mathcal{F}_1$ in more detail. In the above, we mentioned that $\mathcal{F}_g$ in the A-model (B-model) is independent of the complex moduli y (Kähler moduli t). To be precise, this is the case for $g > 1$. At $g = 1$, $\mathcal{F}_1$ does not distinguish the A-model and the B-model. So the precise statement in this case is that $\partial_t \mathcal{F}_1$ is independent of $y, \bar{y}$ and that $\partial_y \mathcal{F}_1$ is independent of $t, \bar{t}$. Namely $\mathcal{F}_1$ is a sum of two terms, one depends on t and $\bar{t}$ and the other depends on y and $\bar{y}$.

If we are interested in the complex moduli dependence of $\mathcal{F}_1$, we may choose any Kähler structure on M to evaluate $\mathcal{F}_1$. Especially we can take the infinite volume limit. Here we employ the argument by Witten [15]. The infinite volume limit for a fixed and finite world–sheet modulus is dominated by constant maps. But as noted in [15], this is not the full story. The reason is that we are integrating over the modulus τ of the torus. No matter how large a volume of Calabi-Yau we choose if we go close enough to the boundary of the moduli space of τ, we can get a finite action. In other words the world–sheets which will have finite actions are the ones concentrated in long thin tubes, which means that the computation reduces to that of an ordinary field theory. We then find that the complex moduli dependent part of $\mathcal{F}_1$ is expressed in terms of the determinants of the Laplacians $\Delta_{p,q}$ acting on $\Omega^{(0,p)}(\wedge^q T^{(1,0)})$, the space of $(0,p)$–forms taking values in $\wedge^q T^{(1,0)}$ on M, as

$$\mathcal{F}_1 = \frac{1}{2}\log\left[\prod_{p,q=0}^{n}(\det\Delta_{p,q})^{(-1)^{p+q}pq}\right] + (\text{independent of } y,\overline{y}) \tag{4.1}$$

where $n = \dim M$.

This particular combination of the determinant can be related to the Ray–Singer holomorphic torsion [16] as

$$\mathcal{F}_1 = \frac{1}{2}\sum_q(-1)^q q\log I(\wedge^q T) + (\text{independent of } y,\overline{y}), \tag{4.2}$$

where

$$I(V) = \prod_p(\det\Delta_{p,V})^{(-1)^p p}, \tag{4.3}$$

V is a vector bundle on M and $\Delta_{p,V}$ is the Laplacian acting on $\Omega^{(0,p)}(V)$. One of the important properties of the Ray–Singer holomorphic torsion is that it obeys the Quillen's anomaly formula [17]

$$\partial\overline{\partial}\log I(V) = \partial\overline{\partial}\sum_p(-1)^p\log g_p + 2\pi i\int_M \mathrm{Td}(T)\mathrm{Ch}(V)\Big|_{(1,1)} \tag{4.4}$$

where g_p is the metric for zero modes of $\Delta_{p,V}$, $\mathrm{Td}(T)$ denotes the Todd class of the tangent bundle over M, and $\mathrm{Ch}(V)$ is the Chern class of the vector bundle V. The symbol $\Big|_{(1,1)}$ in the above formula means that we take a $(n+1,n+1)$–form of the integrand and integrate it over the n–fold M to be left with a $(1,1)$–form on the complex moduli space . Applying this formula to (4.1), we recover the holomorphic anomaly formula (3.6) for $\mathcal{F}_1$ [12]. Thus, in this particular case of the B–model at one–loop, the holomorphic anomaly reduces to the Quillen's formula.

The holomorphic anomaly (3.6) can be easily integrated to obtain

$$\mathcal{F}_1 = \log\left[X(t,\overline{t})|f(t)|^2\right] + \log\left[\widetilde{X}(y,\overline{y})|\widetilde{f}(y)|^2\right] \tag{4.5}$$

where

$$X(t,\overline{t}) = e^{-\frac{\chi}{24}K}\prod_{p,q}(\det g_{p,q})^{(-1)^{p+q}\frac{p+q}{4}}. \tag{4.6}$$

with $g_{p,q}$ being the metric for the (c,c)–primary fields with the $U(1)$ charges (p,q)

$$X(t,\overline{t}) = e^{-\frac{\chi}{24}\widetilde{K}}\prod_{p,q}(\det\widetilde{g}_{p,q})^{(-1)^{p+q}\frac{p+q}{4}}. \tag{4.7}$$

with $\widetilde{g}_{p,q}$ being the metric for the (c,a)–primary fields with the $U(1)$ charges $(p,-q)$. Here $f(t)$ and $\widetilde{f}(y)$ is some holomorphic object which cannot be determined from the anomaly equation alone. In order to fix f and $\tilde{f}$, we need to know the behavior of $\mathcal{F}_1$ near the boundary of the moduli space of M. For $\tilde{f}$, this is not an easy task since the behavior of the sigma–model near the boundary of the complex moduli space of M is not well–known. On the other hand, the behaviour of $\mathcal{F}_1$ near the boundary of the Kähler moduli space can be derived from the sigma–model path integral as we see below. Thus, $f(t)$ can in principle be derived. However in this case, the computation of the metric $g_{p,q}$ is difficult. Both of these difficulties could be avoided if the manifold M has a mirror partner $\widetilde{M}$. If this is the case, we can compute $g_{p,q}$ of M by using the computation of $\tilde{g}_{p,q}$ on $\widetilde{M}$ (which can be done using the variation of Hodge structure), and at the same time we can fix the boundary condition for $\tilde{f}(y)$ on M by using the Kähler moduli space of $\widetilde{M}$. Thus, if M has a mirror partner $\widetilde{M}$, we can compute $\mathcal{F}_1$ explicitly.

Previously we saw that, in the A–model, the Yukawa coupling C_{ijk} is expressed as a sum over rational curves on M. Thus one may expect that the genus–g amplitude $\mathcal{F}_g$ can also be expanded as a sum over genus–g curves. Similarly the Kähler moduli dependent part of $\mathcal{F}_1$ should be given as a sum over holomorphic elliptic curves on M. This is almost true, but a precise statement is as follows [11]. The fact that the instanton approximation is exact for the Yukawa coupling (an instanton in this case is a holomorphic map $S^2 \to M$) is a consequence of the BRST symmetry on the world–sheet generated by G^+ [2], [5]. Since the BRST symmetry is slightly broken at $g \geq 1$ due to the holomorphic anomaly (3.4), (3.6), it is not exactly true that $\mathcal{F}_g$ can be expressed as a sum over holomorphic elliptic curves. In fact, the holomorphic anomaly implies that $\mathcal{F}_g$ depends on the anti–holomorphic Kähler moduli $\bar{t}^i$ as well as t^i. In the sigma–model Lagrangian, the moduli $\bar{t}^i$ appear in the combination

$$\bar{t}^i \int_{\Sigma_g} k_i \bar{\partial} X \partial \overline{X}. \tag{4.8}$$

Thus if the instanton approximation were exact, $\mathcal{F}_g$ should not depend on $\bar{t}^i$. This clearly is inconsistent with the holomorphic anomaly. However we can recover the sum over holomorphic maps from $\mathcal{F}_g$ as follows. By sending $\bar{t}^i \to \infty$, we can suppress non–holomorphic maps $X : \Sigma_g \to M$ in the sigma–model path integral because of the form (4.8) of the Lagrangian.

In the case of $g = 1$, the evaluation of the path integral in the limit $\bar{t}^i \to \infty$ gives the following formula [11].

$$\frac{\partial}{\partial t^i} \mathcal{F}_1 \simeq \frac{(-1)^n}{24} \int_M k_i \wedge c_{n-1} + \sum_s s_i N_s^{(1)} \left(\sum_{N=1}^{\infty} \frac{e^{-N \sum s_j t^j}}{1 - e^{-N \sum s_j t^j}} \right) + \tag{4.9}$$

$$+ \frac{1}{12} \sum_s s_i N_s^{(0)} \frac{e^{-\sum s_j t^j}}{1 - e^{-\sum s_j t^j}}. \tag{4.10}$$

The leading term $(-1)^n \int_M k_i \wedge c_{n-1}$ in the right–hand side is a contribution from the constant map $\Sigma_1 \to M$. This term is multiplied by the factor $\frac{1}{24}$, which is equal a minus–half of the Euler character of the moduli space Σ_1. This is because the constant map exists for any modulus τ of Σ_1, thus we must perform integration over the moduli space of Σ_1. In the second term, $N_s^{(1)}$ is the number of holomorphic elliptic curves on M. The factor

$$\sum_{N=1}^{\infty} \frac{e^{-N\sum s_j t^j}}{1 - e^{-N\sum s_j t^j}} \tag{4.11}$$

is there in order to count multi–covering of a holomorphic elliptic curve in M by the world–sheet torus Σ_1. It is curious that the number of the rational curves $N_s^{(0)}$ also contribute to $\mathcal{F}_1$ as we see in the third term in the right–hand side. This is due to the following two reasons. One of the origins of the third term is the compactification of the moduli space of curves in M (the instanton moduli space). In general, a curve is not necessarily isolated in M and there may be a continuous family of curves. In this case, $N_s^{(1)}$ represent a computation of the Euler character of a certain vector bundle over the instanton moduli space , the Gromov–Witten invariant. In order to compute the Euler character , we must compactify the instanton moduli space . We must then take into account contributions from the boundary of the moduli space , and they are given by $N_s^{(0)} s_i e^{-\sum s_j t^j}$ which gives a leading term in the Taylor expansion of the third term in $e^{-\sum s_j t^j}$. The rest of the third term is explained as follows. The world–sheet Σ_1 may be represented as a branched double cover of S^2 which may be holomorphically embedded in M. Contributions from such configurations are given by $N_s^{(0)} s_i e^{-2\sum s_j t^j}$ (the factor 2 in the exponent is due to the fact that the world–sheet Σ_1 double–covers S^2), and this gives the second term in the Taylor expansion. We must also take into account multi–coverings of these configurations. Combining all these contributions, we have obtained the formula (4.10).

To summarize, in the one–loop amplitude $\mathcal{F}_1$, the complex moduli dependence and the Kähler moduli dependence decouple. The complex moduli dependent part is related to the Ray–Singer torsion as in (4.1) while the Kähler moduli dependent part contains informations on the number of holomorphic elliptic curves $N_s^{(1)}$ on M as in (4.10). If the manifold M has a mirror partner $\widetilde{M}$, we can combine these two formulae and compute $\mathcal{F}_1$. In the case of the quintic 3–fold in $\mathbf{CP}^4$, we have carried out this procedure explicitly and computed the numbers $N_s^{(1)}$ of holomorphic elliptic curves [11].

Table 1. the number of elliptic curves on the quintic 3–fold

Degree	$N_s^{(1)}$
1	0
2	0
3	609250
4	3721431625
5	12129909700200
6	31147299732677250
7	71578406022880761750
8	154990541752957846986500
9	324064464310279585656399500

5   Higher Loops

In [12], the formula (4.10) for $g = 1$ is extended to the case of $g = 2$ as

$$\mathcal{F}_2 \simeq \frac{\chi}{5760} + \sum_s N_s^{(2)} e^{-\sum s_j t^j} + \frac{1}{240} \sum_s N_s^{(0)} \frac{e^{-\sum s_j t^j}}{(1 - e^{-\sum s_j t^j})^2} \tag{5.1}$$

in the $\bar{t}^i \to \infty$ limit of the A–model. The first term in the right–hand side is the contribution from the constant map $\Sigma_2 \to M$ with the coefficient

$$\frac{1}{5760} = \frac{1}{2} \int_{\mathcal{M}_2} (c_1)^3, \tag{5.2}$$

and the second term is from holomorphic maps $\Sigma_2 \to M$. Third term is the contributions from the boundary of the instanton moduli space and from the multi–covering of Σ_2 over S^2. The coefficient $\frac{1}{240}$ in front of the third term is a minus of the Euler character of the moduli space of genus–2 curves. In order to extend this result to still higher genera $g \geq 3$, we need to understand better the contributions from the boundary of the instanton moduli space and from the multi–coverings.

On the other hand, in the B–model, the genus–g amplitude $\mathcal{F}_g$ should be related to a g–loop computation of some quantum field theory for the reason we described in the previous section (the infinite–volume limit is exact in the B–model). Such a quantum field theory should be regarded as a string field theory of the B–model. In fact, we have identified such a quantum field theory. It is related to the theory of Kodaira and Spencer [18] which describes the deformation of the complex structure of M. The moduli space of a Calabi-Yau 3–fold could be regarded as the space of solution of the topological string of the B–model since the consistent propagation of this topological string requires the Ricci flatness of the target manifold [5]. It is then natural that the Kodaira–Spencer theory emerges as a string field theory since it described the moduli space of the complex structure of a Calabi-Yau manifold. For more details, we again refer the reader to [12].

As in the case of $g = 1$ (4.5), there is a systematic procedure to integrate the holomorphic anomaly equation (3.4) for $\mathcal{F}_g$. To do this, we need to introduce a "canonical" prepotential S satisfying

$$\overline{C_{ijk}} = e^{-2K} \overline{D_{\bar{i}}} \overline{D_{\bar{j}}} \overline{D_{\bar{k}}} S. \tag{5.3}$$

That we can find such an object S was shown in [12]. We then introduce some notations.

$$S^i = G^{i\bar{j}} \overline{D_{\bar{j}}} S \ , \quad S^{ij} = G^{i\bar{k}} \overline{D_{\bar{k}}} S^j \tag{5.4}$$

With these preparations, the holomorphic anomaly (3.4) can be integrated, for example in the case of $g = 2$, as

$$\mathcal{F}_2 = \frac{1}{2} S^{ij} D_i D_j \mathcal{F}_1 + \frac{1}{2} D_i \mathcal{F}_1 S^{ij} D_j \mathcal{F}_1 - \frac{1}{8} S^{jk} S^{mn} C_{jkmn} - \tag{5.5}$$

$$-\frac{1}{2} S^{ij} C_{ijm} S^{mn} D_n \mathcal{F}_1 + \frac{\chi}{24} S^i D_i \mathcal{F}_1 + \tag{5.6}$$

$$+\frac{1}{8} S^{ij} C_{ijp} S^{pq} C_{qmn} S^{mn} + \frac{1}{12} S^{ij} S^{pq} S^{mn} C_{ipm} C_{jqn} - \tag{5.7}$$

$$-\frac{\chi}{48} S^i C_{ijk} S^{jk} + \frac{\chi}{24}\left(\frac{\chi}{24} - 1\right) S + f_2(t). \tag{5.8}$$

As in the case of $g = 1$, there is a holomorphic object $f_2(t)$ which cannot be determined from the anomaly equation alone. This can be fixed when the manifold M has a mirror partner $\widetilde{M}$. In this case, by the mirror symmetry , $\mathcal{F}_g$ is also equal to the g–loop amplitude of the Kodaira–Spencer theory on $\widetilde{M}$, and this geometric interpretation gives a strong restriction on the behavior of $f_2(t)$ near the boundary of the moduli space. We can combine this information with the formula (5.1) for the A–model on M to determine $f_2(t)$. In the case when M is the quintic 3–fold in $\mathbf{CP}^4$, we again carried out this procedure and derived the number $N_s^{(2)}$ of genus–2 curves on M by using (5.1).

Table 2. the number of genus–2 curves on the quintic 3–fold

Degree	$N_s^{(2)}$
1	0
2	0
3	0
4	534750
5	75478987900
6	871708139638250
7	5185462556617269625
8	9006736425242423675345000
9	3258596871473582660010240500

We have also derived similar formulae for all g by integrating the holomorphic anomaly [12]. It turned out that the resulting expressions can be reproduced by the Feynman rule of some finite dimensional quantum mechanical system. It would be interesting to understand a geometric origin of this system.

References

[1] E.Witten, Commun. Math. Phys. 117 (1988) 353; and 118 (1988) 411.

[2] M.Dine, N.Seiberg, X.Wen and E.Witten, Nucl. Phys. B278 (1986) 769 and B289 (1987) 319.

[3] A.Strominger and E.Witten, Commun. Math. Phys. 101 (1985) 341.

[4] T.Eguchi and S.-K.Yang, Mod. Phys. Lett. A5 (1990) 59.

[5] E.Witten, in *Essay on Mirror Manifolds,* ed. by S.-T. Yau, International Press (1992).

[6] L.Dixon, in Proceedings of the ICTP Summer Workshop in High Energy Physics and Cosmology, Trieste 1987.

[7] T.Eguchi, H.Ooguri, A.Taormina and S.-K.Yang, Nucl. Phys. B315 (1989) 193.

[8] W.Lerche, C.Vafa and N.Warner, Nucl. Phys. B324 (1989) 427.

[9] B.R.Greene and M.R.Plesser, Nucl. Phys. B338 (1990) 15.

[10] P.Candelas, X. de la Ossa, Green and Parkes, Nucl. Phys. B359 (1991) 21.

[11] M.Bershadsky, S.Cecotti, H.Ooguri and C.Vafa, *Holomorphic Anomalies in Topological Field Theories*, HUTP–93/A008, RIMS–915, hep-th/9302103.

[12] M.Bershadsky, S.Cecotti, H.Ooguri and C.Vafa, *Kodaira–Spencer Theory of Gravity and Exact Results for Quantum String Theories*, HUTP–93/A025, RIMS–946, SISSA–142/93/EP.

[13] I.Antoniadis, E.Gava, K.S.Narain and T.R.Taylor, *Topological Amplitudes in String Theory*, hep-th/9307158.

[14] V.Kaplunovsky, Nucl.Phys. B307 (1988) 36;
L.J.Dixon, V.S.Kaplunovsky and J.Louis, Nucl.Phys. B355 (1991) 649;
S.Ferrara, C.Kounnas, D.Lüst and F.Zwirner, Nucl. Phys. B365 (1991) 431;
I.Antoniadis, E.Gava and K.S.Narain, Nucl. Phys. B383 (1992) 93 and Phys. Lett. B283 (1992) 209;
J.-P. Derendinger, S. Ferrara, C. Kounnas and F. Zwirner, Nucl. Phys. B372 (1992) 145.

[15] E.Witten, *Chern–Simons Gauge Theory as a String Theory*, IASSNS-HEP-92/45, hep-th/9207094.

[16] D.B.Ray and I.M.Singer, Ann. Math. 98 (1973) 154.

[17] J.M.Bismut and D.S.Freed, Commun. Math. Phys. 106 (1986) 159 and 107 (1986) 103.

[18] K.Kodaira and D.C.Spencer, Ann. Math. 67 (1958) 328 and 71 (1960) 43;
K.Kodaira, L.Niremberg and D.C.Spencer, Ann. Math. 68 (1958) 450;
K.Kodaira and D.C.Spencer, Acta. Math 100 (1958) 281.

INTEGRABLE QFT$_2$ ENCODED ON PRODUCTS OF DYNKIN DIAGRAMS

Ennio QUATTRINI

Dip. di Fisica, Università di Bologna
Via Irnerio 46, I-40126 Bologna, Italy

Francesco RAVANINI

INFN - Sez. di Bologna
Via Irnerio 46, I-40126 Bologna, Italy

and

Roberto TATEO

Dip. di Fisica Teorica, Università di Torino
via P.Giuria 1, I-10125 Torino, Italy

Abstract: A large class of Thermodynamic Bethe Ansatz equations governing the Renormalization Group evolution of the Casimir energy of the vacuum on the cylinder for an integrable two-dimensional field theory, can often be encoded on a tensor product of two graphs. We demonstrate here that in this case the two graphs can only be of ADE type. We also give strong numerical evidence for a new large set of Dilogarithm sum Rules connected to $ADE \times ADE$ and a simple formula for the ultraviolet perturbing operator conformal dimensions only in terms of rank and Coxeter numbers of $ADE \times ADE$. We conclude with some remarks on the curious case $ADE \times D$.

1   Introduction

The space of two-dimensional Quantum field Theories is in connection with that of Conformal Field Theories (CFT$_2$) in the sense that the ultraviolet (UV) and infrared (IR) limit of the former must lie in the space of theories of the latter. In other words, the space of two-dimensional QFT consists of conformal submanifolds (fixed points) of given central charge and of Renormalization Group (RG) flows connecting them. Many of these flows are not integrable, but there exists a subclass of integrable flows that defines the space of two-dimensional (euclidean) Integrable Quantum Field Theories (IQFT$_2$).

Quantum Field Theory and String Theory, Edited by
L. Baulieu *et al.*, Plenum Press, New York, 1995

The challenge to classify all IQFT$_2$ is one of the most attractive of modern Quantum Field Theory. However, this task is very far from completion and at the present stage we can only hope to get partial classifications similar to what happens in CFT$_2$. There also, the whole classification is far from completed, but for example, the subclass of rational Conformal field theories (RCFT) is (strongly) conjectured to be organized in series of coset models [1] G/H, thus putting their description in correspondence with general properties of the Lie algebras and their affine generalizations.

Analogously in IQFT$_2$ it would be interesting to identify some kind of general patterns organizing the vaste zoology of integrable models, in other words some kind of "Mendeleev table", able to show hierarchical structures to be later interpreted as effect of some symmetry underlying the integrable theories. To recognize this organizing criterion can be useful as the first step to figure out some general property of the yet misterious symmetries governing IQFT$_2$'s.

In the present paper we deal with deformations of some coset CFT G_k/H_{Ik} by one of its relevant scalar operators, say $\Phi(x)$ of (left) conformal dimension $\Delta_P < 1$. Here G is a compact Lie Algebra and H a proper subalgebra embedded in G with index I. The subscripts on G and H denote the level of the corresponding Kac-Moody algebra. The theories we are interested in are defined by the action

$$\mathcal{A} = [G_k/H_{Ik}] + \lambda \int d^2x \Phi(x) \tag{1.1}$$

where λ is a bare perturbing parameter of dimension $y = 2(1 - \Delta_P)$ and the symbol $[G_k/H_{Ik}]$ stands for the action of the G_k/H_{Ik} coset CFT. Such a theory can develop a mass gap and flow to a massive IR theory as the scale increases, or alternatively keep some of its states massless and flow towards a non-trivial infrared CFT. Zamolodchikov's c-theorem [2] implies, for unitary theories, that the central charge of the IR CFT must be smaller than that of the UV one.

An IQFT$_2$ possesses by definition an infinite set of local conserved charges. This implies [3] that the N-particle S-matrix is factorizable in terms of 2-particle ones. There are various ways to conjecture the form of the S-matrix once the conserved local and non-local charges are known. If the theory is massless, one can circumvent the problems with the definition of the S-matrix by considering asymptotic left and right movers, as recently proposed in a series of papers [4, 5, 63, 7].

The so far most effective method to recover the UV behaviour of a theory whose factorizable S-matrix is given is the so called Thermodynamic Bethe Ansatz (TBA). That the thermodynamics of a scattering theory can be reconstructed completely from its S-matrix was known since the end of the sixties [8], and the use of Bethe Ansatz techniques to implement this program for integrable theories was issued in a seminal paper by Yang and Yang [9] for a non-relativistic scattering problem. More recently Al.Zamolodchikov has proposed this method to investigate factorized scattering theories corresponding to IQFT$_2$ [10].

Therefore, any IQFT$_2$ must have a (possibly infinite) set of TBA equations. It can be seen as a set of equations governing non-perturbatively and exactly the evolution of the Casimir vacuum energy of the theory put on a cylinder along the Renormalization

274

Group flow. In sect.2 we present various interesting systems of TBA equations for some classes of well known coset CFT perturbed by relevant operators in integrable directions and we shall illustrate how in most cases the resulting TBA equations can be encoded in a sort of tensor product of two graphs. This encoding is even more evident by passing from TBA equations to an equivalent system of functional equations (the so-called *Y-system*) (sect.3). For the Y-system to be consistent with TBA interpretation, the two graphs cannot be chosen arbitrarily. We shall see in sect.3 how one can prove that they must be both restricted to the set of Dynkin diagrams of ADE type (or their foldings and their extensions). We also mention two properties of the $ADE \times ADE$ Y-system, that we checked numerically, namely a set of Dilogarithm sum rules and a periodicity formula. The amusing fact is that they can be expressed in terms of properties of Dynkin diagrams only.

The next obvious step is to explore (sect. 4) the set of would be flows suggested by the various ADE choices for the two graphs in the product, and recognize them with known flows or interpret them as signal of new, yet unexplored IQFT$_2$'s. We are conscious that the observation of a consistent set of equations does not prove the effective existence of the flow it seems to describe (although many reasonable checks like those of sect.4 can be proposed in this direction). However, it can be used as a leading tool to investigate, conjecture and then, when possible, verify the existence of new IQFT$_2$'s and better understand the relations among them and the pattern organizing IQFT$_2$. In this respect we trust it as a very productive method.

Of course the set of TBA-like equations we explore here does not exhaust even the set of IQFT$_2$ of the form (1.1). Our goal is to understand as far as possible the relation between the G and H algebras in the definition of the coset UV CFT, and the two Dynkin graphs appearing in the TBA equations, and possibly find extensions to this set able to incorporate other models.

2   TBA Equations on $\mathcal{G} \times \mathcal{H}$

In this section we want to recall some facts known in the literature about TBA equations and their encoding on products of Dynkin diagrams.

For a clear exposition of the deduction of TBA equations from the diagonal S-matrix of a purely elastic scattering theory we refer the reader to the origianl paper of Al.Zamolodchikov [10] and also to ref. [12, 13]. In this case, the equations turn out to have the form

$$\nu_a(\theta) = \varepsilon_a(\theta) + \frac{1}{2\pi} \sum_b [\phi_{ab} * \log(1 + e^{-\varepsilon_b})](\theta) \tag{2.1}$$

where the indices $a, b = 1, ...r$ label the different species of particles of masses m_a respectively, the symbol * stands for convolution and the kernel $\phi_{ab}(\theta)$ is determined from the S-matrix via the formula

$$\phi_{ab}(\theta) = -i\frac{d}{d\theta} \log S_{ab}(\theta) \tag{2.2}$$

θ is the rapidity parametrizing the energy (momentum) of a particle as $m\cosh\theta$ ($m\sinh\theta$). The massive behaviour of the theory is encoded in the *energy terms* $\nu_a(\theta) = m_a R\cosh\theta$

(R being the radius of the cylinder on which the theory is put, or, by modular invariance, the inverse temperature). The unknowns of this set of integral equations are the *pseudoenergies* $\varepsilon_a(\theta)$. These in turn determine the Casimir energy of the vacuum on the cylinder via the equations

$$E(R) = -\frac{\pi c(r)}{6R} \quad \text{and} \quad c(r) = \frac{3}{\pi^2} \sum_a \int_{-\infty}^{+\infty} d\theta \nu_a(\theta) L_a(\theta) \tag{2.3}$$

where $L_a(\theta)$ is short for $\log(1 + e^{-\varepsilon_a(\theta)})$, and $r = m_1 R$ (m_1 is the mass of the lightest particle) is such that the UV limit corresponds to $r \to 0$ and the IR one to $r \to \infty$. The function $c(r)$ is even more interesting than $E(R)$, as it interpolates directly the central charges of the UV and IR theories: $c(0) = c_{UV}$, $\lim_{r\to\infty} c(r) = c_{IR}$ (if the theory is massive $c_{IR} = 0$).

Klassen and Melzer [12] explored with this method the whole set of $\mathcal{G} = A, D, E$ purely elastic minimal S-matrices, confirming that they describe the set of theories with action

$$\mathcal{A} = \left[\frac{\mathcal{G}_1 \times \mathcal{G}_1}{\mathcal{G}_2^{diag}} \right] + \lambda \int d^2 x \phi_{adj}^{id,id}(x) \tag{2.4}$$

where the perturbing operator is here labelled by representations of $\mathcal{G} \times \mathcal{G}$ (upper indices) and $\mathcal{G}^{diag}$ (lower index) as usual in coset CFT. The link between S-matrix and ADE Lie algebras can be made explicit in the TBA equations thanks to a transformation proposed by Al.Zamolodchikov in the very interesting paper ref. [11]. For all the ADE theories explored in [12] the TBA can be recast in the following form

$$\nu_a(\theta) = \varepsilon_a(\theta) + \frac{1}{2\pi}\{\phi_g * \sum_a \mathcal{G}_{ab} * [\nu_b - \Lambda_b]\}(\theta) \tag{2.5}$$

where $\Lambda_b = \log(1 + e^{\varepsilon_b})$, $\phi_g(\theta) = g/2\cosh(g\theta/2)$ is a universal kernel depending on the Coxeter number g of $\mathcal{G}$ only, and all the dependence of the kernel from the indices a, b has been confined in the incidence matrix of the Dynkin diagram of the Lie algebra $\mathcal{G}$, that we shall denote in this paper as $\mathcal{G}_{ab}$. Hence we can think of (2.5) as a set of equations where the unknowns couple to each other obeying the connectivity of the $\mathcal{G}$ Dynkin diagram, and in this sense they can be thought as uniquely defined once the diagram is given. The masses of the particles are known to be encoded on the Perron-Frobenius eigenvector $\psi_{\mathcal{G}}$ of $\mathcal{G}$: $m_a = m_1 \psi_{\mathcal{G}}^a$.

A large set of IQFT$_2$ is known where the S-matrix is not diagonal. The deduction of TBA equations in this case is much more cumbersome, and one has to resort to Higher Level Bethe Ansatz techniques to diagonalize the "color" transfer matrix appearing in the Bethe equations. In spite of these difficulties, Al. Zamolodchikov was able to work out some simple case of TBA and then to conjecture a generalization valid for the whole set of minimal models (the $(A_1)_k \times (A_1)_1/(A_1^{diag})_{k+1}$ coset CFT's) perturbed by their least relevant ϕ_{13} operator [15, 4]. These equtions are of the form

$$\nu_i(\theta) = \varepsilon_i(\theta) + \frac{1}{2\pi}\sum_j (A_k)_{ij}(\phi_2 * L_j)(\theta) \tag{2.6}$$

where the indices $i, j = 1, ... k$ run on the incidence matrix of the A_k Dynkin diagram. The most curious feature of this system is the form of the energy terms. For negative λ the behaviour is massive and can be reproduced by choosing $\nu_i = \delta_i^1 mR\cosh\theta$. For positive λ these theories describe massless interpolating flows between the k-th and $k-1$-th minimal models. In spite of the conceptual difficulties to define an S-matrix for scattering of massless objects, one can still present consistent sets of TBA equations with the concept of particle substituted by that of left and right movers, of energy $mRe^\theta/2$ and $mRe^{-\theta}/2$ respectively. The sets of TBA equations are basically the same as (2.6), but now $\nu_i = \frac{mR}{2}(\delta_i^1 e^\theta + \delta_i^k e^{-\theta})$. In both cases, almost all nodes on the diagram are attached vanishing energy terms. This is a quite general feature of many IQFT$_2$. From the few known examples, one can argue that the Bethe Ansatz procedure usually performs a sort of "factorization" of the scattering process in two subprocesses: first a pure elastic scattering of the particles as they had no internal indices (colors), then a pure exchange of color between particles, that one is free to think as mediated by virtual objects with no energy nor momentum called *magnons* , somehow in connection with the nodes of the diagram having zero energy term, that are therefore called *magnonic nodes*.

Other progresses have been made on the same path. Putting masses on the k-th node or left/right movers on k-th/l-th nodes of the A_{k+l-1} diagram, one obtains a TBA set of equations describing the massive (negative λ) and massless (positive λ) behaviour of all the coset theories $(A_1)_k \times (A_1)_l/(A_1^{diag})_{k+l}$ perturbed by their $\phi_3^{1,1}$ relevant operator of dimension $\Delta_P = 1 - \frac{2}{k+l+1}$ [16]

One can also ask if it is possible to use diagrams different from A_k for this "magnonic" TBA's. In ref. [17] the $\mathbf{Z}_k$ parafermions (in coset language the $(A_1)_k/U(1)$ RCFT's) deformed by their $\psi_1\overline{\psi}_1 + c.c.$ operator were studied. The TBA system proposed there is similar to (2.6) but encoded on D_k. See ref. [17] for details. There is an elegant way to write in a unified form all the TBA systems cited above and also other generalizations. The idea is to introduce **two** graphs, one encoding the particle (or kink) species ($\mathcal{G}$), the other the magnonic structure of the colors ($\mathcal{H}$). The TBA is then fixed by giving the graph tensor product $\mathcal{G} \times \mathcal{H}$ and specifying the form of the energy term $\nu_a^i(\theta)$ attached to each node (a, i) (the labels $a, b = 1, \ldots r = rank(\mathcal{G})$ run here on the $\mathcal{G}$ graph and $i, j = 1, \ldots k = rank(\mathcal{H})$ run on the $\mathcal{H}$ graph)

$$\nu_a^i(\theta) = \varepsilon_a^i(\theta) + \frac{1}{2\pi}\phi_g * \left\{ \sum_b \mathcal{G}_{ab}[\nu_b^i - \Lambda_b^i] - \sum_j \mathcal{H}_{ij}L_a^j \right\}(\theta) \qquad (2.7)$$

Here g is simply some real number whose connection with the graph $\mathcal{G}$ shall specified better below. The rationale under this generalization comes from the study of the $\phi_{adj}^{id,id}$ perturbations of the large class of coset models $\mathcal{G}_k \times \mathcal{G}_l/\mathcal{G}_{k+l}(\mathcal{G} = A_n, D_n, E_{6,7,8})$, which are the most starightforward generalizations of the minimal models perturbed by ϕ_{13}. They are [18] the first evident example where one has to introduce both particle and magnonic indices. The TBA equations can be encoded on the *product* of diagrams $\mathcal{G} \times A_{k+l-1}$, the first describing the connectivity of the particle indices, the second that of the magnonic ones. The cases previously discussed can all be encoded in this general class (2.7), with suitable choices of $\mathcal{G}$ and $\mathcal{H}$. One is naturally lead to ask how large is the scope of integrable flows described by system (2.7) and if there are restrictions on the choice of $\mathcal{G}$ and $\mathcal{H}$. We deal with this subject in the next section.

3   The Y-System

Here we wish to recall a very important result of paper [11] (or better its generalization [18] to the more general system (2.7)), that perhaps has not sufficiently been emphasised and appreciated by part of the S-matrix community, but that will be the main instrument of our investigation, i.e. the fact that all solutions of the system (2.7) are also solutions of the system of functional equations

$$Y_a^i\left(\theta - \frac{i\pi}{g}\right) Y_a^i\left(\theta + \frac{i\pi}{g}\right) = \prod_b (1 + Y_b^i(\theta))^{\mathcal{G}_{ab}} \prod_j (1 + Y_a^j(\theta)^{-1})^{-\mathcal{H}_{ij}} \tag{3.1}$$

often referred as the *Y-system*. The relation with the TBA equations is given by $Y_a^i(\theta) = e^{\epsilon_a^i(\theta)}$. Here the encoding on the $\mathcal{G}$ and $\mathcal{H}$ graphs is even more evident. This system has been encountered in many applications of TBA, but such kind of objects appear in other areas of physics and mathematics too, for example in lattice integrable models, τ-functions for Toda lattices, etc..., so the importance to study such mathematical object is larger than the TBA problem we are considering here.

The derivation from TBA (2.7) given in [18] is valid for $\mathcal{G}$ any ADE Dynkin diagram, $g = cox(\mathcal{G})$ and $\mathcal{H}$ any connected unoriented graph. One can however consider the Y-system for any $\mathcal{G} \times \mathcal{H}$ graph and any $g \in R$ and then ask when it is consistent with a TBA interpretation, in other words when it can be used to represent an integrable Renormalization Group flow (this does not mean that the flow *exists*, simply that it is *mathematically* possible). In this connection, we shall prove the following two basic statements:

1. The Y-system (3.1) can be consistent with a TBA interpretation only if the graph $\mathcal{G}$ is a Dynkin diagram of a simply-laced simple Lie algebra or a folding of it ($ADET$). This algebra encodes the mass ratios of the particles (or the relative crossover scales of the left and right movers) of the integrable QFT underlying it.

2. if the $\mathcal{G}$ graph is $ADET$ then the $\mathcal{H}$ graph (that is known to encode the "color" structure of the theory) can only be $ADET$ or at least an extended Dynkin diagram $\widehat{ADE}$ or a folding of it.

1   Proof of Statement 1

The Y-system (3.1) can be depicted on the *tensor product* diagram $\mathcal{G} \times \mathcal{H}$, organized in $\mathcal{G}$-*rows* reproducing the $\mathcal{G}$ diagram, and $\mathcal{H}$-*columns* reproducing the $\mathcal{H}$ diagram. We wish to prove that a physically relevant system of this kind can only exist for $\mathcal{G}$ in the set of simply-laced Dynkin diagrams or foldings of them $ADET$. *Physically relevant* means that we are interested in systems of the kind (3.1) that allow an interpretation in terms of a scattering theory; i.e. they must come from some TBA, able to reproduce sensible physical behaviours at UV and IR. As $Y_a^i(\theta) = e^{\varepsilon_a^i(\theta)}$ and the asymptotic behaviours for large R and large θ of the pseudoenergies are driven by the corresponding energy terms $\nu_a^i(\theta)$, the possible behaviours of $Y_a^i(\theta)$ are of the following two types:

$$Y_a^i(\theta) \underset{R,\theta\to+\infty}{\sim} \begin{cases} e^{m_a^i R e^\theta/2} & \text{if the node } (a,i) \text{ is massive or left mover} \\ e^{const.} & \text{if the node } (a,i) \text{ is magnonic or right mover} \end{cases} \tag{3.2}$$

In what follows we denote the nodes having the massive or left mover behaviour as *black*; those with magnonic or right mover behaviour as *white*.

If a system of type (3.1) has white nodes only, i.e. if the corresponding TBA has all the $\nu_a^i(\theta) \equiv 0$, it degenerates completely to have the totally trivial solution $c(r) \equiv 0$. Therefore a physically sensible Y-system must have at least one black node. We now prove that if a $\mathcal{G}$-row contains a black node, then all the nodes in that row must be black. Indeed, consider a would-be white node a on the same $\mathcal{G}$-row, connected to the black node b. Connected means that $\mathcal{G}_{ab} \neq 0$. On the left hand side of (3.1) we then have a constant asymptotic behaviour. On the right hand side the product over $\mathcal{H}$ also behaves as a constant irrespective of the color of the nodes adjacent to a in the $\mathcal{H}$-column direction. However, the product over $\mathcal{G}$ contains at least one term (the one connecting a with b) that has the exponential behaviour typical of black nodes. The constant behaviour is exponentially depressed compared to the "black" one, therefore, comparing terms of order e^θ in the asymptotic of (3.1) we would get the condition

$$0 = m_b^i R e^\theta/2 \tag{3.3}$$

which would imply that the b node has zero mass, in contradicion with the hypotesis that it is black. Hence we conclude that for a given $\mathcal{G}$-row the nodes are all black or all white.

We said that the Y-system must have at least one black node: this implies that indeed there must be at least one black $\mathcal{G}$-row. With this result at hand, we proceed further by picking up a black $\mathcal{G}$-row for a fixed $\mathcal{H}$ index i, and considering the asymptotics of system (3.1) for this row. As before, the $\mathcal{H}$ product on the r.h.s. drops having constant behaviour, while the $e^\theta/2$ terms give the asymptotic consistency condition

$$2m_a^i \cos\frac{\pi}{g} = \sum_{b=1}^r \mathcal{G}_{ab} m_b^i \tag{3.4}$$

¿From this equation some very important facts follow:

1. if $\mathcal{G} = A_1$, i.e. the 1×1 matrix 0, then the r.h.s. of (3.4) vanishes. For the l.h.s to vanish too, we must have $g = 2$.

2. if $\mathcal{G}$ is the incidence matrix of a non-trivial connected graph, then the r.h.s. of (3.4) is positive, therefore, to have a positive l.h.s. we must require $g > 2$.

3. for $2 < g < \infty$, as $\cos \frac{\pi}{g} < 1$ we have the condition

$$\sum_{b \in \mathcal{G}} \mathcal{G}_{ab} m_b^i < 2m_a^i \tag{3.5}$$

where the vector $m^i = (m_1^i, m_2^i, ..., m_r^i)$ has all positive components. This is exactly the condition that selects, among all possible connected graphs the list of simply-laced Dynkin diagrams and their possible foldings,

$$\mathcal{G} = A_n \,, \; D_n \,, \; E_{6,7,8} \,, \; T_n = A_{2n}/Z_2 \tag{3.6}$$

4. As the vector m^i is an eigenvector of $\mathcal{G}$ with all non negative components, it must be proportional to the (unique) Perron-Frobenius eigenvector $\psi^{(\mathcal{G})}$ of $\mathcal{G}$

$$m_a^i = m^i \psi_a^{(\mathcal{G})} \tag{3.7}$$

for each black $\mathcal{G}$-row. Moreover, g plays the role of the (dual) Coxeter number of $\mathcal{G}$, hence it is a positive integer.

2 Proof of Statement 2

Now we prove another statement, i.e. if $\mathcal{G}$ is a simply-laced $(ADET)$ Dynkin diagram then $\mathcal{H}$ is or in the set $ADET$ too, or in the set of extended simply-laced Dynkin diagrams or their foldings. This can be done by generalizing an argument given in [14]. Consider stationary solutions (i.e. indipendent on θ) of the system (3.1). We know that these solutions enter the calculation of the UV central charge and in particular that at least one solution with real ε_a^i must occur (for the simplest cases one can prove that this real solution exists and is unique, we shall assume this as true for all cases in what follows). Reality and finiteness of the ε_a^i implies that $y_a^i > 0$. Then

$$2 \log y_a^i = \sum_b \mathcal{G}_{ab} \log(1 + y_b^i) - \sum_j \mathcal{H}_{ij} \log(1 + 1/y_a^j) \tag{3.8}$$

The quantity $z_a^i = \log(1 + y_a^i)$ is real and strictly positive, which implies that it satisfies $\sum_b \mathcal{G}_{ab} z_b^i < 2z_a^i$, as $\mathcal{G}$ is in the list of $ADET$ Dynkin diagrams. This allows, after few manipulations, to obtain the inequality

$$\sum_j \mathcal{H}_{ij} w_a^j < 2w_a^i \tag{3.9}$$

where $w_a^i = \log(1 + 1/y_a^i) > 0$. This proves that $\mathcal{H}$ must also be in the list of $ADET$ Dynkin diagrams. It can happen that the system degenerates to have some $y_i^a = 0$ i.e. $\epsilon_i^a = -\infty$. In this case one can prove, using tecniques similar to those of the proof of statement 1, that if one node has $y_i^a = 0$, then all nodes have $y_i^a = 0$. In this case instead of (3.9) we get the equality $\sum_j \mathcal{H}_{ij} w_a^j = 2w_a^i$ selecting the extended Dynkin diagram $\widehat{A}_n, \widehat{D}_n, \widehat{E}_{6,7,8}$ or the vaste set of their foldings. We have to mention here also this somehow degenerate case, as it has physical applications, e.g. in [19].

3   Some Useful Formulas on the Y-System

Many authors converge on the opinion that the Y-system seems to encode a great amount of information about the $IQFT_2$ it is attached to. In particular the connection between the Y-system and the Rogers Dilogarithm function seems to be very productive in reproducing the properties of the UV and IR limits of the integrable model under consideration [20]. One can extremize this point and think to a *reconstruction program*, i.e. to an answer to the question: given a Y-system and the asymptotic behaviour of its solutions for $\theta, R \to +\infty$, can we reconstruct completely an $IQFT_2$ underlying it? Is this theory unique? This program is up to now very far from completion. However many facts indicate that it is not hopeless. We shall not deal with this very interesting aspect in this paper; we shall be content to comment here about the two simplest quantities that can be extracted from a Y-system and their relation to the graphs $\mathcal{G} \times \mathcal{H}$.

1. The stationary solutions Y_a^i of the Y-system allow to compute the central charge via Dilogarithm sum rules. The main ingredient is the sum

$$s(\mathcal{G} \times \mathcal{H}) = \sum_{a \in \mathcal{G}} \sum_{i \in \mathcal{H}} \mathcal{L}\left(\frac{1}{1+Y_a^i}\right) = \frac{\pi^2}{6} \frac{rk}{g+h} h \qquad (3.10)$$

   Here $\mathcal{L}(z)$ denotes the Rogers Dilogarithm function, $g = cox(\mathcal{G})$ and $h = cox(\mathcal{H})$. For $\mathcal{G} = A_r$ and $\mathcal{H} = A_k$ this sum rule is proven ([21]), we checked it numerically to very high precision up to $r = rank(\mathcal{G})$ and $k = rank(\mathcal{H})$ both equal to 50 for the remaining cases, for which this sum rule is new and is proposed to the mathematicians for a rigorous proof.

2. The solutions of the Y-system possess a periodicity $Y_a^i(\theta+i\pi P) = Y_a^i(\theta)$ $(P \in \mathbf{Z}/g))$ In [11, 4] arguments are given to relate P to the conformal dimension $\Delta_P = 1 - 1/P$ of the UV perturbing operator and that of the IR attracting operator if it is the case. We checked numerically the following formula for the periodicity to be valid for all the $\mathcal{G} \times \mathcal{H}$ Y-sytems

$$P = \frac{g+h}{g} \qquad (3.11)$$

4   Zoology of $\mathcal{G} \times \mathcal{H}$ $IQFT_2$

Armed with the properties and formulas of the preceeding section, one can explore the whole set of $ADET \times ADET$ cases. Here we only briefly report on a very preliminary exploration where we only computed UV and IR central charges, thanks to (3.10) and conformal dimensions of UV perturbing and IR attracting operators, thanks to (3.11). The standard way to extract these data from the $\mathcal{G} \times \mathcal{H}$ Y-system (or TBA) is widely explained in the literature, see e.g. [12, 4]. In the following we just comment on some interesting cases. Here $\mathcal{G} = ADE$.

- $\mathcal{G} \times A_{k+l-1}$ is the case discussed in [18]. It corresponds to the best studied class of coset models, namely $\mathcal{G}_k \times \mathcal{G}_l/\mathcal{G}_{k+l}$, perturbed by the operator $\phi_{adj}^{id,id}$. We refer to [18] for the details.

- A complete list of all the $A_1 \times ADET$ cases is given in [14].

- A set of interesting cases where one is able to reconstruct the S-matrix, that turns out to be diagonal, is $\mathcal{G} \times T_1$ [22]. All the cases involving the T_n diagrams correspond to non-unitary theories. There are curious developments in this direction [23], in particular for $T_1 \times A_k$.

- A partially new and interesting case is $\mathcal{G} \times D_k$ with a massive energy term on one node of the fork, or left and right movers on the two nodes of the fork. This case generalizes to higher $\mathcal{G}$ what has been discussed in [17] for $A_1 \times D_k$. In that paper this TBA has been used to describe Z_k parafermionic theories perturbed by their $\psi\overline{\psi}+\psi^\dagger\overline{\psi}^\dagger$ operators (ψ being the generating parafermion). In the massless direction these theories flow to an IR limit given by the minimal models $(A_1)_k \times (A_1)_1/(A_1)_{k+1}$. Taking the limit for large k one defines a flow from $c = 2$ to $c = 1$ which is interpreted as the $O(3)$ massless sigma model with theta term $\theta = \pi$, which renormalizes at IR to the $SU(2)_{k=1}$ WZW model. One would search for a generalization of this class of flows to higher $\mathcal{G}$. In a certain sense the most obvious candidate is exactly this set $\mathcal{G} \times D_k$. With the choice of left and right movers on the two nodes of the fork, this TBA does indeed reproduce an IR limit towards $\mathcal{G}_k \times \mathcal{G}_1/\mathcal{G}_{k+1}$. The UV CFTs, however, are not the $\mathcal{G}_k/U(1)^r$ as one would naively expect by generalizing the A_1 case, but better the list of so called *Normal Forms* over $\mathcal{G}$ (for a definition and a list, see the appendix in [24]). This suggests that the corresponding sigma models defined as the $k \to \infty$ limit of these series are *quantum* integrable and that it is possible to add some kind of topological term transforming the model in a massless one with IR limit the $\mathcal{G}_{k=1}$ WZW model. However, it seems that these sigma models (apart the case $A_1/U(1)$) do not allow topological terms [25] and, to our knowledge, even the quantum integrability is an unclear issue here. Moreover, another fact seems to invite to deepen the study of these models. In the A_1 case, it has recently proposed a so called *sausage model*, i.e. a deformation of the target space of the $O(3)$ sigma model defining a one parameter family of integrable sigma models. A TBA has been proposed for a discrete set of values of this parameter, encoded on $A_1 \times \widehat{D}_k$. The main feature is that all the UV central charges of these TBA's are equal to 1, irrespective of the rank of the D diagram, thus allowing the sausage interpretation. Now, if this interpretation can be extended to higher $\mathcal{G}$, one should expect constant central charges, usually equal to some integer. This is not the case for all $\mathcal{G} \times \widehat{D}_k$ with

$\mathcal{G}$ other than A_1. The generalized sausage models, if they exist, have nothing to do with the sigma models on the normal forms of $\mathcal{G}$. We intend to return on this puzzling problem in the future.

To conclude, we have explored here a wide class of sets of TBA equations, that incorporate almost all examples shown so far. We have determined definitively the scope of this class, and, exploring the zoology of the models encoded in it, we have found repeated regularities but also some unclear problems to be investigated further. Of course, we are conscious that this class of IQFT$_2$'s does not exahust all the possible deformations of RCFT's. A well known counterexample is the set of $\mathcal{G}_k \times \mathcal{G}_l / \mathcal{G}_{k+l}$ CFT deformed by their $\phi_{adj}^{id,id}$ operator, where $\mathcal{G}$ is a non-simply-laced algebra. This case needs a TBA, and then a Y-system, which is not included in the set described above. The most reasonable generalizations one can think are

- to allow *different* diagrams $\mathcal{H}_a$ on different nodes of $\mathcal{G}$

- to allow for shift terms on the right-hand-side of the Y-system too.

This is indeed the case for the non-simply-laced $\mathcal{G} \times \mathcal{G}/\mathcal{G}$. We shall report on this in a forthcoming publication.

Acknowledgements – We are grateful to P.Dorey, P.Frè F.Gliozzi, A.Kirillov, G.Sotkov, M.Stanishkov and A.Valleriani for useful discussions and help. R.T. thanks the Theory Group at Bologna University for the kind hospitality during various stages of this work.

References

[1] P.Goddard, A.Kent and D.Olive, *Comm. Math. Phys.* 103 (1986) 105

[2] A.B.Zamolodchikov, JETP Lett. 43 (1986) 730

[3] A.B.Zamolodchikov and Al.B.Zamolodchikov, Ann. Phys. 120 (1979) 253

[4] Al.B.Zamolodchikov, *Nucl. Phys.* B358 (1991) 524

[5] A.B.Zamolodchikov and Al.B.Zamolodchikov, *Nucl. Phys.* B379 (1992) 602

[6] P.Fendley, H.Saleur and Al.B.Zamolodchikov, *Massless Flows I* and *II*, hepth 9304050-51, to appear in Int. J. Mod. Phys. **A**

[7] P.Fendley and H.Saleur, *Massless integrable quantum filed theories and massless scattering in* $1 + 1$ *dimensions*, Los Angeles preprint USC-93-022

[8] R.Dashen, S.-K.Ma and H.J.Bernstein, Phys.Rev. 187 (1969) 345

[9] C.N.Yang and C.P.Yang, J. Math. Phys. **10** (1969) 1115

[10] Al.B.Zamolodchikov, *Nucl. Phys.* B342 (1990) 695

[11] Al.B.Zamolodchikov, *Phys. Lett.* B253 (1991) 391

[12] T.Klassen and E.Melzer, *Nucl. Phys.* B338 (1990) 485

[13] T.Klassen and E.Melzer, *Nucl. Phys.* B350 (1991) 635

[14] F.Ravanini, R.Tateo and A.Valleriani, Int. J. Mod. Phys. **A8** (1993) 1707

[15] Al.B.Zamolodchikov, *Nucl. Phys.* B358 (1991) 497

[16] Al.B.Zamolodchikov, *Nucl. Phys.* B366 (1991) 122

[17] V.A.Fateev and Al.B.Zamolodchikov, *Phys. Lett.* B271 (1991) 91

[18] F.Ravanini, *Phys. Lett.* B282 (1992) 73

[19] V.Fateev, E.Onofri and Al.B.Zamolodchikov, LPTHE preprint 92-46

[20] W.Nahm, A.Recknagel and M.Therhoeven, Bonn preprint, hep-th/9211034

[21] A.Kirillov, Newton inst. preprints, hepth 9211137 and 9212150

[22] F.Ravanini, R.Tateo and A.Valleriani, *Phys. Lett.* B293 (1992) 361

[23] F.Ravanini, M.Stanishkov and R.Tateo, in preparation

[24] M.A.Olshanetskii and A.M.Perelomov, Phys. Rep. **94** (1983) 313

[25] A.M.Perelomov, Physica **4D** (1981) 1

REMARKS ON TOPOLOGICAL STRING THEORIES[a]

E. RABINOVICI

University of Jerusalem
Jerusalem, Israel

Abstract: We discuss various issues in topological string theory. These include the relation between world sheet and target space topology and the emergence of a rigid metric and signature in a 2-d topological theory of induced gravity. This is a consequence of a vacuum choice.

1 Motivation

The study of universal properties of gauge theories has led to a qualitative understanding of the interactions appearing in the standard model. The confining, higgs and coloumb phases of gauge theories manifest themselves in the color, weak and electromagnetic interactions (oblique confinement has yet to find a place). This understanding was achieved before an exact solution of either system is available. Universal properties of systems invariant under general coordinate invariance are much less known. One is aware that the symmetry can be realized locally in some cases without introducing a metric tensor [1]. It is not clear if one may view theories with and without metrics in a Wilsonian manner as different phases of a larger system.

In gauge theories the softer short distance properties of massive gauge theories enables the renormalization of the weak-interaction. An analogous structure in gravitational theories would provide a relief to the short distance behavior of gravity.

Unfortunately while it is very easy to identify topological theories as extreme infra-red limits of systems (such as 2-d QCD), similar results do not emerge in the ultra-violet. String theory provides hints that more symmetries appear at short distances however

[a]Work supported in part by the Israel Research foundation and the American-Israel Bi National Science foundation.

this has not yet been understood in terms of the phases structure of some effective field theory. In this talk I will describe various attempts [2, 3, 4] to learn something about this problem. I will mainly emphasize a study of induced gravity in topological systems.

1 World Sheet Topology vs. Target Space Topology

While topological theories exist in many dimensions, there properties are extensively analyzed in 2-d dimensions. String theory actually promotes the importance of 2-d systems as they can be used on the world sheet in order to generate target space theories. In particular the question was posed [2] if topology on the world-sheet implies topology in the target space. To answer this question one studies [2] generic theories with N observables termed topological sigma models. Theories with N observables, $O_\alpha\{\alpha = 1,\ldots,N\}$, have an operator product expansion of the form

$$O_\alpha O_\beta = C_{\alpha\beta}^\gamma O_\gamma \tag{1.1}$$

with $C_{\alpha\beta}^\gamma$ obeying commutativety and associativity constraints. Generically one can diagonalize this basis and bring it to the form

$$O_i O_j = \delta_{ij} O_j \tag{1.2}$$

(This can not be done in the presence of a charge for the operators O_α, for example $n = 2$ susy systems posses such a charge). In any case one can draw the N dimensional moduli space of theories with N observables by adding operators of the form

$$\sum_i \mu_i \int d^2 \times \sqrt{-g} R O_i \tag{1.3}$$

to a topological lagrangian. Note the analogy and difference to the marginal operators in conformed field theory (non-topological) obtained by

$$\sum_i \mu_i \int d^2 \times \sqrt{-g} O_i \tag{1.4}$$

One solves the generic theory on a world sheet with a general genus g to obtain "exact" correlation functions in a string theory by summing over all g.

$$\langle\langle (O_i)^n \rangle\rangle_{\text{exact}} = \frac{\mu_i^2}{\mu_i - 1} \tag{1.5}$$

one can show that:

1. These exact correlation functions obey the axioms of a topological theory.

2. These exact correlation functions can be reproduced by some 2-d, $g = 0$, topological theory (or actually any single g theory) with N observables. While the original theory was characterized by the point $(\mu_1,\ldots,\mu_N)$ in moduli space, the $g = 0$ theory is situated at the point $(\frac{\mu_1^2}{\mu_1-1},\ldots,\frac{\mu_N^2}{\mu_N-1})$ in the same moduli space.

3. For $N = 2$ one can construct explicitly the action of the effective 2-d topological target space theory.

4. One can also show that different pair of points in moduli space correspond to the same theory (an embryo duality). This is true genus by genus for some pairs, additional pairing occurs for the exact theory.

All in all we have shown that topology on the world sheet corresponds to topology in target space for these models. Moreover any such model could be obtained itself as an effective target space theory. There is a caveat to these results, they assume the exact correlation functions are obtained by summing over connected Riemann surfaces. Allowing for disconnected surfaces results in correlation functions which do not obey the axioms of topological field theory. The topology can be restored if the target space theory itself includes a sum over disconnected spaces. Requiring this leads to third quantization in target space. One should also note that the correlation functions have only a finite radius of convergence in moduli space. The significance of the singularity is still unclear.

It was next found that also the topological Chern–Simmons theory is an effective string theory [11]. The corresponding world sheet theories were related to type A and type B $N = 2$ super symmetric systems on Calabi–Yau manifolds with boundaries. This is an open string theory.

2 Induced Gravity in 2-d Topological Theories [3]

A central problem is the search for circumstances in which a metric will pop-out. Induced gravity models [5, 6] are a suitable arena for such a search. These models posses ab-initio the general coordinate invariance symmetry without the necessity to introduce a metric. By some definition this would be a topological theory. One then proceeds, classically, by assuming a flat potential permits an expansion around a non-symmetric classical configuration. The spontaneous breaking of a higher symmetry (including a gauge group G) results in the formation of a condensate which has the appropriate index structure to serve as a metric. On such configurations a general coordinate transformation is implemented by a gauge one and actually the metric on the gauge group G induces a space-time metric.

This semi-classical approach is in general not sufficient to investigate the full quantum structure. Some effort has been devoted to investigating 2 and 3-d models - mainly CS for 3-d [6] and ϕF for 2-d [7] where the theory may be solved exactly at the quantum level. Here we reconsider the topological 2-d ϕF theory, discussing in some detail its gauge structure and determining the full quantum wave functional. In topological theories all wave functionals have zero energy and thus they will indeed have a flat potential. One can ask as to what are the metric properties of each wave functional. For some special groups the ϕF has a gravitational interpretation. We start by discussing the ϕF as a gauge theory.

2 General Two Dimensional ϕF Theory - Gauge Theory Aspects

1 The Classical ϕF Model

The topological two dimensional ϕF gauge theory of a Lie group G consists of a gauge connection A in the adjoint representation as well as $dim(G)$ real scalar fields transforming under the adjoint representation of G.

The topological action is given by:

$$S = \int_\Sigma Tr(\phi \cdot F) d^2 x \tag{2.1}$$

where $F = dA + A^2$ is the field strength corresponding to the gauge connection A and $tr(XY) = \eta_{ab} X^a Y^b$ is a real invariant bi-linear form on the Lie algebra of G. Σ is the two dimensional world sheet.

The equation of motion are:

$$F^a_{\mu\nu} = 0 \tag{2.2}$$

$$D_\mu \phi^a = 0 \tag{2.3}$$

where $D_\mu \phi^a$ is the covariant derivative of ϕ

$$(D_\mu \phi)^a = \partial_\mu \phi^a + f^a_{bc} A^b_\mu \phi^c \tag{2.4}$$

f^a_{bc} being the structure constants of G.

The topological theory can be studied by considering various properties of the moduli space of the equations of motion [7]. In particular the partition function measures some appropriately chosen volume of the moduli space (which depends on the group G and the genus of Σ).

2 Quantization of the $\phi \cdot F$ Model

We perform a canonical quantization in the $A^a_0 = 0$ gauge. From now on A will denote A_1.

In order to be able to apply a Hamiltonian formulation we chose Σ to be $R^1 \times S^1$.

¿From eq. 2.1 it follows that the gauge fields A and the scalar fields ϕ are canonical conjugate to each other, the equal time commutation relations they satisfy are:

$$[\phi^a(x), A^b(y)] = -i\hbar \eta^{ab} \delta(x - y) \tag{2.5}$$

we can thus choose a complete set among the (ϕ^a, A^a) variables.

Some of the results of quantization can be anticipated by recalling that the two dimensional $\phi \cdot F$ theory eq. (2.1) is a dimensionally reduced 3-dimensional Chern-Simon theory (CS) of the same gauge group G defined by the action

$$S = \frac{k}{4\pi} \int_M Tr(A \wedge dA - \frac{2}{3} A \wedge A \wedge A) \tag{2.6}$$

on a three-manifold $M = \Sigma \times S^1$ (Σ is the 2-d manifold of eq. (2.1)). The theories described by eq. (2.5) and (2.6) are both topological, the reason a scale dependent procedure such as dimensional reduction retained the topological nature of the theory is that dimensional reduction results actually by considering the large k limit of the CS theory in eq. (2.6), [8].

In that limit, only configurations of the type eq. (2.5) contribute as long as G has no center. For a general G

$$Z_{cs}(G, k) \to Z_{\phi \cdot F}(G) exp(a\chi(\Sigma)) \# Z(G) \qquad \text{for k} \to \infty \qquad (2.7)$$

where a does not depend on Σ, $\chi(\Sigma)$ is the Euler number of Σ and $\# Z(G)$ the number of elements in the center of G. Now, it is known [9] that the Hilbert space of the CS model eq. (2.6) consists (on $\Sigma = T^2$) of the conformal blocks of the affine Lie algebra of G. In particular in the large k limit, to each representation of G corresponds one state in the Hilbert space.

Thus the Hilbert space of the $\phi \cdot F$ theory eq. (2.1) should also be in one to one correspondence to the representations of G. This is verified by an explicit calculation of the Hilbert space. In addition, in the CS case the phase of the wave functional had an interesting structure, it was identical to the action of the chiral $WZNW$ model [9], [10]. In the $\phi \cdot F$ case the phase will also have a geometrical interpretation.

Another consequence of the relation to the CS model is that $\phi \cdot F$ models are also effective string theories. This follows from the fact that CS is a string theory [11] for any k in particular for $k \to \infty$. This theory has an infinite number of observables.

The ϕF theory being topological requires all wave functionals to have zero energy. The wave functionals themselves are chosen so that they are invariant under the residual gauge transformations in the $A_0^a = 0$ gauge. This imposes constraints on the states.

It is possible [7] to represent the wave functionals in an A^a basis. A complete basis for the allowed wave functionals are the characters $Tr_r(w)$, where r is an irreducible representation of G, w is the Wilson loop constructed from the connection A by:

$$w = Pexp \oint dx' A(x') \qquad (2.8)$$

where P denotes a path ordered exponential taken around the circumference of space S^1.

For our goal to connect the theory to gravity and to the emergence of a target space, we find it instructive to construct the wave functional in the ϕ^a basis.

While in CS theories the two possible conjugate coordinates A_1^a and A_2^a had the same properties under gauge transformations, A^a and ϕ^a in 2.5 have different transformation properties under the gauge group G. If we denote $\Psi(\phi)$ the wave functional in the $\phi^a(x)$ basis the operators $A_a(x) = \eta_{ab} A^b(x)$ act upon $\Psi(\phi(x))$ as $i\frac{\delta}{\delta \phi^a(x)}$.

The generators of the residual gauge symmetries are given by the spatial component of eq. (2.4)

$$D_1 \phi^a = \partial_1 \phi^a + f_{bc}^a A^b \phi^c. \qquad (2.9)$$

They are obtained by varying eq. (2.5) with respect to A_0^a. We will suppress the spatial index in the derivative term.

When expressed in terms of the basis $\phi^a(x)$, the fact that physical states should obey Gauss's law is represented by:

$$G^a(x)\Psi(\phi(x)) = [\partial\phi^a(x) + if_b^{ca}\phi^b\frac{\delta}{\delta\phi^c}]\Psi(\phi(x)) = 0. \tag{2.10}$$

Eq. (2.10) constrains the configurations $\phi(x)$ for which $\Psi(\phi(x))$ does not vanish.

It turns out $\Psi(\phi(x))$ does not vanish only for configurations $\phi(x)$ which are of the form

$$\phi(x) = h^{-1}(x)\phi_D h(x) \tag{2.11}$$

where ϕ_D are constant in space and belong to the Cartan sub algebra C of G or rather, to a specific Weyl chamber in C. The metric $h(x)$ stands for a general element of G. In particular, every Casimir invariant polynomial of the Lie algebra, when calculated in terms of $\phi(x)$, which are in the support of $\Psi(\phi(x))$, is x independent. This is reasonable for a topological theory. The Casimir invariant polynomials as observables in a topological theory should have space independent correlation functions. One can solve explicitly the remaining dependence of $\Psi(\phi(x))$ on $\phi(x)$. For example for the case $G = SU(2)$, one can show that the wave functional

$$\Psi(\rho,\sigma,\alpha) = exp(\oint \frac{\rho\partial\sigma - \sigma\partial\rho}{M^2 - \alpha^2}\alpha dx) \quad F(M^2) \tag{2.12}$$

satisfies eq. (2.10) where

$$\phi = \phi_1\tau_1 + \phi_2\tau_2 + \phi_3\tau_3$$
$$\rho = \frac{\phi_1 + i\phi_2}{\sqrt{2}}$$
$$\sigma = \frac{\phi_1 - i\phi_2}{\sqrt{2}}$$
$$\alpha = \phi_3$$
$$M^2 = \phi_1^2 + \phi_2^2 + \phi_3^2 = 2\rho\sigma + \alpha^2$$

τ_i being the Pauli matrices (in this case $\frac{1}{2}Tr\phi_D^2 = M^2$ an d thus M is constant). F is an arbitrary function of M^2 (arbitrary up to some normalization requirement on the wave function).

In order to proceed for a general G it will be useful to be a little less explicit on the dependence of Ψ on the angular field variables. The solution will be obtained along the lines of the determination of the wave functional in CS theory [10] respecting the differences between the cases.

To solve eq. (2.10) on admissible configurations (of the type eq. (2.11)) recall that Gauss' laws are the infinitesimal generators of time independent gauge transformations. Indeed, under a gauge transformation $g(x)$, the operators $\phi(x)$ and $A(x)$ transform as:

$$\phi(x) \rightarrow g(x)\phi(x)g^{-1}(x) \tag{2.13}$$

$$A(x) = i\frac{\delta}{\delta\phi(x)} \rightarrow g(x)i\frac{\delta}{\delta\phi(x)}g^{-1} + g(x)\partial g^{-1}(x). \tag{2.14}$$

To implement these transformations, the unitary operator, U_g, acting on Ψ, corresponding to $g(x)$ should be defined by:

$$(U_g(x)\Psi)(\phi(x)) = exp(-iTr \int gdg^{-1}\phi)\Psi(g^{-1}(x)\phi(x)g(x)). \qquad (2.15)$$

In eq. (2.15) the transformation of the argument of Ψ gives rise to the homogeneous part of eq. (2.13), while the phase factor in front is needed to produce the inhomogeneous part in eq. (2.14).

In particular, for a configuration of the type of eq. (2.11) one obtains:

$$\Psi(\phi(x) = h^{-1}(x)\phi_D h(x)) = exp(iTr \int (hdh^{-1}\phi_D))\Psi(\phi_D). \qquad (2.16)$$

This $\Psi(\phi(x))$ fulfills eq. (2.10).

At this stage it seems that the only constraint on the function $\Psi(\phi_D)$ is that it be normalizable in some appropriate measure. We will proceed to find additional constraints on $\Psi(\phi_D)$.

The solution as expressed in eq. (2.16) is less explicit than that of eq. (2.12), it requires to find the appropriate $h(x)$ for each $\phi(x)$. In fact $h(x)$ is defined by $\phi(x)$ only up to a left multiplication by an element of $T \times W$, where T is the subgroup generated by the Cartan sub algebra C and W is the Weyl group. If ϕ_D is fixed to be in a particular Weyl chamber there is left only an arbitrarily in T: $h(x)$ and $t(x)h(x)$ with $t(x)\varepsilon T$ correspond to the same $\phi(x)$. Therefore the phase factor in eq. (2.16) should be invariant under $h \to th$.

Comparing the phases of eq. (2.16) in both cases:

$$Tr(thd(th)^{-1}\phi_D) = Tr(hdh^{-1}\phi_D) + Tr(tdt^{-1}\phi_D) \qquad (2.17)$$

$t(x)$ can be written as

$$t(x) = exp(\sum_{i=1}^{r} \rho^i(x)T_i) \qquad (2.18)$$

thus invariance would result from eqs. (2.17) and (2.18) if

$$exp(iTr \int tdt^{-1}\phi_D) = exp(i\sum_i \int \partial\rho^i Tr(T_i\phi_D)dx) \qquad (2.19)$$

would be equal to the unity matrix. As ϕ_D is constant this would occur in particular for any periodic function $\rho^i(x)$, for any value of the constant ϕ_D.

However, the function ρ^i appearing in the parameterization of eq. (2.18) need not be periodic. All that is required is that $t(x)$ be a periodic solution.

For example, for a compact simply connected group, $t(x)$ is continuous whenever

$$\rho^i(2\pi R) = \rho^i(0) + \alpha^i \qquad (2.20)$$

where the r dimensional vector α belongs to the dual of the weight lattice of G and R is the radius of the spatial world sheet circle. (For non compact groups, restrictions hold, although there are not always lattices present). Therefore, for $\Psi(\phi)$ to be single valued under two different representations of the same $\phi(x)$ in terms of h and in terms of th,

ϕ_D has to belong to the weight lattice of G, modulo W, i.e. it corresponds to some highest weight. In particular, for $G = SU(2)$, for ϕ_D, which is of the form $\phi_D = \frac{iM}{2}\tau_3$ M has to be an integer. For that group $\Psi(\phi)$ is restricted to configurations for which

$$\phi_1^2 + \phi_2^2 + \phi_3^2 = M^2, \tag{2.21}$$

M is an integer.

The general observations is that, as anticipated, to each representation of G corresponds one state in the Hilbert space. For an allowed configuration $\phi(x) = h(x)\phi_D h^{-1}(x)$ the dependence on the angular variables $h(x)$ is fixed by the phase as appearing in eq. (2.16). To appreciate the geometrical significance of the phase, it can be recast as a two dimensional integral on a surface, D, whose boundary is s^1. The general form is:

$$\Psi(\phi(x) = h^{-1}\phi_D h) = exp(-i \int_{D(\partial D=L)} Tr(hdh^{-1} \wedge hdh^{-1}\phi_D))\Psi(\phi_D) \tag{2.22}$$

where this time the phase depends only on the values of $\phi(x)$ as continued into the disc.

For example for $G = SU(2)$ one obtains, writing $\phi(x)$ for which $\Psi(\phi(x))$ has support as:

$$\phi(x) = \frac{i}{2}(M \sin\theta(x)\cos\varphi(x)\tau_1 + M \sin\theta(x)\sin\varphi(x)\tau_2 + M \cos\theta(x)\tau_3). \tag{2.23}$$

$$\Psi(\theta(x), \phi(x)) = exp(-i\frac{M}{2}\int(1 - \cos\theta)d\phi)\Psi(\frac{i}{2}M\tau_3). \tag{2.24}$$

The wave functional is thus:

$$\Psi(\phi(x)) = exp(-i\frac{M}{2}\,(\text{area enclosed on the unit sphere}S^2 \tag{2.25}$$

$$\text{by the curve}\phi(x)))\Psi(\frac{i}{2}M\tau_3).$$

There are many ways to define a disc on S^2 with a fixed boundary. Their areas differ from each other by integer multiplies of the total area of the sphere. We thus reobtain the quantization condition

$$\frac{M}{2}4\pi = 2\pi \quad \text{integer} \quad \Rightarrow \text{M}\varepsilon\text{Z}. \tag{2.26}$$

Therefore, in general, the phase is just the appropriate area bounded by the configuration $\phi(x)$.

In the 3-d Chern-Simons case [9], [10] the phase of the wave functional itself correspond to a 2-d action, that of the chiral WZNW. In the present case the phase is again an action, this time of a quantum mechanical system [12] containing also only first order derivatives.

3 $\phi \cdot F$ Theory for Some Non-Compact Groups

The above analysis holds also for non-compact groups.

In particular we will consider $G = SO(2,1), SO(2,1) \times U(1)$, a contracted version of $SO(2,1) \times U(1)$ and $SO(2,2)$.

These theories are related, in some sense, classically to various two dimensional gravitational systems. This relationship will be discussed in greater detail.

Starting with $G = SO(2,1)$, one notes that G has three types of adjoint orbits, the generators being elliptic, hyperbolic and parabolic. Denoting $\phi(x)$ by

$$\phi(x) = i\phi_1\tau_1 + i\phi_2\tau_2 + \phi_3\tau_3. \tag{2.27}$$

A configuration $\phi(x)$ can be brought to the form ϕ_D of eq. (2.11) depending on the sign of M^2, where now:

$$\frac{1}{2}tr\vec{\phi}^2 = \phi_3^2 - \phi_1^2 - \phi_2^2 = M^2 \tag{2.28}$$

M^2 remains x independent as we have shown. For a positive value of M^2, $\phi(x)$ can be brought to the form $M\tau_3$ (M real). The orbit given by eq. (2.28) represents a two sheeted hyperboloid, as the stability groups of such orbits are compact for $M^2 > 0$, the analysis in the compact case $G = SU(2)$ applies here as well, rendering M to be an integer. As $SO(2,1)$ contains no operation relating the two disjoint parts of the two sheeted hyperboloid, there exist two different classes of representations denoted by M positive or negative. They are in one to one correspondence to the discrete series representations, D_+, D_-, if $SO(2,1)$. Representations for which the invariant Casimir is negative and the eigenvalue of the Cartan sub algebra generator is bounded from either above or below.

Negative values of M^2, lead to orbits on the one sheeted hyperboloid, in which case the generator of the stability group is a hyperbolic, non compact operator. Thus $\phi(x)$ can be brought to the form $iM\tau_1$, no quantization is needed to ensure that $t = exp(\alpha\tau_1)$ be continuous. To every real value of M corresponds a basis element of the wave functional. The states are in one to one correspondence to the so-called principal series representations of $SO(2,1)$, which have no restriction on the value of the real Casimir. The orbits are classified by a fixed radius of the one sheeted hyperboloid.

For $M^2 = 0$ ϕ can be brought to the form $\phi = i\phi_1\tau_1 + \phi\tau_3$ and one distinguishes two cases. The first orbit is $\vec{\phi} = 0$, which is represented by the tip of the light one, the second orbit is the light cone itself. The first is related to the singlet representation of $SO(2,1)$, we have not found an orbit associated with the complementary representations of $SO(2,1)$.

The area phase, for positive M^2, can be chosen to be the finite area contained by $\phi(x)$. For negative values of M^2, the cases can be classified according to their winding number around the axis of the hyperboloid.

As for $SU(2)$, one can obtain the explicit solution of eq. (2.10) in terms of ϕ_1, ϕ_2, ϕ_3. It is with the definitions of eq. (2.12)

$$\Psi(\rho, \sigma, \alpha) = exp(\oint \frac{\rho\partial\sigma - \sigma\partial\rho}{M^2 + \alpha^2}\alpha dx)F(M^2). \tag{2.29}$$

Again to the phase one has to add a part $M \oint dx$ for $M^2 > 0$. For configurations with other values of M^2 it is redundant.

Let us next consider the group $G = SO(2,1) \times U(1)$, in that case it is enough to treat the $U(1)$ case, as the problem factorizes and the $SO(2,1)$ has just been studied.

For $U(1)$ the solution of eq. (2.10) is immediate, the field ϕ_4 associated to the $U(1)$ generator must be x independent. If the $U(1)$ gauge group is non compact, ϕ_4 may assume any value, for a compact $U(1)$, ϕ_4 will be appropriately quantized in units of the $U(1)$ radius.

Finally let us consider the twisted $SO(2,1) \times U(1)$ algebra (a Euclidean version of this was first treated in [13]).

One can supplement the $SO(2,1)$ algebra

$$[J_0, P_\pm] = \pm P_\pm$$
$$[P_+, P_-] = J_0$$

by another generator I of a $U(1)$ group making it an $SO(2,1) \times U(1)$ algebra, enabling a study of a twisted algebra of these four generators

$$[J_0, P_\pm] = \pm P_\pm$$
$$[P_+, P_-] = \lambda I.$$

I commutes with all generators. This is still a Lie algebra which may actually be reached by a continuous deformation of the $SO(2,1) \times U(1)$ case. The two operators $C_1 = \lambda I$, $C_2 = P_+ P_- + \lambda I J_0$ are Casimir operators commuting with the algebra. Consider again the action $S = Tr \int \phi F$ built on that algebra. Here F is the curvature of the gauge field based on the above algebra and ϕ is a four component Higgs field in the adjoint representation:

$$\phi = \phi^+ P_+ + \phi^- P_- + \phi^0 J_0 + \phi^I I. \tag{2.30}$$

The symbol Tr denotes here the invariant bilinear form of the algebra [13]

$$Tr(\phi F) = \phi^+ F^- + \phi^- F^+ + \frac{1}{\lambda}(\phi^I F^0 + \phi^0 F^I). \tag{2.31}$$

Under a gauge transformation ϕ^0 is unchanged and the combination $\phi^+ \phi^- + \frac{1}{\lambda}\phi^0 \phi^I$ is also invariant. Gauss's law thus implies that the wave functional obtains support only on configurations for which

$$\phi^+ \phi^- + \frac{C}{\lambda}\phi^I = M^2 \tag{2.32}$$

C and M^2 (the defining radius of the parabolic hyperboloid) are constant in space. An invariant metric on our Lie algebra is

$$ds^2 = a^2(d\phi^+ d\phi^- + \frac{1}{\lambda}d\phi^0 d\phi^I) + b^2(d\phi^0)^2 \tag{2.33}$$

a and b are arbitrary constants. On a given orbit the metric will be just $a^2 d\phi^+ d\phi^-$. Every $\phi(x)$ configuration on such an orbit can be brought by a gauge transformation $h(x)$ to the x independent form: $h(x)\phi(x)h^{-1}(x) = \phi$ with, say, $\phi_D^\pm = 0$,

$\phi_D^0 = C$, $\phi_D^I = \frac{\lambda M^2}{C}$. The transformation h is defined modulo right multiplication by a transformation t which preserves ϕ_0 i.e. t is generated by J_0 and I with no $P_\pm$. The corresponding phase in the wave functional $Tr \int (hdh^{-1}\phi_D)$ is ambiguous by a phase

$$Tr \int (tdt^{-1}\phi_0) = Tr \int (d\alpha^0 J_0 + d\alpha^I I)(C J_0 + \frac{\lambda M^2}{C} I) \qquad (2.34)$$
$$= \int \frac{\lambda^2 M^2}{C} d\alpha^0 + \int \lambda C d\alpha^I.$$

Since J_0 generates a non compact group, the first term vanishes. If however, I generates a compact $U(1)$, then the second term may give $\int d\alpha^I = $ integer which imposes quantization on possible values of $\lambda \phi^0 = \lambda C$.

4 Some Gauge Theory Aspects

The $\phi \cdot F$ theory is a gauge theory, as it is also a topological theory all of the states in the theory have quantum mechanically zero energy thus falling in the category of flat potentials. For all groups G the symmetry group remains G if the system is chosen to be in the state corresponding to the singlet. (For example for $SU(2)$ this implies: $\Psi(\phi) = \delta(\phi)$). A generic result is that for a wave function supported on any other orbit the symmetry is reduced at most to $\prod_{i=1}^r H_i$. H_i are various one dimensional groups ($U(1)$'s for compact groups) and r is the rank of the group. Indeed, for compact gauge groups, a scalar as the ϕ field in the adjoint representation reduces G to a product of groups whose ranks add up to the rank of G, generically reducing G to $(U(1))^r$. For a flat potential G can either be maintained or broken down to the above factors.

The wave function supported on a fixed orbit allows $\phi(x)$ in general to vary in space. The identification of the residual H_i symmetries is more complex. For example one may employ the methods used for the monopole configuration [14]. This will be further discussed in relation to gravity.

Treating $\phi \cdot F$ as a gauge theory one may calculate it's order parameters. We have already demonstrated that the operator O, defined by

$$O = tr\phi^2(x) \qquad (2.35)$$

is x independent (as it should be in a topological theory) and it's expectation value is given by

$$< 0 > = \int |\Psi(\phi_D)|^2 (tr\phi_D^2) D\phi_D. \qquad (2.36)$$

Note that the phase of $\Psi(\phi(x))$ and with it all $h(x)$ dependence is eliminated. For $\Psi(\phi_D)$ on a fixed orbit ϕ_D^f

$$< 0 > = tr(\phi_D^f)^2. \qquad (2.37)$$

Another gauge theory order parameter is the value of the Wilson loop. In the Hamiltonian formulation the calculation of the average value of the Wilson loop reduces to an addition of two sources to the system.

This is done in the following manner:

Suppose the expectation value of a Wilson loop in the representation R is inserted. One deals then with the path integral:

$$Z(A) = \int D\phi DA_0 DA e^{Tr\phi F} Tr_R Pe^{\int_C A}. \qquad (2.38)$$

Choose the Wilson line C to run along the t axis at $x = 0$. Consider, first the Abelian case. The representation R is fixed then by the charge q of the loop and one computes the expectation of $exp(q \int A)$. In the gauge $A_0 = 0$, varying with respect to A_0 Gauss' law is changed into

$$[d\phi + q\delta(x)] = 0. \qquad (2.39)$$

Naturally the flux of the charged particle enters into Gauss' law and the wave function is supported on configurations of ϕ which is constant everywhere except at the position of the particle where it makes a jump of height q which represents the flux of the particle. If space is a compact circle, it must contain zero total charge and another particle of charge $-q$ has to be added, say, at the point $x = L$. The allowed configurations are then those with a constant $\phi = M$ everywhere except for the interval between the two opposite charges where the flux line between them is manifested as a shift in ϕ to the value $M - q$. Notice however that in this topological theory the flux line carries no energy density and does not really confine the sources.

For the non-abelian case a similar analysis holds. Tensionless strings forming as long as the sources have non zero "N-ality". The tension can be provided.

Actually recall that the theory described by:

$$S = \int_\Sigma d^2x Tr(\phi \cdot F + e^2 \sqrt{g}\phi^2) \qquad (2.40)$$

would give pure QCD in two dimensions once the ϕ degrees of freedom are integrated upon to yield

$$S = \int_\Sigma d^2x \sqrt{g} \frac{1}{e^2} Tr F^2 \qquad (2.41)$$

e^2 being a dimensionfull coupling.

The Hilbert space of the action given by eq. (2.40) is identical to the Hilbert space described the action of eq. (2.1) in the $A_0^i = 0$ gauge. This follows as the additional term $e^2 tr\phi^2$ is A_0 independent. It is however metric dependent and thus modifies the energy of the system in a way similar to a magnetic field in the Zeeman-effect as $tr\phi^2$ is constant the action is only area dependent. For any group G, the infinite degeneracy is lifted by the $e^2 tr\phi^2$ term.

The exact value of the energy can be renormalized, one expects, from pure $(QCD)_2$, that the renormalized energy is this case is not $tr\phi^2$, but the second Casimir. In any case the wave functionals themself are left unchanged.

For the $\phi \cdot F$ theory we have found that the wave functionals behave as in strongly coupled lattice QCD, in particular QCD is manifestly confining in that region of parameter space.

The formation of the flux lines, in the case when the inserted representation is sensitive to the group center, is the clear geometrical (stringy) feature of confinement.

The next step in any such evaluation is to measure the dependence of the energy difference between the energy of the system with inserted charges and without such charges on the distance between the sources. For confining systems it grows linearly with that distance. In particular when a chromo-electric flux line is formed among the sources confinement is manifest.

For the topological theory the energy of all states, with or without sources remains zero (if the sources can be coupled in a topological manner, which is possible in the $\phi \cdot F$ case) and thus no dependence of the distance between the sources can emerge. Indeed any such dependence would violate the topological nature. However the moment the $tr\phi^2$ term will be switched on, the flux line will have a fixed energy per unit length and the wave functionals in the presence of sources (which remain unchanged when $e^2 tr\phi^2$ was turned on) will now indeed signify that the confinement properties of pure $(QCD)_2$.

The scalar field ϕ playing the role of a chromoelectric field, whose constancy in $(QCD)_2$ is well known. The close relation between $\phi \cdot F$ and $(QCD)_2$ gives a hint to search the string representation of $(QCD)_2$ in a deformed CS theory which leads to the $Tr\phi^2$ term in the large k limit.

Let us now put forward a speculation on how a transition could occur which would increase the number of states. One of the difficulties in imagining a transition between a topological gravitational theory and a non-topological phase of gravity is that a non topological phase of gravity seems to consist of many more states. Imagine that a transition on the world sheet would result in tearing the world sheet, in particular in turning closed space to many open spaces. If this would occur in a theory of the type $\phi \cdot F$, it would result in the formation of many more states [3]. Exactly how many more is determined by the general nature of the group (compact or non compact) and on its exact specification. Is there a mechanism which would allow to tear the world sheet. It was pointed out [15] that one aspect of crossing the $c > 1$ barrier is the rapture of the world sheet. Similar phenomena occur at the K.T. transition. Above $c = 1$ the 2-d gravitational system changes from a zero number of local degrees of freedom system to one with a positive number of such degrees of freedom. Recall also that while for the 3-d CS theory on a closed manifold each conformal block is represented by one state in the Hilbert space, for an open manifold the whole conformed block is liberated. The Hilbert space contains an infinite number of states [16, 10].

3   Gravitational Aspects for the Gauge Theory

1   Review

We have solved quantum mechanically the topological gauge theory, the promise of an emergence of a metric actually appears classically. Thus, we retreat to the classical description and later return to search for a fulfillment of these classical promises in the quantum setting. In particular we discuss classically the case of the non-compact gauge group $SO(2,1)$. The theory has three generators which are denoted by P_1, P_2, J they fulfill the algebra:

$$[J, P_a] = 2i\varepsilon_a^b P_b$$
$$[P_a, P_b] = -i\Lambda\varepsilon_{ab}J.$$

For now we suppress the scale Λ and set $\Lambda = 1$.

As the theory has three gauge generators, it was suggested [7] to relate them to the parameters associated with the two dimensional diffeomorphisms and the local Lorentz transformation similarly to what was proposed earlier for a 3-d case [6]. This was shown [7] in the following manner.

Under a gauge transformation $\lambda = \lambda^i J_i$ (J_i denote here the group generators), the gauge fields, $A_\mu = A_\mu^i J_i$, and the scalar field, $\phi = \phi^i J_i$, transform infinitesimally respectively as:

$$\delta A_\mu = -\partial_\mu \lambda - [\lambda, A_\mu] \equiv -D_\mu \lambda$$
$$\delta \phi = [\lambda, \phi]$$

under a diffeomorphism $\delta x^\mu = \varepsilon^\mu(x)$ the fields transform as

$$\delta A_\mu = \varepsilon^\rho F_{\rho\mu} + D_\mu(\varepsilon^\rho A_\rho)$$
$$\delta \phi = \varepsilon^\rho D_\rho \phi - [\varepsilon^\rho A_\rho, \phi].$$

As the equation of motions are

$$F^{\mu\nu} = 0 \tag{3.1}$$

$$D_\mu \phi = 0 \tag{3.2}$$

gauge invariance guarantees general coordinate transformations on configurations which are solutions of the equations of motions; the gauge and coordinate transformations are related by:

$$\lambda^i = \varepsilon^\rho A_\rho^i. \tag{3.3}$$

That is, general coordinate transformations define configuration dependent gauge transformation (on the configuration $A_\rho^i = 0$, for example, $\lambda^i = 0$). This may be implemented for any gauge group G. Choosing two group directions ($i = 1, 2$) one may determine ε_ρ from λ^i if $det_{\rho i} A_\rho^i \neq 0$.

Moreover, it was suggested to define

$$A_\mu = e_\mu^a P_a + w_\mu J \tag{3.4}$$

such that e_μ^a, w_μ play the role of zweibeins and spin connections respectively. In general a metric $g_{\mu\nu}$ can be defined from the zweibeins by:

$$g_{\mu\nu} = e_\mu^a e_\nu^b B_{ab} \tag{3.5}$$

where B_{ab} is the $SO(2,1)$ group metric restricted to the transverse direction:

$$\begin{pmatrix} 1 & 0 \\ 0 & -1 \end{pmatrix}. \tag{3.6}$$

Introducing $P_\pm = P_1 \pm P_2$, we find from eq. (C1-1)

$$[J, P_+] = -2i P_+$$
$$[J, P_-] = 2i P_-$$
$$[P_+, P_-] = 2i J.$$

The field components are defined by

$$2\phi = \rho P_+ + \sigma P_- + \alpha J$$
$$2A_\mu = e_\mu P_+ + \lambda_\mu P_- + \beta_\mu J.$$

Thus also:

$$2F_{\mu\nu} = ((\partial_\mu e_\nu - \partial_\nu e_\mu) - i\beta_\mu e_\nu + i\beta_\nu e_\mu)P_+$$
$$+((\partial_\mu \lambda_\nu - \partial_\nu \lambda_\mu) + i\beta_\mu \lambda_\nu - i\beta_\nu \lambda_\mu)P_- \tag{3.7}$$
$$((\partial_\mu \beta_\nu - \partial_\nu \beta_\mu) + ie_\mu \lambda_\nu - ie_\nu \lambda_\mu)J = 0.$$

The equations of motions eq. (2.3) are explicitly

$$-\partial_\mu \rho + i\rho\beta_\mu - i\alpha e_\mu = 0$$
$$-\partial_\mu \sigma - i\sigma\beta_\mu + i\alpha \lambda_\mu = 0 \tag{3.8}$$
$$\partial_\mu \alpha - i\rho\lambda_\mu + i\sigma e_\mu = 0.$$

The two equations requiring the vanishing of the P_+ and P_- components of the electric field can be used to express the gauge field β_μ in terms of the other two gauge fields λ_μ and e_μ by:

$$\beta_\mu = \frac{e_\mu \varepsilon^{\rho\sigma}\partial_\rho \lambda_\sigma + \lambda_\mu \varepsilon^{\rho\sigma}\partial_\rho e_\sigma}{det(e,\lambda)} \qquad det(e,\lambda) \equiv \varepsilon^{\mu\nu} e_\mu \lambda_\nu. \tag{3.9}$$

Assigning to e_μ, λ_μ the zweibein roles, the J component in eq. (3.7) can be rewritten as:

$$R = 1 \tag{3.10}$$

where R is defined by analogy to be:

$$R \equiv \frac{\varepsilon^{\mu\nu}\partial_\mu \beta_\nu}{det(e,\lambda)}. \tag{3.11}$$

R looks as a Ricci curvature scalar. The gauge equations of motion can be reinterpreted as an equation for constant curvature which could have been derived from a purely gravitational theory of the form

$$S = \int \eta(R-1)\sqrt{g}d^2x \tag{3.12}$$

where η is a Lagrange multiplier [17]. The reinterpretation of the symmetries of the equations of motion and the possibility to recast the gauge equations in a gravitational form suggest the emergence of a metric in the topological gauge theory.

Before trying to understand what this classical metric emergence means quantum mechanically we wish to reinstate a role for the scalar fields ϕ^i in the analysis. First note that the equations of motion eq. (3.8) require that

$$tr\phi^2 \equiv M^2 = 2\rho\sigma - \alpha^2 \tag{3.13}$$

be constant in both space and time. In section B we have seen that the spatial components of eq. (3.8) were the Gauss's law in the $A_0 = 0$ gauge so that

$$\partial_1(M^2) = 0 \tag{3.14}$$

was also valid at the quantum level. The vanishing of $\partial_t M^2$ follows also quantum mechanically from the vanishing of the Hamiltonian.

2 Constructing an H Invariant Metric

Following the gauge theory point of view, as expressed in the former section, any classical choice of a configuration ϕ of a given "length" M^2, would result in the "breaking" of the group $G = SO(2,1)$ down to a one generator sub group H of $SO(2,1)$. As the Hamiltonian vanishes identically, the classical "potential" is flat and any configuration ϕ^a of given length M^2 is allowed. For $M^2 > 0$ $SO(2,1)$ could be viewed as broken to a compact $U(1)$, while for $M^2 < 0$ the residual symmetry is a non compact $O(1,1)$.

In both cases the axis of the remaining symmetry is determined by the configuration ϕ^a. (The $M^2 = 0$ cases will be treated separately). Realizing that residual $SO(2,1)$ gauge invariance will require a well defined superposition of all ϕ^a on an orbit of a given allowed M^2, we follow the original suggestion as used for the monopole configuration [14].

The gauge fields are decomposed in terms of a longitudinal coordinate, parallel to the ϕ^a field and two transverse coordinates in the orthonormal directions.

Making use of Gauss' law one obtains an H invariant metric

$$g_{\mu\nu} = \frac{Tr(\partial_\mu\phi\partial_\nu\phi)}{2M^2}. \tag{3.15}$$

Let us comment on this result. We notice first that the induced metric is not only locally invariant under H but also globally under G. Since the surface $Tr\phi^2 = M^2$ embedded in the 3 dimensional ϕ space with invariant metric, has a constant curvature proportional to $\frac{1}{M^2}$, the metric in eq. (3.15) induced on the world sheet has a constant, M independent curvature. Notice that for $M^2 > 0$ it has Euclidean signature while Minkowski signature emerges for $M^2 < 0$.

We could have defined the zweibeins, or the metric, by multiplying $\partial_\mu\phi$ by an arbitrary function of M (the only H invariant function) to obtain any (possibly M dependent) constant value for the curvature.

Let us now pause to consider what actually is the extent of the classical promise for a formation of a metric.

As $M^2 \equiv \frac{1}{2}tr\phi^2$ is constant, one may use it to classify ϕ^a configurations. For a given value of M^2, all non constant configurations lead, by the equations of motion, to induced world sheet metrics which differ only by a gauge transformation, for non singular metrics they differ only by a diffeomorphism as well. These metrics, when invertible, lead all to the same constant curvature, that induced by the group. Singular metrics lead to locally ill-defined curvatures, for constant configurations, the metric vanishes globally as well.

Thus classically, up to diffeomorphism and to topological considerations, there exists only one induced metric, the theory allows the existence of a fixed metric, a metric with some singularities as well as of no metric.

The quantum theory can either wash out or maintain that structure, it may also add quantum fluctuations to the metric based configurations. We will find out that the quantum theory does maintain the structure; however as the theory is topological many of the equations of motion are valid quantum mechanically limiting severely the allowed quantum fluctuations of the induced metric. Put differently, pure 2-d quantum gravity is in any case at best a topological theory, one can't expect much more than the appearance of a space with very rigid curvature.

3 Quantum Mechanical Analysis-Flat Potential

As we have stated several times, an intriguing structure that emerges in topological theories is a flat potential even at the quantum level. One may choose any wave functional obeying Gauss' law and inquire as to the type of universe encoded in it.

We first consider physical states confined to a fixed value of M^2. They can be classified in many respects according to the sign of M^2

Among all orbits a special role is played by the one defined by

$$\phi^a = (0, 0, 0); M^2 = 0. \tag{3.16}$$

This orbit consists of a single ϕ configuration. It maintains the full $SO(2, 1)$ gauge symmetry and all the world sheet is mapped into a single point in target space - the tip of the light-cone. By eq. (3.15) this constant map does not induce a metric, the theory is purely topological. The wave functional carries also no phase. (In the $\Psi(A_i)$ picture, the choice of a singlet representation allows all holonomies with equal amplitude). This wave functional carries some analogy to the $\phi^a = 0$ special point of flat potentials. It represents a group singlet and its choice breaks no symmetry.

Symmetry is broken on the other orbits. Let's first consider an orbit with $M^2 > 0$.

In a sense all these M^2 orbits are equivalent (just as different values of $(\phi^a)^2$ in the case of a flat potential), they describe two sheeted hyperboloids and the euclidean metric on these hyperboloids is pulled back to the world sheet.

The support of the wave functional lies in $\frac{SO(2,1)}{T}$, for any non-constant value of ϕ^a, eq. (3.15) leads in target space to a metric which has a Euclidean target space signature and to the same target space constant curvature.

Fixed ϕ^a configurations on the other hand map the world sheet into a single point target space, not allowing therefore any target space as well as world sheet metric interpretation. Constant maps do not separate between different space points. Their totality maps the full target space, but one can't associated metric to any of them. For constant map the phase of the wave functional associated with a target space area also vanishes. Points for which a configuration $\phi(x)$ has vanishing first derivative cannot participate in a metric space.

The wave functional describes a very rigid gravity. All non-constant configurations are general coordinate equivalent (or equivalently gauge equivalent) and thus describe closed loops on a single manifold of fixed curvature (up to topological obstructions). The singular configurations are gauge but not diffeomorphic equivalent and describe topological pockets, their measure is not clearly defined and thus we can only state their presence. The above is just the restatement of the relation between gauge and co-ordinate transformations.

For $M^2 < 0$ the manifold described is the one sheeted hyperboloid with a Minkowski metric induced by the group metric. For $M^2 = 0$ one obtains the light like metric. A wave function which is a superposition of different values of M^2 can also be considered, including the superposition of different metric signatures (following from different sign of M^2). One touches here upon two fantasies, both the emergence of a metric and of

the space-time signature are a matter of vacuum choice. The most symmetric state
does not allow a metric interpretation. One should note that the system has actually
only four different choices. The tip of the light-cone the light-cone ($M^2 = 0$), $M^2 > 0$
and $M^2 > 0$ orbits.

Calculating a suggested [18] order parameter one can [3] verify from a different point
of view an emergence of a dimension two space-time for 3 of the later choices. This
order parameter is related to an average length of a loop with a given topology. Another
candidate for an order parameter is the action of the BRST charge on the physical states.
It is difficult to expect a global BRST operator not to annitilate physical states in non-
anomalous theories. Nevertheless one finds that in a special topological system, related
to conformed quantum mechanics, the spectrum of states is continous and bounded
from below and excludes the edge of the spectrum which is the zero energy state. In
such a system one can calculate topological quantities even among state with non-zero
energy [4].

Acknowledgements: I wish to thank my collaborators D. Amati, L. Baulieu, S. Elitzur
and A. Forge with whom the work reported here was done.

References

[1] E. Witten, Communication. Math. Phys. 117 (1988) 353; 118 (1988)
411, Nucl. Phys. B340 (1990) 281. L. Baulieu and C. M. Singer, Nucl.
Phys. B (Proc. Suppl.) 5B (1988) 12; R. Brooks, D. Montano and J.
Sonnenschein, Phys. Lett. B214 (1988) 91.

[2] S. Elitzur, A. Forge and E. Rabinovici, Nucl. Phys. B388 (1992) 131.

[3] D. Amati, S. Elitzur and E. Rabinovici, SISSA preprint (1993), IAS -
SNS - Hep 93/ , Racha Institute preprint submitted for publication.

[4] L. Baulieu and E. Rabinovici, to be published in Physics. Lett. B.

[5] A.D.Sakharov, Sov. Phys. Dokl. 12 (1968), 1040; D.Amati and
G.Veneziano, Nucl. Phys. B204 (1982), 451.

[6] E.Witten, Comm. Math. Phys. 117 (1988), 353; Nucl. Phys. B311
(1988/89) 46.

[7] A.Chamseddine and D.Wyler, Phys. Lett. B228 (1989) 75; Nuclear
Physics B 340 (1990) 595; K.Isler and C.A.Trugenberger, Phys. Rev.
Lett. 63 (1989) 834.

[8] E.Witten, Commun. Math. Phys. 141 (1991) 153; H.Verlinde, String the-
ory and quantum gravity, ICTP, 1991; World Scient. p. 178 1992, Sixth
Marcel Grossman Conference Meeting on General Relativity M.Sato, ed.
(World Scientific, Singapore, 1992).

[9] E.Witten, Commun. Math. Phys. 121 (1989) 351.

[10] S.Elitzur, G.Moore, A.Schwimmer and N.Seiberg, Nucl. Phys. B326 (1989) 104.

[11] E. Witten, IAS preprint 92/45 (1992).

[12] A.Alekseev, L.Fadeev and S.Shatashvili, Journal of Geometry and Phys. 5 (1989) 391.

[13] D.Cangemi and R.J.Jackiw, Phys. Rev. Lett. 69 (1992) 233; R.Jackiw, MIT preprint, CTP 2105 (1993).

[14] G. 't Hooft, Nucl. Phys. B79 (1974) 276; A.M.Polyakov, JEPT Lett. 20 (1974) 194.

[15] N. Seiberg, Notes on Quantum Liouville Theory and Quantum Gravity. In: Common Trends in Mathematics and Quantum Field Theory, Proc. of the 1990 Yu' Progress of Theo. Phys. Sup. 102 319 (1990).Edited by T.Eguchi, T. Inami and T. Miwa. To appear in the Proc. of the Cargese meeting, Random Surfaces, Quantum Gravity and Strings, 1990. RU-90-29.

[16] E.Witten, Nucl. Phys. B311 (1988/89) 46.

[17] C.Teitelbaum, Phys. Lett. B126 (1983): in Quantum theory of gravity, S.Christensen ed. (Adam Hilger, Bristol, 1984); R.Jackiw, ibid; Nucl. Phys. B252 (1985) 343.

[18] A.M.Polyakov, Lecture presented at Les-Houches 1992, Princeton University, preprint.

HAMILTONIAN REDUCTION OF BRST COMPLEX
AND N=2 SUSY

Vladimir SADOV

Lyman Laboratory of Physics
Harvard University
Cambridge, MA 02138
and
L.D. Landau Institute for Theoretical Physics, Moscow

We study the nonunitary representations of $N = 2$ Super Virasoro algebra for the
rational central charges $\hat{c} < 1$. The resolutions for the irreducible representations of
$N = 2$ $SVir$ in terms of the "2-d gravity modules" are obtained and their characters
are computed. The correspondence between $N = 2$ nonunitary minimal models and the
Virasoro (q, p) minimal models coupled to 2-d gravity is shown at the level of states.
We also define the Hamiltonian reduction of the BRST complex of $\widehat{sl}(N)/\widehat{sl}(N)$ coset
to the BRST complex of the W-gravity coupled to the W matter. The case $\widehat{sl}(2)$ is
considered explicitly. Finally, we trace down the mechanism of the correspondence
between $\widehat{sl}(2)/\widehat{sl}(2)$ coset and 2-d gravity models.

1 Introduction

(The reader who is only interested in $N = 2$ nonunitary representations and don't care
for cosets for the first reading may skip the first four paragraphs of the Introduction
and the whole Section 2.)

In the first part of this paper we continue studying the relationships between the
noncritical W_N strings and $\widehat{sl}(N)_k/\widehat{sl}(N)_k$ topological cosets started in [1]. Our main
goal here is to explain the role of the Hamiltonian reduction in the story. It is well
known, that the Hamiltonian reduction [2], [3] maps the representations of $\widehat{sl}(N)_k$ to
that of W_N. (In a sense, we can consider this *as a definition* of W_N). Technically
speaking, to a given module $M_{sl(N)}$ it relates another module M_W. The latter is formed
by the semi-infinite BRST cohomology (twisted by a character) of $M_{sl(N)}$ over the
current algebra $\widehat{N_+}(sl(N))$ taking values in the maximal nilpotent subalgebra of $sl(N)$.
So defined, M_W has a natural structure of a W- algebra module [3].

Quantum Field Theory and String Theory, Edited by
L. Baulieu *et al.*, Plenum Press, New York, 1995

On the other hand, when computing the spectrum of physical states in the topological coset $\widehat{sl}(N)_k/\widehat{sl}(N)_k$, we take the semi-infinite BRST cohomology of the *whole* current algebra $\widehat{sl}(N)$ acting on the tensor product of two $\widehat{sl}(N)$ representations[a]. There is a suspicion, backed by the explicit calculation for $\widehat{sl}(2)$ coset and Vir-gravity respectively, that the spectra of the physical (BRST) states, including the discrete states, coincide in $\widehat{sl}(N)_k/\widehat{sl}(N)_k$ coset and in W_N gravity [4] [5] [6][1]. To make the statement more precise, let us consider two modules, M_1 and M_2 of $\widehat{sl}(2)$, and two modules M_1^{DS} and M_2^{DS} of the Virasoro algebra corresponding to the first pair by the (quantum Drinfeld-Sokolov) Hamiltonian reduction. For definiteness, let M_1 be an "admissible" irreducible representation ($k_1 + 2 = \frac{p}{q}$ — rational) and M_2 be a Wakimoto (free fields) representation with the value of the central charge $k_2 = -k_1 - 4$.

Then, the $\widehat{sl}(2)$-BRST homology of $M_1 \otimes M_2$ do coincide with $W_2 = Vir$-BRST homology of $M_1^{DS} \otimes M_2^{DS}$. The similar result is true if we take two Wakimoto's /it or one Wakimoto and one "transposed" Wakimoto modules for $\widehat{sl}(2)$ and two free boson Fock modules for Vir respectively.

Whereas the spectra of topological cosets can be found explicitly [4] [5] [6][1], it is quite difficult to do for W_N-gravity. Although it is really possible to construct a BRST complex in some cases [7][8], from the algebraic point of view it is not at all clear why it is possible. Then, it appears that even when exists this complex is not very convenient for the direct computations [26].

Taken together, all these facts motivate a desire to define a procedure of Hamiltonian reduction not only for one $\widehat{sl}(N)$ module (quantum Drinfeld-Sokolov reduction), but also for the whole BRST complex. In Section 2 we address this issue and give a proper definition of the reduction procedure. Then we outline how it works for $\widehat{sl}(2)$.

At this stage, quite naturally, the (topologically twisted) $N = 2$ superconformal algebra appears. It is $N = 2$ Super Virasoro for the basic example of $\widehat{sl}(2)/\widehat{sl}(2)$ coset. We show that the "reduced" $\widehat{sl}_k(2)$-BRST complex, with an irreducible representation in the matter sector is just a direct sum of two copies of the irreducible representation of $N = 2$ $SVir$ with the central charge $\widehat{c} = \frac{k}{k+2}$ as a vector space. The differential is given by the zero mode G_0^+ of the superconformal current. (This is proven in Section 4.1).

In Section 3 we present the necessary background material on the nonunitary representations of $N = 2$ $SVir$, following [13]. In fact, in [13] only the case of irrational $\widehat{c}$ was considered, so here we have to generalize that results to the more complicated situation when $\widehat{c}$ is rational. We obtain a "2-d gravity" resolution for the irreducible representations of $N = 2$ $SVir$ in terms of the triple products $L(Vir) \otimes F(Liouv.) \otimes F_{gh}$, where $L(Vir)$, $F(Liouv.)$ and F_{gh} denote respectively a Virasoro irreducible representation ("the matter"), a free bosonic Fock space ("the Liouville field") and a two fermions

---

[a]Usually one of them is irreducible (the matter sector) and another is a Wakimoto representation (Toda sector).

306

Fock space ("the diffeomorphisms ghosts"). Using this resolution, we find in particular a character formula for the irreducible representations of $N = 2$ $SVir$.

This formula explicitly incorporates the Lian-Zuckerman states of the Virasoro (p, q) (with $\frac{p}{q} = k + 2$) minimal model coupled to gravity. Thus it explains *for the representations* the relations between $N = 2$ $SVir$ and 2d-gravity, found in [9],[10] *for the chiral algebras*. It also gives a piece of evidence in favour of the idea [11] that $N = 2$ minimal theory "may already know about 2-d gravity". Also it is nice to have an object which puts together an infinite number of LZ states for the given matter field.

Finally, in Section 4.2 we combine the BRST complex Hamiltonian reduction of Sec. 2 and the resolutions of Sec. 3 to trace explicitly the mechanism identifying the physical states in $\widehat{sl}(2)$ coset and 2-d gravity. From the point of view of the $N = 2$ $SVir$ representations theory this last section may be viewed as the consistency check for the results we obtained in the Section 3, for the "2-d gravity resolution" in particular.

2 Hamiltonian reduction of $\widehat{sl}(N)$ BRST complex

1 Some motivations

In a sense, this section is the second Introduction. We wish to informally explain here what we mean by the Hamiltonian reduction of the BRST complex. The technical details can be found in the next two sections. First, it seems natural to repeat some motivations from [1].

In [2],[3] the quantum Drinfeld-Sokolov (DS) Hamiltonian reduction for Wakimoto and irreducible representations has been defined as the homology H^0 of the DS BRST complex, associated with the constraints

$$e^{\alpha}(z) = 1 \text{ if } \alpha \text{ is a simple root} \tag{2.1}$$

$$e^{\alpha}(z) = 0 \text{ if } \alpha \text{ is not a simple root}$$

Suppose now that we have a $\widehat{sl}(N)$-BRST complex which computes the cohomology of the tensor product of two modules M_1, M_2. Here M_1 can be either an irreducible representation $L_k(\widehat{sl}(N))$ or a Wakimoto representation Wak_k (or possibly a "transposed" Wakimoto representation $\widehat{Wak}_k$) and M_2 is a Wakimoto representation. In principle, it is also interesting to consider other combinations. In the language of the $\widehat{sl}(N)/\widehat{sl}(N)$ coset model the modules M_1, M_2 represent respectively the fields of *the matter* and of *the Liouville-Toda* sectors of the theory. The cohomology $H^*_{Q_{BRST}}(M_1 \otimes M_2)$, where M_1 runs over some specified set of representations with the fixed level k and M_2 runs over all Wakimoto representations with level $-2N - k$ form the spectrum of physical states of the theory. Usually we consider rational levels $k + N = \frac{p}{q}$ and restrict M_1 to the "admissible" irreducible representations of $\widehat{sl}(N)$.

We wish to have for *the whole BRST complex* something like what the DS reduction is for the single module. More precisely, ultimately we wish to obtain a W_N BRST complex by this "something". From the first sight it seems natural to try reduction independently on each factor. So we would have to add two sets of the DS ghost-antighost pairs, labeled by the positive roots of $sl(N)$ and consider the DS BRST operators Q_1, Q_2 acting respectively on M_1 and M_2, then take the product $M_1^{DS} \otimes M_2^{DS} \otimes \{\widehat{sl}(N)\ ghosts\}$ where $M_i^{DS} = H_{Q_i}^0(M_i \otimes \{DS\ ghosts\}_i)$ are the reduced modules. M_i^{DS} are the representations of W_N algebra.

This procedure does not work because we have to require that the reduction BRST operator, which is $Q_1 + Q_2$ here, commute with $\widehat{sl}(N)$-BRST operator, and it is not difficult to check, that there is no proper modification of $Q_1 + Q_2$, commuting with Q_{BRST} [b].

The other problem to address here is that of $\widehat{sl}(N)$ ghosts. They form a representation of $\widehat{sl}(N)$ (at level $2N$) and the general ideology requires to reduce this representation also — after all we must somehow obtain the W_N ghosts! After a short reasoning it seems very natural *not* to introduce special reduction ghosts at all and to try to make the reduction of the BRST complex using the $\widehat{sl}(2)$ ghosts themselves.

2 The basic definition

(If the reader is not interested in "general nonsense" and only wants to learn about $\widehat{sl}(2)$ and 2-d gravity, he may skip this subsection and start from 2.3.)

Let us give a definition.

Definition-Hypothesis 1 *Let M_1 be either the irreducible or the Wakimoto representation at level k, and M_2 be the Wakimoto representation at level $-k - 2N$ of the $\widehat{sl}(N)$ algebra. Then there exists a spectral sequence, converging (at the second term) to the $\widehat{sl}(N)$-BRST cohomology $H^*_{Q_{BRST}}(M_1 \otimes M_2)$ of the module $M_1 \otimes M_2$. Denote by Q_R a differential in the first term of the spectral sequence. Then the second term of the spectral sequence, i.e. the complex $(H^*_{Q_R}, Q_W)$ with the differential Q_W is quasiisomorphic (has the same cohomology) to the W_N-BRST complex of the module $M_1^{DS} \otimes M_2^{DS}$ where the superscript "DS" denotes the standard (Drinfeld-Sokolov) Hamiltonian reduction.*

We say, that Q_R "makes the Hamiltonian reduction of $\widehat{sl}(N)$-BRST complex" and call $(H^*_{Q_R}, Q_W)$ "the reduced BRST complex".

Comment Because of the required quasiisomorphism the cohomology computed by the spectral sequence are determined by the cohomology of W_N-BRST. One the other hand this is the $\widehat{sl}(n)$-BRST cohomology just by the definition. Thus we see that the equivalence of spectra of cosets and W-gravity should follow from the hypothesis above.

---

[b] mainly because in $Q_1 + Q_2$ necessarily participate not only the symmetric combinations of currents like $J_1^a + J_2^a$ but also the antisymmetric ones like $J_1^a - J_2^a$. The commutators of the latter with Q_{BRST} cannot be compensated by adding extra terms with ghosts

3 An example of the construction

We have formulated a general hypothesis. It may seem a little bit complicated and not very explicit. Now we wish to show how it can be proved for the algebra $\widehat{sl}(2)$. In this example the coincidence of spectra was known for some time from the explicit computation, which was fairly straightforward in this case. However we shall see that our construction is nontrivial already for this simplest example and gives rise to the interesting N=2 supersymmetric structure.

Let us we consider a decomposition $Q_{BRST} = Q_R + Q_W$ with

$$Q_R = \oint c^+(E_1 + E_2) + c^0(H_1 + H_2 + 2(b_+c^+ - b_-c^-)) \tag{2.2}$$
$$Q_W = \oint c^-(F_1 + F_2 + c^+b_-)$$

It is convenient to rephrase our hypothesis in this particular case in the form of the

Theorem 1 *Let M_1 be either the irreducible or the Wakimoto representation at level k, and M_2 be the Wakimoto representation at level $-k - 4$ of the $\widehat{sl}(2)$ algebra. Then $\widehat{sl}(2)$-BRST complex for $M_1 \otimes M_2$ has a structure of the double complex with differentials (Q_R, Q_W) such that $Q_R + Q_W = Q_{BRST}$. Consider a spectral sequence of this double complex whose first term has a differential Q_R. Then the second term of the spectral sequence, i.e. the complex $(H^*_{Q_R}, Q_W)$ with the differential Q_W is quasiisomorphic (has the same cohomology) to the Virasoro-BRST complex of the module $M_1^{DS} \otimes M_2^{DS}$ where the superscript "DS" denotes the standard (Drinfeld-Sokolov) Hamiltonian reduction.*

In other words, there exists a Hamiltonian reduction of $\widehat{sl}(2)$-BRST complex".

It is easy to check that the operators (2.2) do define the structure of the double complex on the $\widehat{sl}_k(2)$ BRST complex. What is $H^*_{Q_R}(\widehat{sl}_k(2) \otimes Wakim_{-4-k})$ — the Q_R cohomology of the chiral algebra $\widehat{sl}_k(2) \otimes Wakim_{-4-k}$ here?

Doing the direct computation[c] one convinces oneself that two currents

$$G^+(z) = \tfrac{1}{2(k+2)}(c^-(F_1 + F_2 + c^+b_0) + 2\partial(\gamma c^-)) \tag{2.3}$$
$$G^-(z) = b_-(E_1 - E_2) \tag{2.4}$$

belong to $H^*_{Q_R}$. Their OPE is

$$G^+(z)G^-(0) = \frac{k/k + 2}{z^3} + \frac{J(z)}{z^2} + \frac{T(z) + \partial J(z)}{z} + regular\ terms \tag{2.5}$$

where

$$J(z) = :c_-b^: - (z) - \frac{2}{\sqrt{2(k+2)}}(\partial\varphi(z) + \frac{1}{\sqrt{2(k+2)}}([E_1 + \beta]\gamma + :b_+c^+ :)) \tag{2.6}$$
$$+ \frac{1}{2(k+2)}\{Q_R, b_0\}$$

---

[c]It can be a good idea to use the Matematica OPE package [14] to do this!

The operators $J(z)$ and $T(z)$ also are the nontrivial elements of

$$H^*_{Q_R}(\widehat{sl}_k(2) \otimes Wakim_{-4-k})$$

Moreover, it is true that $T(z)$ is equivalent modulo Q_R-exact term to the twisted stress-energy of the topological $\widehat{sl}(2)/\widehat{sl}(2)$ coset and that four currents $G^+(z)$, $G^-(z)$, $T(z)$, $J(z)$ form a closed chiral algebra which is just a (topologically twisted) N=2 SuperVirasoro with the central charge

$$\widehat{c} = \frac{k}{k+2} \tag{2.7}$$

This is not very surprising. For example, the similar phenomena was observed in [12] for the Kazama-Suzuki coset with $k = -3$.

Thus $SVir \subset H^*_{Q_R}(\widehat{sl}_k(2) \otimes Wakim_{-4-k})$. In fact, there is also an operator c^0_0 — the zero mode of the ghost $c^0(z)$ — which belongs to H^*_R. In the next sections we shall see using more complicated technique that actually

$$H^*_{Q_R}(\widehat{sl}_k(2) \otimes Wakim_{-4-k}) = [N = 2\ SVir) \oplus c^0_0(N = 2\ SVir] \tag{2.8}$$

as a *chiral algebra*. For now let us assume this is true.

Similarly it can be shown (see Sec.4.1) that *for the representations* the cohomology $H^*_{Q_R}(L_k(\widehat{sl}(2)) \otimes Wak_{-4-k})$ are given just by the direct sum of two copies (one is again shifted by c^0_0) of the irreducible representation $L(N = 2\ SVir)$ of N=2 $SVir$:

$$H^*_{Q_R}(L_k(\widehat{sl}_k(2)) \otimes Wak_{-4-k}) = [L(N = 2\ SVir) \oplus c^0_0 L(N = 2\ SVir)] \tag{2.9}$$

The crucial for the following observation is that the second differential Q_W (2.2) of our double complex is just a zero mode of one of the superconformal currents:

$$Q_W = G^+_0 \tag{2.10}$$

Thus the whole *reduced BRST complex for $\widehat{sl}_k(2)/\widehat{sl}(2)$* coset can be expressed solely in terms of N=2 Super Virasoro algebra:

$$(H^*_{Q_R}(L_k(\widehat{sl}_k(2) \otimes Wak_{-4-k}), Q_W) = (L(N = 2\ SVir) \oplus c^0_0 L(N = 2\ SVir), G^+_0) \tag{2.11}$$

Now we shall use the relation between N=2 $SVir$ and 2d gravity coupled to the minimal matter found in [9], [10] to establish the quasiisomorphism the main hypothesis claims. Let me remind here this relation.

Consider the BRST complex of 2d gravity coupled to the minimal matter with the central charge c_M. The chiral algebra of the matter sector is Vir, and that of the Liouville and ghost sectors are correspondingly $Heis$ and $Clif$ with the total central charge $-c_M$. Then

There is an embedding

$$N = 2\ SVir \to Vir \otimes Heis \otimes Clif \tag{2.12}$$

of the chiral algebras and the corresponding map on the representations such that $G^-(z) \to b(z)$, $G^+(z) \to J_{BRST}(z)$, *where* $J_{BRST}(z)$ *is a Vir-BRST current plus a total derivative term. In particular, a zero mode of* $G^+(z)$ *is mapped:* $G_0^+ \to Q_{Vir}$ *to a BRST operator of 2-d gravity.*

We shall show in Sections 3,4 that the similar relation exists at the level of representations. In particular, for the irreducible representation $L(N = 2\ SVir)$ there exists a resolution in terms of the "2-d gravity" modules $L(Vir) \otimes F(Liouv.) \otimes F_{gh}$, where $L(Vir)$ denotes the irreducible representation of the Virasoro algebra, $F(Liouv.)$ is a free boson Fock space and F_{gh} is the diffeomorphisms ghosts Fock space:

$$0 \to L(N = 2\ SVir) \to L(Vir) \otimes F(Liouv.) \otimes F_{gh} \to \tag{2.13}$$
$$\to L'(Vir) \otimes F'(Liouv.) \otimes F_{gh} \to\to L''(Vir) \otimes F''(Liouv.) \otimes F_{gh} \to \cdots$$

(This resolution is infinite in the most interesting cases, see Section 3.)

To complete the proof, we should substitute this resolution into the reduced BRST complex (2.11) to obtain the double complex of "2-d gravity" modules with two differentials. One of them comes from the resolution (2.13). The other one is just

$$Q_W = G_0^+ = Q_{Vir} \tag{2.14}$$

We are almost done now. The cohomology of the double complex are computed in the Sec. 4.2 where it is shown that they are equal to $H^*_{Q_{Vir}}(L(Vir) \otimes F(Liouv.) \otimes F_{gh})$ — the 2-d gravity BRST cohomology of the first term in the resolution (2.13).

Before passing to the details of the proof let us summarize what we have learned and propose the possible generalization for the algebras $\widehat{sl}(N)$. First, we decomposed the Lie algebra BRST operator into two pieces, corresponding to the Borel subalgebra of $sl(2)$ and its compliment (+some ghost terms corrections) to obtain a structure of a double complex. This decomposition for $\widehat{sl}(N)$ goes through if we take as a Q_R a proper piece of $\widehat{sl}(N) - Q_{BRST}$, corresponding to the maximal parabolic subalgebra of (the finite dimensional) $sl(N)$.

Then we computed the cohomology of Q_R which turned to be the irreducible representation of N=2 $SVir$. For $\widehat{sl}(N)$ it is likely the representation of N=2 SuperW algebra. The cohomology has the structure of a complex with differential $Q_W = G_0^+$.

Finally we used the map (2.12) from [9], [10] to relate this complex to the BRST complex of 2d gravity coupled to the minimal matter. Such map into W-gravity+W minimal matter also exists for any N=2 SuperW algebra [10]. Then we show that the cohomology of the reduced BRST complex are the same as of the W-system.

3 Representations of N=2 SuperVirasoro

1 The bosonization formulas

In this subsection we recall the results from the representation theory of $N = 2SVir$, obtained in [13]. It is important, that we actually need the *nonunitary* representations of $N = 2SVir$, it follows from the formula (2.7) for the central charge (remember that k is just rational, not necessarily integer).

Let us introduce the basic notations. We have a system consisting of the Virasoro algebra (Vir) — a matter sector, a free bosonic field ϕ with the background charge $(Heis)$ — a Liouville sector and a pair of fermions b, c of spins 2,-1 $(Clif)$ — the diffeomorphism ghosts (in the brackets are the names of the corresponding chiral algebras). We require the total central charge be equal to zero. Then the currents

$$J(z) =: cb : +\alpha_-\partial\phi \tag{3.1}$$

$$G^+(z) =: c[T_{Vir} + T_\phi + \tfrac{1}{2}T_{bc}] : -2\alpha_-\partial(c\partial\phi) + \tfrac{1}{2}(1 - 2\alpha_-^2)\partial^2 c \tag{3.2}$$

$$G^-(z) = b(z) \tag{3.3}$$

$$T = T_{Vir} + T_\phi + T_{bc} \tag{3.4}$$

$$T_\phi = -\tfrac{1}{4} : (\partial\phi)^2 : +\beta_0\partial^2\phi \tag{3.5}$$

satisfy the OPE of (twisted) N=2 $SVir$ chiral algebra. In fact these formulas give the embedding of N=2 $SVir$ into the tensor product of three other chiral algebras $Vir \otimes Heis \otimes Clif$ (we used this fact in the previous section). To describe the properties of this map it is technically convenient to bosonize the Virasoro algebra by the free field X with the background charge α_0. In other words we embed Vir into the Heisenberg algebra generated by ∂X which we denote by $Heis'$ to distinguish it from the Liouville $Heis$. Substituting the bosonized matter stress energy

$$T_{Vir}(z) = \tfrac{1}{4} : (\partial X)^2 : +\alpha_0 X \tag{3.6}$$

$$\beta_0^2 - \alpha_0^2 = 1, \quad \alpha_\pm = \alpha_0 \pm \beta_0 \tag{3.7}$$

$$c_{Vir} = 1 - 24\alpha_0^2$$

into (2.14) we finally obtain the bosonization prescriptions for N=2 $SVir$ we need. Unlike the standard bosonization [17],[18], [19] the formulas for $G^+(x)$ and $G^-(z)$ are very asymmetric. Comparing (3.6) with (2.7) we see that

$$\alpha_- = \frac{-1}{\sqrt{k+2}} \tag{3.8}$$

(it is a nonstandard notation for $\widehat{sl}(2)_k$!). For the representations, we take a Fock space

$$F_{\alpha\beta} = F_\alpha(X) \otimes F_\beta(\phi) \otimes F_{gh} \tag{3.9}$$

$F_\alpha(X)$ and $F_\beta(\phi)$ here are the standard Fock modules of $Heis'$ and $Heis$ with vacuums $|\alpha>$ and $|\beta>$ respectively and F_{gh} is a ghosts Fock space (a $Clif$ Verma module) with the vacuum vector $|0>$ annihilated by

$$c_n|0>=0 \; n>1, \quad b_n|0>=0 \; n>-2 \tag{3.10}$$

In F_{gh} we take a vector $|0>_{phys}= c_1|0>$ and define the N=2 vacuum as $\Omega = |\alpha> \otimes|\beta> \otimes|0>_{phys}$. This procedure is well known in string theory. Here we use it to endow the free field Fock space $F_{\alpha\beta}$ with a structure of a highest weight N=2 $SVir$ module:

$$L_n\Omega = J_n\Omega = G_n^-\Omega = 0, \; n \geq 0 \tag{3.11}$$

$$G_n^+\Omega = 0, \; n > 0 \tag{3.12}$$

$$L_0\Omega = \Delta\Omega = (-1 + \alpha(\alpha - 2\alpha_0) - \beta(\beta - 2\beta_0)) \tag{3.13}$$

$$J_0\Omega = Q\Omega = (1 + 2\alpha_-\beta)\Omega \tag{3.14}$$

It is convenient to rewrite the formula for the conformal weight as

$$\Delta(\alpha, \beta) = (\alpha + \beta - \alpha_+)(\alpha - \beta - \alpha_-) \tag{3.15}$$

There are two *screening operators* in our bosonization. One of them is just $E = \oint : e^{\alpha_+ X(z)} :$. It comes from the bosonization of the Vir matter. The other one is $F_1 = \oint : b(z)e^{-\frac{\alpha_+}{2}(X(z)+\phi(z))} :$ [10],[20],[21]. It is fermionic and local to itself:

$$F_1^2 = 0 \tag{3.16}$$

Together, E and F_1 form a quantum superalgebra $u_q(n_+(sl(2|1)))$ with $q = e^{\pi i\alpha_+^2}$. Namely they satisfy the Serre relation:

$$E^2 F_1 - (q + q^{-1})EF_1E + F_1E^2 = 0 \tag{3.17}$$

(As usual, the left hand side of (3.17) is to be understood as a part of some formal polynomial in screenings acting on the appropriate state.)

There is a simple but important remark to me made here. The basic object for the correspondence N=2 $SVir \rightarrow \{2-d\ gravity\}$ above is the Virasoro algebra itself, not the free field $X(z)$ which is just a useful tool for describing this correspondence. Bosonizing the Virasoro algebra, we could choose a screening operator $E^{(-)} = \oint : e^{\alpha_- X(z)} :$ instead of $E^{(+)} = \oint : e^{\alpha_+ X(z)} :$. The vertex operators corresponding to $E^{(-)}$, F are local with respect to each other, so we may simply set

$$E^{(-)}F - FE^{(-)} = 0 \tag{3.18}$$

Hence the algebraic structure produced by the pair $E^{(-)}$, F is much simpler and therefore much less powerful for the purposes of the representation theory then the structure produced by $E^{(+)}$, F. Therefore in this subsection we'd better stick to the latter. In fact, we will be able to take advantage of the simplicity of (3.18) later, *when we already know the representation theory of* N=2 *SVir*.

Using very rigid conditions (3.16), (3.17) we can classify the irreducible representations of N=2 *SVir* according to the types of the free fields resolution they have. By such resolution we mean the complex of the free field Fock spaces with the cohomology being nontrivial only in the zero degree where it is represented by the irreducible representation. Having a complex we compute its Euler characteristics (character valued, as usual) which turns to be a character of the irreducible $L_{\alpha\beta}$. It is convenient to deal with the normalized characters

$$\tilde{\chi}_{\alpha\beta} = \frac{\chi_{\alpha\beta}}{\chi(F_{\alpha,\beta})} \; , \; \chi(F_{\alpha,\beta}) = Tr_{F_{\alpha,\beta}}(q^{L_0} x^{2J_0}) \tag{3.19}$$

2 The case of irrational $\hat{c}$

First let me describe the classification for the generic (irrational) values of $\hat{c}$. It was originally obtained in the works [15],[17],[18], [19],[16], but it is convenient to get it using our bosonization (3.1-3.6) as a warm up for the more complicated case when $\hat{c}$ is rational. Depending on (α, β), the irreducible representation $L_{\alpha,\beta}$ may belong to either of the four following types.

Case I. (α, β) is generic, the module $F_{\alpha\beta}$ is irreducible. The corresponding complex is therefore trivial, $\tilde{\chi}_{\alpha\beta} = 1$.

Case II.

$$\alpha_{n\ m} = \alpha_+ \frac{1-n}{2} + \alpha_- \frac{1-m}{2} \tag{3.20}$$

β is generic. This case essentially reduces to the well known theory for the Virasoro algebra. The map E^n is surjective, its kernel is a submodule in $F_{\alpha\beta}$ generated by the highest weight vector. It gives

$$L(N = 2\ SVir) = L_{m\ n}(Vir) \otimes F_\beta \otimes F_{gh} \tag{3.21}$$

This is the only map *from* $F_{\alpha\beta}$ and the only map *to* $F_{\alpha+n\alpha_+\ \beta}$; there is no maps *to* $F_{\alpha\beta}$ or *from* $F_{\alpha+\alpha_+\ \beta}$. Therefore the submodule in $F_{\alpha\beta}$ generated by the highest weight vector is irreducible. The quotient of $F_{\alpha\beta}$ by this submodule is also irreducible and coincides with $F_{\alpha+n\alpha_+\ \beta}$. The character is

$$\tilde{\chi}_{\alpha\beta} = 1 - q^{nm} \tag{3.22}$$

This was the typical example of the argument to be used in this sort of constructions.

Case III_-.

$$\alpha - \beta = -\alpha_- l \tag{3.23}$$

When $l \geq -1$ the map F_1 sends the highest weight vector of $F_{\alpha,\beta}$ to a nonzero element, which generate in $F_{\alpha-\frac{\alpha_+}{2},\beta-\frac{\alpha_+}{2}}$ a proper submodule $SF_{\alpha-\frac{\alpha_+}{2},\beta-\frac{\alpha_+}{2}}$. There is no other maps into $F_{\alpha-\frac{\alpha_+}{2},\beta-\frac{\alpha_+}{2}}$, so $SF_{\alpha-\frac{\alpha_+}{2},\beta-\frac{\alpha_+}{2}}$ is the only proper submodule. Therefore it must coincide with the kernel of the map F_1 from $F_{\alpha-\frac{\alpha_+}{2},\beta-\frac{\alpha_+}{2}}$ to $F_{\alpha-\alpha_+,\beta-\alpha_+}$. This means that the diagram III_- is *exact* — the image of the incoming arrow coincide with the kernel of the outgoing arrow.

The diagram III_- has already a natural structure of the complex. The graded components are just the Fock spaces at the vertices and the differentials are given by the arrows. This complex is infinite in both directions. It is exact, so its cohomology is trivial. To obtain the resolution of $L_{\alpha\beta}$ one cuts the diagram by the arrow going *from* $F_{\alpha\beta}$ to obtain two complexes with the equal cohomology (so there are two resolutions in fact), one can use either of them. The character is

$$\widetilde{\chi}_{\alpha\beta} = \frac{1}{1 + x^{-1}q^{l+1}} \tag{3.24}$$

We see that for $l \geq -1$ the formula (3.24) can naturally be interpreted as a character of the representation with the highest weight $(\Delta_{\alpha\beta}, q_{\alpha\beta})$. But for $l < -1$ the identical transformation $(3.24) \rightarrow \frac{xq-l-1}{1+xq^{-l-1}}$ shows that the character we compute now correspond to the weight $(\Delta_{\alpha+\frac{\alpha_+}{2},\beta+\frac{\alpha_+}{2}}, q_{\alpha+\frac{\alpha_+}{2},\beta+\frac{\alpha_+}{2}})$. The reason for this phenomena is simple. For $l < -1$ the map F_1 kills the highest weight vector of $F_{\alpha\beta}$ and sends some vector $w \in F_{\alpha\beta}$ to the highest vector of $F_{\alpha-\frac{\alpha_+}{2},\beta-\frac{\alpha_+}{2}}$. Hence each Fock space has one cosingular vector $w_{\alpha\beta}$ and the irreducible representation is a *submodule* of $F_{\alpha\beta}$ generated by the highest weight vector of the Fock space[d]. Now, to obtain a resolution of $L_{\alpha\beta}$ we should cut the diagram III_- by the arrow coming *into* $F_{\alpha\beta}$. The character is

$$\widetilde{\chi}_{\alpha\beta} = \frac{1}{1 + xq^{-l-1}} \tag{3.25}$$

Case III_+

$$\alpha + \beta = \alpha_-(l+1) + \alpha_+ \tag{3.26}$$

One can repeat everything that have been said about III_-. The only subtlety here is to check that the composition of two consequent maps in the diagram is zero. The reader should convince oneself it is true using (3.17) and simple q-polynomial identities. The formulas for the characters are given by the same formulas (3.24) ,(3.24).

Case IV_- — the conditions II and III_- (IV_-) are met simultaneously.

$$\alpha_{nm} = \alpha_+\frac{1-n}{2} + \alpha_-\frac{1-m}{2} \tag{3.27}$$
$$\beta_{nml} = \alpha_+\frac{1-n}{2} + \alpha_-\frac{1-m+2l}{2}$$

---

[d]Compare to $l \geq -1$ when the irreducible representation was a *quotient* of the Fock space.

We denote $F_{\alpha_{nm}\beta_{nml}}$ by F_{nml}. We know everything already about the maps in the diagram. First consider the resolution of the representation with $n \geq 0$, i.e. belonging to the left column in Fig.1. To obtain a resolution we should again cut the diagram by the horizontal line crossing the arrow *above* $(\alpha\beta)$ for $l \geq -1$ or *below* $(\alpha\beta)$ for $k < -1$. Keeping the upper half we end up with a "ladder" shown in the Fig.2. The structure of the complex $\{C^r, d_{(r)}\}_{r \geq 0}$ is given by

$$C^0 = F_{n+1\ m\ l}, \quad C^r = F_{n+1+r\ m\ l} \oplus F_{-(n+r)\ m\ l} \tag{3.28}$$

$$d_{(0)} = E^{n+1} \oplus F_1, \quad d_{(r)} = \begin{pmatrix} F_1 & 0 \\ E^{n+1+r} & x_{n+r}EF_1 - F_1E \end{pmatrix}$$

$$x_{l+1} = (q + q^{-1}) - \frac{1}{x_l}, \quad x_0 = q + q^{-1} \tag{3.29}$$

The character is

$$l \geq -1 \quad \tilde{\chi}_{\alpha\beta} = \frac{1 - q^{m(n+1)} + q^{m-l-1}(1 - q^{mn})}{(1 + x^{-1}q^{l+1})(1 + xq^{m-l-1})} \tag{3.30}$$

$$l < -1 \quad \tilde{\chi}_{\alpha\beta} = \frac{1 - q^{mn} + q^{m-l-1}(1 - q^{m(n-1)})}{(1 + xq^{-l-1})(1 + xq^{m-l-1})} \tag{3.31}$$

Case IV_+ — the conditions II and III_+ are met simultaneously.

$$\alpha_{nm} = \alpha_+ \frac{1-n}{2} + \alpha_- \frac{1-m}{2} \tag{3.32}$$

$$\beta_{nml} = \alpha_+ \frac{1+n}{2} + \alpha_- \frac{-1+m+2(l+1)}{2} \tag{3.33}$$

One just repeats what was said about IV_+ (probably it is better to take a bottom half of the cut diagram to construct a resolution of the representations with $n > 0$ in this case). The formulas for the characters (3.30) are applicable.

Note that the equation

$$\Delta(\alpha, \beta) = (\alpha + \beta - \alpha_+)(\alpha - \beta - \alpha_-) = 0 \tag{3.34}$$

has two branches of solutions corresponding to either $III_\pm$ or $IV_\pm$ with $l = -1$.

It is important to note that the resolutions in the cases $I - IV$ above can be rewritten in terms of modules $L_\alpha \otimes F_\beta \otimes F_{gh}$ instead of $F_\alpha \otimes F_\beta \otimes F_{gh}$, where L_α is the irreducible representation of the Virasoro algebra.

Indeed, for the case II we have shown this explicitly in (3.21). For the cases I, III this is trivial because the irreducible and free field representations are the same object: $L_\alpha = F_\alpha$. In the case IV we can compute the cohomology of the free field resolution using the spectral sequence, associated with the "vertical" filtration, shown in Fig.2. The first term of this spectral sequence computes the "horizontal" (in Fig.2) cohomology, which gives exactly $L_\alpha \otimes F_\beta \otimes F_{gh}$. Then the second term represents the resolution we are after:

$$0 \to L_{nml} \to L_{nm} \otimes F(\beta_{nml}) \otimes F_{gh} \to L_{n+1\ m} \otimes F(\beta_{n+1\ ml}) \otimes F_{gh} \to \cdots \tag{3.35}$$

Existence of such resolution in terms of irreducible representations of Virasoro is not surprising at all, because it is the Virasoro algebra, not a free field $X(z)$ which is basic in the correspondence between Supervirasoro and 2-d gravity, so everything has to be expressible in terms of it. This fact should be viewed as a counterpart for the *representations* of the map (2.12) between the *chiral algebras*.

In such form, the resolutions above can be generalized to the rational values of $\hat{c}$.

3 The case of rational $\hat{c}$

Up to now we dealt with irrational k. But as (2.7) shows, to consider the most interesting "minimal" cosets we must take k rational! The free field resolution becomes more complicated in this case, comparing to what we had in $I - IV$. The reason for it is that for $k + 2 = \frac{p}{q}$, p, q — integer numbers, the bosonic screening E becomes nilpotent:

$$E^q = 0 \tag{3.36}$$

It results essentially in that there appear more maps among the free Fock spaces than there were for irrational central charges. Easy to see, that it changes the diagrams only for the (α, β) pairs where α satisfies the integrality condition given by (3.20), i.e., for the cases II, $IV_\pm$. Thus the diagrams of maps in Fig.1 remain the same for the cases I, III. The proper modification of the diagram IV in Fig.1 is shown in the Fig.3. Comparing Fig.1, Fig.3. and the Felder resolution for the irreducible "discrete" representation of the Virasoro algebra we see that passing to the rational values of $\hat{c}$ in N=2 $SVir$ effectively amounts to using the correct input for the Virasoro piece for the rational central charge c_M.

Unlike the diagrams shown in Fig.1, the diagram in Fig.3 does not admit a natural structure of the complex. Essentially this is because of presence of the (shaded in Fig.3) rows, corresponding to the boundary of the Kac table for Virasoro. To obtain the complex (actually, the double complex), one throws away these lines to come up with a picture, shown in Fig.4 — it is a commutative diagram and the composition of any two consequent vertical or horizontal arrows is zero. Now we can use a spectral sequence, similar to what we used before for the complex in the Fig.2. The "horizontal" cohomology again are nontrivial in only one column and give the irreducible representations of Virasoro, via the Felder resolution. The vertical cohomology give the resolution (here L_{nml} denotes the irreducible representation of $N = 2$ $SVir$, corresponding to $(\alpha_{nm}, \beta_{nml})$, whereas L_{nm} denotes the irreducible representation of Vir, corresponding to α_{nm})

$$0 \to L_{nml} \to L_{nm} \otimes F(\beta_{nml}) \otimes F_{gh} \to L_{(n+1)m} \otimes F(\beta_{(n+1)ml}) \otimes F_{gh} \to \cdots \tag{3.37}$$

$$\to L_{(q-1)m} \otimes F(\beta_{(q-1)ml}) \otimes F_{gh} \to L_{(q-1)m} \otimes F(\beta_{(q-1)m(p-m+l)}) \otimes F_{gh} \to$$

$$\to L_{(q-2)m} \otimes F(\beta_{(q-2)m(p-m+l)}) \otimes F_{gh} \to \cdots$$

$$\to L_{1m} \otimes F(\beta_{1m(p-m+l)}) \otimes F_{gh} \to L_{1m} \otimes F(\beta_{1m(l+p)}) \otimes F_{gh} \to \cdots$$

$$\cdots \qquad\qquad \cdots$$

$$\uparrow F_1 \qquad\qquad \uparrow x_{\alpha+\alpha_+}EF_1-F_1E$$

$$(\alpha-\alpha_+,\beta-\alpha_+) \qquad\qquad (\alpha+\alpha_+,\beta-\alpha_+)$$

$$\uparrow F_1 \qquad\qquad \uparrow x_{\alpha+\frac{\alpha_+}{2}}EF_1-F_1E$$

$$(\alpha_{nm},\beta) \xrightarrow{\ E^n\ } (\alpha_{-nm},\beta) \qquad (\alpha-\tfrac{\alpha_+}{2},\beta-\tfrac{\alpha_+}{2}) \qquad (\alpha+\tfrac{\alpha_+}{2},\beta-\tfrac{\alpha_+}{2})$$

$$\uparrow F_1 \qquad\qquad \uparrow x_\alpha EF_1-F_1E$$

$$(\alpha,\beta) \qquad\qquad (\alpha,\beta)$$

$$\uparrow F_1 \qquad\qquad \uparrow x_{\alpha-\frac{\alpha_+}{2}}EF_1-F_1E$$

$$\cdots \qquad\qquad \cdots$$

Case II   Case III$_-$ Case III$_+$

$$\cdots \qquad\qquad \cdots$$

$$\uparrow F_1 \qquad\qquad \uparrow x_n EF_1-F_1E$$

$$(n,m,l)_- \xrightarrow{\ E^n\ } (-n,m,l)_+$$

$$\uparrow F_1 \qquad\qquad \uparrow x_{n-1}EF_1-F_1E$$

$$\cdots \qquad\qquad \cdots$$

$$\uparrow F_1 \qquad\qquad \uparrow x_0 EF_1-F_1E$$

$$(1,m,l)_- \xrightarrow{\ E\ } (-1,m,l)_+$$

$$\uparrow F_1$$

$$(0,m,l)$$

$$\uparrow F_1$$

$$(1,m,l)_+ \xrightarrow{\ E\ } (-1,m,l)_-$$

$$x_0 EF_1-F_1E \uparrow \qquad\qquad \uparrow F_1$$

$$\cdots \qquad\qquad \cdots$$

Case IV$_-$

Figure 1

The mapping diagrams for the irrational $\widehat{c}$

$$\cdots \qquad\qquad \cdots$$

$$\uparrow F_1 \qquad\qquad\qquad \uparrow x_{n+2}EF_1 - F_1E$$

$$(n+2, m, l)_- \quad \xrightarrow{E^{n+2}} \quad (-n-2, m, l)_+$$

$$\uparrow F_1 \qquad\qquad\qquad \uparrow x_{n+1}EF_1 - F_1E$$

$$(n+1, m, l)_- \quad \xrightarrow{E^{n+1}} \quad (-n-1, m, l)_+$$

$$\uparrow F_1 \qquad\qquad\qquad \uparrow x_nEF_1 - F_1E$$

$$(n, m, l)_- \quad \xrightarrow{E^n} \quad (-n, m, l)_+$$

Figure 2

This bicomplex gives a resolution in case IV_-

$$
\begin{array}{ccccccc}
& & \cdots & & \cdots & & \cdots \\
& & \uparrow F_2 & & \uparrow F_1 & & \uparrow F_2 \\
\cdots \xrightarrow{E} & & (2q-1,m,l+2p-m)_+ & \xrightarrow{E^{q-1}} & (1,m,l-p)_- & \xrightarrow{E} & (-1,m,l+p-m)_+ & \xrightarrow{E^{q-1}} \\
& & \uparrow F_2 & & \uparrow F_1 & & \uparrow F_2 \\
\cdots \longrightarrow & & (2q,m,l)_- & \longrightarrow & (0,m,l+p-m)_+ & = & (0,m,l-p)_- & \longrightarrow \\
& & \uparrow F_1 & & \uparrow F_2 & & \uparrow F_1 \\
\cdots \xrightarrow{E} & & (2q-1,m,l)_- & \xrightarrow{E^{q-1}} & (1,m,l+p-m)_+ & \xrightarrow{E} & (-1,m,l-p)_- & \xrightarrow{E^{q-1}} \\
& & \uparrow F_1 & & \uparrow F_2 & & \uparrow F_1 \\
& & \cdots & & \cdots & & \cdots \\
& & \uparrow F_1 & & \uparrow F_2 & & \uparrow F_1 \\
\cdots \xrightarrow{E^{q-1}} & & (q+1,m,l)_- & \xrightarrow{E} & (q-1,m,l+p-m)_+ & \xrightarrow{E^{q-1}} & (-q+1,m,l-p)_- & \xrightarrow{E} \\
& & \uparrow F_1 & & \uparrow F_2 & & \uparrow F_1 \\
\cdots \longrightarrow & & (q,m,l+p-m)_+ & = & (q,m,l)_- & \longrightarrow & (-q,m,l-m)_+ & = \\
& & \uparrow F_2 & & \uparrow F_1 & & \uparrow F_2 \\
\cdots \xrightarrow{E^{q-1}} & & (q+1,m,l+p-m)_+ & \xrightarrow{E} & (q-1,m,l)_- & \xrightarrow{E^{q-1}} & (-q+1,m,l-m)_+ & \xrightarrow{E} \\
& & \uparrow F_2 & & \uparrow F_1 & & \uparrow F_2 \\
& & \cdots & & \cdots & & \cdots \\
& & \uparrow F_2 & & \uparrow F_1 & & \uparrow F_2 \\
\cdots \xrightarrow{E^{n+1}} & & (2q-n-1,m,l+p-m)_+ & \xrightarrow{E^{q-n-1}} & (n+1,m,l)_- & \xrightarrow{E^{n+1}} & (-n-1,m,l-m)_+ & \xrightarrow{E^{q-n-1}} \\
& & \uparrow F_2 & & \uparrow F_1 & & \uparrow F_2 \\
\cdots \xrightarrow{E^{n}} & & (2q-n,m,l+p-m)_+ & \xrightarrow{E^{q-n}} & (n,m,l)_- & \xrightarrow{E^{n}} & (-n,m,l-m)_+ & \xrightarrow{E^{q-n}} \\
& & \uparrow F_2 & & \uparrow F_1 & & \uparrow F_2 \\
& & \cdots & & \cdots & & \cdots
\end{array}
$$

Figure 3

$(i,j,k)_\pm$ denotes the module $F_\alpha(X) \otimes F_{\beta_\pm}(\phi) \otimes F_{gh}$

$$\alpha = \alpha_{ij} = \frac{1-i}{2}\alpha_+ + \frac{1-j}{2}\alpha_-$$

$$\beta_\pm = \beta_{ijk}^\pm = \frac{1 \pm i}{2}\alpha_+ + \frac{\pm j + 2k + 1}{2}\alpha_-$$

Shaded are the "boundary rows"

$$
\begin{array}{ccccccccc}
& & \cdots & & & & \cdots & & & & \cdots \\
& & \uparrow F_2 & & & & \uparrow F_1 & & & & \uparrow F_2 \\
\cdots \xrightarrow{E} & & (2q-1,m,l+2p-m)_+ & \xrightarrow{E^{q-1}} & & (1,m,l-p)_- & \xrightarrow{E} & & (-1,m,l+p-m)_+ & \xrightarrow{E^{q-1}} \\
& & \uparrow F_2 F_1 & & & & \uparrow F_1 F_2 & & & & \uparrow F_2 F_1 \\
\cdots \xrightarrow{E} & & (2q-1,m,l)_- & \xrightarrow{E^{q-1}} & & (1,m,l+p-m)_+ & \xrightarrow{E} & & (-1,m,l-p)_- & \xrightarrow{E^{q-1}} \\
& & \uparrow F_1 & & & & \uparrow F_2 & & & & \uparrow F_1 \\
& & \cdots & & & & \cdots & & & & \cdots \\
& & \uparrow F_1 & & & & \uparrow F_2 & & & & \uparrow F_1 \\
\cdots \xrightarrow{E^{q-1}} & & (q+1,m,l)_- & \xrightarrow{E} & & (q-1,m,l+p-m)_+ & \xrightarrow{E^{q-1}} & & (-q+1,m,l-p)_- & \xrightarrow{E} \\
& & \uparrow F_1 F_2 & & & & \uparrow F_2 F_1 & & & & \uparrow F_1 F_2 \\
\cdots \xrightarrow{E^{q-1}} & & (q+1,m,l+p-m)_+ & \xrightarrow{E} & & (q-1,m,l)_- & \xrightarrow{E^{q-1}} & & (-q+1,m,l-m)_+ & \xrightarrow{E} \\
& & \uparrow F_2 & & & & \uparrow F_1 & & & & \uparrow F_2 \\
& & \cdots & & & & \cdots & & & & \cdots \\
& & \uparrow F_2 & & & & \uparrow F_1 & & & & \uparrow F_2 \\
\cdots \xrightarrow{E^{n+1}} & & (2q-n-1,m,l+p-m)_+ & \xrightarrow{E^{q-n-1}} & & (n+1,m,l)_- & \xrightarrow{E^{n+1}} & & (-n-1,m,l-m)_+ & \xrightarrow{E^{q-n-1}} \\
& & \uparrow F_2 & & & & \uparrow F_1 & & & & \uparrow F_2 \\
\cdots \xrightarrow{E^{n}} & & (2q-n,m,l+p-m)_+ & \xrightarrow{E^{q-n}} & & (n,m,l)_- & \xrightarrow{E^{n}} & & (-n,m,l-m)_+ & \xrightarrow{E^{q-n}} \\
& & \uparrow F_2 & & & & \uparrow F_1 & & & & \uparrow F_2 \\
& & \cdots & & & & \cdots & & & & \cdots
\end{array}
$$

Figure 4

The same as in Figure 3 with the boundary rows dropped. There is a
bicomplex structructure now.

(This is IV_- case, the similar resolution exists for IV_+.) Starting from $L_{(q-1)m} \otimes F(\beta_{(q-1)m(p-m+l)}) \otimes F_{gh}$ the resolution becomes periodic (with the period $2(q-1)$) in its Virasoro sector. Passing to the next period shifts the Liouville charge β by the amount $\alpha_- p$ (cf. the charges of the first and the last a Fock spaces in (3.37)). Note also that in (3.37) participate only the representations L_{nm} from the principal Kac table (i.e. with $0 < n < q$, $0 < m < p$).

The "throwing away" procedure that we use seems a little bit *ad hoc*. In fact, it can be understood using the same logic as we used in the previous section for irrational $\hat{c}$. Another way to obtain the resolution (3.37) is explained in the next section.

Let us compute the characters of the representations of the type IV here, using the resolution (3.37). The formulas are (unlike the previous subsection, these are just the usual nonnormalized characters):

$$\chi_{nml} = x^{-2\alpha - \beta_{nml}} \chi_{bc}(t, x) \sum_{r=1}^{q-1} (-1)^{r-n} x^{r-n} \chi_{rm}^{Vir}(t) \left(\sum_{\substack{s=0 \ if \ r \geq n \\ s=1 \ if \ r < n}}^{\infty} x^{2sq} t^{\Delta^+(s)} \right.$$

$$\left. + x^{2(q-r)} \sum_{s=1}^{\infty} x^{2sq} t^{\Delta^-(s)} \right) \tag{3.38}$$

$$\Delta^+(s) = -\beta_{rm(l+sp)}(\beta_{rm(l+sp)} - \alpha_+ + \alpha_-) \tag{3.39}$$

$$\Delta^-(s) = -\beta_{rm(l-m+sp)}(\beta_{rm(l-m+sp)} - \alpha_+ + \alpha_-) \tag{3.40}$$

— for the IV_- case and

$$\chi_{nml} = x^{-2\alpha - \beta_{nml}} \chi_{bc}(t, x) \sum_{r=1}^{q-1} (-1)^{r-n} x^{r-n} \chi_{rm}^{Vir}(t) \left(\sum_{\substack{s=0 \ if \ r \leq n \\ s=1 \ if \ r > n}}^{\infty} x^{2sq} t^{\Delta^+(s)} \right.$$

$$\left. + x^{2(q-r)} \sum_{s=0}^{\infty} x^{2sq} t^{\Delta^-(s)} \right) \tag{3.41}$$

$$\Delta^+(s) = -\beta_{rm(l+sp)}(\beta_{rm(l+sp)} - \alpha_+ + \alpha_-) \tag{3.42}$$

$$\Delta^-(s) = -\beta_{rm(l+m+sp)}(\beta_{rm(l+m+sp)} - \alpha_+ + \alpha_-) \tag{3.43}$$

— for the IV_+ case. Because the symbol q is reserved already for the index of the minimal model, we denoted the modular parameter by t. The functions $\chi_{rm}^{Vir}(t)$ and $\chi_{bc}(t, x)$ are the characters of respectively the Virasoro irreducible representation L_{rm} and the (b, c) ghosts Fock space (the latter is independent of n, m, l).

It should be noted that the formulas for the characters were also obtained by Kac and Wakimoto and, independently, in [19]. Both groups used the method, different from ours, which did not allow to see the role of the Virasoro characters.

It is particularly interesting to consider the representations IV_+ with $l = jp$ and IV_- with $l = -1 + jp$, where j is any integer number. For definiteness let us consider the latter.

Note, that for such l the charges β of the Liouville Fock spaces coupled to the Virasoro irreducible representation $L_{mn}(Vir)$ in the resolution (3.37) are exactly the same as in the Lian-Zuckerman's papers [22],[23],[24] on the spectrum of 2-d gravity; it means that these liouvilles are the "dressing fields" of the discrete states. More precisely, the weights $\Delta^{\pm}(s)$ in (3.38) are related to the weights of the singular vectors in *the Verma module* $M_{mn}(Vir)$ (whose quotient is $L_{mn}(Vir)$) by the formula:

$$\Delta^{\pm}(s) + \Delta_{nm}(-j-s) = 1 \tag{3.44}$$

In this formula, $\Delta_{nm}(i)$ denotes the weight of the singular vector in $M_{mn}(Vir)$, and i is the ghost number of the LZ state, corresponding to this singular vector.

It is interesting, that the the infinite number of LZ states (with the ghost numbers $\leq -j$) are combined in the single object — the character of the irreducible representation of $N = 2$ SuperVirasoro algebra. It demonstrates the relationship between 2-d gravity and $N = 2$ $SVir$ very explicitly at the level of representations. In Section 4.2 we shall learn more about this.

4 Details of computations and proof of the equivalence theorem

1 Reduction of the BRST complex of $\widehat{sl}(2)/\widehat{sl}(2)$ coset

Now we can compute the cohomology $H^*_{Q_R}(L_k \otimes Wak_{-k-4})$. It is convenient to take the Felder resolution of the irreducible representation L_k of $\widehat{sl}(2)$ in terms of free Fock modules generated by the currents $\partial x(z)$, $\beta_M(z)$, $\gamma_M(z)^e$, using a usual screening operator

$$E^{(-)}_{sl(2)} = \oint \; : \beta_M(z) e^{\sqrt{2}\alpha_- x(z)} : \tag{4.1}$$

In principle, there is another screening

$$E^{(+)}_{sl(2)} = \oint \; : (\beta_M(z))^{-(k+2)} e^{\sqrt{2}\alpha_+ x(z)} : \tag{4.2}$$

Although the notion of the rational powers of $\beta_M(z)$ can be justified if we make bosonization of the β_M, γ_M pair, we prefer to deal with the "conventional" choice (4.2).

We know that under the standard (Drinfeld-Sokolov) Hamiltonian reduction these two screenings go to the two screenings

$$E^{(\pm)}_{Vir} = \oint \; : e^{\alpha_{\pm} x(z)} : \tag{4.3}$$

of the Virasoro algebra. Thus we can anticipate that it is also so in *our* reduction scheme, which means that the "conventional" screening $E^{(-)}_{sl(2)}$ must go to $E^{(-)}_{Vir}$ and the "unconventional" one $E^{(+)}_{sl(2)}$ must go to $E^{(+)}_{Vir}$.

---

[e] As the Toda-Liouville sector is already bosonized, we use the subscript "T" for its β, γ system, so this sector is generated by the currents $\partial\varphi(z), \beta_T(z), \gamma_T(z)$.

Let us decompose the reduction BRST operator Q_R as

$$Q_R = \hat{Q}_R + c^0(0)\mathcal{H}_0 \qquad (4.4)$$
$$\mathcal{H}_0 = (H_1 + H_2 + 2 : b_+ c^+ : -2 : b_- c^- :)_0$$

and note that there is a relation

$$\mathcal{H}_0 = \{Q_R, b_0(0)\}$$

Then the usual argument shows that on the cohomology

$$\mathcal{H}_0|_{H^*_{Q_R}} = 0 \qquad (4.5)$$

must hold. Moreover, we see that the operator $c^0(0)$ is Q_R-nontrivial so it maps the Q_R cohomology to itself:

$$|\lambda > \in H^*_{Q_R} \longrightarrow c^0(0)|\lambda > \in H^*_{Q_R} \qquad (4.6)$$

(This "doubling" of Q_R cohomology will be important in relations to 2-d gravity.)

It is convenient to choose the "light-cone" coordinates in the β, γ sector:

$$\beta_+ = \tfrac{1}{\sqrt{2}}(\beta_M + \beta_T) \qquad (4.7)$$
$$\beta_- = \tfrac{1}{\sqrt{2}}(\beta_M - \beta_T) \qquad (4.8)$$
$$\gamma_+ = \tfrac{1}{\sqrt{2}}(\gamma_M + \gamma_T) \qquad (4.9)$$
$$\gamma_- = \tfrac{1}{\sqrt{2}}(\gamma_M - \gamma_T) \qquad (4.10)$$

To compute $H^*_{Q_R}$ let us use a double complex with differentials

$$d_1 = \sqrt{2} \oint (c^+ \beta_+ - \sqrt{2} c^0(: \beta_+ \gamma_+ : - : b_+ c^+ :)) \qquad (4.11)$$
$$d_2 = \oint c^0(-\sqrt{2}\alpha_+(\partial\varphi + \partial x) - 2 : \beta_- \gamma_- : -2 : b_- c^- :) \qquad (4.12)$$

Thus we have managed to completely separate in d_1 and d_2 two subsets of fields: $\{c^+(z), b_-(z), \beta_+(z), \gamma_+(z)\}$ and $\{c^0(z), b_0(z), c^-(z), b_-(z), x(z), \varphi(z), \beta_-(z), \gamma_-(z)\}$. It means that our double complex is in fact a direct product of two complexes with differentials respectively d_1 and d_2. Their cohomology can be computed independently of each other. Let us compute the cohomology of d_1 first. It is easy to see that in $H^*_{d_1}$ two systems β_+, γ_+ and b_+, c^+ "cancel" each other. It means that there are no excitations along these four directions in the "physical" space (i.e. $H^*_{d_1}$). This is an example of the famous Kugo-Ojima quartet decoupling mechanism. To compute $H^*_{d_2}$ we need to bosonize the β_-, γ_- system in terms of two free bosons $\psi(z)$, $\chi(z)$:

$$: \beta_- \gamma_- : (z) = \partial\psi(z) \qquad (4.13)$$
$$\beta_-(z) = e^{(\psi + i\chi)} \qquad (4.14)$$

To be precise, following [25], the Hilbert space $F_{\beta,\gamma}$ of the β_-, γ_- system is represented by the (0-) cohomology of the screening operator

$$F = \oint \; : exp(-i\chi(z)) : \tag{4.15}$$

acting on the free Fock modules

$$\bigoplus_{\cdots} F(\psi) \otimes F(\chi) \tag{4.16}$$

(so it is a free field resolution). Now we have a double complex again: one differential is d_2 which acts now by the "free field" formula

$$d_2 = \oint \left(-\sqrt{2}\alpha_+(\partial\varphi(z) + \partial x(z)) - 2\partial\psi(z) : -2 : b_- c^- : \right) \tag{4.17}$$

and the other one is F. To compute the cohomology of this double complex let us find $H^*_{d_2}$ first. It is easy to do because it is actually the $U(1)$ cohomology. The representatives of $H^*_{d_2}$ can be written as:

$$X(z) = \sqrt{2}x(z) - \alpha_+(\psi + i\chi) \tag{4.18}$$
$$\phi(z) = \sqrt{2}\varphi(z) + \alpha_+(\psi + i\chi) \tag{4.19}$$
$$B(z) = b_- e^{(\psi + i\chi)}(z) \tag{4.20}$$
$$C(z) = c^- e^{-(\psi + i\chi)}(z) \tag{4.21}$$

Together they form a space

$$\bigoplus F(X) \otimes F(\phi) \otimes F(B, C) \tag{4.22}$$

Now we should compute $H^*_F(H^*_{d_2})$ (the second term of the spectral sequence of the double complex). If we only need the Q_R cohomology of two Wakimoto modules, we are done, and the answer is represented by the free field resolution

$$F_{\alpha\beta} \to F_{\alpha - \frac{\alpha_+}{2} \beta - \frac{\alpha_+}{2}} \to F_{\alpha - \alpha_+ \beta - \alpha_+} \to \cdots \tag{4.23}$$

The charges α, β in terms of the weights J_M, J_T of the Wakimoto modules are given by $\alpha = \alpha_- J_M$, $\beta = -\alpha_- J_T$. The differential in (4.23 is F).

However, if we are interested in the Q_R cohomology of the product of the irreducible representation times Wakimoto module, we should remember about the screening $E^{(-)}_{sl(2)}$ and the Felder resolution we made. Note, that the $\widehat{sl}(2)$ screenings $E^{(\pm)}_{sl(2)}$ can be represented by $E^{(\pm)}_{Vir} = \oint \; : e^{\alpha_\pm X(z)} :$ just as we thought (cf. (4.2,4.2)):

$$: e^{\alpha_+ X(z)} := : e^{\sqrt{2}\alpha_+ x(z) - (k+2)(\psi + i\chi)} := (\beta_M)^{-(k+2)} : e^{\sqrt{2(k+2)}x(z)} : \tag{4.24}$$

$$: e^{\alpha_- X(z)} := : e^{\sqrt{2}\alpha_- x(z) + (\psi + i\chi)} := \beta_M : e^{-\frac{2}{\sqrt{2(k+2)}}x(z)} : \tag{4.25}$$

(By the triviality of $H^*_{d_1}$ we may simply set $\beta_-(z) = \beta_M(z)$ at the level of the chiral algebra. In (4.24) we use this relation.)

Also, adding the trivial piece

$$\{d_2, b_0(z)\} = -\sqrt{2}\alpha_+(\partial\varphi(z) + \partial x(z)) - 2\partial\psi(z) : -2 : b_- c^- : \qquad (4.26)$$

to $i\chi(z)$ and using (4.13,4.18) we see that F (4.15) can be represented as

$$F = \oint : exp(-i\chi(z)) := \oint B(z) : e^{-\frac{\alpha_\pm}{2}(X(z)+\phi(z))} \qquad (4.27)$$

so actually there is no abuse of notations in (4.15) and we may think of F just as of the fermionic screening from the representation theory of N=2 $SVir$ which we introduced in Sec 3.

Thus, finally, we are left with the cohomology of the double complex (the last one in this story) with two differentials. One of them is just F and another one comes from the differential of the Felder resolution and is given by the certain powers of the screening E^-. This "ultimate" double complex is shown in the Fig.5.

It is interesting to compare this complex with the double complex in the Fig.4 because we anticipate that their cohomology are the same (and give the irreducible representation of N=2 Supervirasoro algebra). Although these two look very similar, there are some differences. The basic distinction between them is that the "horizontal" differentials are "made" of two different Vir-screenings: $E^{(+)}$ for the complex in the Fig.4 and $E^{(-)}$ for that one in the Fig.5.

They represent two different choices of the Felder resolutions for the same irreducible representation of Virasoro. As $E^{(-)}$ and F commute (cf. (3.18)), the vertical differential in Fig.5 is always F. The relations between $E^{(+)}$ and F are more involved. As a result in the Fig.4 the vertical differential is either F (denoted by F_1) or a combination like $xE^{(+)}F + FE^{(+)}$ (denoted by F_2), or $FE^{(+)}F$, depending on the place in the complex. Then, the weights α corresponding to the principal Virasoro Kac table representations are concentrated along one column in Fig.4. There are no weights from the boundary of the Kac table (we "threw them away"). On the other hand, the boundary weights are present in the Fig.5 and the weights from the interior of the principal Kac table are situated along the "shifted" vertical segments, shaded in the picture. (A shift occurs each time we pass through the horizontal line corresponding to the boundary weight.)

These differences result in the difference in computation of the cohomology. In both cases one can use the "vertical" filtration of the double complex to compute the "horizontal" cohomology first. In the case of the complex in the Fig.5 we immediately had the Kac table irreducible representations of Virasoro times some Fock spaces of ghosts and Liouville field all along one column. The "vertical" differential was induced either from F or $xE^{(+)}F + FE^{(+)}$ or $FE^{(+)}F$. The latter combination appeared between the rows where we have thrown away the "boundary row".

$$
\begin{array}{ccccccccc}
 & & \cdots & & \cdots & & \cdots & & \\
 & & \uparrow F_1 & & \uparrow F_1 & & \uparrow F_1 & & \\
\cdots & \xrightarrow{E^{m}_{(-)}} & (2q-1, 2p-m, l+p-m)_- & \xrightarrow{E^{p-m}_{(-)}} & (2q-1, m, l)_- & \xrightarrow{E^{m}_{(-)}} & (2q-1, -m, l-m)_- & \xrightarrow{E^{p-m}_{(-)}} & \\
 & & \uparrow F_1 & & \uparrow F_1 & & \uparrow F_1 & & \\
 & & \cdots & & \cdots & & \cdots & & \\
 & & \uparrow F_1 & & \uparrow F_1 & & \uparrow F_1 & & \\
\cdots & \xrightarrow{E^{m}_{(-)}} & (q+1, 2p-m, l+p-m)_- & \xrightarrow{E^{p-m}_{(-)}} & (q+1, m, l)_- & \xrightarrow{E^{m}_{(-)}} & (q+1, -m, l-m)_- & \xrightarrow{E^{p-m}_{(-)}} & \\
 & & \uparrow F_1 & & \uparrow F_1 & & \uparrow F_1 & & \\
\cdots & \xrightarrow{E^{m}_{(-)}} & (q, 2p-m, l+p-m)_- & \xrightarrow{E^{p-m}_{(-)}} & (q, m, l)_- & \xrightarrow{E^{m}_{(-)}} & (q, -m, l-m)_- & \xrightarrow{E^{p-m}_{(-)}} & \\
 & & \uparrow F_1 & & \uparrow F_1 & & \uparrow F_1 & & \\
\cdots & \xrightarrow{E^{m}_{(-)}} & (q-1, 2p-m, l+p-m)_- & \xrightarrow{E^{p-m}_{(-)}} & (q-1, m, l)_- & \xrightarrow{E^{m}_{(-)}} & (q-1, -m, l-m)_- & \xrightarrow{E^{p-m}_{(-)}} & \\
 & & \uparrow F_1 & & \uparrow F_1 & & \uparrow F_1 & & \\
 & & \cdots & & \cdots & & \cdots & & \\
 & & \uparrow F_1 & & \uparrow F_1 & & \uparrow F_1 & & \\
\cdots & \xrightarrow{E^{m}_{(-)}} & (n+1, 2p-m, l+p-m)_- & \xrightarrow{E^{p-m}_{(-)}} & (n+1, m, l)_- & \xrightarrow{E^{m}_{(-)}} & (n+1, -m, l-m)_- & \xrightarrow{E^{p-m}_{(-)}} & \\
 & & \uparrow F_1 & & \uparrow F_1 & & \uparrow F_1 & & \\
\cdots & \xrightarrow{E^{m}_{(-)}} & (n, 2p-m, l+p-m)_- & \xrightarrow{E^{p-m}_{(-)}} & (n, m, l)_- & \xrightarrow{E^{m}_{(-)}} & (n, -m, l-m)_- & \xrightarrow{E^{p-m}_{(-)}} & \\
 & & \uparrow F_1 & & \uparrow F_1 & & \uparrow F_1 & & \\
 & & \cdots & & \cdots & & \cdots & &
\end{array}
$$

Figure 5

$$E_{(-)} = E^{(-)}_{Vir} = \oint : e^{\alpha_- X(z)} :$$

Shaded are the "shifted segments" mentioned in Section 4.1

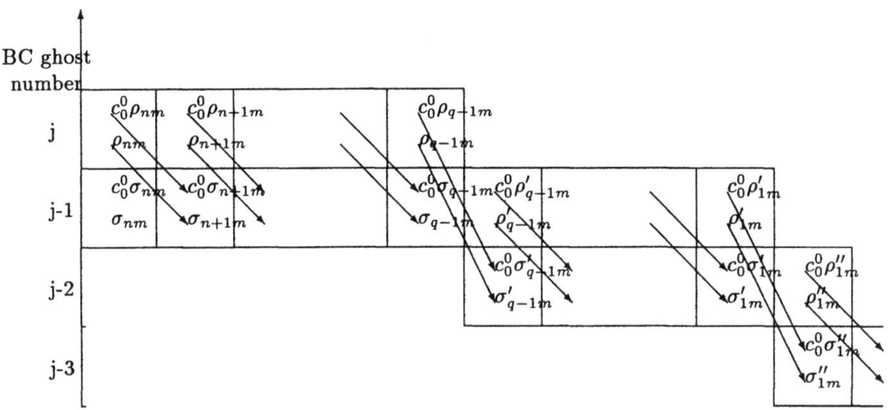

Figure 6

Arrows show the action of the differential d_2.

$(\sigma_{nm}, \sigma'_{nm}, \sigma''_{nm}, \ldots)$ and $(\rho_{nm}, \rho'_{nm}, \rho''_{nm}, \ldots)$ are the LZ states in the irreducible representation L_{nm} with all possible Liouville dressings.

The leftmost lower cell is the only one containing nontrivial cohomology $H^*_{d_2}$.

On the other hand, in the case shown in the Fig.6, the horizontal cohomology give the irreducible representations for the rows corresponding to the weights from the interior of the principal Kac table and zeros for the rows corresponding to the weights from the boundary of the Kac table. We obtain zero cohomology for the "boundary" rows because each such row represents two (left- and right-sided) glued together resolutions of the same irreducible representation, so the resulting horizontal complex is exact. The nontrivial cohomology are concentrated along the "shifted segments".

The "vertical" cohomology gives again the resolution (3.37). To see this we should first recall that the three pairs of indices (n, m), $(n + q, m + p)$ and $(q - n, p - m)$ describe the same Virasoro weight. Note then that the "knight move" differential d_2 is nontrivial for this spectral sequence. It is a "connecting differential" — it acts from the top of one "segment" to the bottom of another "shifted segment", mending all these segments together in one complex which is just (3.37).

Now to get $H^*_{Q_R}(L_k \otimes Wak_{-k-4})$ we recall about the c^0 zero mode "doubling". (Here also, as in the case of $H^*_{Q_R}(Wak_k \otimes Wak_{-k-4}$ there could be a problem with the "knight move" differential of the main spectral sequence, but it is zero here for the same reason as there.) Thus we have

$$H^*_{Q_R}(L_k \otimes Wak_{-k-4}) = [L(N = 2\ SVir) \oplus c^0_0 L(N = 2\ SVir)] \qquad (4.28)$$

Let the $sl(2)$ weights (spins) of the modules L_k (the matter) and Wak_{-k-4} (the "Toda-Liouville") in (4.28) be respectively J_M and J_T. Then the parameters (α, β) of the irreducible representation $L_{\alpha,\beta}(N = 2\ SVir)$ there are given by

$$\alpha = -\alpha_- J_M \qquad (4.29)$$

$$\beta = \alpha_- J_T \qquad (4.30)$$

(remember that $\alpha_- = -\frac{1}{\sqrt{k+2}}$)

In particular, let $L_k(J_M)$ be the admissible irreducible representation of $\widehat{sl}(2)$, i.e.

$$J_M = \frac{m - 1}{2} + (k + 2)\frac{1 - n}{2} \qquad (4.31)$$

and let the weight J_T of $Wak_{-4-k}(J_T)$ be such that it dresses properly a singular vector (having weight J'_M) in the Verma module $M_k(J_M)$, which means that $J_T + J'_M = -1$. Then the representation $L_{\alpha\beta}(N = 2\ SVir)$ in (4.28) in the notations of Section 3 is L_{nml} of IV_- type with $l = -jp$, where j is the ghost number of the corresponding $\widehat{sl}(2)$-BRST state.

Now, to prove (2.8) (it was promised in the Sec.2), we take the vacuum representation of $\widehat{sl}(2)$ which is a VOA for $\widehat{sl}(2)$ and, looking at (4.28), recall that the vacuum representation of N=2 $SVir$ algebra we see in the right hand side is a VOA for the latter. (In the more physical language it just means that by the $Operators \rightarrow States$ correspondence the chiral algebras $N = 2\ SVir$ and $\widehat{sl}(2)$ are related to the "descendants of the unity operator" which form the corresponding vacuum representations.)

2 Equivalence of the spectra of topological $\widehat{sl}(2)/\widehat{sl}(2)$ coset and minimal matter coupled to gravity

Finally, we compute the cohomology $H^*_{Q_W}$ of the reduced $\widehat{sl}(2)$ — BRST complex $H^*_{Q_R}(L_k \otimes Wak_{-k-4})$ and show they coincide with the 2-d gravity BRST cohomology of $L_k^{DS} \otimes Wak_{-k-4}^{DS}$ [f]. For definiteness, we do it for this choice of $\widehat{sl}(2)$ representations, but it can easily be done also for other choices, mentioned in Section 2.

To do the actual computation, let us use the equivalence of two complexes:

$$(H^*_{Q_R}(L_k(\widehat{sl_k}(2)) \otimes Wak_{-4-k}), Q_W) = (L(N = 2\ SVir) \oplus c_0^0 L(N = 2\ SVir), G_0^+) \quad (4.32)$$

Suppose that the spin of $L_k(\widehat{sl_k}(2))$ is given by (4.31) and the spin of the Wakimoto module Wak_{-4-k} is "dressing" (see the discussion after (4.31)). Then the $N = 2$ irreducible representation $L(N = 2\ SVir)$ in (4.32) is L_{nml} with $l = -jp$.

Now, take the "2-d gravity" resolution (3.37) of L_{nml}. On the "2-d gravity" modules $L(Vir) \otimes F(Liouv.) \otimes F_{gh}$ the action of the differential G_0^+ is given by the formula (2.14): $Q_W = G_0^+ = Q_{Vir}$. So we have a double complex. One differential there is Q_{Vir} and the other one comes from the resolution (3.37).

The cohomology of this double complex, by our construction, compute the $\widehat{sl}(2)$ BRST cohomology of $L_k(\widehat{sl_k}(2) \otimes Wak_{-4-k}$ — the physical states of the coset model. At this stage we have to assume that *we know* all the physical states *either* for the coset model *or* for 2-d gravity. Then we shall be able to find the the spectrum of states for the other theory. Suppose for definiteness that we know the spectrum of 2-d gravity coupled to (q, p) – minimal matter[g].

It means that in our double complex we can compute the "gravitational" cohomology $H^*_{Q_{Vir}}$ first — it gives the first term of the spectral sequence. Recalling that in (3.37) for the representations IV_- with $l = -jp$ the charges of the liouvilles are just right to dress the "discrete states", we end up with the situation shown in the Fig.6. It is important to remember that we compute the usual, i.e. "absolute" cohomology. It is well known that the BRST cohomology of the properly dressed irreducible representation are given by *two* elements at the adjacent ghost numbers. This "Virasoro doubling" is due to the zero mode of the diffeomorphisms ghost C_0. Thus all nontrivial cohomology states in the first term of the spectral sequence (the Lian-Zuckerman states) are concentrated along the "shifted segments" according to their ghost numbers. Each time we pass half-period $p - 1$ (recall the periodic structure of (3.37)), we get a shift by -1 of the ghost number. At the ghost number zero there are two states, and at each positive ghost number, there are four states. Note the "$\widehat{sl}(2)$ doubling" due to the zero mode c_0^0.

---

[f] As above, the superscript "DS" denotes the standard (Drinfeld-Sokolov) reduction.

[g] Using the standard homological algebra, it is easy to go the other way — i.e. to obtain the 2-d gravity spectrum *from* the spectrum of the coset.

To compute the second term of the spectral sequence, we need the properties of the differential d of the resolution (3.37). This operator was studied in [21]. Remember that it is induced either from the fermionic screening F (when it acts "within one half-period") or from FEF (when it acts "between two half-periods"). Thus it changes the ghost number by -1 or by -2 units respectively. Its action on the LZ states is shown by the arrows. We see that the cohomology are concentrated in the same two degrees as the cohomology $H^*_{Q_{Vir}}(L_{nm} \otimes F(\beta_{nml}))$ and also have two elements, now due to the "$\widehat{sl}(2)$ doubling". Of course, this is the correct answer for $H^*_{Q_{BRST}}(L_k \otimes Wak_{-k-4})$. The nontrivial thing that happens is that the "Virasoro doubling" gets transformed into the "$\widehat{sl}(2)$ doubling".

Now we should only note that $L_{nm} \otimes F(\beta_{nml})$ — the first term of the resolution (3.37), — is nothing else but the Drinfeld-Sokolov reduced $(L_k)^{DS} \otimes (Wak_{-k-4})^{DS}$. Thus we have shown, that the coset- and 2-d gravity cohomology are the same thing, basically because they both describe the G_0^+ cohomology of the $N = 2$ $SVir$ irreducible representation L_{nml}.

Acknowledgements

I am grateful to M. Bershadsky, E. Frenkel, C. Vafa for the interesting discussions. I thank S. Chung for bringing to my attention the paper [19].

Research supported in part by the Packard Foundation and by NSF grant PHY-87-14654

References

[1] V. Sadov *On the spectra of $\widehat{sl}(N)_k/\widehat{sl}(N)_k$-cosets and W_N gravities.* Harvard preprint HUTP-92/A055

[2] M. Bershadsky, H. Ooguri *Comm. Math. Phys.* **126***(1992) 49*

[3] B. Feigin, E. Frenkel *Phys. Letts.* **B246**(1990) 75

[4] O. Aharony, O. Ganor, N. Sochen, J. Sonnenschein, S. Yankielowicz *Physical states in G/G Models and 2d Gravity* TAUP-1961-92

[5] O. Aharony, J. Sonnenschein, S. Yankielowicz *G/G Models and W_N Strings* TAUP-1977-92

[6] H. L. Hu, M. Yu *On BRST cohomology of SL(2)/SL(2) gauged WZNW models* AS-ITP-92-32

[7] W. Lerche, D. Nemeshansky, M. Bershadsky, N. Warner *A BRST Operator for non-critical W-Strings* HUTP-A034/92

[8] E. Bershgoeff, A. Sevrin, X. Shen *A derivation of the BRST operator for noncritical strings* preprint

[9] B. Gato-Rivera, A. Semikhatov *Phys. Letts.***B293** (1992) 72

[10] M. Bershadsky, W. Lerche, D. Nemeshansky, N. Warner *N=2 Extended superconformal structure of Gravity and W Gravity coupled to Matter* HUTP-A034/92

[11] A. Losev *Descendants constructed from matter field and K. Saito higher residue pairing in Landau-Ginzburg theories coupled to topological gravity* preprint TPI-MINN-92-40-T

[12] S. Mukhi, C. Vafa *Two dimensional Black hole, c=1 Non-Critical Strings and a Topological Coset Model.* Harvard preprint HUTP-93/A002

[13] V. Sadov *Free field resolution for nonunitary representations of N=2 SuperVirasoro* Harvard preprint HUTP-92/A070

[14] K. Thielemans *Int. J. Mod. Phys.* **C** Vol.2 No.3, 787 (1991)

[15] V. Dobrev *Phys. Lett.* **186B** (1987) 43

[16] V. Dobrev *Structure of Verma modules and characters of irreducible highest weight modules over $N = 2$ superconformal algebras* in *Proc. of the XV Int. Conf. on Diff. Geom. Meth. in Theor. Physics, Clausthal (July 1986)*, Eds. H. D. Doebner and J. Hennig (World Sci, Singapore, 1987) pp. 289-307

[17] G. Mussardo, G. Sotkov, M. Stanichkov *Int. J. Mod. Phys***A4** (1989) 1135

[18] K. Ito, *Nuclear Physics***B332** (1990) 566

[19] C. Ahn,S. Chung, S. H. H. Tye "New Parafermion, SU(2) Coset and N=2 Superconformal Field Theories", *Nucl. Phys.* **B365**, 191-242 (1991)

[20] L. Rozansky *a letter to M. Bershadsky*, 1989

[21] Vl. Dotsenko *Mod. Phys. Lett.***A7** (1992) 2505

[22] B. H. Lian, G. J. Zuckerman *Phys. Lett* **254B***(1991) 417*

[23] B. H. Lian, G. J. Zuckerman *Phys. Lett* **266B***(1991) 21*

[24] B. H. Lian, G. J. Zuckerman *Comm. Math. Phys.* **145***(1992) 54*

[25] B. Feigin, E. Frenkel *Comm. Math. Phys.* **128***(1990) 161*

[26] P. Bouwknegt, J. McCarthy, K. Pilch *On the BRST structure of W_3 gravity coupled to c=2 matter* preprint USC-93-14

LATTICE MODELS
AND $N = 2$ SUPERSYMMETRY

Hubert SALEUR and Nicholas P. WARNER

Physics Department
University of Southern California
University Park
Los Angeles, CA 90089-0484
USA

Abstract: We review the construction of exactly solvable lattice models whose continuum limits are $N = 2$ supersymmetric models. Both critical and off-critical models are discussed. The approach we take is to first find lattice models with natural topological sectors, and then identify the continuum limits of these sectors with topologically twisted $N = 2$ supersymmetric field theories. From this, we then describe how to recover the complete lattice versions of the $N = 2$ supersymmetric field theories. We discuss a number of simple physical examples and we describe how to construct a broad class of models. We also give a brief review of the scattering matrices for the excitations of these models.

1 Introduction

$N = 2$ supersymmetric theories in two dimensions are in some ways much simpler than their non-supersymmetric kin. This is essentially because the $N = 2$ supersymmetry implies the presence of a topological sector for which semi-classical analysis yields exact quantum results. (In the language of supersymmetry, the "F-terms" are not renormalized.) This does not mean that the model is semi-classically rigid, but only that a key (topological) subsector of the theory is semi-classically determined: the complete theory is very rich and has all of the complexity of a non-supersymmetric theory. The topological subsector has thus provided a "bridgehead" from which many of the non-trivial quantum aspects of the model can be explored, and usually with greater facility, and often in more detail than is possible for non-supersymmetric theories. (For reviews, see [1, 2, 3].) For example, $N = 2$ Landau-Ginzburg theories have a superpotential that is exact. For $N = 2$ superconformal models this means that the operator algebra and the anomalous dimensions (conformal weights) of the Landau-Ginzburg fields are trivially computable [4, 5]. On the other hand, it is still somewhat unclear as to how much of the rest of the theory, and in particular, how much of the complete operator content is

Quantum Field Theory and String Theory, Edited by
L. Baulieu *et al.*, Plenum Press, New York, 1995

determined by the Landau-Ginzburg potential. Indicative results are known: For example, the ADE classification of modular invariants of $N = 2$ minimal models collapses to the ADE classification of modality zero singularities [6, 4]. Ramond sector characters can be obtained from the Landau-Ginzburg potential by computing the elliptic genus [7, 8, 9, 10].

For $N = 2$ supersymmetric quantum integrable field theories (QIFT's [a]) the Landau-Ginzburg description leads to transparent analysis of the soliton structure [11, 12, 13]. In $N = 2$ QIFT's, and even non-integrable $N = 2$ QFT's, differential equations can also be obtained for some of the scaling functions [14]. For the $N = 2$ QIFT's these scaling functions can also be obtained from the thermodynamic Bethe ansatz, but instead of differential equations, one obtains complicated integral equations [15].

Over the last decade it has also become evident that many of the structures of quantum integrable field theories have analogues in exactly solvable lattice models (see, for example [16, 17, 18, 19, 20]). Using this technology, lattice models have been constructed in which the continuum limits give rise to many of the non-supersymmetric conformal field theories. It is therefore natural to expect that there should be exactly solvable lattice models whose continuum limits are $N = 2$ supersymmetric QIFT's. We shall henceforth refer to such lattice models as $N = 2$ lattice models[b]. It is also to be hoped that such lattice regularizations may have especially simple features. If so, they may in turn help us to understand some mysterious issues like the appearance of branching functions in local height probabilities [21]. There are also practical reasons for constructing $N = 2$ lattice models. The simplicity of $N = 2$ QFT's has led to more progress than in any other kind of field theory. It is especially attractive to try to compare some of the advanced results with real or computer experiments. To do so, the construction of a "physically reasonable" lattice model whose continuum limit is the $N = 2$ QFT of interest is a natural way to proceed. We also hope that the $N = 2$ lattice models will be "more universal" than their non-supersymmetric counterparts. Indeed, consider a general non-supersymmetric model, and suppose that we do not impose the requirement of exact solvability. Since there are generically very many relevant operators in such a model, the corresponding coupling constants would need to be very finely tuned in order to find a particular second order phase transition and hence a particular conformal field theory. Thus one of the side-effects of exact solvability is to automatically make such a fine-tuning of the couplings. The $N = 2$ lattice models have a further special, and robust, identifying feature: the presence of a topological sector. Thus, not only can one construct them by imposing exact solvability, but one can also identify them by virtue of their topological sector. In fact, we will argue in this review that any "trivial" statistical mechanics model will probably give rise to a non-trivial $N = 2$ lattice model, and thus the $N = 2$ models will be of relevance in a large variety of situations. For instance the polymer and percolation problems, the statistics of Bloch

---

[a] In this review, conformal field theory will be abbreviated as CFT, quantum field theory as QFT, and quantum integrable field theory as QIFT.

[b] This nomenclature is somewhat misleading in that it suggests that the supersymmetry is realized on the lattice, whereas it still remains unclear whether this can be done in most of the $N = 2$ lattice models thus far constructed.

walls in high temperature Ising model, Brownian motion, are all described by $N = 2$ supersymmetric theories.

The investigation of lattice models associated with supersymmetric QFT's has a rather long history. To our knowledge[c] it started with the study of $N = 1$ supersymmetry in the tricritical Ising model [22, 23, 24] and the fully frustrated XY-model [25] (the latter related to Josephson junctions arrays).

Then $N = 1$ supersymmetry together with $N = 2$ supersymmetry were observed at some special points of the Ashkin-Teller model and the six-vertex (or XXZ) model [26, 27, 28]. By analysis of torus partition functions it was easy to identify more generally special points of the higher spin $U_q(SU(2))$ vertex models [29, 30] that were $N = 2$ supersymmetric. A physical explanation of this came later [31]. In the simplest case of the six-vertex model it leads to $N = 2$ supersymmetry in the percolation problem. Similarly, by analysis of the related Izergin-Korepin ($O(n)$) model [32, 33], another family of lattice models with $N = 2$ supersymmetry was found (or conjectured). The simplest of this family leads to $N = 2$ supersymmetry in the self-avoiding walk (polymer) problem. This turns out to be a very favorable case for comparison with real and numerical experiments [34]. More complete study of the $SU(2)$ based models was carried out in [35]. Recently, it has been shown how one can go beyond models based upon $SU(2)$. That is, $N = 2$ lattice models based on any simply-laced Lie algebra have been constructed by using "partial restriction" of modified solid-on-solid (SOS) models, or equivalently by twisting and performing *partial* quantum group truncation of the corresponding vertex models. The continuum limits of these lattice models are the $N = 2$ superconformal coset models based on hermitian symmetric space [36, 37, 38].

At the present time, there is a respectable number of known $N = 2$ lattice models. They resemble in some respects the corresponding QFT, although the structures are not yet as closely linked as one would wish. They have already led to comparison with real and numerical experiments, and very good agreement has been found. The purpose of this review is to describe in some detail the current state of knowledge and to point to new directions of research.

The first section is rather qualitative and collects observations about the expected structure of $N = 2$ lattice models. A variety of simple examples is worked out, starting from simple geometrical ideas and introducing the fundamental tools that will be used extensively later. One of the basic themes underlying the $N = 2$ lattice constructions is the role of the "trivial" topological sector. This will be used directly in later sections to construct and analyse the $N = 2$ lattice models based on general Lie algebras. In the fourth section we will briefly discuss the scattering matrices of excitations in the $N = 2$ lattice models and their continuum limits. In section five we discuss further physically interesting aspects of $N = 2$ lattice models, and in the last section we conclude by summarizing what we believe are the important open problems in the subject.

---

[c]We apologize for any reference missing in this short history.

2 Features of $N = 2$ lattice models

Our purpose in this section is to study a number of closely related and simple models that exhibit the basic features of the $N = 2$ lattice models. Our aim is to try to give some intuitive understanding of $N = 2$ lattice models, to show how one can analyse their various features, and ultimately to evolve a strategy for finding such models.

1 The clues from the topological sector

If the theory has $N = 2$ supersymmetry then, as mentioned earlier, there is a topological sector, with the following characteristics. The topological sector of the $N = 2$ QFT consists of operators that are annihilated by two of the four supercharges. The topological physical[d] states consist of only the ground states of the non-topological $N = 2$ lattice model. The operators of the topological model are order parameters of the non-topological sector, and the correlation functions of these operators are constant in the topological sector.

The presence of some sort of "topological" sector is one of the main attributes of an $N = 2$ lattice model, and because of the links between the topological and non-topological sectors in the continuum, the identification of the lattice topological sector proves an important tool in the construction of the complete $N = 2$ lattice model. More precisely, the idea is to study representations of the lattice algebras and lattice symmetries (such as Temperley-Lieb and quantum group representations). For certain choices of parameters some of these will be very simple or trivial, and the corresponding lattice models will be "topological." The idea is to define the topological model precisely, fix the parameters to their topological values, but then modify the choice of representations or modify the choice of lattice variables in such a manner that the $N = 2$ lattice model emerges from its topological limit. We will describe two "dual" methods for accomplishing this. The first relies on the fact that a lattice model can have very different properties when it is described by different sets of observables that are not mutually local. One thus obtains some of the $N = 2$ lattice models by considering a "trivial" model where the given lattice variables do not interact, and then one makes a "change of variables" to new set of geometrical observables that have highly non-trivial properties. The non-interacting observables describe the topological model, while the new observables describe the non-trivial $N = 2$ lattice model. The second method is to consider the restriction process in solid-on-solid (SOS) models (or quantum group truncation in the equivalent vertex formulation), and start with a restriction that results in a "frozen" or completely rigid topological model. One then "gets something from nothing" by relaxing the restrictions imposed on the SOS model, while at the same time one "topologically untwists" the Boltzmann weights to avoid singularities in the transfer matrix. It turns out that the first method is *a priori* a little more intuitive, while the second method is much easier to generalize. We therefore start by describing a simple example of the first procedure and then trace a connection to the second method.

---

[d]The definition of "physical" for the lattice model can be quite different that of the field theory.

2 Examples of lattice topological sectors

1 Percolation: an example of a geometrical model

Consider a square lattice and put on every edge a variable σ that can take two possible values, 0 or 1. Suppose that the edge variables do not interact with each other, but put the whole system in a magnetic field. Since the spins do not interact, all correlation functions are trivially constant (or, at least are delta functions). This appears to be a completely trivial model and the partition function is $\mathcal{Z} = \left(1 + e^{-H}\right)^{N_E}$, where N_E is the number of edges. We now pass to a new set of observables (called geometrical observables in what follows): Call edges with variable $\sigma = 0$ "empty" and edges with variable $\sigma = 1$ "occupied" (we represent occupied edges with heavy lines in figure 1) and consider the geometrical properties of clusters, that is, of the connected sets of occupied edges. These cluster observables are non-local with respect to the original σ variables, and the "trivial" model now describes the bond percolation problem. Since the sum over variables $\sigma = 0, 1$ is unconstrained, there is a probability of $p = \frac{e^{-H}}{1+e^{-H}}$ of having an occupied edge, and a probability of $p = \frac{1}{1+e^{-H}}$ of having an empty edge. The correlation functions of new variables, such as the probability that two vertices belong to the same cluster, are non-trivial. The original trivial model can, of course, be recovered by deciding to study only local questions like "is this edge occupied" and forgetting about non local questions like "are these two edges part of the same cluster". As we will see, the distinction between these two questions is clearer from a representation theory point of view, and easy to implement, for instance, in a transfer matrix formalism.

To determine the supersymmetry point of this model we consider it upon a torus. The new observables can be given various boundary conditions: for instance, one can sum over non-contractible clusters giving each some weight, or fugacity. The presence of unbroken supersymmetry means that, in the thermodynamic limit, the free energy per edge cannot depend upon the boundary conditions on the torus. Intuitively, if $p > 1/2$ or $p < 1/2$, one will find large clusters of occupied or unoccupied edges. Thus non-contractible clusters will make significant contributions to the free energy, and thus the latter will depend upon the boundary conditions. However, when one has $p = 1/2$, or $H = 0$, one is at the percolation point, a critical point where the geometrical correlation functions decay algebraically and there is a vanishing fraction of big clusters of occupied or unoccupied edges. The partition function is $\mathcal{Z} = 2^{N_E}$. This lattice partition function is, as usual, defined up to a non-universal term of the form e^{fN_E}. In the continuum, the partition function simply counts the number of "different observables" of the model. It is rather reasonable that there is only one such observable (beside the identity): the variable σ. One thus expects $Z = 2$. This is also consistent with the identification of this percolation problem as an $N = 2$ lattice model. As we will see, the foregoing lattice partition function can be interpreted as the Witten index of the $N = 2$ lattice model. We will soon present rather more direct evidence that for $p = 1/2$ this is indeed an $N = 2$ lattice model.

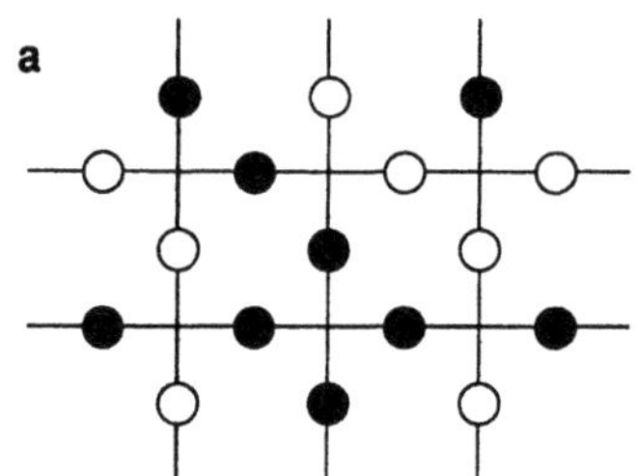

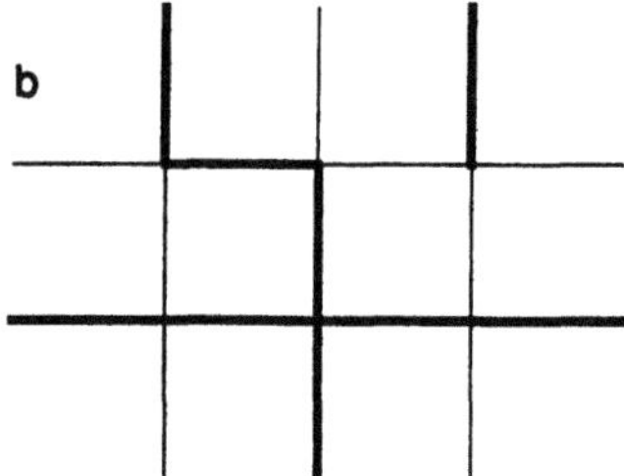

Figure 1. a) A configuration of "up" (black dot) or "down" (white dot) spins on the edges of the square lattice. b) The corresponding configuration of "occupied" or "empty" edges.

2 The dual point of view: the "frozen" Potts model

The foregoing non-trivial bond percolation problem can also be described using the Q-state Potts model in the limit $Q \to 1$. The general Q-state Potts model is defined by putting a spin variable, σ, on each *vertex* of the square lattice. These spin variables take values in $1, \ldots, Q$, and the lattice partition function is

$$\mathcal{Z}_{\text{Potts}} \;=\; \sum_{\{\sigma\}} \prod_{<i;j>} e^{K\,\delta(\sigma_i, \sigma_j)} \;, \tag{2.1}$$

where the sum is over all spins and the product is over is over nearest neighbours. The connection with the percolation problem becomes evident upon making a high temperature expansion of the Potts model: The partition function can be written

$$Z_{\text{Potts}} \;=\; \text{tr} \left(\prod_{<i;j>} \left\{ 1 \,+\, (e^K - 1)\delta(\sigma_i, \sigma_j) \right\} \right) \;, \tag{2.2}$$

which can then be expanded graphically by connecting all vertices with the same spin σ. One then finds that

$$Z \;=\; \sum_{\text{graphs}} (e^K - 1)^{N_B}\, Q^{N_C} \;, \tag{2.3}$$

where N_B is the number of bonds in the graph, and N_C is the number of clusters. This high temperature expansion coincides with the bond percolation problem only if the $[Q^{N_C}$ term is equal to one, *i.e.* if and only if $Q = 1$. One must also make the following identification:

$$e^K - 1 \;=\; e^{-H} \;. \tag{2.4}$$

However, if one considers the Potts model with $Q = 1$, one finds that it is a trivial, completely frozen, "topological" model. The partition function (2.1) collapses to $\mathcal{Z}_{\text{Potts}} = e^{KN_E}$. The bond percolation problem should really be viewed as the $Q \to 1$ limit of the Potts model. For example, the derivatives of $\mathcal{Z}_{\text{Potts}}$ at $Q = 1$ describe generating functions for the percolation problem [39]. Similarly, derivatives of the Green functions correspond to geometrical correlation functions; for example the spin-spin two point function of the Potts model yields the probability that two vertices belong to the same cluster.

We therefore see that the percolation problem is closely associated with a frozen model. To obtain the non-trivial model from the frozen model we need to find models that are "close to the frozen model." The natural way of doing this is to use an algebraic approach.

3 Representations of the Temperley-Lieb algebra

To proceed further it is convenient to think in algebraic terms and turn to a transfer matrix formalism [40]. Consider thus a rectangle of size $L \times T$ and propagation in the $\hat{y}$ direction. We define more precisely the geometrical observables in terms of connectivities as follows. Number the vertices on a given time slice by $1, 2, \ldots, L$, running from left to right. The space of states $\mathcal{H}_L^{\text{geom}}$ is the set of all possible partitions of $1, 2, \ldots, L$. In a partition of L at time t, two vertices are grouped together if at time t they are connected through a cluster that extends into their present and past. For instance, for $L = 4$ the partition $(13)(2)(4)$ correspond to the fact that there is connectivity between 1 and 3, while 2 and 4 are not currently connected to any other vertex by connections through clusters in their past (see figure 2). Introduce operators, e_k, that act on $\mathcal{H}_L^{\text{geom}}$. For odd subscripts, e_{2j-1} transforms connectivities at time t to connectivities at time $t + 1$ by inserting vertical "occupied" bonds at all but the jth vertex. Thus one has $e_3|(13)(2)(4) >= |(13)(2)(4) >$, and $e_1|(13)(2)(4) >= |(1)(2)(3)(4) >$. For even subscripts, e_{2j}, modifies connectivities at the same time, and simply inserts a horizontal edge between the jth and $(j + 1)^{\text{th}}$ vertices. For example, one has $e_2|(13)(2)(4) >= |(123)(4) >$.

These operators satisfy the relations

$$e_j^2 = \sqrt{Q}\, e_j, \qquad e_j e_{j\pm1} e_j = e_j, \qquad [e_j, e_k] = 0, \; |j - k| \geq 2 \qquad (2.5)$$

with $\sqrt{Q} = 1$, and hence furnish a particular representation of the Temperley-Lieb algebra [41, 42, 33]. The general form of this algebra can be realized in a number of simple models, and, in particular, in the Q-state Potts model.

Define $\mathcal{H}_L^{\text{Potts}}$ to be the space of all possible (Potts model) spin states on the L vertices across a time slice of the lattice. Introduce the following operators on $\mathcal{H}_L^{\text{Potts}}$:

$$(e_{2j})_{\sigma,\sigma'} = \sqrt{Q} \prod_k \delta(\sigma_k, \sigma_k')\, \delta(\sigma_j, \sigma_{j+1}) \qquad (2.6)$$

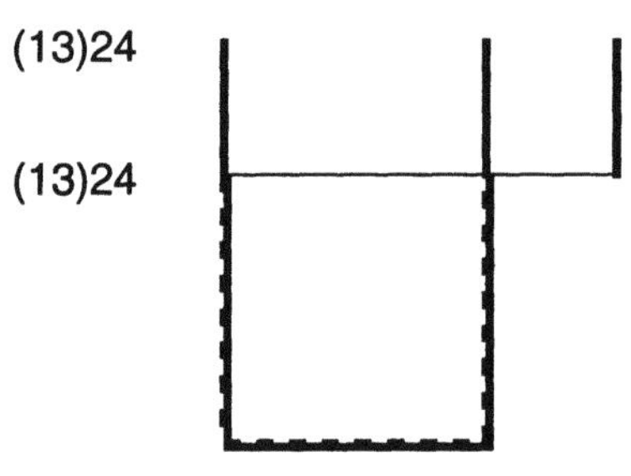

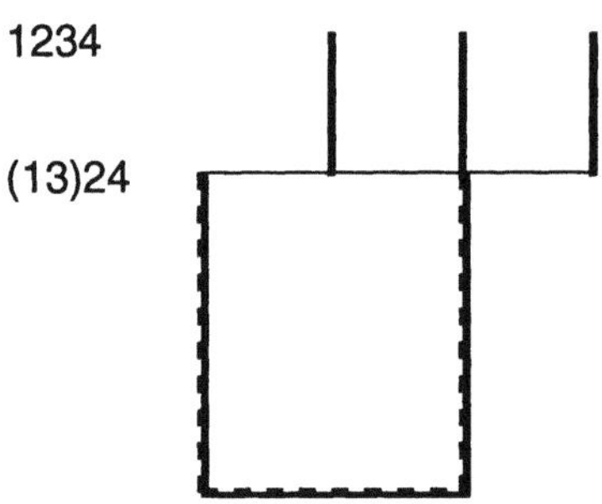

Figure 2. The partition (13)24 means that vertices 1 and 3 are connected through a path of occupied edges at previous times (by convention past means edges below $y = t$). Acting with e_3 corresponds to adding an occupied vertical edge in all but the second column between $y = t$ and $y = t+1$ resulting again in (13)24. Acting with e_1 corresponds to adding an occupied vertical edge in all but the first column between $y = t$ and $y = t + 1$ and this destroys the connection between vertices 1 and 3, resulting in the partition 1234.

and

$$(e_{2j-1})_{\sigma,\sigma'} = \frac{1}{\sqrt{Q}} \prod_{k \neq j} \delta(\sigma_k, \sigma'_k) . \tag{2.7}$$

The operators e_j satisfy the Temperley-Lieb algebra for general Q. The transfer matrix of the Potts model may be written:

$$\tau = Q^{L/2} \prod_{j=1}^{L} X_{2j} \prod_{j=1}^{L} X_{2j-1} \tag{2.8}$$

where

$$X_{2j-1} = \frac{e^K - 1}{\sqrt{Q}} + e_{2j-1}, \quad X_{2j} = 1 + \frac{e^K - 1}{\sqrt{Q}} e_{2j} . \tag{2.9}$$

The decoupled spin model in the magnetic field is related to this by setting $Q = 1$ and $e^K = 1 + e^{-H}$ as in (2.4). The transfer matrix of the bond percolation-problem is also given by (2.8), (2.9), but with the Temperley-Lieb representation given at the beginning of this subsection. More generally the transfer matrices of the vertex and

342

Temperley-Lieb algebra. The various possible weights associated to non-contractible clusters and the various natural geometrical sectors have precise meaning in terms of traces and representation theory [43].

3 Unfreezing frozen models

We believe that the foregoing features are generic for an $N = 2$ lattice model. Such a model should be closely related to two trivial models: one rigid or frozen, and the other some kind of high temperature dual in which the lattice variables are decoupled from one another. In between these extremes is the Hilbert space of the $N = 2$ lattice model. Common to all of these models is the representation theory of some underlying lattice algebra. The problem of making the $N = 2$ lattice model is to find the proper observables in the decoupled model, or to find the proper "unfreezing" of the frozen model. As we will discuss, this may all be thought of in terms of choices of the representations of the lattice algebra. We will also find that while the "decoupled" description of the model is more intuitive, it is the "unfreezing" process that is easiest to implement more generally. To define more precisely the unfreezing procedure, we first want to describe two other representations of the Temperley-Lieb algebra, and their frozen limits.

1 Yet more frozen models

Consider the representation of the Temperley-Lieb algebra provided by restricted solid on solid (RSOS) models. (We refer to more general extensions of such models as interaction-round-a-face (IRF) models.) These representations have $\sqrt{Q} = 2\cos\pi/(m+1)$, where m is an integer. The model is defined by introducing heights, $\ell = 1, \ldots, m$, on the vertices and faces of the original square lattice (figure 3). Represent these heights as the nodes of the Dynkin diagram of A_m. The transfer matrix now acts on a configuration space $\mathcal{H}_{2L}^{\mathrm{RSOS}}$ whose basis is given by elements $\{\ell_j, j = 1, \ldots, L\}$ with the constraint that neighbouring vertices on the lattice carry heights that are neighbours on the A_m diagram. Let v_ℓ be the components of the Perron-Frobenius eigenvector of the Cartan matrix of A_m (or of the incidence matrix of the Dynkin diagram). That is, $v_\ell = sin(\pi\ell/(m+1))$. The representation of the Temperley-Lieb algebra is then given by:

$$(e_j)_{\ell\ell'} = \delta(\ell_j, \ell_{j+2}) \frac{\left(v_{\ell_{j+1}} v_{\ell'_{j+1}}\right)^{1/2}}{v_{\ell_j}} \prod_{k \neq j+1} \delta(\ell_j, \ell'_j). \qquad (2.10)$$

The RSOS transfer matrix is then given by using this in (2.8) and (2.9). To get the representation with $Q = 1$, one simply takes $m + 1 = 3$. The model is rigid, with heights alternating between the value 1 on one sublattice, and the value 2 on the other. This model thus has two states, depending upon which sublattice takes the value 1 and which takes the value 2.

This model has some form of duality with the original non-interaction spin model. The non-interacting spin model can be considered as the $T \to \infty$ limit of an Ising model on the medial lattice. The dual [41] is an Ising model defined on the faces and vertices

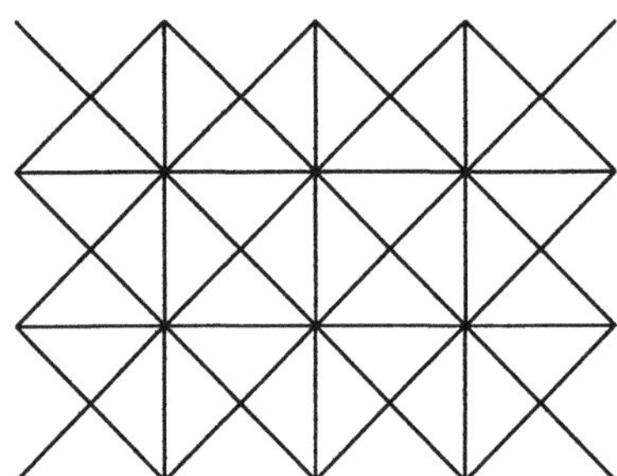

Figure 3. The RSOS variables are defined on the vertices of the diagonal lattice, that is the vertices and the faces of the original square lattice.

of the original lattice and at $T = 0$, which is completely "frozen," and has exactly two states.

More importantly for our current purposes, this model has an equivalent vertex formulation whose continuum limit is a Gaussian model. This will finally yield more direct evidence of the supersymmetry.

2   The vertex model and its continuum limits

We now introduce our final representation of the Temperley-Lieb algebra: the one provided by the six-vertex model. Introduce spin variables $\pm$ (usually represented by arrows) on the edges of the medial graph of the original square lattice (figure 4). Think of these spin variables as being basis vectors in $\mathbb{C}^2$. The configuration space of the transfer matrix is now $\mathcal{H}_{2L}^{\mathrm{vertex}} = \left(\mathbb{C}^2\right)^{2L}$. To define the operators e_j, introduce 4×4 matrices $E_{ij,kl}$, where i, j, k, l are either $+$ or $-$, by setting all the entries to zero, except the ij,kl entry which is set equal to 1. Then e_j is equal to the identity in all but the jth and $(j+1)^{\mathrm{th}}$ copies of $\mathbb{C}^2$, where it is given by

$$e_j = q^{-1}E_{+-,+-} + qE_{-+,-+} - E_{+-,-+} - E_{-+,+-} \, . \tag{2.11}$$

The parameter Q of the Potts model is related to the parameter q of the six-vertex model by:

$$\sqrt{Q} \ = \ q + q^{-1} \, , \tag{2.12}$$

and so the supersymmetric model corresponds to setting $q = \exp(i\pi/3)$.

The continuum limit of this model is a Gaussian model [45, 18, 19]. More precisely, introduce variables ϕ on the vertices of the original lattice and in the middle of its faces, with values defined recursively by $\phi_u = \phi_d \pm 1$, that is, the value of ϕ above one edge of the medial equal to the value under it plus the value of the spin carried by the edge.

344

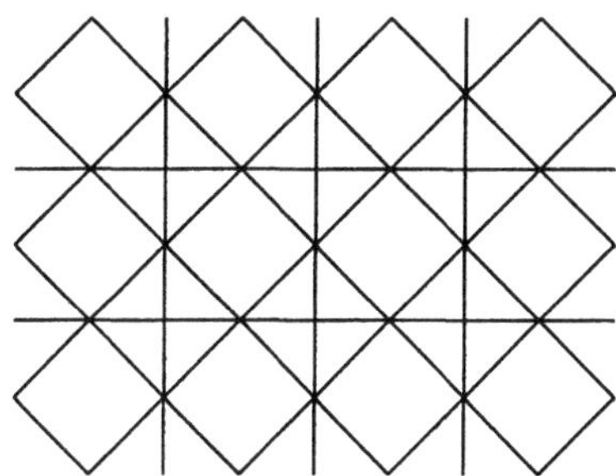

Figure 4. The six-vertex arrows are defined on the edges of the diagonal (medial) lattice. They are also interpreted as domain walls for a SOS model whose heights are defined on the vertices and faces of the square lattice (see figure 3).

The dynamics of these variables is described by a Gaussian action at large distance, and in the conventions where the topological defects are not renormalized, the action is

$$A = \frac{g}{4\pi} \int (\partial\phi)^2 \tag{2.13}$$

with $g = 2/3$. As a result, the partition function on a torus with doubly periodic boundary conditions,

$$\mathcal{Z} = \mathrm{tr}\left(\tau^{\mathrm{vertex}}\right)^T , \tag{2.14}$$

is given, in the continuum, by

$$Z = Z_G(3/2) , \tag{2.15}$$

where Z_G is a Gaussian partition function. That is,

$$Z_G(g) = \frac{1}{\eta\bar{\eta}} \sum_{e,m \in Z} p^{h_{em}/4} \bar{p}^{\bar{h}_{em}/4} , \tag{2.16}$$

where

$$h_{em} = \frac{1}{4}\left(\frac{e}{\sqrt{g}} + m\sqrt{g}\right)^2 , \qquad \bar{h}_{em} = \frac{1}{4}\left(\frac{e}{\sqrt{g}} - m\sqrt{g}\right)^2 \tag{2.17}$$

and $\eta(p) = p^{1/24} \prod(1 - p^n)$ is the Dedekind eta function, and the elliptic nome is given by $p = \exp - 2\pi T/L$. Equation (2.15) coincides with the partition function of the first minimal (central charge $c = 1$) $N = 2$ superconformal model with projection on odd fermion number. The holomorphic generators of the $N = 2$ superconformal algebra can be written (after normalizing the boson appropriately):

$$\begin{aligned} G^+(z) &= e^{i\sqrt{3}\phi(z)} ; \quad G^-(z) = e^{-i\sqrt{3}\phi(z)} ; \\ J(z) &= \tfrac{i}{\sqrt{3}}\partial\phi(z) ; \quad T(z) = -\tfrac{1}{2}(\partial\phi(z))^2 . \end{aligned} \tag{2.18}$$

If one used the trivial representation of the Temperley-Lieb algebra given by equations (2.6) and (2.7), instead of the representation given above, it would lead to

$$\mathcal{Z} = \mathrm{tr}\left(\tau^{\mathrm{Potts}}\right)^T = 2^{N_E} . \tag{2.19}$$

number[e]. It is therefore constant and corresponds to the Witten index of the model. It may also be thought of as the partition function of the continuum topological matter model.

Beside partition functions, various physical observables in the percolation problem can be conveniently studied in the $N = 2$ supersymmetry formalism. Usually the critical percolation problem is considered as a $c = 0$ CFT, *i.e.* a twisted $N = 2$ superconformal theory. The physical observables of statistical mechanics are then the ones that are usually discarded as unphysical from string theory point of view. The probability that two edges belong to the same cluster corresponds to an operator with half-integer labels in the Kac table of the $c = 0$ CFT. Only by turning to an $N = 2$ formalism can its correlation functions be studied. This operator belongs in the $N = 2$ formalism to the sector "in between" Ramond and Neveu-Schwarz, with spatial boundary conditions twisted by $(\sqrt{-1})^F$. Its dimension in the twisted theory is $h = \frac{5}{96}$.

3   A practical use of the topological sector

The identification of a topological sector is not only useful for supersymmetry purposes. If supersymmetry is unbroken, the free energy of a non-trivial lattice model, like the six-vertex model with $q = e^{i\pi/3}$, can be readily computed by turning to the topological frozen model. At the special supersymmetric point, it is not necessary to perform a Bethe ansatz computation to determine f. This will be discussed further in section 3.

4   Quantum group truncation, restriction and freezing

Because the six-vertex model (with free boundary conditions) has a quantum group symmetry, one can use this to reduce the Hilbert space of the model to obtain a new truncated model. In the language of height models this correspond to making the RSOS restriction. It may also be viewed as the lattice version of the BRST reduction of the Gaussian model to the Virasoro minimal models [44]. The first step is to take q to be a root of unity. Here we will take $q = e^{i\pi/m+1}$, for some integer m. One then replaces the traces of operators by Markov traces, that is, the trace of an operator $\mathcal{O}$ is defined by:

$$tr_M(\mathcal{O}) \; = \; \mathrm{tr}\left(\mathcal{O} \, q^{2H} \right) \, , \tag{2.20}$$

where H is the Cartan subalgebra generator of the $U_q(SU(2))$ symmetry of the model [17, 55] For example, the partition function of the truncated vertex model is defined by $\mathcal{Z} - tr_M(\tau^T)$.

To see how this truncation works, consider the Hilbert space of the vertex model, and imagine decomposing it into representations of the $U_q(SU(2))$ symmetry. The effect of the factor of q^{2H} in (2.20) is to weight the contribution of each representation by its q-dimension. In particular, this means that only the type II representations contribute to the trace[f]. This means that we can restrict the trace in (2.20) to a trace over type

---

[e]In the purely bosonic formulation the fermion number is defined by $F = e^{\pi i (J_0 + \widetilde{J}_0)}$, where J and $\widetilde{J}$ are the holomorphic and anti-holomorphic $U(1)$ currents.

[f]The other representations are called type I and are those representations that are reducible, but indecomposable. Such representations also have vanishing q-dimension [17].

$\mathcal{II}$ representations. The result of this truncation in the continuum limit is that the Gaussian model becomes the minimal model with central charge $c = 1 - \frac{6}{m(m+1)}$.

To relate the vertex model to a height model one makes a change of basis in the vertex model Hilbert space [16, 46, 17, 47]. At the vertices and faces of the original lattice one introduces positive integer heights, ℓ, in the following manner. One views the arrows of the vertex model as defining the basis elements of the spin-$\frac{1}{2}$ representation of $U_q(SU(2))$. One starts at one side of the lattice with a fixed element of some spin-j_0 representation of $U_q(SU(2))$; if one tensors this with an element of the spin-$\frac{1}{2}$ representation then one obtains a combination of vectors in the spin-$(j_0 + \frac{1}{2})$ and spin-$(j_0 - \frac{1}{2})$ representations. Perform this tensoring successively with the spin-$\frac{1}{2}$ states on the edges across the lattice. After each tensoring with a spin-$\frac{1}{2}$ state, associate the spin of the resulting $U_q(SU(2))$ representation to the next vertex. At the same time, keep track of the total Cartan subalgebra eigenvalue of the state. A sequence of such spins across the lattice, along with Cartan subalgebra eigenvalue is a new basis for the vertex model Hilbert space (figure 5). The new basis is related to the old by a huge collection of Clebsch-Gordon coefficients. The quantum group symmetry means that the transfer matrix is independent of the Cartan subalgebra eigenvalue, and so we may discard it. The heights, ℓ, are then related to the spins, j, by $\ell = 2j + 1$.

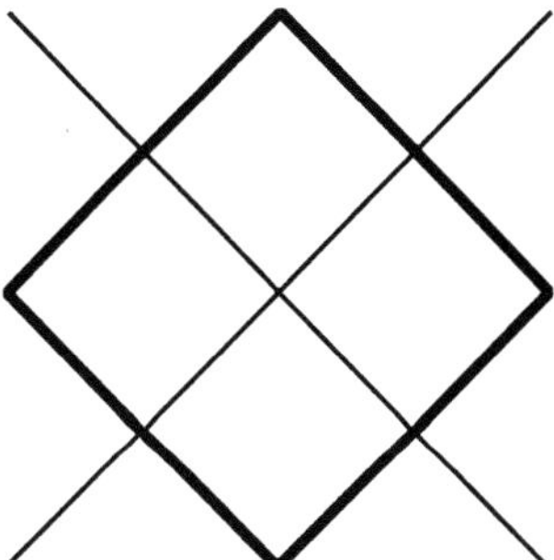

Figure 5. The geometry used in the vertex-RSOS transformation. Heights sit on the vertices of the heavy square. The dotted edges carry the spins of the six-vertex model. Heights are viewed as $U_q(SU(2))$ highest weights and neighbours are connected by tensor product with the fundamental representation.

One can rewrite the transfer matrix, and indeed the individual Boltzmann weights, in this new basis. They are related to the vertex model Boltzmann weights by quantum $6j$-symbols [46, 47]. The effect of performing the quantum group truncation in the vertex model is equivalent to restricting the spins, j, of the representations on each vertex to those of type $\mathcal{II}$ representations. That is, for $q = e^{i\pi/m+1}$, one restricts the heights happen if one made the change of basis but did not perform the truncation. One would find that the some other $6j$-symbols would vanish, and as a result some Boltzmann weights would become singular. Thus, when q is a root of unity, this particular height formulation is necessarily the restricted (RSOS) model.

For $q = e^{i\pi/3}$, the only type $\mathcal{II}$ representations of $U_q(SU(2))$ are spin-0 and spin-$\frac{1}{2}$. Therefore, as has already been observed, the heights in the RSOS model alternate between 1 and 2, and the model is frozen.

From this perspective it is somewhat clearer what we must do in order to recover an $N = 2$ model from a frozen model. We must generalize a process that takes us back to the six-vertex model from the RSOS model. Morally speaking, this should involve some form of releasing the height restriction, however it is not quite this simple since, as we mentioned above, doing this in the RSOS model will result in both vanishing and singular Boltzmann weights. The key point is to realize that the solution to this problem of singular Boltzmann weights is directly related to the process of untwisting the energy-momentum tensor of the continuum topological theory into the energy-momentum tensor of the $N = 2$ superconformal theory. In more concrete terms, the fact that the transfer matrix in the six vertex model with free boundary conditions commutes with the quantum group means that the transfer matrix and the corresponding spin-chain Hamiltonian necessarily contain special boundary terms [17]. The continuum limit of this Hamiltonian is that of the topological theory. Thus to obtain the $N = 2$ lattice model one has to twist the transfer matrix of the vertex model with free boundary conditions so as to remove these boundary terms [36]. Doing this also breaks the quantum group symmetry and thus modifies the spectrum of the spin-chain Hamiltonian [37]. The result is the pure Gaussian model described earlier. An alternative way of removing these boundary terms is simply to use periodic boundary conditions (this was implicitly used earlier). This works for the six vertex model, but not for its generalizations, in which we only want to break part of the quantum group symmetry. We will discuss this more extensively in the next section.

5   Where is the supersymmetry on the lattice?

As stressed above, one of the indications of supersymmetry in any model is the existence of a trivial sector. We have also seen that the $N = 2$ superconformal generators can be explicitly constructed in the continuum limit. One would, however, like to see more direct evidence of the supersymmetry on the lattice. For example, one would like to be able to exhibit lattice quantities that reproduce the supersymmetry algebra. More generally, one would like to find some lattice fermion operators.

Unfortunately, although some progress has recently been made [56] in the identification of lattice quantities whose continuum limit is the Virasoro algebra, the situation is still very unclear for supersymmetry generators. An obvious related difficulty is that on the lattice we only see the symmetry $U_q(SU(2))$ with $q = \exp(i\pi/3)$, while the continuum theory is characterized by a pair of quantum groups, the other one being $U_{q'}(SU(2))$, $q' = \exp(i\pi/2)$, and it is this latter quantum group that has much to do with the supersymmetry algebra. For the more general models of next section, we will again observe the "wrong" quantum group on the lattice.

For the $c = 1$ superconformal model there is, however, an $N = 2$ lattice model based on polymers that exhibits a $U_{q'}(SU(2))$ symmetry. Although no explicit realization of supersymmetry is known in this model, it at least allows a rather satisfying identification of lattice fermionic degrees of freedom. Polymers, or self-avoiding walks, are better described as the limit of the $O(n)$ model [62] as $n \to 0$. The topological model is simply a model with no degrees of freedom at all, and with $\mathcal{Z} = 1$. The non-local geometrical observables are obtained by considering properties of self-avoiding mutually avoiding walks on the lattice. Typical correlation functions at coupling β have the form

$$\sum_N \beta^N \Omega_N \ , \tag{2.21}$$

where the sum is taken over all self-avoiding walks connecting two points (see figure 6), and Ω_N, is the number of such walks of length N. The critical point, where the correlators decay algebraically, occurs when $\beta = \beta_c$, where β_c^{-1} is the "effective connectivity constant," which is defined for large N by $\Omega_N \approx \beta_c^{-N}$, $N >> 1$. The proper algebraic setting for this model is $A_2^{(2)}$. It has however a $U_q(SU(2))$ symmetry with $q = i$ [63]. A simple way of identifying fermionic degrees of freedom is to think of the zero weight given to closed loops as being the sum of statistical factors $+1$, the -1 corresponding to the loop carrying a bosonic or fermionic variable. On a torus, the choice of antiperiodic boundary conditions for fermions gives a weight $1 + 1 = 2$ to some families of loops: this allows a simple recovery of the various sectors of the $N = 2$ theory from the lattice. This fermion interpretation allows also a transparent interpretation of the new index of [14, 15] in polymer terms.

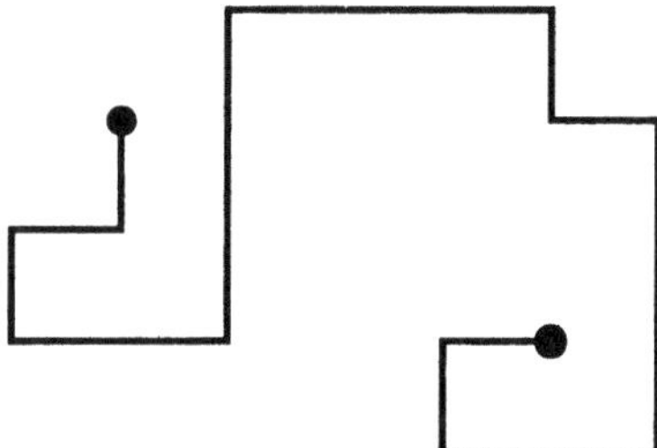

Figure 6: The derivative of the spin two-point function in the $O(n)$ model, evaluated at $n = 0$, becomes a generating function for the self-avoiding walks that connect two points.

Off-critical $N = 2$ supersymmetry

The first, and most obvious property guaranteed by (unbroken) supersymmetry in a field theory is that the ground state energy will be exactly zero. This does not immediately imply that the bulk free energy of the lattice model is zero since the partition function, $\mathcal{Z}$, is ambiguous up to non-universal factors of the form e^{fN}, where N is the number of "sites" of the model. Equivalently, one can always multiply all of the Boltzmann weights by some overall function that is analytic and nowhere vanishing in the regime of interest. This will produce the foregoing ambiguity in the partition function. Thus, in a supersymmetric lattice model we should expect that the free energy will vanish up to logarithms of such analytic functions.

As we remarked in the introduction, the percolation problem has $N = 2$ supersymmetry in the continuum limit only at its critical point. To have unbroken supersymmetry in the scaling region it is necessary that the trivial sector remains trivial in the perturbation, and moreover that the free energy per vertex continues to be sector independent. If this is so, then the free energy cannot have any singular part. Recall that in general the free energy has two parts:

$$f \; = \; f_{\text{reg}} \; + \; f_{\text{sing}} \; , \tag{2.22}$$

where f_{reg} is analytic in the variable measuring the distance away from criticality; and

$$f_{\text{sing}} \; = \; A_\pm \, \xi^{-2} \; , \tag{2.23}$$

where $A_\pm$ are amplitudes above and below the critical point, and ξ is the correlation length. The singular part of the free energy, f_{sing}, can be identified with E_0/L where E_0 is the ground state energy of the quantum theory. If supersymmetry is unbroken then one has $E_0 = 0$, and therefore $f_{\text{sing}} = 0$.

This lattice criterion can immediately be used to claim that for the polymer problem the flow to the dilute region (where $\beta < \beta_c$ in (2.21)) does not break supersymmetry. On the other hand the flow to the dense region $\beta > \beta_c$ does break supersymmetry. This is because for bigger weights, the walks fill a finite fraction of the available space (figure 6) [65], so as long as one loop is allowed there is a non-trivial free energy: $\mathcal{Z}_R = 1$ but $\mathcal{Z} \approx e^{fTL}$ otherwise. Supersymmetry turns out to be broken spontaneously, a fact that is possible because of the non-unitarity of the perturbed polymer problem. We will discuss this more in a later section.

If one starts with the six vertex model at criticality, then the eight vertex model is the obvious off-critical generalization. In the regime where the elliptic nome is positive, and for $q = e^{i\pi/3}$, the supersymmetry is not spontaneously broken. (This will be established as a consequence of the results in the next section.) Indeed, it has been known for some time that the singular part of the free energy vanishes for the eight-vertex model for this value of q [41, 48]. In the continuum limit, the eight-vertex model can be thought of as a theory with Landau-Ginzburg potential $W = X^3 - \lambda X$. In the other regimes, where the theory is non-unitary, we expect that supersymmetry will be broken.

We believe that the foregoing is a generic property of $N = 2$ lattice models: there will be an off-critical regime in which supersymmetry is unbroken, and a non-unitary regime where its spontaneously broken.

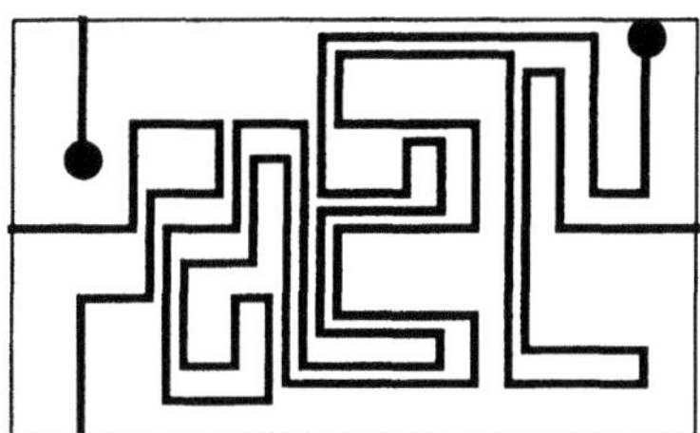

Figure 7: If the variable β of the generating functions is real and bigger than the radius of convergence β_c in infinite volume, the self-avoiding walks start to fill the system and properties depend very much on the boundary conditions. The free extremities of the walks repel one another algebraically.

6 Summary

At the end of this long, and perhaps somewhat confusing, section we would like to recall the properties that we expect of an $N = 2$ lattice model and summarize them in the form of a strategy that can be used in seeking out such lattice models.

For any Lie algebra, spin and quantum group parameter q there is a natural lattice algebra with which solutions of the Yang-Baxter equations are built. This algebra has several types of representations, including a RSOS one, a vertex one, and some others of mixed type [64]. Choose the value of q such that the RSOS representation is trivial. One obtains then a candidate for a topological sector. Study then the complete model obtained from the vertex representation or maybe the mixed one. Compute in particular its torus partition function and compare it with known $N = 2$ ones. To investigate unbroken supersymmetry away from the critical point, study integrable deformations of the critical Boltzmann weights, and (i) check that the free energy has no singular part (ii) compute the scaling dimension and $U(1)$ charge of the lattice operators corresponding to the integrable perturbation and verify that the continuum limit of these operators preserves the supersymmetry.

3 The coset construction of $N = 2$ lattice models

Thus far, we have considered only the simplest of the $N = 2$ lattice models: those whose quantum group structure is $U_q(SU(2))$ and whose Hilbert space forms a representation of the Temperley-Lieb algebra. The critical, continuum limits of these models are the $N = 2$ superconformal minimal models. To go beyond this and obtain theories from models with larger quantum group symmetries, and whose continuum limits are the $N = 2$ superconformal coset models, one first finds a formulation of the coset models in which the topological sector has a simple lattice formulation.

Consider the $N = 2$ superconformal coset model of the form [66]:

$$\mathcal{M}_k(G;H) \equiv \frac{G_k \times SO_1(dim(G/H))}{H} \, , \tag{3.1}$$

where G is simply laced and G/H is a hermitian symmetric space. It was observed in [67] that there is a natural perturbation of this model leading to an $N = 2$ supersymmetric quantum integrable field theory, $\mathcal{M}_k^*(G; H)$. Moreover, the topological subsector of this quantum integrable model can be identified with the topological coset conformal field theory:

$$\frac{G_k \times G_0}{G_{k+0}} \,. \tag{3.2}$$

In particular, the correlators of the topological sector of $\mathcal{M}_k^*(G; H)$ are all constant, and can be written in terms of the structure constants of the fusion algebra of G_k. Since lattice analogues of the $G_k \times G_\ell / G_{k+\ell}$ are well known, the basic approach should now be evident. To see more precisely how the N=2 lattice model is constructed we need to elaborate some of the details of the superconformal model.

1 A Coulomb Gas Formulation

For simplicity, take the level of G in (3.1) to be one (*i.e.* $k = 1$)[9]. Let r be the rank of G. Since G and H have the same rank, the representations of the current algebra of G_1 are finitely decomposable as representations of the current algebra of H_1. Because G/H is a symmetric space, the representations of the current algebra of $SO_1(dim(G/H))$ are finitely decomposable into representations of the current algebra of H_{g-h}, where g and h are the dual Coxeter numbers of G and H respectively. It follows that $\mathcal{M}_k(G; H)$ can be thought of as a coset model of the form:

$$H_1 \times H_{g-h}/H_{g-h+1} \,, \tag{3.3}$$

but with a special choice of modular invariant. Because of this equivalence, one can find a Coulomb gas formulation that directly generalizes the $SU(2)$ Coulomb gas formulation of subsection 2.3.2. For $SU(2)$, this was a simple Gaussian model, here one gets a Gaussian model with r free bosons compactified on a scaled version of the weight lattice of G. The model has a quantum group symmetry of $U_q(H_0)$ where H_0 is the semi-simple factor of H. (For hermitian symmetric spaces G/H, the group H has the form $H = H_0 \times U(1)$, where H_0 is semi-simple.) The quantum group parameter, q, is the one appropriate to the denominator factor of the coset (3.3), that is, one takes

$$q = e^{\frac{i\pi}{(g-h+1)+h}} = e^{\frac{i\pi}{g+1}} \,. \tag{3.4}$$

This choice fixes the radius of compactification of the Gaussian model to the "supersymmetric radius."

To reduce the Gaussian model to the requisite coset model one must perform the BRST reduction of the field theory, which, on the lattice, is the quantum group truncation with respect to $U_q(H_0)$. The model knows it origins as $\mathcal{M}_k(G; H)$ essentially because the Gaussian model is compactified upon the scaled weight lattice of G. The

---

[9]The field theories with higher levels, k, can be obtained by including generalized parafermions, and the lattice models with higher values of k can be obtained by fusion.

field theory also contains two operators, $\mathcal{X}^+$ and $\mathcal{X}^-$, that in the Coulomb gas formulation extend the generators of $U_q(H_0)$ to a twisted form of the affine quantum group $U_q(\widehat{G})$. In the $N = 2$ superconformal model, these operators are hermitian conjugates of each other, they are relevant, and together provide a perturbation that yields a unitary $N = 2$ supersymmetric quantum integrable field theory [68, 67]. This is directly parallel to the situation for non-supersymmetric quantum integrable models obtained by conformal perturbation theory: the perturbation is usually one that leads to an affine extension of any underlying quantum group structure. Here, however, one needs two perturbing operators to make the quantum integrable model, one of which extend $U_q(H_0)$ to $U_q(G)$ and the other extends this to $U_q(\widehat{G})$.

The topological twist of any $N = 2$ supersymmetric field theory can be implemented by first replacing the energy momentum tensor by [69, 70]:

$$T^{\text{top}}_{\mu\nu} = T^{N=2}_{\mu\nu} + \frac{1}{2}\epsilon_{\mu\rho}\, g^{\rho\sigma}\partial_\sigma J_\nu \, , \tag{3.5}$$

where J_ν is the conserved $U(1)$ charge of the $N = 2$ supersymmetric theory. With this energy momentum tensor, two of the supercharges become dimension zero charges and can be used as BRST charges to truncate the original Hilbert space down to the topological Hilbert space.

The topological twist shifts the dimension of the operators $\mathcal{X}^+$ and $\mathcal{X}^-$ so that they provide exactly the correct charges to extend $U_q(H_0)$ to $U_q(\widehat{G})$. The topological energy momentum tensor thus commutes with $U_q(G)$. The type $\mathcal{II}$ representations of $U_q(G)$ at the value of q that we have chosen in (3.4) are all trivial, and the BRST reduction, or quantum group truncation, will result in a rigid topological model.

In the foregoing discussion we used a minor sleight of hand that we wish to bring into the open. In the continuum field theory formulation of a coset model there are always two quantum groups, one associated with a numerator factor and the other associated with the denominator factor. Thus in (3.3) there are two quantum groups, the one that we have been discussing with q given by (3.4), and one associated with the numerator factor of H_{g-h} with $q = e^{i\pi/g}$. Two supercharges can be added to the latter quantum group extending it to a twisted form of $U_q(\widehat{G})$, while the extension of the denominator quantum group is accomplished by the relevant perturbing operators $\mathcal{X}^+$ and $\mathcal{X}^-$ as described above. To define the topological theory one is supposed to use one of the supercharges as a BRST charge, and this implicitly suggests that one gets the topological theory using the numerator copy of $U_q(G)$. However it should be remembered that the BRST reduction of a Coulomb gas formulation of a coset model can be done using either one of the two quantum groups, and we have used, and will need to use, the denominator copy of the quantum group.

2 Formulating the $N = 2$ Lattice Models

To build the $N = 2$ lattice models one simply reverses the foregoing course. One starts either with a vertex model built using the fundamental representation, $\mathcal{V}$, of G, or with the IRF model whose heights are the weights of G. The "topologically twisted" transfer matrix is built from the $\check{R}$-matrix of $U_q(G)$. If one performs the quantum group truncation of the vertex model [17, 16], or the restriction of the height model, using the complete $U_q(G)$ with q given by (3.4) the result is a rigid lattice model. In the IRF formulation each successive height will be the unique level one fusion product of the previous height and the representation $\mathcal{V}$.

To get the $N = 2$ lattice model one must topologically untwist the foregoing transfer matrix, breaking the $U_q(G)$ symmetry to $U_q(H_0)$ in such a way that the two extra generators in $U_q(\widehat{G})$ are given the same spin. One then only performs the quantum group truncation with respect to $U_q(H_0)$, or, in the IRF formulation, only restricts the heights to be affine highest weights of H_0 at level $g - h + 1$, leaving the $U(1)$ direction unrestricted. This is referred to as partial quantum group truncation, or a partial RSOS model. This is the procedure as it was originally implemented in [36]. In [36] the Coulomb gas formulation was extensively analysed to precisely define the topological untwisting of the transfer matrix, and to provide further evidence that the result was indeed an $N = 2$ lattice model. In this review we will use more recent results to define the off-critical Boltzmann weights in the IRF formulation of these models with $G = SU(N)$. It turns out that in the off-critical formulation the "untwisting" is easy to describe and is rather intuitive. We begin by reminding the reader about the construction of RSOS models based on the weight lattice of $SU(N)$.

Introduce an orthonormal basis, f_j, $j = 1, \ldots, N$, in $\mathbb{R}^N$ and define vectors e_j by $e_j = f_j - \frac{1}{N}(f_1 + \ldots + f_N)$. The vectors e_j can be thought of as weights of the fundamental of $SU(N)$. Consider the oriented lattice shown in figure 8. To each vertex assign a height of the form:

$$\Lambda = \Lambda_0 + \sum_{j=1}^{N-1} n_j e_j , \tag{3.6}$$

where $n_j \in \mathbb{Z}$ and Λ_0 is an as yet arbitrary vector. This "initial vector," Λ_0, will play a major role in the forthcoming discussion.

If Λ is a height at the beginning of an oriented edge and Λ' is the height at the other end of the oriented edge, then we require that $\Lambda' - \Lambda = e_j$, for some j. For the typical plaquette shown in figure 9, the evolution is defined by the Boltzmann weights $w(\Lambda, \Lambda + e_i, \Lambda + e_j + e_j, \Lambda + e_k|u)$. Introduce the shorthand notation:

$$[\nu] \equiv \vartheta_1(\gamma\nu|\tau) \tag{3.7}$$

$$\equiv 2p^{\frac{1}{8}} sin(\pi\gamma\nu) \prod_{n=1}^{\infty}(1 - p^n)(1 - e^{2\pi i\gamma\nu}p^n)(1 - e^{-2\pi i\gamma\nu}p^n) , \tag{3.8}$$

where $q = e^{i\pi\gamma}$ is the quantum group parameter and $p \equiv e^{2\pi i\tau}$ is the elliptic nome. Let ρ denote the Weyl vector of $SU(N)$:

$$\rho = \frac{1}{2}\Big((N - 1)e_1 + (N - 3)e_2 + \ldots\ldots - (N - 1)e_N\Big) . \tag{3.9}$$

The non-vanishing Boltzmann weights are as follows [64]:

$$w(\Lambda, \Lambda + e_i, \Lambda + 2e_i; \Lambda + e_i|u) = \frac{[u + 1]}{[1]}$$

$$w(\Lambda, \Lambda + e_i, \Lambda + e_i + e_j; \Lambda + e_i|u) = \frac{[(\Lambda + \rho) \cdot (e_i - e_j) - u]}{[(\Lambda + \rho) \cdot (e_i - e_j)]}$$

$$w(\Lambda, \Lambda + e_i, \Lambda + e_i + e_j; \Lambda + e_j|u) =$$

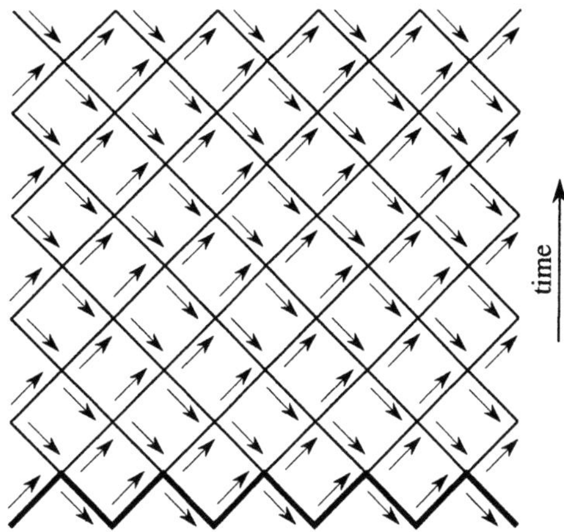

Figure 8. A section of the lattice upon which the model is defined. The bold zig-zag is the initial time slice, and the arrow indicate orientations of edges.

$$\frac{[u]}{[1]} \left(\frac{[(\Lambda + \rho) \cdot (e_i - e_j) + 1][(\Lambda + \rho) \cdot (e_i - e_j) - 1]}{[(\Lambda + \rho) \cdot (e_i - e_j)]^2} \right)^{\frac{1}{2}}, \qquad (3.10)$$

where $i \neq j$.

To obtain the restricted (RSOS) models corresponding to $G_\ell \times G_1/G_{\ell+1}$, one takes $\Lambda_0 = 0$, $\gamma = 1/(\ell + g + 1)$ and restricts the heights to the fundamental affine Weyl chamber of $G_{\ell+1}$. That is, the heights must be affine highest weights of $G_{\ell+1}$. If one were to make this choice of Λ_0 and γ, but not make this restriction on the heights, then the Boltzmann weights would be singular for certain combinations of heights. It is elementary to verify that the transfer matrix arising from the Boltzmann weights in (3.10) preserves this restriction. The topological, or rigid model corresponds to setting $\ell = 0$.

The lattice analogues [36] of the $N = 2$ superconformal grassmannian models:

$$\mathcal{G}_{1,m,n} \equiv \frac{SU_1(m + n) \times SO_1(2mn)}{SU_{n+1}(m) \times SU_{m+1}(n) \times U(1)}, \qquad (3.11)$$

are constructed by once again taking the heights as in (3.6) (with $N = n+m$). However one now performs the partial restriction by requiring that the heights be affine highest weights of both of the subgroups $SU_{n+1}(m)$ and $SU_{m+1}(n)$ of $SU(m + n)$, but one does not restrict the heights in the $U(1)$ direction. The value of q is taken to be the "topological value" (3.4). The topological untwisting is accomplished by choosing

$$\Lambda_0 = \frac{1}{\gamma g}(\rho_G - \rho_H) \, \tau , \qquad (3.12)$$

where ρ_G and ρ_H are the Weyl vectors of G and H. For the grassmannian model (3.11), this is simply

$$\Lambda_0 = \frac{m + n + 1}{2(m + n)}[n(e_1 + \ldots + e_m) - m(e_{m+1} + \ldots e_{m+n})]\tau . \qquad (3.13)$$

355

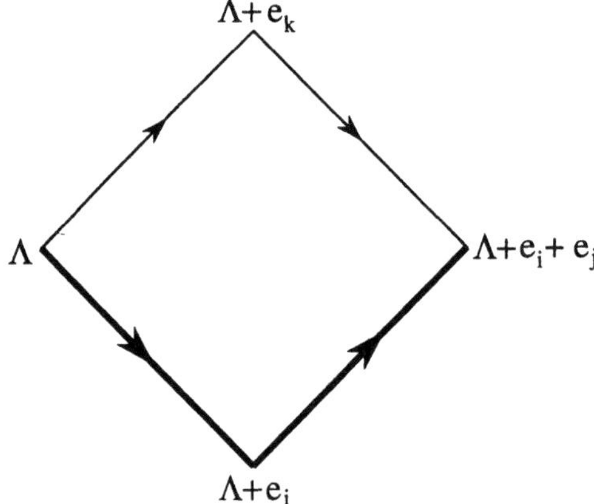

Figure 9. In the IRF formulation the heights, Λ, are associated to vertices as shown.

There are several things to note about this choice. First, the proof in [64, 71] that the Boltzmann weights (3.10) satisfy the star-triangle relations works for a general value of Λ_0 so this model is indeed still exactly solvable. Secondly, Λ_0 is orthogonal to the $SU(m)$ and $SU(n)$ directions and so these sectors of the model behave as if Λ_0 were zero. This means that the transfer matrix preserves the partial restriction on the subgroup $SU(m) \times SU(n)$. Hence this model is an *IRF* formulation of a particular form of (3.3). Because the $U(1)$ direction is unrestricted, there is, in principle, a danger that the Boltzmann weights could be singular for some configurations. In addition, in the physical model, the vector (3.12) is purely imaginary, and so one might be concerned that the Boltzmann weights may cease to be real in the physical regime. As we will see momentarily the Boltzmann weights are still real and non-singular.

It is convenient to think of the foregoing models from slightly different perspective: the heights can be taken on the partially restricted weight lattice of $G = SU(m + n)$ with $\Lambda_0 = 0$, but the Boltzmann weights must now incorporate the shift (3.12). We will henceforth adopt this perspective. The shift by Λ_0 does not modify the Boltzmann weights in the $H_0 = SU(m) \times SU(n)$ direction, but the other Boltzmann weights are, significantly modified. The effect of this shift is to replace certain strategic ϑ_1's by ϑ_4's, and this neatly removes all the potential singularities in the Boltzmann weights owing to the unrestricted $U(1)$. In order to describe the results explicitly let $[\nu]$ be given by (3.8) and let $\{\nu\}$ is defined by:

$$\{\nu\} \equiv \vartheta_4(\gamma\nu|\tau) \tag{3.14}$$

$$\equiv \prod_{p=1}^{\infty}(1 - p^n)(1 - e^{2\pi i \gamma\nu}p^{n-\frac{1}{2}})\left(1 - e^{-2\pi i \gamma\nu}p^{n-\frac{1}{2}}\right). \tag{3.15}$$

Set the parameter γ to $\frac{1}{m+n+1}$, and take the vector Λ_0 to be zero. The non-vanishing Boltzmann weights are then as follows:

If $1 \leq i, j \leq m$ *or* $m + 1 \leq i, j \leq m + n$ then, for $i \neq j$, one has:

$$w(\Lambda, \Lambda + e_i, \Lambda + 2e_i; \Lambda + e_i|u) \;=\; \frac{[u+1]}{[1]}$$

$$w(\Lambda, \Lambda + e_i, \Lambda + e_i + e_j; \Lambda + e_i|u) \;=\; \frac{[(\Lambda + \rho) \cdot (e_i - e_j) \;-\; u]}{[(\Lambda + \rho) \cdot (e_i - e_j)]}$$

$$w(\Lambda, \Lambda + e_i, \Lambda + e_i + e_j; \Lambda + e_j|u) \;=\;$$

$$\frac{[u]}{[1]} \left(\frac{[(\Lambda + \rho) \cdot (e_i - e_j) + 1][(\Lambda + \rho) \cdot (e_i - e_j) - 1]}{[(\Lambda + \rho) \cdot (e_i - e_j)]^2} \right)^{\frac{1}{2}} . \tag{3.16}$$

If $1 \leq i \leq m$ *or* $m + 1 \leq j \leq m + n$, or vice-versa, then one has

$$w(\Lambda, \Lambda + e_i, \Lambda + e_i + e_j; \Lambda + e_i|u) \;=\; \frac{\{(\Lambda + \rho) \cdot (e_i - e_j) \;-\; u\}}{\{(\Lambda + \rho) \cdot (e_i - e_j)\}}$$

$$w(\Lambda, \Lambda + e_i, \Lambda + e_i + e_j; \Lambda + e_j|u) \;=\;$$

$$\frac{[u]}{[1]} \left(\frac{\{(\Lambda + \rho) \cdot (e_i - e_j) + 1\}\{(\Lambda + \rho) \cdot (e_i - e_j) - 1\}}{\{(\Lambda + \rho) \cdot (e_i - e_j)\}^2} \right)^{\frac{1}{2}} . \tag{3.17}$$

From the Coulomb gas analysis of [36] and as confirmed by the Bethe Ansatz computations in [37], the foregoing height model at criticality yields the Neveu-Schwarz sector of the $N = 2$ superconformal model. The Ramond sector can be obtained by a uniform spectral flow in the $U(1)$ direction. This is easily implemented: one shifts all the lattice height appropriately, or equivalently one takes $\tau \to \tau + 1$ in (3.12).

3 Some simple properties of these models

As has already been mentioned, a signal that a model is supersymmetric is that the free energy per lattice site is analytic when the supersymmetry is unbroken. For the models discussed in the last subsection, this computation is very revealing in that it also exhibits and utilizes the connection between the supersymmetric model and its topological sector. We will begin, however, by illustrating the foregoing construction for $G = SU(2)$, and relate it to the constructions of section 2.

1 A simple example

For $G = SU(2)$ the foregoing lattice model construction collapses to what is basically the IRF version of the eight-vertex model [41, 48]. The weight vectors e_1 and e_2 of the foregoing subsection satisfy $e_1 = -e_2$, and the Weyl vector is given by $\rho = e_1$. The unrestricted heights lie on the weight lattice of $SU(2)$, which we will parametrize by an integer, ℓ, where $\Lambda = \Lambda_0 + (\ell - 1)e_1$. The integer, $(\ell - 1)$, is equal to twice the spin of the corresponding $SU(2)$ weight. These are exactly the heights discussed in section 2.3.1. If one takes $\Lambda_0 = 0$ and $\gamma = 1/3$ then the quantum group truncated model is completely rigid with ℓ alternating between 1 and 2. There are thus two states in this model depending upon whether a given site has height 1 or 2. This is the $SU_1(2)/SU_1(2)$ lattice model. To get the lattice analogue of the $N = 2$ supersymmetric theory, one should only quantum group truncate with respect to the H subgroup, or more precisely, with respect to the semi-simple part, $H - 0$, of H. In this instance, this means that one performs no quantum group truncation at all, and one therefore has a Gaussian model. The height, ℓ, takes the values $0, 1$ or 2 modulo 3, and values of γ and Λ_0 are:

$$\gamma = \frac{1}{3} \quad \text{and} \quad \Lambda_0 = \frac{3}{2} e_1 \tau \,. \tag{3.18}$$

We have thus "unfrozen" the heights ℓ of the RSOS model and have avoided the problem of singular Boltzmann weights through the choice of Λ_0. As described in section 2.3.2, this value of γ means that at criticality and in the continuum the Gaussian model is the $N = 2$ superconformal minimal model, with central charge, $c = 1$. The off-critical $N = 2$ supersymmetric, quantum integrable model corresponds to the most relevant chiral primary perturbation of the conformal model [11], and may be thought of as sine-Gordon at the supersymmetric value of the coupling constant. From (3.16) and (3.17) the elliptic Boltzmann weights are the following:

$$w(\ell, \ell \pm 1, \ell \pm 2; \ell \pm 1 | u) = \frac{\vartheta_1(\frac{1}{3}(u+1)|\tau)}{\vartheta_1(\frac{1}{3}|\tau)}$$

$$w(\ell, \ell \pm 1, \ell; \ell \pm 1 | u) = \frac{\vartheta_4(\frac{1}{3}(\ell \mp u)|\tau)}{\vartheta_4(\frac{\ell}{3}|\tau)}$$

$$w(\ell, \ell \pm 1, \ell; \ell \mp 1 | u) = \frac{\vartheta_1(\frac{u}{3}|\tau)}{\vartheta_1(\frac{1}{3}|\tau)} \left(\frac{\vartheta_4(\frac{1}{3}(\ell-1)|\tau)\, \vartheta_4(\frac{1}{3}(\ell+1)|\tau)}{(\vartheta_4(\frac{\ell}{3}|\tau))^2} \right)^{\frac{1}{2}} \,.$$

These Boltzmann weights are precisely the same as those of the $A_2^{(1)}$ cyclic solid-on-solid models described in [41, 49]. In this context the labelling of the model by $A_2^{(1)}$ refers to the fact that, because of the periodicity of the Boltzmann weights, the unrestricted $U(1)$ is, in fact, cyclic and so rather than taking the heights to be in $\mathbb{Z}$ one can view them as living on the extended Dynkin diagram of A_2.

2 The free energy

The easiest way to compute the free energy per unit volume, $\mathcal{F}$, is to use its analytic and inversion properties (see chapter 13 of [41]). In this review we will only consider regime *III* of the $SU(N)$ model, *i.e.* $0 < p \equiv e^{2\pi i \tau} < 1$, $-\frac{1}{2}N < Re(u) < 0$. (This regime corresponds to the *unitary*, quantum integrable field theory of interest – other regimes either have different conformal limits or correspond to perturbations with imaginary coupling.) We will also consider the models defined by (3.15) – (3.17) where γ is now an arbitrary, positive, real parameter.

To compute $\mathcal{F}$ it is first convenient to multiply the Boltzmann weights (3.15) – (3.17) by a factor of

$$e^{\frac{i\pi \gamma^2}{\tau}(u^2 + \frac{2u}{N})} \,, \tag{3.19}$$

and perform the modular inversion $\tau \to -1/\tau$. After making a gauge transformation one then finds that the Boltzmann weights are manifestly periodic under:

$$u \to u + \frac{2\tau}{\gamma} \,. \tag{3.20}$$

One then writes

$$\mathcal{F} = -\log(\kappa(u)) \,,$$

and requires that $\log(\kappa)$ be analytic in a region containing regime III and also be periodic under (3.20). Note that one does not impose the other periodicity of the theta functions $(u \to u + \frac{2}{\gamma})$ because such shifts of u would go outside the domain of analyticity of the free energy.

Using the properties of the corner transfer matrix one can show that $\kappa(u)$ must also satisfy:

$$\kappa(u)\,\kappa(-u) = h(1-u)\,h(1+u) \,, \tag{3.21}$$

$$\kappa(\lambda+u)\,\kappa(\lambda-u) = h(\lambda-u)\,h(\lambda+u) \,, \tag{3.22}$$

where $\lambda = -N/2$ for $SU(N)$, and

$$h(u) \equiv \frac{\vartheta_1\left(\frac{u\gamma}{\tau}\Big|-\frac{1}{\tau}\right)}{\vartheta_1\left(\frac{\gamma}{\tau}\Big|-\frac{1}{\tau}\right)} \,. \tag{3.23}$$

The function $h(u)$ that appears on the right hand side of these equations depends only upon the inversion relations of the elliptic Boltzmann weights. These equations, along with analyticity and periodicity, determine $\log(\kappa(u))$ completely. One simply writes $\log(\kappa(u))$ as a general Fourier series in $e^{i\pi\gamma u/\tau}$ and uses (3.21) and (3.22) to determine the coefficients. This is a little tedious but it is straightforward.

At this point one should note that the foregoing equations and constraints *do not depend on the choice of* Λ_0. This means that the free energy for all of of the lattice analogues of the grassmannian models (3.11) only depends upon $N = m + n$, and this free energy is exactly that of the topological $SU_1(N) \times SU_0(N)/SU_1(N)$ model. This will be a general feature of these models, the free energy of the models will be equal to that of the rigid topological model, and thus it must be possible to normalize the elliptic Boltzmann weights analytically so that the free energy is zero.

The fact that the free energy is independent of Λ_0 also means that we can use the known results for $\Lambda_0 = 0$ for the models based on $SU(N)$ [72]. One therefore has

$$\log(\kappa(u)) = \log\left(\frac{\vartheta_1(\gamma(u+1)|\tau)}{\vartheta_1(\gamma|\tau)}\right) - \sum_{k=-\infty}^{k=\infty} f(k;u,\gamma,\tau) \,, \tag{3.24}$$

where

$$f(x;u,\gamma,\tau) \equiv \frac{\sinh(\frac{\pi i}{\tau}(1-\gamma)x)\,\sinh(\frac{2\pi i}{\tau}\gamma u x)\,\sinh(\frac{\pi i(N-1)}{\tau}\gamma x)}{x\,\sinh(\frac{\pi i}{\tau}x)\,\sinh(\frac{\pi i N}{\tau}\gamma x)} \,. \tag{3.25}$$

This expression differs slightly from that of [72] in that we have subtracted the logarithm of the phase (3.19) so that the result is no longer exactly periodic under (3.20), but so that it does give the free energy for the model defined by the Boltzmann weights in (3.16) and (3.17).

To obtain the result as a function of $p = e^{2\pi i\tau}$ one needs to perform the modular inversion of the second term in (3.24). This can be done by Poisson resummation (see, for example, chapter 10 of [41], or appendix D of [73]). That is, one defines

$$\hat{f}(\zeta) \equiv \int_{-\infty}^{+\infty} e^{2\pi i\zeta x}\,f(x)\,dx \tag{3.26}$$

and uses the equality:

$$\sum_{k=-\infty}^{k=\infty} f(k) \;=\; \sum_{k=-\infty}^{k=\infty} \widehat{f}(k) \;. \tag{3.27}$$

The Fourier transform of (3.25) can be performed by closing the contour above or below the real axis, depending upon the sign of ζ, and then summing the residues. This sum over residues and the sum in (3.24) generates an expansion in powers of p. The vanishing of $sinh(\frac{\pi i N}{\tau}\gamma x)$ in denominator of (3.24) gives rise to residues that are proportional to $p^{\frac{1}{N\gamma}}$. This is the source of the non-analytic behaviour of the free energy as a function of p. The residues coming from the vanishing of the other $sinh$ function in the denominator of (3.24) are all proportional to integral powers of p. It is easy to extract the leading behaviour as $\tau \to i\infty$ from this sum over residues. One finds

$$\mathcal{F} \;\sim\; \frac{4\, sin(\frac{1}{N}(\frac{1}{\gamma}-1)\pi)\, sin(\frac{2\pi u}{N})\, sin(\frac{\pi}{N})}{sin(\frac{\pi}{N\gamma})}\; p^{\frac{1}{N\gamma}} \;. \tag{3.28}$$

For generic values of γ this means that $\mathcal{F} \sim p^{\frac{1}{N\gamma}}$, while from hyperscaling one has $\mathcal{F} \sim \frac{1}{\xi^2}$, where ξ is the correlation length. Therefore, we have

$$\tau \;\sim\; \frac{iN\gamma}{\pi}\, log(\xi) \;, \tag{3.29}$$

as $\tau \to i\infty$.

One should note that for

$$\gamma^{-1} \;=\; Nj + 1, \qquad j = 1,2,3,\dots \tag{3.30}$$

the numerator of (3.25) vanishes whenever $sinh(\frac{\pi i N}{\tau}\gamma x)$ vanishes. As a result, the Poisson resummation of (3.24) gives rise to an expansion in integral powers of p. That is, if γ satisfies (3.30) then the free energy is an analytic function of p. The choice $\gamma^{-1} = N+1$ corresponds to our supersymmetric models.

For these special values of γ it is, of course, no longer true that $\mathcal{F} \sim p^{\frac{1}{N\gamma}}$ as $\tau \to i\infty$. However, continuity in γ means that the relation (3.29) is true even at these special values.

It turns out that for γ given by (3.30) one can easily express the free energy in terms of theta functions. Here we will give the result for the supersymmetric model: $\gamma^{-1} = N+1$. The general case is similar. For $\gamma = 1/(N+1)$, the second term in (3.24) can be written:

$$\frac{2\pi i}{\tau}\, u\gamma(1-2\gamma) \;-\; \sum_{k=1}^{\infty} \frac{1}{k(1-\tilde{p}^k)}\, (z^k - z^{-k})\,(w^{-k} - w^{-kN}) \;, \tag{3.31}$$

where $\tilde{p} = e^{-2\pi i/\tau}$, $z = e^{2\pi i\gamma u/\tau}$ and $w = e^{2\pi i\gamma/\tau}$. This is a standard expansion of the logarithm of the ratio of two theta functions[h]. Finally one can perform modular inversions on the theta functions and the final result is

$$log(\kappa(u)) \;=\; log\left(\frac{\vartheta_1(\gamma(u-1)|\tau)}{\vartheta_1(\gamma|\tau)}\right) \;=\; log\left(\frac{[u-1]}{[1]}\right) \;. \tag{3.32}$$

---

[h]One simply takes the logarithm of the product formula for the theta functions, expands all of the $log(1 - q^n x^{\pm 1})$ terms into power series in $q^n x^{\pm 1}$, and then one can perform the sum over n to obtain (3.31).

Observe that, as promised, this free energy is analytic in the whole of regime *III*. Indeed, if one multiplies all of the Boltzmann weights in (3.16) and (3.17) by $\frac{[1]}{[u-1]}$ then the Boltzmann weights are still analytic in regime *III*, and the free energy, $\mathcal{F}$, vanishes identically.

3 The off-critical perturbing operators

For the critical limit of these models one can use Coulomb gas methods to show that the continuum limit is $N = 2$ supersymmetric. The free energy computations suggest that the off-critical models are also $N = 2$ supersymmetric in the continuum, but we would like to demonstrate this more directly.

From the analysis of the perturbations of $N = 2$ superconformal coset models, we know that the are natural perturbations that lead to $N = 2$ supersymmetric quantum integrable models. These operators have the form:

$$\psi \;\equiv\; G^-_{-\frac{1}{2}}\widetilde{G}^-_{-\frac{1}{2}}\phi \qquad \text{and} \qquad \widetilde{\psi} \;\equiv\; G^+_{-\frac{1}{2}}\widetilde{G}^+_{-\frac{1}{2}}\widetilde{\phi} \;, \tag{3.33}$$

where ϕ is a very particular chiral primary field, and $\widetilde{\phi}$ is its anti-chiral conjugate. In particular, the fields ψ and $\widetilde{\psi}$ have conformal weights and $U(1)$ quantum numbers:

$$h_\psi \;=\; \overline{h}_\psi \;=\; h_{\widetilde{\psi}} \;=\; \overline{h}_{\widetilde{\psi}} \;=\; \frac{1}{2} \;+\; \frac{1}{2(N+1)}$$

$$Q_\psi \;=\; \overline{Q}_\psi \;=\; -Q_{\widetilde{\psi}} \;=\; -\overline{Q}_{\widetilde{\psi}} \;=\; 1 \;-\; \frac{1}{(N+1)} \tag{3.34}$$

We remarked above that in the vertex form of the model at criticality, these operators extend $U_q(H_0)$ to $U_q(G)$. Thus, based on general expectations about lattice models, one would expect these operators to be the ones that correspond to the elliptic "deformation" of the critical model. We can see this connection much more explicitly as follows.

Consider the continuous family of models where the initial height vector is taken to be

$$\Lambda_0 \;=\; \frac{2\mu}{\gamma g}(\rho_G - \rho_H)\,\tau \;, \tag{3.35}$$

where μ is a parameter. When $\mu = 0$ we have the topological model and the lowest powers of p in the Boltzmann weights (3.10) are p^0 and p^1. If we increase μ, then the lowest powers of p change smoothly to p^μ and $p^{(1-\mu)}$. Now recall that in the scaling limit, $\tau \to i\infty$, we know that τ is related to the correlation length by (3.29), and thus $p \sim \xi^{-\frac{2N}{N+1}}$. It follows that the elliptic perturbation must involve two operators, and that their coupling constants must scale as $p^\mu \sim \xi^{-\frac{2N\mu}{N+1}}$ and $p^{(1-\mu)} \sim \xi^{-\frac{2N(1-\mu)}{N+1}}$. The corresponding operators therefore have conformal weights:

$$h_1 \;=\; \overline{h}_1 \;=\; 1 - \frac{N\mu}{(N+1)} \;; \qquad h_2 \;=\; \overline{h}_2 \;=\; 1 - \frac{N(1-\mu)}{(N+1)} \;. \tag{3.36}$$

When $\mu = \frac{1}{2}$ these two operators have precisely the same conformal dimension and this is the conformal dimension given in (3.34). If we now consider the gradual untwisting of the energy momentum tensor of the quantum field theory:

$$T(z) \;=\; T^{\text{top}}(z) \;-\; \mu\,\partial J(z) \;; \qquad \widetilde{T}(\overline{z}) \;=\; \widetilde{T}^{\text{top}}(\overline{z}) \;+\; \mu\,\overline{\partial}\widetilde{J}(\overline{z}) \;, \tag{3.37}$$

we see that the conformal weights of the operators (3.33) depend upon the parameter μ exactly as in (3.36).

In this way one can not only identify the dimensions of the perturbations that lead to the elliptic Boltzmann weights, but one can also determine their $U(1)$ charges. Moreover, the foregoing also demonstrates that smoothly changing the initial lattice vector, Λ_0, causes the dimensions of the perturbing operators to change in exactly the way that they should under topological untwisting. As a result one also obtains further confirmation that one has correctly determined the lattice analogue of topological untwisting. A more explicit relation between Λ_0 and perturbed conformal theories can be obtained using the method of [59].

4 Scattering matrices

Since we are considering critical and off-critical $N = 2$ lattice models, there will of course be massless and massive S-matrices associated with the excitations of these models. There have been quite a number of papers written on this subject (see, for example, [11, 50, 12, 13, 51, 52, 53]), and we will not try perform even a partial survey. Our purpose here is to simply make a few remarks about some of the known S-matrices, and how their construction fits in with the general strategy of the construction of $N = 2$ lattice models.

It is first useful to recall some general observations about conformal coset models and their perturbations leading to quantum integrable models. In a generalized Coulomb gas description of the conformal model, there are always two quantum groups that usually have the same underlying Lie algebra, but have different roots of unity. One of these quantum groups is generically associated with the numerator of the coset, while the other quantum group is associated with the denominator. The corresponding lattice model only exhibits one of these quantum group symmetries, and it is usually the one associated with the denominator of the coset model. The perturbing operators that lead to quantum integrable models usually extend this lattice (or denominator) quantum group to a larger, affine quantum group. The other (numerator) quantum group usually plays a major role in the scattering theory of the quantum integrable model. That is, the scattering matrices are usually built out of the $\check{R}$-matrices of this numerator quantum group. The corresponding affine extension of the scattering quantum group leads to operators in the theory that commute with the S-matrix. Such non-local conserved charges have been used extensively in the construction of the S-matrices. The intuitive reason as to why the scattering theory and lattice quantum groups are distinct is that the solitons of the scattering theory must be local with respect to the perturbing operators leading to the quantum integrable model [54].

The supersymmetric models also fit this mould. The denominator quantum group is indeed that of the lattice model, while numerator quantum group is that of the scattering theory. The non-local conserved charges that extend the scattering quantum group to the affine quantum group are simply the supersymmetry generators. Thus we find that while almost all of our lattice models do not have explicit supersymmetry, the corresponding scattering theories do. In other words supersymmetry appears only as a dynamical symmetry.

1 Massless S-matrices and the six-vertex model

The six-vertex model is diagonalizable by Bethe ansatz [41], and the solutions are classified in terms of $1, 2$ strings and the anti-string 1^-, with a coupling diagram as in figure 10 (see [57]). To study the scattering theory simply observe that for a relativistic QFT the thermodynamic Bethe ansatz [58], in addition to the pseudo-particles appearing in the solution of the Bethe equations, involves an additional physical particle, as in figure 10.

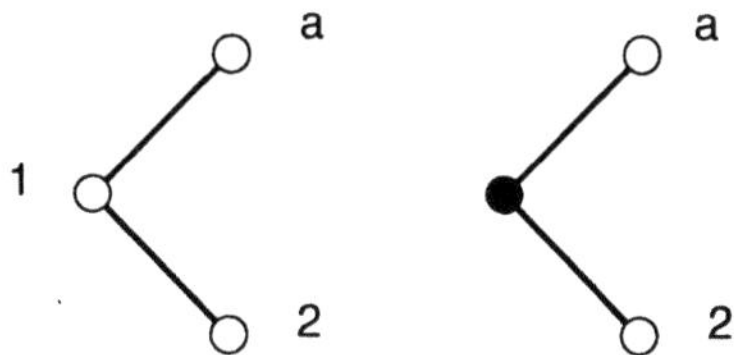

Figure 10. The first diagram represents the pseudo-particles (solutions of the Bethe ansatz) and their couplings at the supersymmetric point of the lattice vertex model. Recall that for $q = e^{i\pi/t}$ one has strings $(1, 2, \ldots, t-1)$ and the anti-string a. In the continuum theory there is an additional node for the physical particles, indicated by a heavy dot on the second diagram. Therefore the number of Bethe ansatz solutions must differ by one from the lattice case, hence the denominator of the quantum group parameter must also differ by one from the lattice case.

Therefore, if the lattice model has $q = \exp(i\pi/3)$, the scattering theory has $q' = \exp(i\pi/2)$. From the lattice model, a massless scattering theory [60] can be extracted which has the following content (see [59, 61]). One has a pair of right moving particles which we parametrize as

$$e_R = p_R = \frac{M}{2}e^{\theta} \tag{4.1}$$

and a pair of left moving particles with

$$e_L = -p_L = \frac{M}{2}e^{-\theta} \tag{4.2}$$

where M is a mass scale and θ is a rapidity. These particles have fermion number equal to ± 1 and they have factorized scattering with

$$S_{LL} = S_{RR} = Z(\theta) \begin{pmatrix} \cosh\theta/2 & 0 & 0 & 0 \\ 0 & i\sinh\theta/2 & 1 & 0 \\ 0 & 1 & i\sinh\theta/2 & 0 \\ 0 & 0 & 0 & \cosh\theta/2 \end{pmatrix} \tag{4.3}$$

where $Z(\theta)$ is a normalization factor ensuring unitarity and crossing symmetry, while the LR scattering is trivial

$$S_{LR} = S_{RL} = 1 \tag{4.4}$$

Acting on the right particles u, d we have

$$Q_-|u(\theta)> = \sqrt{M/2}e^{\theta/2}|d(\theta)> \tag{4.5}$$

$$Q_+|d(\theta)> = \sqrt{M/2}e^{\theta/2}|u(\theta)> \tag{4.6}$$

and same type of relations with left particles and $\overline{Q}^{\pm}$ generators. The supersymmetry algebra is

$$Q_+^2 = Q_-^2 = \overline{Q}_+^2 = \overline{Q}_-^2 = \{Q_+, \overline{Q}_-\} = \{Q_-, \overline{Q}_+\} = 0 \tag{4.7}$$

$$\{Q_+, \overline{Q}_+\} = \{Q_-, \overline{Q}_-\} = 0 \tag{4.8}$$

$$\{Q_+, Q_-\} = Me^{\theta}, \quad \{\overline{Q}_+, \overline{Q}_-\} = 2Me^{-\theta} \tag{4.9}$$

2 Massive S-matrices

There are several ways in which one can extract the S-matrices of the massive quantum integrable models that appear in the continuum limits of the $N = 2$ lattice models. We will briefly discuss two of them here. The first approach is that taken in [11, 12], whose starting point is to use the Landau-Ginzburg description of the model to determine the ground-state and soliton structure. Since the solitons are generically doublets of the superalgebra, one can then use the Bogomolnyi bounds in the superalgebra to determine the soliton masses. The solitons also have fractional fermion numbers that can be determined from the Landau-Ginzburg potential. This information, combined with the fact that the S-matrix must commute with the supersymmetry, be unitary, satisfy the Yang-Baxter equations, and obey crossing symmetry is more than enough to determine the S-matrix completely. This has been done for a large class of models [12], but there is an even larger class for $N = 2$ quantum integrable models for which the S-matrices are not known. The first reason for this is that one can easily show that in this larger class the solitons alone cannot form a kinematically closed scattering theory [11, 51]. This almost certainly means that breather states must be included in the spectrum. This does not pose, a priori, any obstacle to the construction of the S-matrices, but as we will describe below, there could be some further conceptual problems to be solved if one is to determine the S-matrices for these more general models.

We will describe the second method of obtaining the S-matrices in more detail here since it is rather more in the spirit of our approach to $N = 2$ lattice models. The basic idea is to once again use topological models and a modification of quantum group truncation.

The scattering matrices are well known for the quantum integrable models obtained from the natural perturbations of the $G_k \times G_\ell / G_{k+\ell}$ coset models (see, for example, [74, 75, 76, 77, 78, 79, 80]). For simplicity we will only consider the models with with $G = SU(N)$ and the level k equal to one. The S-matrices can then be obtained from RSOS restrictions of affine Toda S-matrices. Let W_j, $j = 1, \ldots, N - 1$, denote the fundamental representations of $SU(N)$ (i.e. the anti-symmetric tensors of rank j).

Each of these representations defines a multiplet of solitons of affine $\widehat{SU(N)}$-Toda, and
these solitons can be labelled by the weights of the representations.

Define $S_{ji}^{(\widehat{G})}(\theta, q)$ to be the two-body S-matrix for the scattering of particles in W_i
with those in W_j:

$$S_{ji}^{(\widehat{G})}(\theta, q): \quad W_i \otimes W_j \;\to\; W_j \otimes W_i \,, \tag{4.10}$$

where, as usual, θ is the relative rapidity of the solitons and q is the quantum group
parameter. Requiring the S-matrix to commute with the $U_q(\widehat{SU(N)})$ symmetry leads
to the following result:

$$S_{ji}^{(\widehat{G})}(\theta, q) \;=\; X_{ji}(\theta)\, v_{ji}(\theta, q)\, R_{ji}(\theta, q) \,. \tag{4.11}$$

The last term, $R_{ji}(\theta, q)$, is the standard R-matrix for the quantum group $U_q(\widehat{SU(N)})$ in
the principal gradation[i]. The spectral parameter, x, and the quantum group parameter,
q, of the R-matrix are related to the rapidity, θ and the Toda coupling constant, β, by:

$$q \;=\; exp\left(-\frac{i\pi}{\beta^2}\right) \,, \quad x \;=\; exp\left(\frac{\theta}{\gamma}\right) \quad \text{where} \quad \gamma \;=\; \frac{\beta^2}{1 - \beta^2} \,. \tag{4.12}$$

The scalar factor, $v_{ji}(\theta, q)$, in (4.11) is the minimal factor that makes the product $v_{ji}R_{ji}$
crossing symmetric and unitary. This is relatively easy to compute and can be found
in [75, 76]. The additional scalar factor $X_{ji}(\theta)$ is a CDD factor, and it satisfies crossing
and unitarity by itself. This factor contains all of the necessary poles for closure of the
bootstrap with the spectrum of soliton masses. It turns out that this CDD factor does
not depend upon the value of the Toda coupling and so can be evaluated by going to
the limit ($\beta \to 1$) in which the Toda model can be related to a Gross-Neveu model [75].

To get the model corresponding to the perturbed coset model one must first incorporate the background charge, which is done in an analogous manner to the twist in the
Markov trace of subsection 2.3.4. Let H be some subgroup of $G = SU(N)$, where H
also has rank $N - 1$, and let ρ_H be the Weyl vector of H. Define $S_{ji}^{(\widehat{G}/H)}(\theta, q)$ by

$$S_{ji}^{(\widehat{G}/H)}(\theta, q) \;=\; \left(x^{-\rho_H \cdot h} \otimes x^{-\rho_H \cdot h}\right) S_{ji}^{(\widehat{G})} \left(x^{\rho_H \cdot h} \otimes x^{\rho_H \cdot h}\right) \,, \tag{4.13}$$

where h represents the vector of Cartan subalgebra generators of the quantum group.
This conjugation only affects the R-matrix part of the S-matrix, and for $H = G$, it
converts the principal gradation to the homogeneous gradation.

By construction, the S-matrices $S_{ji}^{(\widehat{G}/H)}(\theta, q)$ commute with the action of the finite
quantum group $U_q(H)$. These generators act in the more familiar rapidity independent
fashion. Therefore one can use the $U_q(H)$ symmetry to restrict the model exactly as in
the lattice model.

For each weight, μ, of W_j we introduce a formal operator $K_\mu^{(j)}(\theta)$. These operators
satisfy an S-matrix exchange relation:

$$K_\mu^{(j)}(\theta_1)\, K_\nu^{(i)}(\theta_2) \;=\; \sum_{\mu', \nu'} \left(S_{ji}^{(\widehat{G}/H)}(\theta, q)\right)_{\mu, \nu}^{\mu', \nu'} K_{\nu'}^{(i)}(\theta_2)\, K_{\mu'}^{(j)}(\theta_1) \,. \tag{4.14}$$

---

[i]The R-matrices that one usually encounters in the literature are written in the homogeneous gradation, which is related to the principal gradation by a trivial automorphism,
which will soon be described.

Let $\mathcal{F}$ denote the multi-particle Fock space generated by the formal action of the operators $K_\mu^{(j)}(\theta)$ on the vacuum. The space $\mathcal{F}$ is an $U_q(H)$ module, and reducible (for q not a root of unity):

$$\mathcal{F} = \bigoplus_a V^{(\lambda_a^{(H)})} \, , \tag{4.15}$$

where $V^{(\lambda_a^{(H)})}$ is an $U_q(H)$ module of highest weight $\lambda_a^{(H)}$. Since the $K_\mu^{(j)}(\theta)$ act on $\mathcal{F}$, one can consider their reduction

$$K^{(n)}_{\lambda_b^{(H)}\lambda_a^{(H)}}(\theta) : \quad V^{(\lambda_a^{(H)})} \longrightarrow V^{(\lambda_b^{(H)})}. \tag{4.16}$$

These operators satisfy the exchange relation:

$$K^{(j)}_{\lambda_b^{(H)}\lambda_a^{(H)}}(\theta_1)\, K^{(i)}_{\lambda_a^{(H)}\lambda_c^{(H)}}(\theta_2) \;=\; \sum_{\lambda_d^{(H)}} \left(S_{ji}^{(\widehat{G}/H)}(\theta,q) \right)^{\lambda_b^{(H)}\lambda_d^{(H)}}_{\lambda_a^{(H)}\lambda_c^{(H)}} K^{(i)}_{\lambda_b^{(H)}\lambda_d^{(H)}}(\theta_2)\, K^{(j)}_{\lambda_d^{(H)}\lambda_c^{(H)}}(\theta_1) \, . \tag{4.17}$$

The S-matrix for the kinks in this equation is the SOS form, and the foregoing construction is once again the vertex/SOS correspondence.

As before, the restriction amounts to taking q to be a root of unity and imposing a limitation on the allowed highest weight labels $\lambda^{(H)}$. To get the S-matrix for the perturbed coset model $G_1 \times H_\ell/H_{\ell+1}$ one must take the Toda coupling so that $\beta^2 = (\ell + h)/(\ell + h + 1)$, where h is the dual Coxeter number of H. Therefore, one has:

$$q \;=\; -exp(-i\pi/(\ell + h)) \, . \tag{4.18}$$

Note that this is the root of unity appropriate to the quantum group structure of the numerator quantum group of the coset model[j].

To get the $N = 2$ supersymmetric models one chooses H so that G/H is hermitian symmetric, and takes $\ell = g - h$. Thus one has $q = -e^{-i\pi/g}$. If one were to take $H = G$ one would once again get the topological coset model, but with the foregoing choice of q, the only $U_q(G)$ representation with non-vanishing q-dimension is the singlet representation. There are thus no solitons in the topological model. Once again one finds the $N = 2$ model by modifying the quantum group truncation procedure. One can also verify that under the twisting process (4.13), the two generators that extend $U_q(H)$ to $U_q(\widehat{G})$ have a rapidity dependence that makes them spin-$\frac{1}{2}$ charges. If the quantum group truncation procedure parallels that of the conformal theory, then these operators will become the supersymmetry generators of the restricted model.

This process works beautifully and simply for $G = SU(N)$ and $H = SU(N-1) \times U(1)$. For this choice, the fundamental representations W_j decompose into a direct sum of two representations of $U_q(SU(N - 1))$. In the RSOS truncation using $U_q(H)$ one therefore finds that each W_j yields two solitons, and these form a supersymmetric doublet. The

---

[j] In the earlier sections of this review, where we were discussing lattice models, we took $q = e^{i\pi/M}$, where M was some positive integer. We could equally well have taken q to be defined by $q = -e^{-i\pi/M}$. Here, however, the affine extension of the quantum group is now playing a more central role and this means that one has to make a specific choice for q. This choice is, of course, convention dependent and we are employing the conventions of [13, 75].

$U(1)$ charge becomes the (fractional) fermion number, and the two generators that extend $U_q(H)$ to $U_q(\widehat{G})$ act as the supersymmetry on this doublet. We have thus incorporated the supersymmetry directly into the affine quantum group. The S-matrices for the $N = 2$ model can then be easily extracted from the Toda S-matrices for $G = SU(N)$ [13].

The surprise comes once one tries this procedure for more general models of the form (3.11), for example, for $G = SU(4)$, $H = SU(2) \times SU(2) \times U(1)$. The problem is that the quantum group truncation does not respect canonical supermultiplet structure. In these more general models all of the W_j's give rise to RSOS solitons with fractional fermion numbers that agree with the Landau-Ginzburg picture. All of the RSOS solitons coming from one of the W_j's are, of course, transformed into each other by the generators of $U_q(\widehat{G})$. The two generators that extend $U_q(H)$ to $U_q(\widehat{G})$ are indeed spin-$\frac{1}{2}$ and commute with the S-matrix. The problem is that there does not seem to be an RSOS truncation upon which the algebra of these generators becomes that of the standard $N = 2$ superalgebra. For example, some W_j's would give rise to a three dimensional "supermultiplet." (This happens in the $SU(4)$ example above for the six-dimensional representation of $SU(4)$.)

There are several possible resolutions of this issue. It is quite conceivable that the Toda approach breaks down, or it might be that there is some supersymmetry anomaly. It is also possible that the integrable model gives rise to two S-matrices, one in which the supersymmetry acts locally, and one in which it does not. It is certainly of interest to resolve this problem. As first sight, the possibility of a supersymmetry anomaly seems the least likely explanation. However this is perhaps not as outrageous as it sounds, after all, the supersymmetry would still be a symmetry of the S-matrix, but it would simply have non-trivial braiding relations with single soliton states. Such a phenomenon is already present in the models that we do understand: the fractional fermion number of single solitons means that the supersymmetry action on single solitons must involve extra phases. It is just conceivable that the more exotic models exhibit a more involved (perhaps non-abelian) form of this. At any rate, it is highly desirable to compute the S-matrices for these models. It is also, perhaps, significant that these "exotic" models are also precisely the ones for which the S-matrices are not yet known because the solitons cannot, by themselves, form a closed scattering theory.

5   Other issues

1   Spontaneous breaking of $N = 2$ supersymmetry

In lattice models there is a natural direction of perturbation that does not break supersymmetry explicitly, but does so spontaneously. This is made possible by the non-unitarity of the perturbation. For example, consider the polymer problem discussed earlier. The critical Landau-Ginzburg potential in this theory is $W = X^3$, and the perturbation that changes the weight β of monomers in the generating functions corresponds exactly to adding a multiple of X to this potential. However if one writes out the F-term term in the supersymmetric action, and properly incorporates the coupling constant, one obtains:

$$S_F = \text{const.} \left(\int d^2z d^2\theta \left(X^3 + (\beta_c - \beta)^{1/2}\, X \right) + \int d^2z d^2\bar{\theta} \left(\overline{X}^3 + (\beta_c - \beta)^{1/2}\, \overline{X} \right) \right). \tag{5.1}$$

If $\beta < \beta_c$ one gets the usual physics of the off-critical quantum integrable model with superpotential $W = X^3 + X$. On the other hand, if $\beta > \beta_c$ one gets very different physics because the perturbation is purely imaginary, and the model is non-unitary. This is easily analyzed qualitatively. For $\beta \leq \beta_c$ polymers are "very tiny" (their fractal dimension is less than two) so the free energy calculated in the Ramond sector, where no polymer at all is allowed, is the same as the free energy calculated in any other sector, where some non-contractible polymers are allowed. For $\beta > \beta_c$ the partition function of the Ramond sector is still $\mathcal{Z}_R = 1$ but as soon as a polymer is allowed it fills most of the available space and $\mathcal{Z}$ grows exponentially with the size of the system. This means that there are infinitely many level crossings between the ultra-violet and infra-red fixed points in the Ramond sector, and although the states of vanishing energy always stay there, they are infinitely far from the true ground state of negative energy in the infra-red. In the infra-red the bosonic degrees of freedom of the theory have become massive and the fermionic ones are a simple (topologically twisted) Dirac fermion, describing the physics of "dense polymers".

Interestingly, the position of the level crossings can be related to the poles of the Painlevé III differential equation [63]. Consider the system on a cylinder of radius R and length T. The renormalization group variable z is proportional to $(\beta_c - \beta)R^{4/3}$. Recall that the generalized supersymmetry index is defined by [15]:

$$Q = i\frac{R}{T}\, tr \left(e^{-RH}\, F(-)^F \right) \tag{5.2}$$

where, from the point of view of the Hamiltonian evolution, the "time" is in the R direction now. This index can be written:

$$Q = \frac{z}{2}\frac{du}{dz} \tag{5.3}$$

where u is the solution of the Painlevé III equation:

$$\frac{d^2u}{dz^2} + \frac{1}{z}\frac{du}{dz} = \frac{9}{16\sqrt{-z}}\cosh u \tag{5.4}$$

with specific asymptotic behaviour [15]. One can show that the level crossings occur at the poles of Q on the negative z axis. One can also show that these poles are simple, and that there are an infinite number of them [63].

2 Ultra-high temperature limits: the non-interacting models revisited

The trivial non-interacting spin model of section 2 has even more to it than was made evident there. Recall that the spins sit on the edges of the original square lattice (figure 1). Suppose that we now draw on the diagonal lattice of figure 3 an occupied edge between spins of *opposite* value. These edges correspond to a Bloch wall, and the loops created by them can traverse each edge of the lattice at most once. (See figure 11.) These are the usual contours of the low temperature Ising model, but considered in the high-temperature phase (actually, in the limit of infinite temperature). We now consider the geometrical properties of these loops. First it should be clear from preceding discussions that they are identical to the properties of the boundaries of percolation clusters (or hulls). The most interesting quantity is the fractal dimension of these loops, which is related to the algebraic decay of the probability that two edges belong to the same loop. That is, if the former decays as r^{-2x} then the fractal dimension of the loop is $D_f = 2 - x$. One can show that the exponent is given by $x = 2h$, where h the conformal weight, in the twisted theory, of the Ramond ground state with the lowest charge. That is:

$$h = \frac{1}{24} - \frac{1}{2}\left(-\frac{1}{6}\right) = \frac{1}{8} \tag{5.5}$$

and thus $D_f = \frac{7}{4}$. This is midway between the fractal dimension for Brownian motion ($D_f = 2$) and the fractal dimension of self-avoiding random walks ($D_f = \frac{4}{3}$). It is interesting to observe that, from the point of view of the low-temperature expansion for the Ising model, another geometrical quantity is very natural: the probability that two edges are extremities of the same open line. This corresponds to the two-point correlation function of disorder operators. This function goes to a constant at large distance, and so the conformal weight is zero. This high-temperature limit is therefore rather interesting because it rather naturally leads to the existence of a non-trivial operator of vanishing dimension in a "lattice topological sector."

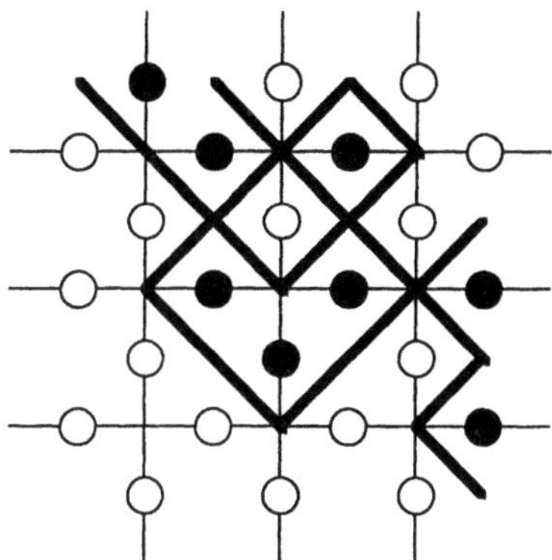

Figure 11. The Bloch walls for Ising model at high temperature. Filled dots represent states of spin +1 and empty dots represent states of spin −1. The walls separate domains of opposite spin.

Now we have a natural direction to study: the properties of Bloch walls in high temperature phases of spin models. The topological sector corresponds to the original spin variables. The non-topological one to the geometry of the Bloch walls. As another example consider the three state Potts model. There are three states per vertex, represented by empty dot, full dot or square in figure 12. The Bloch walls carry one arrow if, when one faces along the arrow, there is either (i) an empty dot on the right and a full dot on the left, or (ii) there is a square on the right and an empty dot on the left. The Bloch walls carry two arrows if there is a square is on the right and a full dot on the left of the arrows. With these rules there is obviously current conservation.

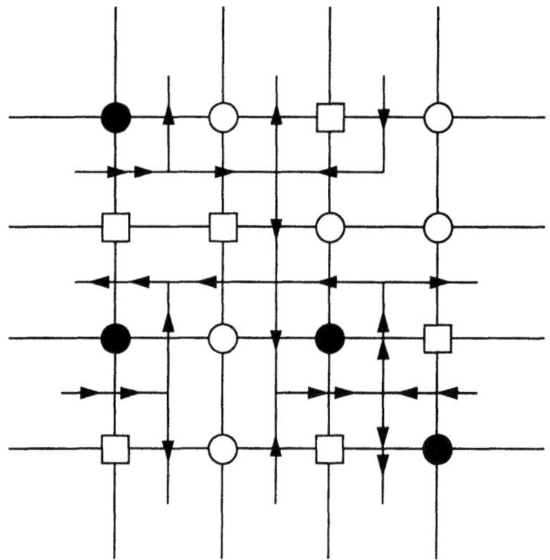

Figure 12. The Bloch walls for the three-state Potts model at high temperature. By convention the full dot, empty dot and square correspond to $\sigma = 1, 2, 3$ respectively. The walls carry one or two arrows depending on the difference of spins between both sides. By convention the highest spin is to the right of the arrow(s).

By analogy with the foregoing Ising model, there are two natural geometrical correlation functions that are constant at large distance: the probabilities that two edges are extremities of the same open line carrying either one or two arrows. Similarly there are two natural correlation functions that decay algebraically: the probabilities that two edges belong to the same loop carrying one or two arrows. In this example there is charge conjugation symmetry that in fact relates simple and double arrows so these pairs of correlation functions are in fact identical. The obvious topological model behind this is the $SU_1(3)/SU_1(3)$ model. The fractal dimension of the contours can be reasonably conjectured to be given by the formula similar to that for the Ising model,

$$h = \frac{1}{16} - \frac{1}{2}\left(-\frac{1}{4}\right) = \frac{3}{16} , \tag{5.6}$$

leading to $D_f = \frac{13}{8}$.

More generally one can conjecture similar properties for the contours in the high temperature phase of the Z_{n+1} models and $SU_1(n+1)/SU_1(n+1)$ theories. The conformal weight h will be $h = \frac{c}{24} - \frac{1}{2}(-\frac{c}{6}) = \frac{c}{8}$, where $c = \frac{3n}{n+2}$ is the central charge of the nth $N = 2$ superconformal minimal model. This leads to a predicted fractal dimension of:

$$D_f = 2 - \frac{c}{4} = \frac{5n + 16}{4(n + 2)} \ . \tag{5.7}$$

6 Conclusions and open problems

The subject of $N = 2$ lattice models has, as yet, not produced the same kind of nice structures as in the continuum $N = 2$ field theories. In particular it is important to emphasize that there is no real algebraic simplification for the computation of most physical properties. This suggests that, despite the large number of models exhibited so far, some better ones still remain to be discovered. In addition, very little is understood about the "lattice supersymmetry" and the way these $N = 2$ lattice models provide discretizations of field theories with fermions. A step in that direction would be to identify supercharges on the lattice and study their anti-commutator in the scaling limit. This might be accomplished in the way that has recently been used for the Virasoro algebra [56].

On the other hand the study of $N = 2$ lattice models has already provided many interesting physical ideas, in particular concerning the topological models. We believe that $N = 2$ theories are the natural framework to describe most geometrical statistical mechanics models and more generally the properties of Bloch walls in non-critical spin models. It is also amusing to notice that what is "physical" in field theory and condensed matter is rather different. For instance spontaneous breaking of $N = 2$ supersymmetry seems to be a very natural feature in lattice models. It is rather remarkable that the associated level crossings are encoded in the poles of differential equations of Painlevé type.

While finding a lattice version of supersymmetry is probably the most important issue to resolve, there are many other open problems in the subject of $N = 2$ lattice models. One can try to establish our observations about Bloch walls in generality, or find evidence to support these observations through numerical simulations. As mentioned in section 4, there are unresolved issues about supersymmetry and the scatering matrices in the more complex $N = 2$ quantum integrable models. (Some suggestions about these scattering matrices have recently been made [81].) Simple corner transfer matrix computations have been done for $N = 2$ lattice models [38, 82]. The results yield $N = 2$ superconformal characters as one would expect, but the computation was done by the traditional method [41]. It would be interesting to see if the remarkable properties of $N = 2$ supersymmetry could be exploited directly to reach this result in a much simpler manner.

There is also the question of more general lattice correlation functions. In the continuum limit, at criticality, one can obtain Knizhnik-Zamolodchikov equations for the correlation functions in $N = 2$ superconformal coset models. At first sight, these equations are no simpler than those for non-supersymmetric coset models. However, once one puts in the special parameter values for the correlators of chiral primary fields one finds that one has saturated some bound and the solution of the differential equation is particularly simple. For example, if the equation is hypergeometric, the correlator of chiral primaries is usually one of the special rational solutions. It would be most interesting to see if similar simplifications occur in the q-deformed Knizhnik-Zamolodchikov equations for the lattice correlators of [83].

Much of the motivation for these open problems is simply to find some $N = 2$ supersymmetry miracles in exactly solvable lattice models. We therefore leave it as a general challenge to find a another formulation of these models, perhaps using the Landau-Ginzburg structure more directly, in which the $N = 2$ supersymmetry leads to dramatic simplifications.

Acknowledgements

One of us (N. W.) would like thank the organizers of the workshop on "New Developments in String Theory, Conformal Models and Topological Field Theory," for their invitation to speak and for creating a very stimulating workshop. The authors also wish to thank the organizers for providing them with the opportunity to write this review, which is the result of many discussions. The authors wish to thank warmly their collaborators, in particular Paul Fendley, Dennis Nemeschansky, Alyosha Zamolodchikov and Ziad Maassarani. This work was partially supported by the DOE under grant No. DE-FG03-84ER40168. H. S. was also supported by the Packard Foundation.

References

[1] E. Martinec, *"Criticality, catastrophes and compactifications,"* in the V.G. Knizhnik memorial volume, L. Brink *et al.* (editors): *"Physics and mathematics of strings."*

[2] N.P. Warner, *Lectures on $N = 2$ superconformal theories and singularity theory",in "Superstrings '89,"* proceedings of the Trieste Spring School, 3–14 April 1989. Editors: M. Green, R. Iengo, S. Randjbar-Daemi, E. Sezgin and A. Strominger. World Scientific (1990); *"$N = 2$ Supersymmetric Integrable Models and Topological Field Theories,"* Lectures given at the Summer School on High Energy Physics and Cosmology, Trieste, Italy, June 15th – July 3rd, 1992. To appear in the proceedings.

[3] C. Vafa, Lectures given at the Summer School on High Energy Physics and Cosmology, Trieste, Italy, June 15th – July 3rd, 1992. To appear in the proceedings.

[4] E. Martinec, *Phys. Lett.* **217B** (1989) 431; C. Vafa and N.P. Warner, *Phys. Lett.* **218B** (1989) 51.

[5] W. Lerche, C. Vafa and N.P. Warner, *Nucl. Phys.* **B324** (1989) 427.

[6] A. Cappelli, C. Itzykson and J.B. Zuber, *Nucl. Phys.* **B280** (1987) 445.

[7] E. Witten, *Nucl. Phys.* **B403** (1993) 159; "On the Landau-Ginzburg description of $N = 2$ minimal models," IASSNS-HEP-93-10, hep-th/9304026.

[8] P. di Francesco, O. Aharony and S. Yankielowicz, "Elliptic Genera and the Landau-Ginzburg approach to $N = 2$ orbifolds," SACLAY-SPHT-93-068, hep-th/9306157.

[9] T. Kawai, Y. Yamada and S.-K. Yang, "Elliptic genera and $N = 2$ superconformal field theory," KEK-TH-362, hep-th/9306096.

[10] M. Henningson, "$N = 2$ gauged WZW models and the elliptic genus," IASSNS-HEP-93-39, hep-th/9307040.

[11] P. Fendley, S. Mathur, C. Vafa and N.P. Warner, *Phys. Lett.* **243B** (1990) 257.

[12] P. Fendley and K. Intriligator, *Nucl. Phys.* **B372** (1992) 533; *Nucl. Phys.* **B380** (1992) 265.

[13] A. LeClair, D. Nemeschansky and N.P. Warner *Nucl. Phys.* **B390** (1993) 653.

[14] S. Cecotti and C. Vafa, *Nucl. Phys.* **367** (1991) 359.

[15] S. Cecotti, P. Fendley, K. Intriligator and C. Vafa, *Nucl. Phys.* **B386** (1992) 405.

[16] V. Pasquier, *Nucl. Phys.* **B295** (1988) 491.

[17] V. Pasquier and H. Saleur, *Nucl. Phys.* **B330** (1990) 523.

[18] B. Nienhuis, *J. Stat. Phys.* **34** (1984) 731.

[19] P. di Francseco, H. Saleur and J.-B. Zuber, *J. Stat. Phys.* **49** (1987) 57.

[20] P. Di Francesco, H. Saleur and J.-B. Zuber, *Nucl. Phys.* **B285** (1987) 454; *Nucl. Phys.* **B300** (1988) 393; I. Kostov, *Nucl. Phys.* **B300** (1988) 559. J.-B. Zuber, "Conformal Field Theories, Coulomb Gas Picture and Integrable models", in *Fields, Strings and Critical Phenomena*, editors E. Brézin and J. Zinn-Justin, Les Houches 1988 Session XLIX.

[21] E. Date, M. Jimbo, T. Miwa and M. Okada, *Phys. Rev.* **B35** (1987) 2105.

[22] D. Friedan, Z. Qiu and S. Shenker, *Phys. Lett.* **151B** (1984) 1575.

[23] D. Kastor, E. Martinec and S. Shenker, *Nucl. Phys.* **B316** (1989) 590.

[24] Z. Qiu, *Nucl. Phys.* **B270** (1986) 205.

[25] O. Foda, *Nucl. Phys.* **B300** (1988) 611.

[26] S.K. Yang, *Nucl. Phys.* **B285** (1987) 183.

[27] S.K. Yang and H.B.Zheng, *Nucl. Phys.* **B285** (1987) 410.

[28] M. Baake and G. von Gehlen, V.Rittenberg, *J. Phys.* **A20** (1987) L479; *J. Phys.* **A20** (1987) L487.

[29] P. di Francesco, H. Saleur and J.-B. Zuber, *Nucl. Phys.* **B300** (1988) 393.

[30] S.-K. Yang, *Phys. Lett.* **B209** (1988) 242.

[31] H. Saleur, *Nucl. Phys.* **B382**(1992) 486, 532.

[32] A.G. Izergin and V.E. Korepin, *Commun. Math. Phys.* **79** (1981) 303.

[33] S.O. Warnaar, M.T. Batchelor and B. Nienhuis, *J. Phys.* **A25** (1992) 3077.

[34] P. Fendley and H. Saleur, *Nucl. Phys.* **B388** (1992) 609.

[35] C. Gomez and G. Sierra, "Spin anisotropy commensurable chains: quantum group symmetries and $N = 2$ Susy," *Nucl. Phys.* **B**, to appear.

[36] Z. Maassarani, D. Nemeschansky and N.P. Warner, *Nucl. Phys.* **B393** (1993) 523.

[37] Z. Maassarani, "Conformal Weights via Bethe Ansatz for $N = 2$ Superconformal Theories," USC preprint USC-93/019.

[38] D. Nemeschansky and N.P. Warner, "Off-Critical Lattice Analogues of $N = 2$ Supersymmetric Quantum Integrable Models," preprint USC-93/018, to appear in *Nucl. Phys. B*

[39] P. Kasteleyn and C. Fortuin, *J.Phys. Soc. Japan Suppl.* **26** (1969) 11.

[40] H.W.J. Blote and M.P.M. Nightingale, *Physica* **112A** (1982) 405.

[41] R.J. Baxter, *"Exactly solved models in statistical mechanics,"* Academic Press, London 1982.

[42] P. Martin, "Potts models," World Scientific.

[43] P. Martin and H. Saleur,"The blob algebra and the periodic Temperley-Lieb algebra", preprint USC-93-009, to appear in *Lett. Math. Phys.*.

[44] G. Felder, *Nucl. Phys.* **B317** (1989), 215.

[45] L.P. Kadanoff and A.C. Brown, *Ann. Phys.* **121** (1979) 318.

[46] V. Pasquier, *Commun. Math. Phys.* **118** (1988) 335.

[47] H. Saleur and J.-B. Zuber, "Integrable Lattice Models and Quantum Groups," in the proceedings of the 1990 Trieste Spring School on String Theory and Quantum Gravity.

[48] G.E. Andrews, R.J. Baxter and P.J. Forrester, *J. Stat. Phys.* **35** (1984) 193.

[49] A. Kuniba and T. Yajima, *J. Stat. Phys.* **52** (1987) 829; P. Ginsparg, *Nucl. Phys.* **B295** (1988) 153; P.A. Pierce and K. Seaton, *Ann. Phys.* **193** (1989) 326; P.A. Pierce and M.T. Batchelor, *J. Stat. Phys.* **60** (1990) 77.

[50] P. Mathieu and M.A. Walton *Phys. Lett.* **254B** (1991) 106.

[51] W. Lerche and N.P. Warner, *Nucl. Phys.* **B358** (1991) 571.

[52] M.T. Grisaru, S. Penati and D. Zanon, *Phys. Lett.* **253B** (1991) 357; G.W. Delius, M.T. Grisaru, S. Penati and D. Zanon, *Phys. Lett.* **256B** (1991) 164; *Nucl. Phys.* **B359** (1991) 125.

[53] J. Evans and T.J. Hollowood *Nucl. Phys.* **B352** (1991) 723; *Nucl. Phys.* **B382** (1992) 662; *Phys. Lett.* **293B** (1992) 100.

[54] G. Felder and A. LeClair, *Int. J. Mod. Phys.* **A7** (1992) 239.

[55] F. M. Goodman, P. de la Harpe, V. F. R. Jones, *"Coxeter graphs and towers of algebras"*, MSRI Publications number 14, Springer Verlag, and references therein.

[56] W.M. Koo and H. Saleur, "Representations of the Virasoro algebra from lattice models," preprint USC-93-025.

[57] M. Takahashi and M. Suzuki, *Prog.Th.Phys.* **48** (1972) 2187.

[58] Al.B. Zamolodchikov, *Nucl. Phys.* **B366** (1991) 122; *Nucl. Phys.* **B358** (1991) 524.

[59] N. Yu Reshetikhin and H. Saleur, "Lattice regularization of massive and massless integrable field theories", preprint USC-93/-20.

[60] L.D. Faddeev and L.A. Takhtajan, *Phys.Lett.* **A85** (1981) 375.

[61] P. Fendley and H. Saleur, "Massless integrable QFT and massless scattering in $1 + 1$ dimensions," preprint USC-93-022.

[62] P.G. de Gennes, *"Scaling concepts in polymer physics"*, Cornell University Press.

[63] P. Fendley, H. Saleur and Al.B. Zamolodchikov, "Massless flows I: the sine-Gordon and $O(n)$ models" and "Massless flows II: the exact S-matrix approach", *Int. J. Mod. Phys.* **A**, to appear.

[64] M. Jimbo, T. Miwa and M. Okado, *Commun. Math. Phys.* **116** (1988) 507.

[65] B. Duplantier and H. Saleur, *Nucl. Phys.* **B290** (1987) 291.

[66] Y. Kazama and H. Suzuki, *Nucl. Phys.* **B234** (1989) 73.

[67] D. Nemeschansky and N.P. Warner, *Nucl. Phys.* **B380** (1992) 241.

[68] P. Fendley, W. Lerche, S.D. Mathur and N.P. Warner, *Nucl. Phys.* **B348** (1991) 66.

[69] E. Witten, *Commun. Math. Phys.* **117** (1988) 353; *Commun. Math. Phys.* **118** (1988) 411; *Nucl. Phys.* **B340** (1990) 281.

[70] T. Eguchi and S.-K. Yang, *Mod. Phys. Lett.* **A4** (1990) 1693.

[71] T. Miwa, M. Jimbo and M. Okado, "Symmetric tensors of the $A_{n-1}^{(1)}$ family," Kyoto preprint (1987), in *Algebraic Analysis,* Festschrift for M. Sato's 60th birthday, Academic Press (1988).

[72] M.P. Richey and C.A. Tracy, *J. Stat. Phys.* **42** (1986) 311.

[73] E. Date, M. Jimbo, A. Kuniba, T. Miwa and M. Okado, *Advanced Studies in Pure Mathematics,* **16** (1988) 17.

[74] T.R. Klassen and E. Melzer, *Nucl. Phys.* **B338** (1990) 485.

[75] D. Bernard and A. LeClair, *Phys. Lett.* **247** (1990) 309; C. Ahn, D. Bernard, and A. LeClair, *Nucl. Phys.* **B346** (1990) 409; D. Bernard and A. LeClair, *Commun. Math. Phys.* **142** (1991) 99.

[76] T. Hollowood, "A Quantum Group Approach to Constructing Factorizable S-Matrices," Oxford Preprint OUTP-90-15P, June 1990.

[77] T. Nakatsu, *Nucl. Phys.* **B356** (1991) 499; H.J. de Vega and V.A. Fateev, *Int. J. Mod. Phys.* **A6** (1991) 3221.

[78] P. Christe and G. Mussardo, *Nucl. Phys.* **B330** (1990) 465; *Int. J. Mod. Phys.* **A5** (1990) 4581; G. Mussardo, *Int. J. Mod. Phys.* **B6** (1992) 2061.

[79] M. Karowski and H. J. Thun, *Nucl. Phys.* **B190** (1981) 61; E. Ogievetsky and P. Wiegmann, *Phys. Lett.* **168** (1986), 360; E. Ogievetsky, N.Yu. Reshetikhin and P. Wiegmann, *Nucl. Phys.* **B280** (1987) 45.

[80] V.V. Bazhanov, N.Yu. Reshetikhin, *Progress in Theor. Phys.* **102** (1990) 301.

[81] C. Gomez and G. Sierra, "On the integrability of $N = 2$ Landau-Ginzburg models: a graph generalization of the Yang-Baxter equation," CERN preprint TH-6963/93, hep-th/9309007.

[82] D. Nemeschansky and N.P. Warner, in preparation.

[83] M. Jimbo, K. Miki, T. Miwa and A. Nakayashiki, "Correlation functions of the XXZ-model for $\Delta < -1$," kyoto PRINT-92-0191, hep-th/9205055; M. Jimbo, T. Miwa and A. Nakayashiki, *J. Phys.* **A26** (1993) 2199; O. Foda, M. Jimbo, T. Miwa, K. Miki and A. Nakayashiki, "Vertex operators in solvable lattice models," RIMS-922, hep-th/9305100; M. Jimbo, T. Kojima, T. Miwa, Y.-H. Quano, "Smirnov's integrals and quantum Knizhnik-Zamolodchikov equation of level 0," RIMS-945, hep-th/9309118.

CANONICAL CONSTRUCTION OF LIOUVILLE FIELD OPERATORS WITH ARBITRARY SPIN

Jens SCHNITTGER[a]

Laboratoire de Physique Théorique de
l'École Normale Supérieure
24 rue Lhomond, 75231 Paris CEDEX 05, France

Abstract: These notes contain an account of some recent work by J.-L. Gervais and
the author on the operator approach to Liouville theory. In the last section the author
presents some qualitative reasoning concerning the structure of the three-point function.

1 Introduction

Recently there has been considerable progress in understanding the structure of two-dimensional gravity from the continuum point of view[1]-[8]. It was brought about by the realization of the underlying quantum group structure of the theory which turns out to govern not only the chiral operator algebra[1][2][6][8], but also the reconstruction of (exponentials of) the Liouville field[7]. The quantum group structure was, however, discovered only after the quantization of the theory had, to some extent, already been carried out by other methods, e.g. by employing a quantum Bäcklund transformation[9], expanding in powers of the cosmological constant[10] or, in the framework considered here[11], by making use of null vector decoupling equations. The latter method is the only one that allowed to establish control over the algebra of the chiral primaries, though only in the sector of the theory consisting solely of degenerate fields, where the null vector decoupling equations apply. The degenerate sector turns out to be described by standard double-sided representations of the quantum group $U_q(sl(2))$ with positive half-integer spin[2], and the corresponding Liouville exponentials $e^{-J\Phi}$ can be obtained as singlets formed from these representations[7]. Of course, to describe the full integrability structure of the theory, it is necessary to construct the complete set of Liouville exponentials with arbitrary continous J, which leads out of the degenerate sector of the theory. In particular, the cosmological constant term with $J = -1$

---

[a] supported by DFG"

appearing in the action, and the Liouville field $\Phi = -\frac{d}{dJ}|_{J=0}e^{-J\Phi}$ itself should possess meaning as quantum operators. The construction of these operators and their correlation functions will provide information about the structure of Liouville theory as a local conformal theory in its own right — not just as a "gravitational dressing" for some particular matter theory — which seems hardly accessible from the matrix model approach. Such insights appear particularly useful to understand 2 dimensional gravity and its W generalizations in their strong-coupling regimes where, so far, only the quantum-group approach has led to significant progress[3] [6] [12] [13]. The presence of the quantum group structure in the degenerate sector of course suggests that actually the full theory should be governed by this symmetry. Indeed, the structural analogy of $U_q(sl(2))$ with its classical counterpart together with the group-theoretical decomposition of the classical Liouville exponentials suggests that in the general situation, the chiral primaries should fall into one-sided highest/lowest weight representations of $U_q(sl(2))[14]$. Furthermore one expects that they should obey a closed algebra under fusion and braiding, determined by the q-Clebsch-Gordan coefficients resp. R-matrix of $U_q(sl(2))$ relevant for these representations. The main point of the present article is to take a first important step in establishing this picture by explicit construction. By setting up a Coulomb-gas type representation for the chiral vertex operators with arbitrary real J, their exchange algebra becomes accessible to free field techniques, and we can prove that the braiding matrix is given by a natural analytic continuation of the positive half-integer spin case, defined in terms of Askey-Wilson polynomials . Passing to suitable linear combinations of these vertex operators which are quantum group covariant, as in the degenerate case, one finds that indeed in this basis, the braiding matrix coincides with the universal $U_q(sl(2))$ R-matrix, specified to the one-sided representations. As a corollary, we obtain that the Liouville exponentials $e^{-J\Phi}$ can again by defined as local operators by forming the quantum group singlet from the left and right moving chiral spin J representations. However, the chiral decomposition is now given by an infinite sum. This leads to a rather drastic modification of the properties of the Liouville exponentials, as charge conservation can no longer be used in the evaluation of their correlation functions. This point will be discussed in the last part of the article, together with some speculative reasoning within the framework of a "minisuperspace" type toy model about the proper evaluation of the three-point function.

2   Gervais-Neveu Quantization

We start with a brief introduction to Gervais-Neveu quantization of Liouville theory. The solutions of the classical Liouville theory,

$$S = \frac{1}{8\pi} \int d^2\xi \{(\partial\Phi)^2 + \mu^2 e^{\alpha\Phi}\}, \tag{2.1}$$

are given by

$$\alpha\Phi = \ln\left[\frac{8A'(u)B'(v)}{\mu^2(A(u) - B(v))^2}\right] \quad , \qquad u = \tau + \sigma, \quad v = \tau - \sigma \tag{2.2}$$

with A and B arbitrary functions, and $\sigma \in [0, 2\pi]$. (We have redefined $\Phi \to \alpha\Phi$ in order to have the standard normalization for the kinetic term). Eq.2.2 is invariant under the projective transformations

$$A \to \frac{aA+b}{cA+d} \qquad B \to \frac{aB+b}{cB+d} \qquad , \qquad (2.3)$$

a, b, c, d complex, which on the quantum level gives rise to the $U_q(sl(2))$ quantum group symmetry. Introducing the chiral fields

$$\psi_m^{(J)} = (A'^{-1/2})^{J-m}(AA'^{-1/2})^{J+m} \quad , \quad \overline{\psi}_m^{(J)} = (B'^{-1/2})^{J+m}(BB'^{-1/2})^{J-m} \qquad , \qquad (2.4)$$

we can write the Liouville exponentials as

$$e^{-J\alpha\Phi} = \sum_{m=-J}^{J} \binom{2J}{J+m}(-1)^{J+m}\psi_m^{(J)}\overline{\psi}_m^{(J)} \qquad (2.5)$$

for any positive half-integer J. It is easy to verify that the $\psi_m^{(J)}, \overline{\psi}_m^{(J)}$ transform as (double-sided) spin J representations of $sl(2, \mathbf{C})$, and Eq.2.5 is the corresponding singlet. For general real J, Eq.2.5 is still valid formally, with the $m-$sum extending to $+\infty$ resp. to $-\infty$, depending on whether we work with the one-sided representations with $J+m$ positive integer , or $J-m$ positive integer; both choices must yield the same result, since they are just related by the particular $sl(2)$-transformation

$$A \to -1/A \qquad B \to -1/B \qquad (2.6)$$

If Φ is periodic, then by making use of the "gauge" symmetry Eq.2.3, we can choose a gauge for A and B such that

$$X_1(u) := \ln A'^{-1/2}(u) \quad , \quad \overline{X}_1(v) := \ln B'^{-1/2}(v)$$

resp.

$$X_2(u) := \ln AA'^{-1/2}(u) \quad , \quad \overline{X}_2(v) := \ln BB'^{-1/2}(v) \qquad (2.7)$$

form the chiral components of two equivalent *periodic* free fields $X_1 + \overline{X}_1$ resp. $X_2 + \overline{X}_2$. The complete symmetry of the treatment of the theory under the exchange of $X_1, \overline{X}_1$ and $X_2, \overline{X}_2$ even on the quantum level is a hallmark of Gervais-Neveu quantization and guarantees the quantum-mechanical preservation of the "residual gauge symmetry" Eq.2.6 which relates the two sets of free fields. Their mode expansions are given by

$$X_j(\sigma) = q_0^{(j)} + p_0^{(j)}\sigma + i\sum_{n \neq 0} e^{-in\sigma} p_n^{(j)}/n, \quad j = 1, 2 \qquad (2.8)$$

The starting point of the quantization is to replace Eqs.2.8 by their quantum counterparts, so that we now have

$$\left[X_1'(\sigma_1), X_1'(\sigma_2)\right] = \left[X_2'(\sigma_1), X_2'(\sigma_2)\right] = 2\pi i\, \delta'(\sigma_1 - \sigma_2) \qquad (2.9)$$

The relation between the modes of X_1 and X_2 is very complicated in general[15], but the zero mode momenta are equal up to sign:

$$p_0^{(1)} = -p_0^{(2)} \qquad (2.10)$$

The functions $(A'^{-1/2})^{J-m}$ and $(AA'^{-1/2})^{J+m}$ are then represented by the normal-ordered exponentials

$$(A'^{-1/2})_{qu}^{J-m} =\, :_1 e^{(J-m)\sqrt{h/2\pi}X_1}:_1 \quad , \quad (AA'^{-1/2})_{qu}^{J+m} =\, :_2 e^{(J+m)\sqrt{h/2\pi}X_2}:_2 \qquad (2.11)$$

where h is related to the central charge C of the theory by

$$h = \frac{\pi}{12}\left(C - 13 - \sqrt{(C-25)(C-1)}\right) \qquad (2.12)$$

and $:_1\ :_1, :_2\ :_2$ denote the normal orderings with respect to the nonzero modes of X_1, X_2. Eqs.2.11 are in fact defined for arbitrary J and m and describe primary conformal fields. The dimension Δ_ρ of $(A'^{-1/2})_{qu}^{2\rho}$ and $(AA'^{-1/2})_{qu}^{2\rho}$ is given by

$$\Delta_\rho = -\rho - \frac{h}{\pi}\rho(\rho + 1) \qquad (2.13)$$

The term linear in ρ is due to the linear "background charge" piece in the energy-momentum tensor, which in terms of the free fields takes the form

$$T_{++} = \frac{1}{2}\, :_1 (\partial_u X_1)^2 :_1 + Q\partial_u^2 X_1 = \frac{1}{2}\, :_2 (\partial_u X_2)^2 :_2 + Q\partial_u^2 X_2,$$

$$Q = \sqrt{(h/2\pi + \pi/2h + 1)} \qquad (2.14)$$

The linear term describes the nonscalar transformation behaviour of the Liouville field under coordinate transformations. One can show that, to leading order in $\sigma - \sigma'$,

$$:_1 e^{(J-m)\sqrt{h/2\pi}X_1(\sigma)}:_1 :_2 e^{(J+m)\sqrt{h/2\pi}X_2(\sigma')}:_2 \sim \left(1 - \frac{z'}{z}\right)^{(m^2-J^2)h/2\pi}\psi_m^{(J)}(\sigma'),$$

where

$$z := e^{i\sigma}, z' := e^{i\sigma'} \qquad (2.15)$$

Here, $\psi_m^{(J)}(\sigma')$ is a conformal primary field of dimension Δ_J fulfilling

$$\psi_m^{(J)}\varpi = (\varpi + 2m)\psi_m^{(J)}, \qquad \varpi := i\sqrt{2\pi/h}p_0^{(1)} \qquad (2.16)$$

The latter two properties in fact determine a primary field uniquely up to normalization. The right-moving vertex operator $\overline{\psi}_m^{(J)}$ is defined analogously. We note that the hermiticity of energy-momentum allows for $\varpi, \overline{\varpi}$ real or purely imaginary, corresponding to the elliptic resp. hyperbolic sector of the theory[16]. In the former case, the boundary conditions imply

$$h\varpi - h\overline{\varpi} = k\pi, \qquad k \in \mathbf{Z} \qquad (2.17)$$

whereas in the latter one must have $\varpi = \overline{\varpi}$. For simplicity, we will consider explicitly only the case $k = 0$ in Eq.2.17, though the analysis is essentially unchanged for general k. Instead of the $\psi_m^{(J)}$, it is convenient to consider the normalized vertex operators $V_m^{(J)}$ with

$$\langle \varpi | V_m^{(J)}(\tau = \sigma = 0) | \varpi + 2m \rangle = 1 \qquad (2.18)$$

which differ from $\psi_m^{(J)}$ only by a ϖ-dependent normalization factor. The ground states $|\varpi\rangle$ are annihilated by all $p_n^{(j)}$ with $n > 0$ and are of course eigenstates of ϖ.

382

3 Derivation of the Chiral Algebra for Continous J

As we have seen, for the reconstruction of the Liouville exponentials with arbitrary J, one needs only the chiral primaries with $J+m$ (positive) integer, or equivalently, those with $J-m$ integer. We shall only discuss the case of integer $J+m$ explicitly here. The other case may be obtained by the trivial substitution $\phi_1 \to \phi_2$ and $\varpi \to -\varpi$ in all the formulae. The purpose of this section is to derive that the $V_m^{(J)}$ with $J+m = 0, 1, 2, \ldots$ obey a closed exchange algebra and thus are in fact the proper quantum generalizations of the classical $\psi_m^{(J)}$. In view of the very complicated relation between X_1 and X_2, it is not very useful to work with the representation Eq.2.15. Classically, it is easy to derive the relation

$$\psi_{1/2}^{(1/2)}(\sigma) = \psi_{-1/2}^{(1/2)}(\sigma)\{e^{2ih\varpi}\int_0^\sigma (\psi_{-1/2}^{(1/2)}(x))^{-2}dx + \int_\sigma^{2\pi}(\psi_{-1/2}^{(1/2)}(x))^{-2}dx\} \tag{3.1}$$

It expresses $\psi_{1/2}^{(1/2)} \equiv e^{\sqrt{h/2\pi}X_2}$ as a function of X_1, by making use of the dimension zero quantity ("screening charge") $A(\sigma) = e^{2ih\varpi}\int_0^\sigma(\psi_{-1/2}^{(1/2)}(x))^{-2}dx + \int_\sigma^{2\pi}(\psi_{-1/2}^{(1/2)}(x))^{-2}dx$. In view of Eq.2.4 we can then express all of the $\psi_m^{(J)}$ in terms of the free field X_1 only. On the quantum level, the situation turns out to be very similar and one finds for the normalized operators $V_m^{(J)}$[17] [18]

$$V_m^{(J)} = (I_m^{(J)}(\varpi))^{-1}U_m^{(J)}, \tag{3.2}$$

where $I_m^{(J)}(\varpi)$ are normalization constants (independent of σ), and

$$U_m^{(J)}(\sigma) = V_{-J}^{(J)}(\sigma)[S(\sigma)]^{J+m} \tag{3.3}$$

$$S(\sigma) = e^{2ih(\varpi+1)}\int_0^\sigma dx V_1^{(-1)}(x) + \int_\sigma^{2\pi} dx V_1^{(-1)}(x). \tag{3.4}$$

Indeed, first it is easily verified that $S(\sigma)$ is a dimension zero operator. Second, using the explicit expression for the Liouville zero-mode $\varpi = ip_0^{(1)}\sqrt{2\pi/h}$ (see Eq.2.16), one verifies that

$$V_m^{(J)}\varpi = (\varpi + 2m)V_m^{(J)} \tag{3.5}$$

as expected. Since the above expressions make sense also for continuous $2J$, provided that $J+m$ is a non-negative integer, we take them as the definition of the generalized vertex operators. The only potential problem comes from divergences of the product of $S(\sigma)$. However it is easily seen that $[S]^{J+m}$ makes sense for small enough h. The singularities which appear when h increases are to be handled by analytic continuation (more about this below). We now turn to the derivation of the algebra of the fields $U_m^{(J)}$. The quantum group picture leads us to expect that there should in fact exist a closed exchange algebra for these operators which is related to the braiding of highest (lowest) weight representations of $U_q(sl(2))$ with continous spin. Hence we start with an ansatz of the form

$$U_m^{(J)}(\sigma)U_{m'}^{(J')}(\sigma') = \sum_{m_1,m_2} R(J,J';\varpi)_{m\ m'}^{m_2m_1} U_{m_2}^{(J')}(\sigma')U_{m_1}^{(J)}(\sigma)$$

for $\quad \pi > \sigma' > \sigma > 0,$

resp.

$$U_m^{(J)}(\sigma)U_{m'}^{(J')}(\sigma') = \sum_{m_1,m_2} \check{R}(J,J';\varpi)_{m\ m'}^{m_2m_1} U_{m_2}^{(J')}(\sigma')U_{m_1}^{(J)}(\sigma)$$

for $\quad \pi > \sigma > \sigma' > 0,$ $\tag{3.6}$

with the sums extending over non-negative integer $J+m_1$ resp. $J'+m_2$. Comparing the monodromy properties of both sides of Eq.3.6, we conclude that, as in the half-integer spin case, the braiding matrix is nonzero only when

$$m_1 + m_2 = m + m' =: M \tag{3.7}$$

In the approach of refs.[17][7], one takes the zero modes of the left-moving and right-moving Liouville modes to commute, so that V and $\overline{V}$ commute. Thus Eq.3.6 is accompanied by a similar relation for the right-moving fields, and locality of the Liouville fields will eventually arise as a consequence of orthogonality relations between left and right braiding matrices. Since one considers the braiding at equal τ one lets $\tau = 0$ once and for all, and works on the unit circle $u = e^{i\sigma}$. In the above equations and in the following we discuss the braiding of the holomorphic components assuming that σ and σ' range between 0 and π for concreteness. As there are no null-vector decoupling equations for continuous J, we have to rely exclusively on the free field techniques just summarized. Fortunately, it turns out that the exchange of two $U_m^{(J)}$ operators can be mapped into an equivalent problem in one-dimensional quantum mechanics, and becomes just finite-dimensional linear algebra. Given the fact that the $U_m^{(J)}$ consist only of the "tachyon operators" $V_{-J}^{(J)}$ resp. $V_1^{(-1)}$, the essential observation is to remember that

$$V_{-J}^{(J)}(\sigma)V_{-J'}^{(J')}(\sigma') = e^{-i2JJ'h\epsilon(\sigma-\sigma')}V_{-J'}^{(J')}(\sigma')V_{-J}^{(J)}(\sigma) \tag{3.8}$$

where $\epsilon(\sigma - \sigma')$ is the sign of $\sigma - \sigma'$. This means that when commuting the tachyon operators in $U_{m'}^{(J')}(\sigma')$ through those of $U_m^{(J)}(\sigma)$, one only encounters phase factors of the form $e^{\pm i2\alpha\beta h}$ resp. $e^{\pm 6i\alpha\beta h}$, with α equal to J or -1, β equal to J' or -1, since we take $\sigma, \sigma' \in [0, \pi]$. Hence we are led to decompose the integrals defining the screening charges S into pieces which commute with each other and with $V_{-J}^{(J)}(\sigma)$, $V_{-J'}^{(J')}(\sigma')$ up to one of the above phase factors. We consider explicitly only the case $0 < \sigma < \sigma' < \pi$ and write

$$S(\sigma) = S_{\sigma\sigma'} + S_\Delta, \quad S(\sigma') = S_{\sigma\sigma'} + k(\varpi)S_\Delta \equiv S_{\sigma\sigma'} + \widehat{S}_\Delta,$$

$$S_{\sigma\sigma'} := k(\varpi)\int_0^\sigma V_1^{(-1)}(x)dx + \int_{\sigma'}^{2\pi} V_1^{(-1)}(x)dx,$$

$$S_\Delta := \int_\sigma^{\sigma'} V_1^{(-1)}(x)dx, \quad k(\varpi) := e^{2ih(\varpi+1)} \tag{3.9}$$

Using Eq.3.8, we then get the following simple algebra for $S_{\sigma\sigma'}, S_\Delta, \widehat{S}_\Delta$:

$$S_{\sigma\sigma'}S_\Delta = q^{-2}S_\Delta S_{\sigma\sigma'}, \quad S_{\sigma\sigma'}\widehat{S}_\Delta = q^2\widehat{S}_\Delta S_{\sigma\sigma'}, \quad S_\Delta\widehat{S}_\Delta = q^4\widehat{S}_\Delta S_\Delta, \tag{3.10}$$

and their commutation properties with $V_{-J}^{(J)}(\sigma), V_{-J'}^{(J')}(\sigma')$ are given by

$$V_{-J}^{(J)}(\sigma)S_{\sigma\sigma'} = q^{-2J}S_{\sigma\sigma'}V_{-J}^{(J)}(\sigma), \; V_{-J'}^{(J')}(\sigma')S_{\sigma\sigma'} = q^{-2J'}S_{\sigma\sigma'}V_{-J'}^{(J')}(\sigma'),$$
$$V_{-J}^{(J)}(\sigma)S_\Delta = q^{-2J}S_\Delta V_{-J}^{(J)}(\sigma), \quad V_{-J}^{(J)}(\sigma)\widehat{S}_\Delta = q^{-6J}\widehat{S}_\Delta V_{-J}^{(J)}(\sigma),$$
$$V_{-J'}^{(J')}(\sigma')S_\Delta = q^{2J'}S_\Delta V_{-J'}^{(J')}(\sigma'), \; V_{-J'}^{(J')}(\sigma')\widehat{S}_\Delta = q^{-2J'}\widehat{S}_\Delta V_{-J'}^{(J')}(\sigma'). \tag{3.11}$$

Finally, all three screening pieces obviously shift the zero mode in the same way:

$$\left.\begin{array}{c} S_{\sigma\sigma'} \\ S_\Delta \\ \widehat{S}_\Delta \end{array}\right\} \varpi = (\varpi + 2)\left\{\begin{array}{c} S_{\sigma\sigma'} \\ S_\Delta \\ \widehat{S}_\Delta \end{array}\right. . \tag{3.12}$$

Using Eqs.3.11 we can commute $V_{-J}^{(J)}(\sigma)$ and $V_{-J'}^{(J')}(\sigma')$ to the left on both sides of Eq.3.6, so that they can be cancelled. Then we are left with

$$(q^{-2J'}S_\Delta + q^{2J'}S_{\sigma\sigma'})^{J+m}(\widehat{S}_\Delta + S_{\sigma\sigma'})^{J'+m'} =$$

$$\sum_{m_1,m_2} R(J, J'; \varpi + 2(J + J'))^{m_2 m_1}_{m\ m'}q^{-2JJ'}(q^{2J}S_{\sigma\sigma'} + q^{6J}\widehat{S}_\Delta)^{J'+m_2}(S_{\sigma\sigma'} + S_\Delta)^{J'+m_1} \tag{3.13}$$

It is apparent from Eq.3.13 that the braiding problem of the $U_m^{(J)}$ operators is governed by the Heisenberg-like algebra Eq.3.10, characteristic of one-dimensional quantum mechanics. However, to see this structure emerge, we had to decompose the screening charges $S(\sigma), S(\sigma')$ in a way which depends on both positions σ, σ'; hence the embedding of this Heisenberg algebra into the 1+1 dimensional field theory is somewhat nontrivial. To evaluate Eq.3.13, we could sort both sides of the equation in powers of $S_{\sigma\sigma'}, S_\Delta$ and then compare coefficients. This indeed can be carried out straightforwardly, upon observing that $S_{\sigma\sigma'}$ and S_Δ resp. $\widehat{S}_\Delta$ behave like the components a, b of a vector in the quantum plane, with $ba = abq^2$, so that one can make use of the q-binomial formula

$$(a + b)^N = \sum_{\nu=0}^N \binom{N}{\nu}q^{(N-\nu)\nu}a^\nu b^{N-\nu}, \qquad \binom{N}{\nu} := \frac{\lfloor N\rfloor!}{\lfloor N - \nu\rfloor!\lfloor \nu\rfloor!}$$

$\binom{N}{\nu}$ is a q-deformed binomial coefficient, with

$$\lfloor\nu\rfloor! = \lfloor 1\rfloor\lfloor 2\rfloor\cdots\lfloor\nu\rfloor, \qquad \lfloor x\rfloor := \sin(hx)/\sin(h) \tag{3.14}$$

denoting q-factorials resp. q-numbers. For our purposes, however, another form of the equations is better suited, which is obtained by choosing the following simple representation of the algebra Eq.3.10 in terms of one-dimensional quantum mechanics (y and y' are arbitrary complex numbers):

$$S_{\sigma\sigma'} = y'e^{2Q}, \quad S_\Delta = ye^{2Q-P}, \quad \widehat{S}_\Delta = ye^{2Q+P}, \qquad [Q, P] = ih. \tag{3.15}$$

The third relation in Eq.3.15 follows from the second one in view of $\widehat{S}_\Delta = k(\varpi)S_\Delta$ (cf. Eq.3.9). This means we are identifying here $P \equiv ih\varpi$ with the zero mode of the original problem. Using $e^{2Q+cP} = e^{cP}e^{2Q}q^c$ we can commute all factors e^{2Q} to the right on both sides of Eq.3.13 and then cancel them. This leaves us with

$$\prod_{j=1}^{J+m}(y'q^{2J'} + yq^{-(\varpi-2J+2j-1)})\prod_{\ell=1}^{J'+m'}(y' + yq^{\varpi-2J'+2m+2\ell-1}) =$$

$$\sum_{m_1}R(J, J'; \varpi)^{m_2 m_1}_{m\ m'}q^{-2JJ'}\prod_{j=1}^{J'+m_2}(y'q^{2J} + yq^{\varpi+4J-2J'+2j-1})\prod_{\ell=1}^{J+m_1}(y' + yq^{-(\varpi-2J+2m_2+2\ell-1)})$$

$$\tag{3.16}$$

where we have shifted back $\varpi + 2(J + J') \to \varpi$ compared to Eq.3.13. Since the overall scaling $y \to \lambda y, y' \to \lambda y'$ only gives back Eq.3.7, we can set $y' = 1$. By putting y equal to the zeros of the first or the second product on the RHS of Eq.3.16, plus one other arbitrary value, we obtain a linear system of equations for R in triangular form, which shows that the solution of Eq.3.16 is unique for any fixed J, J', m, m'. We will now demonstrate that it is given by a straightforward extension of the R-matrix of ref.[8], given in terms of orthogonal polynomials. The R-matrix of [8] reads for the V fields [b] :

$$\hat{R}(J, J'; \varpi)_{m\ m'}^{m_2 m_1} = q^{2mm'+m'^2-m_2^2+\varpi(m'-m_2)} \frac{g_{J,x+M}^{x+m_2} g_{J',x+m_2}^{x}}{g_{J',x+M}^{x+m} g_{J,x+m}^{x}} \left\{ \begin{smallmatrix} J & x+M \\ J' & x \end{smallmatrix} \Big| \begin{smallmatrix} x+m_2 \\ x+m \end{smallmatrix} \right\} \tag{3.17}$$

with

$$M = m + m' = m_1 + m_2, \quad x := (\varpi - \varpi_0)/2, \quad \varpi_0 := 1 + \pi/h, \quad q := e^{ih}. \tag{3.18}$$

Here, $\left\{ \begin{smallmatrix} J & x+M \\ J' & x \end{smallmatrix} \Big| \begin{smallmatrix} x+m_2 \\ x+m \end{smallmatrix} \right\}$ denotes the quantum 6-j symbol, and the g's are coupling constants which can be represented as follows (with $F(z) \equiv \Gamma(z)/\Gamma(1-z)$):

$$g_{J_1 J_2}^{J_{12}} = \prod_{k=1}^{J_1+J_2-J_{12}} \sqrt{F(1 + (2J_1 - k + 1)h/\pi)} \times$$

$$\sqrt{\frac{F(1 + (2J_2 - k + 1)h/\pi)F(-1 - (2J_{12} + k + 1)h/\pi)}{F(1 + kh/\pi)}} \tag{3.19}$$

Since even for the general J case, $J_1 + J_2 - J_{12}$ remains a (positive) integer, the extension of Eq.3.19 is trivial. Thus we only need to specify the proper continuation of the 6j-symbol. Using the relation between the 6j-symbol and the $_4F_3$ q-hypergeometric function[11][21][8], one can write[c]

$$\left\{ \begin{smallmatrix} J & x+M \\ J' & x \end{smallmatrix} \Big| \begin{smallmatrix} x+m_2 \\ x+m \end{smallmatrix} \right\} = \sqrt{\lfloor 2x + 2m_2 + 1 \rfloor \lfloor 2x + 2m + 1 \rfloor} \times$$

$$\Delta(J, x + M, x + m_2)\Delta(J, x, x + m)\Delta(J', x, x + m_2)\Delta(J', x + m, x + M) \times$$

$$\frac{\lfloor 2x + N + 1 \rfloor!}{\lfloor 2x + m_2 + m - J - J' \rfloor! \lfloor J + m_1 \rfloor! \lfloor J - m \rfloor! \lfloor J' + m' \rfloor! \lfloor J' - m_2 \rfloor! \lfloor m_2 - m' \rfloor!} \times$$

$$_4F_3 \left(\begin{smallmatrix} m-J, & m_2-J', & -(J+m_1), & -(J'+m') \\ -(2x+N+1), & m_2-m'+1, & 2x+m_2+m+1-J-J' \end{smallmatrix} ; q, 1 \right) \tag{3.20}$$

with

$$N = J + J' + M, \qquad \Delta(a, b, c) = \sqrt{\frac{\lfloor -a + b + c \rfloor! \lfloor a - b + c \rfloor! \lfloor a + b - c \rfloor!}{\lfloor a + b + c + 1 \rfloor!}}$$

The RHS of Eq.3.20 makes sense for arbitrary J. The q-hypergeometric function is defined as

$$_4F_3 \left(\begin{smallmatrix} a, & b, & c, & d \\ e, & f, & g \end{smallmatrix} ; q, \rho \right) = \sum_{n=0}^{\infty} \frac{\lfloor a \rfloor_n \lfloor b \rfloor_n \lfloor c \rfloor_n \lfloor d \rfloor_n}{\lfloor e \rfloor_n \lfloor f \rfloor_n \lfloor g \rfloor_n \lfloor n \rfloor!} \rho^n ,$$

$$\lfloor a \rfloor_n := \lfloor a \rfloor \lfloor a + 1 \rfloor \cdots \lfloor a + n - 1 \rfloor, \quad \lfloor a \rfloor_0 := 1 \tag{3.21}$$

---

[b]For an earlier derivation in a different form, see also [19]
[c]our conventions for q-hypergeometric functions coincide with the ones of ref.[6].

386

In the present context, we have

$$a = m - J, \; b = m_2 - J', \; c = -J - m_1, \; d = -n' = -(J' + m'), \; e = -2x - N - 1,$$

$$f = m_2 - m' + 1, \; g = 1 + a + b + c + d - e - f. \tag{3.22}$$

To make contact with orthogonal polynomials, we bring the $_4F_3$ into a different form by means of the q-version[d] of a well-known transformation formula for balanced $_4F_3$ functions:

$$_4F_3 \left({a, \, b, \, c, \atop e, \, f, \, 1+a+b+c-n'-e-f} \; {-n' \atop} ; q, 1 \right) = \frac{\lfloor f - c \rfloor_{n'} \lfloor f + e - a - b \rfloor_{n'}}{\lfloor f \rfloor_{n'} \lfloor f + e - a - b - c \rfloor_{n'}} \times$$

$$_4F_3 \left({e-a, \; e-b, \atop e,} \; {c, \atop f+e-a-b, \; 1+c-f-n'} \; {-n' \atop} ; q, 1 \right) \tag{3.23}$$

where n' is a non-negative integer. The $_4F_3$ on the RHS of Eq.3.23 can now be identified with an Askey-Wilson (or Racah) polynomial[29]:

$$p_{n'}(\mu(z); \alpha, \beta, \gamma, \delta; q) = \; _4F_3 \left({e-a, \; e-b, \atop e,} \; {c, \atop f+e-a-b, \; 1+c-f-n'} \; {-n' \atop} ; q, 1 \right)$$

where

$$\mu(z) = q^{-2z} + q^{2(z+c+e-b)}, \; z = -c = J + m_1,$$

and

$$\alpha = e - 1, \; \beta = -n' - a, \; \gamma = c - n' - f, \; \delta = e + f - b + n' - 1. \tag{3.24}$$

$n' = J' + m'$ is the degree in the variable $\mu(z)$. Note that $p_{n'}(\mu(z))$ really depends on z only via $\mu(z)$; the other coefficients are functions only of J, J' and M. The Askey-Wilson polynomials satisfy orthogonality relations of the form [29]:

$$\sum_{z=0}^{N} p_n(\mu(z)) \, p_m(\mu(z)) \, w(z) = 0 \quad \text{for } m \neq n. \tag{3.25}$$

Here $N = J + J' + M$. The weight function $w(z)$ is defined as

$$w(z; \alpha, \beta, \gamma, \delta; q) = \frac{\lfloor 2z + 1 + \gamma + \delta \rfloor \lfloor \gamma + \delta + 1 \rfloor_z \lfloor \alpha + 1 \rfloor_z \lfloor \beta + \delta + 1 \rfloor_z \lfloor \gamma + 1 \rfloor_z}{\lfloor \gamma + \delta + 1 \rfloor \lfloor z \rfloor! \lfloor \gamma + \delta - \alpha + 1 \rfloor_z \lfloor \gamma - \beta + 1 \rfloor_z \lfloor \delta + 1 \rfloor_z}, \tag{3.26}$$

where $\alpha, \beta, \gamma, \delta$ are the same coefficients as in the definition of $p_{n'}$.

Thus we have arrived at defining an extension of the braiding matrix Eq.3.17 for the V fields to general J. In view of Eq.3.2, the corresponding R-matrix for the U fields is then obtained as

$$R(J, J'; \varpi)^{m_2 m_1}_{m \; m'} = \widehat{R}(J, J'; \varpi)^{m_2 m_1}_{m \; m'} \frac{I_m^{(J)}(\varpi) I_{m'}^{(J')}(\varpi + 2m)}{I_{m_2}^{(J')}(\varpi) I_{m_1}^{(J)}(\varpi + 2m_2)} \tag{3.27}$$

The normalizations $I_m^{(J)}$ can be computed by making use of the Fateev-Dotsenko integration formulae[26]. One finds[e], letting $n = J + m$,

$$I_m^{(J)}(\varpi) = i^n \prod_{l=1}^{n} \left\{ e^{i\pi\beta(l-1)} (1 - e^{2\pi i(\gamma + \beta(l-1))}) \right\} \times$$

---

[d]it can be proven [22] exactly along the same lines using the method explained, for instance in ref.[2] of the classical formula[24]

[e]details see [27]

$$\prod_{l=1}^{n}\left\{\frac{\Gamma(1-\beta)\Gamma(1+\gamma+(l-1)\beta)\Gamma(1+\alpha+(l-1)\beta)}{\Gamma(1-l\beta)\Gamma(2+\gamma+\alpha+(n-2+l)\beta)}\right\}$$

$$\alpha=2J\frac{h}{\pi},\quad \beta=-\frac{h}{\pi},\quad \gamma=\frac{h}{\pi}(\varpi+2m-1)-1. \tag{3.28}$$

This relation is valid for arbitrary J,ϖ and $J+m=n$ a non-negative integer. We remark that the divergence of the product $[S]^{J+m}$ appears as the first pole in h in the formula just written. It shows where the integral representation defined by Eqs.3.3,3.4 breaks down. However, Eq.3.28 has meaning beyond this point by the usual analytic continuation of the gamma function. Indeed, the general formula Eq.2.15 is valid for arbitrary h and will give rise to a 3-point function which is analytic in h. Hence Eq.3.28 is valid for all h.

The last step is now to show that the braiding matrix for the U fields defined in this way actually solves Eqs.3.16. For this purpose, we simply insert into Eqs.3.16 and consider first the case where y is equal to one of the zeros of the LHS,

$$y=-q^{\varpi+2J'-2J+2j_0-1},\qquad j_0=1\cdots J+m$$

resp.

$$y=-q^{-(\varpi-2J'+2m+2l_0-1)},\qquad l_0=1\cdots J'+m' \tag{3.29}$$

This gives the following homogeneous relations:

$$0=\sum_{z=0}^{N}\lfloor 2x+j_0+1\rfloor_{N-z}\lfloor J'-J-M+j_0\rfloor_z\frac{(-1)^z\lfloor -2J\rfloor_z}{\lfloor -2x-2M\rfloor_z}p_{n'}(\mu(z))w(z),$$

resp.

$$0=\sum_{z=0}^{N}\lfloor J'-m-M-2x-l_0\rfloor_z\lfloor J-m-l_0+1\rfloor_{N-z}\frac{(-1)^z\lfloor -2J\rfloor_z}{\lfloor -2x-2M\rfloor_z}p_{n'}(\mu(z))w(z),$$

with

$$j_0=1\cdots J+m,\qquad l_0=1\cdots J'+m',\qquad N=J+J'+M, n'=J'+m'$$

$$\tag{3.30}$$

Next, one proves by induction in $n'=J'+m'$ (J,J' and N fixed) that the conditions Eqs.3.30 are equivalent to the orthogonality relations Eqs.3.25. Considering the first relation in Eq.3.30, we have to show that the q-products in front of $p_{n'}(\mu(z))w(z)$ are given by a linear combination of $p_k(\mu(z))$ with $k\neq n'$. Indeed we shall prove that

$$\lfloor 2x+j_0'-n'+1\rfloor_{N-z}\lfloor -J-M-m'+j_0'\rfloor_z\frac{(-1)^z\lfloor -2J\rfloor_z}{\lfloor -2x-2M\rfloor_z}=\sum_{k=n'+1}^{N}a_{j_0'k}p_k(\mu(z)),$$

with

$$j_0'=j_0+n'=n'+1\cdots N \tag{3.31}$$

We start at $n'=N-1$ and go downward. For $n'=N-1$, where $j_0'=N$ is the only allowed value, the q-product on the left must be proportional to $p_N(\mu(z))$. The latter

388

is given by Eq.3.24, with the $_4F_3$ collapsing into a $_3F_2$ since two of the coefficients are equal:

$$p_N(\mu(z)) = {}_3F_2\left(\begin{matrix} -(J'-J+M+2x+1), & -z, & -(2J+2M+2x+1-z) \\ -(N+2x+1), & -(2M+2x), & \end{matrix} ; q, 1\right) \tag{3.32}$$

By means of a well-known relation for the $_3F_2$ functions[24], this can be further reduced to a single product of q-numbers, so that

$$p_N(\mu(z)) = \text{const} \cdot \lfloor 2x + 2 \rfloor_{N-z} \lfloor J' - J - M + 1 \rfloor_z \frac{(-1)^z \lfloor -2J \rfloor_z}{\lfloor -2x - 2M \rfloor_z} \tag{3.33}$$

where the constant is independent of z. In view of Eq.3.31, this proves the induction start. Consider now the induction step $m' \to m' - 1$, $m \to m + 1$. This amounts to changing $j'_0 \to j'_0 + 1$ in Eq.3.31. The case $j'_0 \leq N - 1$ is covered by the induction hypothesis. When $j'_0 = N$, we have to show that

$$\lfloor 2x + 2N + 1 - n' - z \rfloor \lfloor J' - m' + z \rfloor \lfloor 2x + N - n' + 1 \rfloor_{N-z} \times$$

$$\lfloor -J - M - m' + N \rfloor_z \frac{(-1)^z \lfloor -2J \rfloor_z}{\lfloor -2x - 2M \rfloor_z} = \sum_{k=n'}^{N} a_k p_k(\mu(z)) \tag{3.34}$$

for suitable a_k. Using the definition Eq.3.24 of $\mu(z)$, it is easy to check that the first two factors in Eq.3.34 can be written in the form $c_1 + c_2\mu(z)$, whereas the others are known by virtue of the induction hypothesis to be a linear combination of $p_{n'+1} \cdots p_N$. But on the other hand the Askey-Wilson polynomials fulfill a 3-term recurrence relation of the form[29]

$$\mu(z)p_n(\mu(z)) = A_n p_{n+1}(\mu(z)) + B_n p_n(\mu(z)) + C_n p_{n-1}(\mu(z)) \tag{3.35}$$

Since $A_N = 0$, the induction step is complete. The proof of the second relation in Eq.3.30 is completely analogous. So we know now that our conjecture for R is correct up to normalization. The latter can be easily checked by putting $y = -q^{\varpi+2M-1}$ in Eq.3.16. Then the products on the RHS vanish except when $J + m_1 = 0$. Inserting for R as before and using $p_n(\mu(0)) \equiv 1 \ \forall \ n$, it is then a straightforward exercise to check that Eq.3.16 is also fulfilled in this inhomogeneous case. Thus we have established that the braiding of the $V_m^{(J)}$ resp. $U_m^{(J)}$ operators with general J, and $J + m$ a non-negative integer, is given by the extension Eq.3.20 or Eq.3.24 of the R-matrix of [8]. In particular, of course, we reproduce the result for the half-integer case, without having taken recourse to the null-vector decoupling equation. So far, we have worked in the Bloch wave, or Coulomb gas basis where the vertex operators are periodic up to a constant and induce a well-defined shift of the zero mode. However, it is well known from the work of Babelon[1] and Gervais[2] that the quantum group structure becomes manifest only after a change to another basis of chiral primaries $\xi_M^{(J)}$, which are linear combinations of the $V_m^{(J)}$. For the half-integer spin case, Gervais established that the braiding for the $\xi_m^{(J)}$ is given by the universal R-matrix of $U_q(sl(2))$:

$$\xi_M^{(J)}(\sigma)\xi_{M'}^{(J')}(\sigma') = (J, J')_{M\,M'}^{M_2 M_1}\xi_{M_2}^{(J')}(\sigma')\xi_{M_1}^{(J)}(\sigma) \tag{3.36}$$

where

$$(J, J')^{M_2 M_1}_{M\ M'} = \delta_{M+M',M_1+M_2} e^{-2ihMM'} \frac{(-1)^n (2i\sin(h))^n e^{ihn(n+1)/2}}{\lfloor n \rfloor!} e^{-ihnM} e^{ihnM'} \times$$

$$\sqrt{\frac{\lfloor J+M \rfloor! \lfloor J-M_1 \rfloor! \lfloor J'-M' \rfloor! \lfloor J'+M_2 \rfloor!}{\lfloor J-M \rfloor! \lfloor J+M_1 \rfloor! \lfloor J'+M' \rfloor! \lfloor J'-M_2 \rfloor!}}, \tag{3.37}$$

for $n \equiv M_2 - M' \geq 0$, and 0 otherwise. Now Eq.3.37 possesses immediate meaning for noninteger $2J$, $2J'$ if we understand the q-factorials as being defined in terms of q-Gamma-functions; the latter can be obtained by a q-deformation of the classical representation in terms of an infinite product[2]. Thus we expect that if we define

$$\xi^{(J)}_M := \sum_{J+m=0}^{\infty} |J, \widetilde{\varpi})^m_M V^{(J)}_m \tag{3.38}$$

with the transformation coefficients $|J, \widetilde{\varpi})^m_M$ given by a suitable analytic continuation of their expressions for half-integer J, then these $\xi^{(J)}_M$ should obey braiding relations given by (the analytic extension of) Eq.3.37. In order to avoid flooding the reader with technical details, we simply note that this can in fact be shown[27], and hence the braiding for continous J is established in both Coulomb gas and quantum group pictures.

4 Liouville Exponentials

Finally we proceed to the construction of local Liouville operators. We will see that it becomes a rather simple corollary of the above analysis for the chiral vertex operators. Clearly, in view of the $sl(2)$-invariance of the classical solution Eq.2.2, we should accept as Liouville operators only such objects which are quantum group singlets. In fact, it is precisely in this way that one *defines* Liouville theory in the strong coupling regime[3][6][12][13] , where no classical interpretation of the field operators is available. Now there is a natural singlet formed from the left and rightmoving $\xi^{(J)}_M$ operators[7]:

$$e^{-J\alpha\Phi} = \sum_{J+M=0}^{2J} (-1)^{J+M} e^{ih(J-M)} \xi^{(J)}_M \bar{\xi}^{(J)}_{-M} \tag{4.1}$$

for positive half-integer J. It is just the general Liouville exponential we are looking for! Indeed one can easily show that in the classical limit, Eq.4.1 reduces to the classical Liouville exponential. It is clearly a conformal primary of weight Δ_J, and thus the only property left to check is locality,

$$\left[e^{-J\alpha\Phi}(\tau, \sigma) , \ e^{-J'\alpha\Phi}(\tau, \sigma') \right] = 0. \tag{4.2}$$

Eq.4.2 now follows immediately from an orthogonality relation[7] between the left and right-moving R-matrices (the right-moving R-matrix is just the complex conjugate of the left-moving one):

$$\sum_{M,M'} (J, J')^{M_2 M_1}_{M\ M'} (\overline{J, J'})^{-\overline{M}_2 -\overline{M}_1}_{-M\ -M'} = \delta_{M_1, \overline{M}_1} \delta_{M_2, \overline{M}_2} \tag{4.3}$$

To see this, note that the phase factors appearing in Eq.4.1 effectively decouple from the braiding problem. This is because locality holds for each fixed value of $M + M'$ separately, if M and M' are the magnetic quantum numbers appearing in the Liouville exponentials $e^{-J\alpha\Phi}$ and $e^{-J'\alpha\Phi}$. Since the orthogonality relation Eq.4.3 is easily seen to remain valid for general J, it is immediately clear how to extend Eq.4.1:

$$e^{-J\alpha\Phi} = \sum_{J+M=0}^{\infty} (-1)^{J+M} e^{ih(J-M)} \xi_M^{(J)} \overline{\xi}_{-M}^{(J)} \tag{4.4}$$

Hence, the only difference to the half-integer case is that now the sum extends to infinity, corresponding to the fact that the $\xi_M^{(J)}$ now form infinite-dimensional one-sided representations of $U_q(sl(2))$ instead of the finite-dimensional double-sided ones. Thus we see that after the preceding rather technical analysis, an almost miraculously simple picture emerges for the objects of physical interest. Of course, as we have seen, a lot of structure is hidden in the precise definition of the $\xi_M^{(J)}$. Actually there exist two possible deformations of the classical group, and correspondingly one can define more general exponentials involving two $U_q(sl(2))$ quantum numbers J and $\widehat{J}$. We will not discuss this possibility here (see, however,[27]).

For the calculation of correlation functions to be discussed below, it is useful to return to the Coulomb gas basis, and we shall briefly rephrase Eq.4.4 in terms of the Bloch wave operators. Again, one finds that the same expression can be used as for half-integer J. Introducing the rescaled operators

$$\widetilde{V}_m^{(J)} = g_{J,x+m}^x V_m^{(J)} \tag{4.5}$$

the Liouville exponentials can be written in the simple form

$$e^{-J\alpha\Phi} = c_J a(\varpi) \sum_{J+m=0}^{\infty} \widetilde{V}_m^{(J)} \overline{\widetilde{V}}_m^{(J)} / a(\varpi) \tag{4.6}$$

where c_J is a normalization constant. Its value can be changed by constant field redefinitions and is not relevant for our considerations. Likewise, we will not specify the function $a(\varpi)$. In fact the latter drops out of correlation functions and does not affect the local or conformal properties, hence can be ignored (up to some subtleties discussed in refs. [7][27][28]). The locality of the expression Eq.4.6 can be checked directly, as a consequence of orthogonality relations fulfilled by the $6j$ symbol. Indeed, notice first that the braiding of the $\widetilde{V}_m^{(J)}$ is given by Eq.3.17 without the coupling constants, i.e. essentially by the $6j$ symbol alone:

$$\widetilde{R}(J,J';\varpi)_{m\ m'}^{m_2 m_1} = \delta_{m+m',m_1+m_2} q^{2mm'+m'^2-m_2^2+\varpi(m'-m_2)} \left\{ \begin{matrix} J & x+M \\ J' & x \end{matrix} \left| \begin{matrix} x+m_2 \\ x+m \end{matrix} \right. \right\} \tag{4.7}$$

Next, it is straightforward to derive from the orthogonality relations of the Askey-Wilson polynomials that

$$\sum_{J_{23}} \left\{ \begin{matrix} J_1 & J_2 \\ J_3 & J_{123} \end{matrix} \left| \begin{matrix} J_{12} \\ J_{23} \end{matrix} \right. \right\} \left\{ \begin{matrix} J_1 & J_2 \\ J_3 & J_{123} \end{matrix} \left| \begin{matrix} J'_{12} \\ J_{23} \end{matrix} \right. \right\} = \delta_{J_{12},J'_{12}} \tag{4.8}$$

when $J_1 + J_2 - J_{12}, J_2 + J_3 - J_{23}, J_1 + J_{23} - J_{123}, J_{12} + J_3 - J_{123}$ are nonnegative integers. Since again the rightmoving R−matrix is just the complex conjugate of the left-moving one (with ϖ to be treated as real always) this implies

$$\sum_{m,m'} \tilde{R}(J, J'; \varpi)^{m_2 m_1}_{m\ m'} \overline{\tilde{R}(J, J'; \varpi)}^{\overline{m}_2 \overline{m}_1}_{m\ m'} = \delta_{m_1 \overline{m}_1} \delta_{m_2 \overline{m}_2} \tag{4.9}$$

and locality follows. We mention that the canonical approaches of refs.[9],[10] have recently been shown [28] to be equivalent to the one presented above, in the sense that the Liouville exponentials can in all cases be written in the form Eq.4.6, with the same operators $\tilde{V}^{(J)}_m$ appearing, but different functions $a(\varpi)$.

5 Correlation Functions

In this section, we will make some -partly speculative- remarks about the evaluation of correlation functions, in particular the three-point function. Naively, the problem of computing N−point functions of the Liouville exponentials is reduced by means of Eq.4.6 to the calculation of the chiral correlators

$$\langle \tilde{V}^{(J_1)}_{m_1}(\sigma_1) \cdots \tilde{V}^{(J_N)}_{m_N}(\sigma_N) \rangle \tag{5.1}$$

for $m_i \geq -J_i$. (Remember that $\tilde{V}^{(J)}_m$ and $\overline{\tilde{V}}^{(J)}_m$ commute in our approach). The Coulomb gas representation Eqs.3.2-3.4 should then immediately provide us with integral representations of the correlators Eq.5.1 for arbitrary N. However, there are two important subtleties to be taken into account.

The first concerns the question as to what are the left resp. right vacua implied in Eq.5.1. Of course, one expects that they should be given by the $SL(2)$-invariant ground states $|0\rangle$, $\langle 0|$ with

$$L_0|0\rangle = L_{\pm 1}|0\rangle = 0, \quad \langle 0|L_{\pm 1} = \langle 0|L_0 = 0. \tag{5.2}$$

However, it was observed in [17] that the free fields X_1, X_2 cannot simultaneously exist on any ground state $|\varpi\rangle$ with $L_0|\varpi\rangle = 0$. More generally, the same is true for all ground states whose weights are in the Kac's table, i.e. which have null states in their Verma modules. This should be quite alarming since by construction the fields X_1, X_2 are completely equivalent; in particular, they enter on equal footing into the representation Eq.2.15 for the $V^{(J)}_m$. The proper interpretation of the situation is that all the ground states corresponding to the Kac's table actually belong to the socalled parabolic sector of the theory[16], whereas the free field construction described above -as well as any other one known in the literature- applies only to the hyperbolic and elliptic sectors. We will not delve further into this problem and its possible resolutions but simply assume that as in the standard $SL(2)$-invariant case, all information about the N−point function is contained in the "reduced" correlator

$$\langle \varpi_1 | \tilde{V}^{(J_2)}_{m_2}(\sigma_2) \cdots \tilde{V}^{(J_{N-1})}_{m_{N-1}}(\sigma_{N-1}) | \varpi_N \rangle \tag{5.3}$$

(see ref.[7] for an interesting suggestion to justify this further). The expression Eq.5.3 can be computed directly from Eqs.3.2-3.4 by the usual free field techniques, at least when ϖ is real, i.e. in the elliptic sector. Then charge -or momentum- conservation can be used since all $\tilde{V}_m^{(J)}$ introduce real shifts of ϖ, and we get the condition

$$\varpi_N = \varpi_1 + 2m_2 + \cdots + 2m_{N-1} \tag{5.4}$$

for nonzero matrix elements. However, at this point we encounter a second difficulty: Recall that all Liouville exponentials should be invariant under the exchange $X_1 \leftrightarrow X_2$ (cf. Eq.2.7). On the other hand, if we perform this operation, say, in $e^{-J_2\alpha\Phi}$, then all $V_{m_2}^{(J_2)}$ go over into $V_{-m_2}^{(J_2)}$ and the m_2-shifts in Eq.5.4 appear with a minus sign. In the case where J_2 is half-integer positive, this poses no problem since $-J_2 \le m_2 \le J_2$, i.e. positive and negative shifts appear in a symmetric way. However, if $2J_2$ is not an integer and $-J_2 \le m_2 \le \infty$, then the set of shifts $\{m_2\}$ has zero intersection with the set of shifts $\{-m_2\}$. Hence if Eq.5.4 can be fulfilled using the X_1-representation for $e^{-J_2\alpha\Phi}$, then necessarily it can never be fulfilled with the X_2- representation, and we arrive at a contradiction. The only way to avoid this inconsistency is to give up Eq.5.4. In fact, we will now present a very simple argument why exactly in the case where $2J$ is not a positive integer - and the sum representing $e^{-J_2\alpha\Phi}$ is infinite - charge conservation does break down. Let us consider instead of the full-fledged field-theoretical operator $e^{-J\alpha\Phi}$ the following drastically simplified "minisuperspace" type toy version involving only zero modes:

$$
\begin{aligned}
\left(e^{-J\alpha\Phi}\right)_{mini} &= e^{-Ji\alpha(x+\kappa p)}\left(1 + e^{i\alpha(x+\kappa p)}\right)^{2J} \\
&= e^{-Ji\alpha(x+\kappa p)} \sum_{n=0}^{\infty} \binom{2J}{n} e^{i\alpha n(x+\kappa p)}
\end{aligned}
\tag{5.5}
$$

where κ is a real coefficient, and $[x,p] = i$. The structural analogy with Eq.2.5 is obvious, if we factor out $(A'^{-1/2})^{2J}(BB'^{-1/2})^{2J}$. On the quantum level, we use the second equality in Eq.5.5 to give meaning to the operator $(e^{-J\alpha\Phi})_{mini}$, in close analogy with Eq.4.6. To study the elliptic sector, where charge conservation is naively expected, we take α real. However already classically we see that we must add a small imaginary part to α,

$$\alpha \to \alpha - i\epsilon, \qquad \epsilon > 0 \tag{5.6}$$

in order to ensure the convergence of the series expansion. For the sign choice of Eq.5.6, the classical series will converge iff $x + \kappa p < 0$. On the other hand according to Eqs.2.6,2.7 there should be a second equivalent representation of $e^{-J\alpha\Phi}$ obtained by exchanging X_1 and X_2. For our toy version, Eq.2.6 is implemented by replacing $x \to -x, p \to -p$ which is indeed a symmetry of Eq.5.5. We thus obtain for $(e^{-J\alpha\Phi})_{mini}$ in the "X_2-representation",

$$
\begin{aligned}
\left(e^{-J\alpha\Phi}\right)_{mini} &= e^{+Ji\alpha(x+\kappa p)}\left(1 + e^{-i\alpha(x+\kappa p)}\right)^{2J} \\
&= e^{+Ji\alpha(x+\kappa p)} \sum_{n=0}^{\infty} \binom{2J}{n} e^{-i\alpha n(x+\kappa p)}
\end{aligned}
\tag{5.7}
$$

with the classical series converging for $x + \kappa p > 0$. The domains of the series expansions in Eqs.5.5,5.7 are complementary, hence there is always exactly one converging series representation, and both must be used if the full range of x and p values is to be described. In the quantum case, the evaluation of matrix elements of $(e^{-J\alpha\Phi})_{mini}$ involves an integration over the full real axis of x or p. Thus the obvious idea is that we have to split up the integration into two half spaces where the X_1 resp. X_2 representation is applicable. But then a given contribution to the series expansion of, say, $\langle p_2|(e^{-J\alpha\Phi})_{mini}|p_1\rangle$ will no longer be proportional to a charge- conservation delta function enforcing condition Eq.5.4 ! Let us make this a little more precise for the above "reduced three-point function" $\langle p_2|(e^{-J\alpha\Phi})_{mini}|p_1\rangle$. We introduce the new set of variables

$$y := x + \kappa p \quad p_y := \frac{1}{2}(p - x/\kappa) \qquad [y, p_y] = i \tag{5.8}$$

According to the philosophy above, we should define

$$\langle p_2|(e^{-J\alpha\Phi})_{mini}|p_1\rangle = \int_{-\infty}^{0} \langle p_2|y\rangle(e^{-J\alpha\Phi}(y))^{(1)}\langle y|p_1\rangle$$

$$+ \int_{0}^{\infty} \langle p_2|y\rangle(e^{-J\alpha\Phi}(y))^{(2)}\langle y|p_1\rangle \tag{5.9}$$

where the superscripts indicate the use of the X_1-resp. X_2 representation Eqs.5.5, 5.7, with an obvious notation for the kernel $e^{-J\alpha\Phi}(y)$. One has

$$\langle p|y\rangle = \frac{1}{\sqrt{2\pi}}e^{i\kappa(p-y/\kappa)^2/2} \tag{5.10}$$

If $J \leq 0$, then the y-integrations converge term by term, and Eq.5.9 can be explicitly evaluated to give

$$\langle p_2|(e^{-J\alpha\Phi})_{mini}|p_1\rangle = \frac{1}{2\pi|\alpha|}e^{i\kappa(q^2+2p_1q)/2} \times$$

$$\sum_{n=0}^{\infty} \binom{2J}{n} \frac{2(n-J)(\epsilon+i)}{q^2/|\alpha|^2 + (J-n)^2(\epsilon+i)^2}, \qquad q := p_2 - p_1 . \tag{5.11}$$

Obviously, this is nonzero for generic values of p_1 and p_2 -charge conservation is not valid ! Now actually the RHS of Eq.5.11 contains singular pieces,

$$\langle p_2|(e^{-J\alpha\Phi})_{mini}|p_1\rangle|_{\text{sing.}} = \frac{1}{2}e^{i\kappa(q^2+2p_1q)/2} \times$$

$$\sum_{n=0}^{\infty} \binom{2J}{n}(\delta(q - m|\alpha|) + \delta(q + m|\alpha|)), \qquad m := n - J \tag{5.12}$$

Hence there is a charge-conserving part, but it appears in a symmetrized way, invariant under Eq.2.6, so that the contradiction mentioned above is avoided ! The result Eq.5.11 can actually be shown to be valid even for $J > 0$. In particular, we can recover the half-integer positive J case. In fact, in this case we are left only with the singular charge-conserving part, since the regular part vanishes by antisymmetry:

$$\sum_{n=0}^{2J} \binom{2J}{n} \frac{2(n-J)i}{q^2/|\alpha|^2 - (J-n)^2} \equiv \sum_{m=-J}^{J} \binom{2J}{J+m} \frac{2im}{q^2/|\alpha|^2 - m^2} = 0. \tag{5.13}$$

Thus we have arrived at a consistent picture for the three-point function. In view of the conformal structure of the theory, all higher-point functions are then in principle determined as well. Though the situation seeems rather clear conceptually, it is of course still a nontrivial step to work out the above argument for the full theory. However, it should be noted that for the three-point function we are still dealing with a problem involving zero modes only. Indeed, using the representation Eq.3.3 for $V_m^{(J)}$ we see that no matter how the zero mode integration is defined, we can always compute the oscillator part of $\langle \varpi_2 | V_m^{(J)} | \varpi_1 \rangle$, which is equal to 1. The same will then of course be true for $\widetilde{V}_m^{(J)}$. The complication lies in the fact that through the coupling constant $g_{J,x+m}^x$ the full-fledged operator $e^{-J\alpha\Phi}$ acquires a much more involved dependence on ϖ, and it is not so trivial to find a "coherent state" basis as in Eq.5.8. This problem is the subject of ongoing studies. Intriguingly, for the case of negative half-integer J a completely different definition of the three-point function has been proposed in ref.[7], and shown to be able to reproduce the three-point functions of minimal matter coupled to gravity as predicted by the matrix models. The "Liouville exponential" appearing there involves a finite sum only and is not the quantization of the classical expression Eq.2.5 (with upper limit infinity), but rather some kind of analytic continuation $J \to -J - 1$ of the quantum Liouville exponential for positive half-integer J. It is therefore a very interesting open problem whether and in which way the two approaches are related. If they yield different results, then one may speculate that they could describe different phases of two-dimensional gravity, and give rise to different scaling exponents as well as a different position dependence of higher-point functions. The answer to these questions is left as a homework to the dedicated reader.

References

[1] O. Babelon, *Phys. Lett.* **B215** ((1988)) 523.

[2] J.-L. Gervais, *Comm. Math. Phys.* **130** (257) (1990) .

[3] J.-L. Gervais, *Phys. Lett.* **B243** (85) (1990) .

[4] E. Cremmer, J.-L. Gervais, *Comm. Math. Phys.* **144** (279) (1992) .

[5] J.-L. Gervais, *Int. J. Mod. Phys.* **6** (2805) (1991).

[6] J.-L. Gervais, *Comm. Math. Phys.* **138** (301) (1991) .

[7] J.-L. Gervais, "Quantum group derivation of 2D gravity-matter coupling" Invited talk at the Stony Brook meeting *String and Symmetry 1991*, LPTENS preprint 91/22, *Nucl. Phys.* **B391** (287) (1993).

[8] E. Cremmer, J.-L. Gervais, J.-F. Roussel, "The quantum group structure of 2D gravity and minimal models II: The genus-zero chiral bootstrap" LPTENS preprint 93/02, hep-th9302035.

[9] E.Braaten, T.Curtright, C.Thorn, *Phys. Lett.* **B118** (1982) 115; *Phys. Rev. Lett.* **48** (1982) 1309; *Ann. Phys. (N.Y.)* **147** (1983) 365; E.Braaten, T.Curtright, G.Ghandour, C.Thorn, *Phys. Rev. Lett.* **51** (1983) 19; *Ann. Phys. (N.Y.)* **153** (1984) 147.

[10] H.J. Otto, G. Weigt, *Phys. Lett.* **B159** (1985) 341; *Z. Phys.* **C31**, (1986)219 ; G. Weigt, "Critical exponents of conformal fields coupled to two-dimensional quantum gravity in the conformal gauge", talk given at 1989 Karpacz Winter School of Theor. Physics, preprint PHE-90-15; G. Weigt, "Canonical quantization of the Liouville theory, quantum group structures, and correlation functions", talk given at 1992 Johns Hopkins Workshop on Current Problems in Particle Theory, Goteborg, Sweden, hep-th 9208075, print-92-0383 (DESY-IFH).

[11] J.-L. Gervais, A. Neveu, *Nucl. Phys.* **B238** (1984) 125; *Nucl. Phys.* **B238** (1984) 396.

[12] J.-L. Gervais, B. Rostand, *Nucl. Phys.* **B346** (473) (1990) .

[13] J.-L. Gervais, B. Rostand, *Comm. Math. Phys.* **143** (175) (1991) .

[14] F. Smirnoff, L. Takhtajan, "Towards a Quantum Liouville Theory with $c > 1$", Kyoto University, Yukawa Institute Library April 1990.

[15] J.-L. Gervais, A. Neveu, *Nucl. Phys.* **B224** (329) (1983).

[16] L. Johannson, A. Kihlberg, R. Marnelius, *Phys. Rev. D* **29** (1984) 2798; L. Johannson, R. Marnelius, *Nucl. Phys.* **B254** (1985) 201.

[17] D. Lüst, J. Schnittger, *Int. J. Mod. Phys.* **A6** (1991) 3625; J. Schnittger, Ph.D. thesis, Munich 1990.

[18] J.-L. Gervais, J. Schnittger, *Phys. Lett.* **B315** (1993) 258.

[19] G. Felder, J. Fröhlich, J. Keller, *Comm. Math. Phys.* **124** (1989) 646.

[20] A. Kirilov, N. Reshetikhin, *Infinite Dimensional Lie Algebras and Groups, Advanced Study in Mathematical Physics* **vol. 7**, Proceedings of the 1988 Marseille Conference, V. Kac editor, p 285, World scientific.

[21] Bo-yu Hou, Bo-yuan Hou, Zhong-qi Ma, *Comm. Theor. Phys.*, **13**, 181, (1990); *Comm. Theor. Phys.*, **13**, 1990, 341.

[22] J.F. Roussel, private communication.

[23] G. Andrews "*q-series: their development and application in analysis, number theory, combinatorics, physics, and computer algebra*", Conference board of the Mathematical Sciences, Regional Conference in Mathematics, # 66, A.M.S. ed.

[24] L.C. Slater, *"Generalized hypergeometric functions"* Cambridge University presss 1966.

[25] R. Askey, J. Wilson, *"A set of orthogonal polynomials that generalize the Racah coefficients or 6-j symbols"* SIAM J. Math. Anal. 10 (1979) 1008.

[26] Vl. Dotsenko, V. Fateev, *Nucl. Phys.* **B251** (691) (1985) .

[27] J.-L. Gervais, J. Schnittger, "Canonical Construction of Liouville Field Operators with Arbitrary Spin" LPTENS preprint 93/

[28] J.-L. Gervais, J. Schnittger, "The Many Faces of the Quantum Liouville Exponentials", LPTENS preprint 93/30, hep-th 9308134, Nucl. Phys. B to appear.

BETHE ANSATZ FOR THE BLOCH PARTICLE
IN MAGNETIC FIELD

P. B. WIEGMANN

James Franck Institute
and
Enrico Fermi Institute
University of Chicago
5640 S.Ellis Ave.,Chicago Il 60637

A.V. ZABRODIN

Institute of Theoretical Physics
Uppsala University
Box 803 S-75108, Uppsala, Sweden
and
Institute of Chemical Physics
Kosygina St. 4, SU-117334, Moscow, Russia

Abstract: We present a new approach to the problem of Bloch electrons in magnetic field, by making explicit a natural relation between the group of magnetic translations and the quantum group $U_q(sl_2)$. The Hamiltonian is represented as trace of a monodromy matrix of an integrable quantum model. This opens a way for an application of the functional Bethe ansatz technique. The approach allows one to express the "mid" band spectrum of the model and the Bloch wave function by solutions of the Bethe Ansatz equations typical for completely integrable quantum systems. The zero mode wave functions are found explicitly in terms of q-deformed classical orthogonal polynomials . Others quasiperiodic equations related to the Quantum group are discussed.

1   Introduction

The peculiar problem of Bloch electrons in magnetic field[1],[2],[3],[4],[5] often emerges in various branches physics.every time presenting a new face to describe another physical application

   i) it resembles some properties of the integer Hall effect [5],

   ii) its spectrum has an extremely rich structure of Cantor set [4],[6], [7]

   iii) it is a characteristic example of *multifractal* behaviour (see [8] for a review)

Quantum Field Theory and String Theory, Edited by
L. Baulieu *et al.*, Plenum Press, New York, 1995

iv) it is one of the most popular *quasiperiodic* model to study the localization phenomenon in incommensurate potential (see e.g.[8] and references therein), it describes also one dimensional quasicrystal, etc. This list may be continued. Recently it has been conjectured that the symmetry of magnetic group may appear dynamically in strongly correlated electronic systems[9],[10].

In this paper we made explicit a long time anticipated connection of this problem (sometimes called Hofstadter problem), with the *quantum group* $U_q(sl_2)$ and therefore with Quantum Integrable Systems. It allows us to apply the *Bethe- Ansatz* technique to find algebraic equations for the s new spectrum. Although we do not solve the Bethe Ansatz equation here, we hope that they provide a basis for analytical study the fractal properties of the spectrum.

The Hamiltonian of a particle on a two dimensional square lattice in magnetic field is

$$H = \sum_{<n,m>} t_{n,m} e^{iA_{\vec{n},\vec{m}}} c_{\vec{n}}^{\dagger} c_{\vec{m}}, \tag{1.1}$$

$$\prod_{plaquette} e^{iA_{\vec{n},\vec{m}}} = e^{i\Phi} \tag{1.2}$$

where $\Phi = 2\pi \frac{P}{Q}$ is a flux per plaquette and P and Q are mutual prime integers and $t_{n,m}$ is a hoping amplitude between the nearest neighbors.

The wave function of a particle in magnetic field forms a representation of the *group of magnetic translations* [1]: let generators of translations be

$$T_{\vec{\mu}}(\vec{i}) = e^{iA_{\vec{i},\vec{i}+\vec{\mu}}} \mid \vec{i} >< \vec{i} + \vec{\mu} \mid \tag{1.3}$$

They form the algebra

$$T_{\vec{\mu}} = T_{-\vec{\mu}}^{-1}, \ T_{\vec{n}} T_{\vec{m}} = q^{-\vec{n} \times \vec{m}} T_{\vec{n}+\vec{m}},$$
$$T_y T_x = q^2 T_x T_y, \ T_y T_{-x} = q^{-2} T_{-x} T_y \tag{1.4}$$

The Hamiltonian (1.1) therefore is

$$H = t_x(T_x + T_{-x}) + t_y(T_y + T_{-y}) \tag{1.5}$$

This group is also equivalent to the Heisenberg-Weyl group: $[\hat{p}, \hat{q}] = i\Phi$, $T_x = \exp\hat{q}$, $T_y = \exp\hat{p}$.

In the most popular Landau gauge $A_x = A_{\vec{n},\vec{n}+\vec{I}_x} = 0, A_y = \Phi n_x$ the Bloch wave function is

$$\psi(\vec{n}) = e^{i\vec{k}\vec{n}} \psi_{n_x}(\vec{k}), \quad \psi_n = \psi_{n+Q} \tag{1.6}$$

where $n_x \equiv n = 1, ..., Q$ is a coordinate in the magnetic cell. With these substitution the Schroedinger equation turns into a famous one-dimensional quasiperiodic difference equation ("Harper's" equation):

$$t_x(e^{ik_x}\psi_{n+1} + e^{-ik_x}\psi_{n-1}) + 2t_y \cos(k_y + n\Phi)\psi_n = E\psi_n \tag{1.7}$$

Below we shall consider the isotropic case only, setting $t_x = t_y = 1$ The spectrum of this equation has Q bands and feels the difference between rational and irrational numbers

- if flux is irrational, the spectrum is singular continuum - uncountable but measure zero set of points (Cantor set). If flux is rational, then the spectrum has Q bands.

To ease the reference we state the main result now:

It is known that due to the gauge invariance, the energy depends on a single parameter $\Lambda = \cos(Qk_x) + \cos(Qk_y)$. We find that the spectrum at $\Lambda = 0$ ("mid" band spectrum) is given by the sum of roots z_l

$$E = iq^Q(q - q^{-1}) \sum_{l=1}^{Q-1} z_l, \tag{1.8}$$

of the *Bethe Ansatz* equations for the quantum group $U_q(sl_2)$ with

$$\frac{z_l^2 + q}{qz_l^2 + 1} = -q^Q \prod_{m=1, m\neq l}^{Q-1} \frac{qz_l - z_m}{z_l - qz_m}, \quad l = 1, ..., Q-1. \tag{1.9}$$

$$q = e^{\frac{i}{2}\Phi}, \tag{1.10}$$

Another version of the Bethe Ansatz equations is presented in Sect.6.

The paper is organized as follows. In Sect.2 we present the Schroedinger equation in two other gauges in which the quantum group structure of the model is more transparent. In Sect.3 we review some representations of the quantum group $U_q(sl_2)$. Relation between the quantum group and magnetic translations (1.4) is revealed in Sect.4 The Hamiltonian (1.1) can be rewritten entirely via the quantum group generators . We give two equivalent (dual) representation of the Hamiltonian as a linear or quadratic form in the quantum group generators. Relation to integrable models with nonperiodic boundary conditions is presented in Appendix B In Sect.5 we discuss a general quadratic forms in $U_q(sl_2)$ generators which gives an "integrable " class of the second order"quasiperiodic "equations (Hill's equations) related with the quantum group . We present their solution in terms of Bethe Ansatz equations for zeros of the eigenfunctions in Sec.6. Then the spectrum is also expressed through the zeros. The Bethe algebraic equations are found by the method similar to the so called functional Bethe Ansatz. In Sect. 7 we apply this general approach to the Hofstadter Hamiltonian to obtain the "mid" band spectrum of the model. An promising connection between the zero mode wave functions and q-deformations of certain classical orthogonal polynomials is discussed in Sect.8

2 Harper's Equation and "Regular" Gauges

In this Section we write Harper's equation in two specific gauges for which the quantum group structure is more transparent. In these gauges (we call them "regular") the wave function has the form

$$\Psi_n = \prod_{m=1}^{N} (q^n - z_m) \tag{2.1}$$

where n- independent z_m. Eventually they would be roots of the Eq.(8). The Landau gauge of the Introduction is not "regular" and is not convenient for revealing the quantum group structure of the model.

1.Modified Landau gauge

Consider the gauge

$$A_x = -A_y = -\Phi n_x \tag{2.2}$$

Let us call it modified Landau gauge. The Bloch wave function is

$$\phi(\vec{n}) = e^{i\vec{k}'\vec{n}}\phi_{n_x}(\vec{k}'), \quad \phi_n = \phi_{n+Q} \tag{2.3}$$

where $\vec{k}'$ is the wave vector (different from that in (1.6)). The Schroedinger equation turns to

$$e^{ik'_x - i\Phi n}\phi_{n+1} + e^{-ik'_x + i\Phi(n-1)}\phi_{n-1} + 2\cos(k'_y + n\Phi)\phi_n = E\phi_n \tag{2.4}$$

(we drop the argument in ϕ_n).

The "mid" band spectrum ($\Lambda = 0$) corresponds to the values

$$\vec{k} = (\frac{1}{2}\Phi(Q-1), \pi) \tag{2.5}$$

in the Landau gauge and

$$\vec{k}' = (0, \pi) \tag{2.6}$$

in the new gauge.

At the "mid" band all the values of $\vec{k}'$ on the line $\exp i(k'_x - k'_y) = -1$ are physically equivalent i.e. connected by a gauge transformation. We've chosen $\exp i(k'_x + k'_y) = -1$ in (2.6).

2.Chiral gauge

Consider the chiral gauge:

$$A_x = -\frac{\Phi}{2}(n_x + n_y), \quad A_y = \frac{\Phi}{2}(n_x + n_y + 1) \tag{2.7}$$

Take the Bloch wave function in the form

$$\chi(\vec{n}) = e^{i\vec{p}\vec{n}}\chi_n(\vec{p}) \tag{2.8}$$

where $\vec{p} = (p_x, p_y), n = n_+; n_\pm = n_x \pm n_y, p_\pm = (p_x \pm p_y)/2$ are light cone coordinates and momenta.

This wave function is defined in *two* magnetic cells because n runs now from 1 to $2Q$. Accordingly, the light cone momenta are confined to the half of the Brillouin zone $[0, \Phi/2]$. The equivalent form of the Harper's equation (1.7) is

$$2e^{\frac{i}{4}\Phi + ip_+}\cos(\frac{1}{2}\Phi n + \frac{1}{4}\Phi - p_-)\chi_{n+1} +$$
$$+2e^{-\frac{i}{4}\Phi - ip_+}\cos(\frac{1}{2}\Phi n - \frac{1}{4}\Phi - p_-)\chi_{n-1} = E\chi_n \tag{2.9}$$

The wave function χ_n is $2Q$-periodic. This doubling of period in comparison with (2.4) is, of course, artificial. Although the coefficients in (2.9) have period $2Q$ there exists a simple transformation of χ_n which makes them Q-periodic:

$$\chi_n = \exp(\frac{i\Phi}{4}n(n-2))\xi_n \tag{2.10}$$

This new wave function ξ_n can be taken Q-periodic. It satisfies the equation

$$(e^{-\frac{i}{4}\Phi+ip_x} + e^{i\Phi n+\frac{i}{4}\Phi+ip_y})\xi_{n+1} +$$
$$+(e^{\frac{i}{4}\Phi-ip_x} + e^{-i\Phi n+\frac{3i}{4}\Phi-ip_y})\xi_{n-1} = E\xi_n \qquad (2.11)$$

The "mid" band spectrum corresponds to $\vec{p} = (\frac{1}{2}\pi, \frac{1}{2}\pi+\frac{1}{4}\Phi)$ in the chiral gauge (compare with (2.6)). Again all the values $\vec{p} = (\frac{1}{2}\pi, \delta)$ for any real δ are physically equivalent. We choose $\vec{p}$ at the "mid" band to be

$$\vec{p} = (\frac{\pi}{2}, \frac{\pi}{2}) \qquad (2.12)$$

Relations between three gauge are given in the Appendix A

3   Quantum Group

The algebra $U_q(sl_2)$ (a q-deformation of the universal enveloping of the sl_2) is generated by the elements A, B, C, D, with the commutation relations [11],[16],[12],[13],[14]

$$AB = qBA,\ BD = qDB,$$
$$DC = qCD,\ CA = qAC,$$
$$AD = 1,\ [B, C] = \frac{A^2 - D^2}{q - q^{-1}} \qquad (3.1)$$

We shall consider the deformation parameter q is a root of ±1 of degree Q:
$q = exp(i\pi\frac{P}{Q})$ where P and Q are mutually prime integers. The central element of this algebra (for arbitrary q) is a $q-$analog of the Casimir operator

$$c = \left(\frac{q^{-\frac{1}{2}}A - q^{\frac{1}{2}}D}{q - q^{-1}}\right)^2 + BC \qquad (3.2)$$

When q is a root of unity some additional central elements appear.

In the classical limit $q \rightarrow 1+\frac{i}{2}\Phi$, the quantum group turns to the sl_2 algebra: $(A - D)/(q - q^{-1}) \rightarrow S_3,\ B \rightarrow S_+,\ C \rightarrow S_-,\ c \rightarrow \vec{S}^2 + 1/4$.

The commutation relations (3.1) are simply another way to write the intertwining relation for the L - operator:

$$R(u/v)(L(u) \otimes 1)(1 \otimes L(v)) = (1 \otimes L(v))(L(u) \otimes 1)R(u/v) \qquad (3.3)$$

with the trigonometric $R-$matrix

$$R(u) = \frac{1}{2}(q+1)(u-q^{-1}u^{-1})+\frac{1}{2}(q-1)(u+q^{-1}u^{-1})\sigma_3\otimes\sigma_3+(q-q^{-1})(\sigma_+\otimes\sigma_- +\sigma_-\otimes\sigma_+) \qquad (3.4)$$

satisfying the Yang-Baxter relation (σ_j are Pauli matrices; $\sigma_\pm = (\sigma_1 \pm i\sigma_2)/2$. Generators $A,\ B,\ C,\ D$ are matrix elements of the L- operator

$$L(u) = \begin{bmatrix} \frac{ukA-u^{-1}k^{-1}D}{q-q^{-1}} & u^{-1}k^{-1}C \\ ukB & \frac{ukD-u^{-1}k^{-1}A}{q-q^{-1}} \end{bmatrix} \qquad (3.5)$$

Here u is a spectral parameter and k is an additional parameter (rapidity at the site). Note that the R- matrix is the L- operator in the spin 1/2 - representation. It is given by the same matrix (3.5) for $k = q^{1/2}$ with elements: $A = q^{\frac{1}{2}\sigma_3}$, $D = q^{-\frac{1}{2}\sigma_3}$, $B = \sigma_+$, $C = \sigma_-$.

Irreducible finite dimensional representations of the dimension $2j+1$ can be expressed in the weight basis where A and D are diagonal matrices: $A = diag\,(q^j, ..., q^{-j})$. An integer or half integer j is the spin of

representation. The value of the Casimir operator (3.2) in this representation is

$$c = \left(\frac{q^{j+1/2} - q^{-j-1/2}}{q - q^{-1}}\right)^2 = [j + 1/2]_q^2 \tag{3.6}$$

is the q- analog of $(j+1/2)^2$. The representation can be realized by difference operators acting in the space of polynomials $\Psi(z)$ of degree $2j$.

$$A\Psi(z) = q^{-j}\Psi(qz), \; D\Psi(z) = q^j\Psi(q^{-1}z),$$
$$B\Psi(z) = z(q - q^{-1})^{-1}\left(q^{2j}\Psi(q^{-1}z) - q^{-2j}\Psi(qz)\right)$$
$$C\Psi(z) = -z^{-1}(q - q^{-1})^{-1}\left(\Psi(q^{-1}z) - \Psi(qz)\right) \tag{3.7}$$

Then $\Psi_0(z) = 1$ is the lowest weight vector whether $\Psi_{2j}(z) = z^{2j}$ is the highest weight vector,i.e. $C\Psi_0(z) = 0$, $B\Psi_{2j}(z) = 0$

This is the q-analog of the representation of the sl_2 algebra by differential operators:

$$S_3 = z\frac{d}{dz} - j, \; S_+ = z(2j - z\frac{d}{dz}), \; S_- = \frac{d}{dz} \tag{3.8}$$

In addition, in the special dimension Q there is three parametrical family of representations having no lowest nor highest weight (some times called cyclic representations). They can be written in the Weyl basis [24], [16], [15]. Let X and Y

$$qXY = YX \tag{3.9}$$

Then the representation of $U_q(sl_2)$ is

$$A = X, \; D = X^{-1},$$
$$B = (bX + \bar{b}X^{-1})Y^{-1},$$
$$C = Y(cX + \bar{c}X^{-1}) \tag{3.10}$$

and characterized by parameters $b, c, \bar{b}, \bar{c}$ obeying the conditions $qbc = q^{-1}\bar{b}\bar{c} = -(q - q^{-1})^{-2}$. The value of the Casimir operator (3.2) depends on parameters and is $c\bar{b} + \bar{c}b - 2(q - q^{-1})^{-2}$. Compare it with 3.6 we find that at $q^2 b/\bar{b} = \bar{c}/c = \mp q$ (for P-odd, even) cyclic representations become regular, i.e. acquire the lowest and the highest weight . This is a very special dimension when $q^{2j+1} = \mp 1$ for P - odd (even) and the Casimir operator (3.6) is

$$c = -4(q - q^{-1})^{-2}, \; for \; P - odd$$
$$c = 0, \; for \; P - even \tag{3.11}$$

4 Magnetic Translations as a Special Representation of the Quantum Group

For a given value of the Bloch wave vector $\vec{k}$ the dimension of physical Hilbert space of our problem is Q .Representation of the dimension $2j + 1 = Q$ of the quantum group $U_q(sl_2)$ for $q = exp(i\pi\frac{P}{Q})$ naturally acts in the space of Bloch states in the magnetic field. Therefore magnetic translations and the Hamiltonian (1,5) can be expressed through generators of the quantum group ..Let us set

$$T_y = e^{ik_y} Y X^{-1},$$
$$T_x = e^{ik_x} Y X, \qquad (4.1)$$

Then, the Hamiltonian (1.5) could be expressed as a linear form in quantum group generators only at the middle of the band (an integrable point)

$$e^{i(p_x - p_y)} = -q^{-1} \qquad (4.2)$$

We may choose

$$b = e^{\frac{i}{2}(p_y - p_x)}(q - q^{-1})^{-1},$$
$$\bar{b} = \frac{t_x}{t_y}q^2(q - q^{-1})^{-1},$$
$$\bar{c} = \frac{t_y}{t_x}(1 - q^2)^{-1},$$
$$c = e^{\frac{i}{2}(p_y - p_x)}(1 - q^2)^{-1}, \qquad (4.3)$$

Then B and C given by (3.10) form a representation of the quantum group with the value of the Casimir operator

$$c = (-(\mp\frac{t_x}{t_y} + \frac{t_y}{t_x}) - 2)(q - q^{-1})^{-2} \qquad (4.4)$$

Using (3.10) and (4.1) we may identify the quantum group generators and magnetic translations

$$t_x T_{-x} + t_y T_{-y} = t_y(q - q^{-1})e^{-\frac{i}{2}(p_y + p_x)}B,$$
$$t_x T_x + t_y T_y = \mp t_x(q - q^{-1})e^{\frac{i}{2}(p_y + p_x)}C,$$
$$T_{-y}T_x = \pm q^{-1}A^2,$$
$$T_{-x}T_y = \pm qD^2 \qquad (4.5)$$

The Hamiltonian now acquires the form

$$H = (q - q^{-1})(\mp t_x e^{\frac{i}{2}(p_y + p_x)}C + t_y e^{-\frac{i}{2}(p_y + p_x)}B) \qquad (4.6)$$

In fact at the integrable point (4.2) the physical states and their energies (midband states) do not depend on the value of $exp\frac{i}{2}(p_y + p_x)$. It simply can be gauged away. Compare (4.4) and (3.11) we conclude that at isotropic case $t_x = t_y$ the representation of the quantum group (4.3) becomes regular, i.e. acquires the highest and the lowest weight. Below we shall concentrate mostly on this case. As a result for the midband states in the isotropic case the relation between magnetic translations and the quantum group is as follows

$$H = i(q - q^{-1})(C \pm B) \tag{4.7}$$

The Hamiltonian (4.7) is a trace of a modified L- operator (3.5) $\tilde{L}(u)$. Say at P odd

$$H = Tr\tilde{L}(u) = TrL(u)\sigma_1 \tag{4.8}$$

Modified L-operator also obeys the intertwining relation (3.3).[a]

Let us point out another useful realization of the quantum group in terms of magnetic translations. Say, for an odd P we have:

$$
\begin{aligned}
&T_{-x} + T_{-y} = -i(q - q^{-1})q^{-\frac{1}{2}}BD, \\
&T_x + T_y = -i(q - q^{-1})q^{-\frac{1}{2}}CA, \\
&T_{-y}T_x = q^{-1}A^2,\ T_{-x}T_y = qD^2
\end{aligned}
\tag{4.9}
$$

also forms a representation of $U_q(sl_2)$ with the same value of the Casimir operator (3.6) and, therefore, with the same dimension.Q. Now the Hamiltonian turns into quadratic form in the $U_q(sl_2)$ generators:

$$H = -i(q - q^{-1})q^{-\frac{1}{2}}(CA + BD) \tag{4.10}$$

Two different forms of the Hamiltonian are in fact gauge equivalent: they correspond to the two choices of gauge discussed in Sect.2. In the Appendix B we show that the Hamiltonian (4.10) represented as a quadratic form of the quantum group generators also can be understood as a trace of a quantum monodromy matrix , though of a more complicated nature - the monodromy matrix for a system with non-periodic boundary conditions.

Since the main ingredients of the quantum inverse scattering method have appeared in the problem there is a strong evidence that the Hofstadter Hamiltonian can be actually diagonalized for arbitrary Q. i.e. its energy spectrum may be expressed by solutions of the Bethe equations. The ambitious goal of solving the problem for arbitrary parameters p_x, p_y and t_x/t_y is beyond of the scope of this paper. Instead we shall present a solution for the fixed values of the parameters when representation of the quantum group or magnetic translations has the lowest and the highest weight i.e for the middle of bands and isotropic hoping. In this case the so -called functional Bethe Ansatz [19] is applicable.

---

[a]L-operator of such kind has been considered recently for the sine-Gordon model [18]. We are grateful to L.Faddeev and A.Volkov for pointing our attention to the Ref.[18].

5 A Class of Difference and Discrete Equations Related to the Quantum group

Quadratic and linear form in generators of the quantum group provides a class of difference and discrete operators which allow complete or partial algebraization. A problem of the Bloch particle in magnetic field is a particular case (the most physically valuable, perhaps) of the general class.

The point is that the space of representation with the lowest and the highest weight can be realized by the space of polynomials of order $2j$. Any form of generators A, B, C, D preserves the space of polynomials. Therefore eigenfunctions of the form , which by means of the representation (3.7) is the difference operator, are polynomials as well. Vice versa is probably also true - one may look on the quantum group as an algebra of difference operators which preserve the ring of polynomials.

Quadratic forms of the "classical" Sl_2 as a class of differential equations of the second order, solvable in polynomials (so called "quasi exactly solvable" models of quantum mechanics or algebraization of the quantum mechanics) have been extensively studied [20] [21] (see also [27] for a review) in the last few years..They includes second differential equations for all classical polynomials and a set of Schroedinger operators with solvable potentials. Attempt of q-generalization for difference functional equations has been made in Ref.[28] [b]. A general quadratic form in quantum group generators in the regular representation provides the class of difference equations which allows similar algebraization

$$G = aA^2 + dD^2 + (q - q^{-1})(c_2 CA + b_2 BD + b_3 BA + c_3 CD) \tag{5.1}$$
$$+ (q - q^{-1})^2(b_1 B^2 + c_1 C^2) \tag{5.2}$$

Latter we shall set parameters a, d, c_i, b_i ($i = 1, 2, 3$) and dimension of the representation $2j + 1$ to reach our Hamiltonian (4.10).

Under the representation (3.7) the quadratic form is a difference operator. Let us consider its spectral problem.

$$G\Psi(z) = a(z)\Psi(q^2 z) + d(z)\Psi(q^{-2}z) - v(z)\Psi(z) = E\Psi(z) \tag{5.3}$$

where
$$a(z) = b_1 q^{-4j+1}z^2 - b_3 q^{-3j}z + aq^{-2j} + c_2 q^{-j}z^{-1} + c_1 q^{-1}z^{-2} \tag{5.4}$$

$$d(z) = b_1 q^{4j-1}z^2 + b_2 q^{3j}z + dq^{2j} - c_3 q^j z^{-1} + c_1 q z^{-2} \tag{5.5}$$

$$v(z) = (q + q^{-1})(b_1 z^2 + c_1 z^{-2}) + (c_2 q^{-j} - c_3 q^j)z^{-1} + (b_2 q^{-j} - b_3 q^j)z \tag{5.6}$$

If all $b_{1,2,3} = 0$ the difference equation (5.3) is known as q-hypergeometric equation.(see e.g. [29]). Their solutions are certain q-Jacobi polynomials. . At the "classical" limit $q \to 1 + \frac{i}{2}\Phi$ they reproduce classical Jacobi polynomials. A general difference equation (5.3) in the classical limit leads to Mathieu and Lame equations. One may call the equation (5.3) q- analog of the Mathieu (Lame) equations.

---

[b]We are indebted to A.Turbiner who informed us about this work

From the theory of representation of the quantum group ,we know that this equation has $(2j+1)$ polynomial solutions of degree no greater than $2j$.

$$\Psi(z) = \prod_{m=1}^{N} (z - z_m) \tag{5.7}$$

Roots of these polynomials will be found in the next section.

The class of "integrable" equations can be extended if we apply a "gauge" transformation

$$\Psi(z) \to f(z)\Psi(z),\ a(z) \to a(z)f(z)/f(q^2 z),\ d(z) \to d(z)f(z)/f(q^{-2}z),\ v(z) \to v(z)$$

A class of "integrable" quasiperiodic discrete equations can be obtained from (5.3) by setting $z = q^{2n}$, $a(q^{2n}) = a_n$, $d(q^{2n}) = d_n$, $v(q^{2n}) = v_n$, $\Psi(q^{2n}) = \Psi_n$.

$$a_n \Psi_{n+1} + d_n \Psi_{n+1} - v_n \Psi_n = E\Psi_n,\ n = 1, ..., Q \tag{5.8}$$

Solutions of the discrete equations are given by a discrete polynomials

$$\Psi(z) = \prod_{m=1}^{N} (z - z_m) \tag{5.9}$$

with the same roots.

6 Functional Bethe Ansatz

The most suitable method to solve the spectral problem (5.3) or (5.8) is the *functional Bethe Ansatz*[c].

Let us plug (5.7) in (5.3) and divide both sides by $\Psi(z)$. We get

$$a(z) \prod_{m=1}^{N} \frac{q^2 z - z_m}{z - z_m} + d(z) \prod_{m=1}^{N} \frac{q^{-2} z - z_m}{z - z_m} - v(z) = E \tag{6.1}$$

If at least one of the coefficients c_1, c_2, c_3 is nonzero, there are two different cases (i) at least one of b_1, b_2, b_3 is nonzero,

(ii)All $b's$ are zero - quadratic form (5.1) includes only A, D and C (generators of the Borel subalgebra of $U_q(sl_2)$). That type of difference equation (5.3) is known as *q-hypergeometric equation*..Their solutions are certain q-Jacobi polynomials - a class of polynomials orthogonal with a discrete measure (see e.g. [29]).

Let us consider the case (i) first. The l.h.s. of (6.1) is a meromorphic function, whereas the r.h.s. is a constant. To make them equal we must cancel all the singularities of the l.h.s. They appear at singular points of $a(z)$, $d(z)$ and $v(z)$ (double and simple poles at $z = 0$ and $z = \infty$) and at $z = z_m$.

The singular part at $z = 0$ vanishes automatically.

---

[c]We note that this method is the most closest to approach of algebraization of quantum mechanical problem by Ushveridze[21]

408

Vanishing of the singular part at $z = \infty$

$$b_1(q^{2N-4j+1} + q^{-2N+4j-1} - q - q^{-1})z^2 + b_2(q^{-2N+3j} - q^{-j})z + b_3(q^j - q^{2N-3j})z +$$
$$+z(q - q^{-1})b_1(q^{2N-4j} - q^{-2N+4j}) \sum_{m=1}^{N} z_m \tag{6.2}$$

determines the degree of the polynomial: $N = 2j$.

Comparing the constant terms in the both sides of (6.1) we find the energy spectrum:

$$E = b_1(q - q^{-1})(q^2 - q^{-2}) \sum_{n<m}^{N} z_n z_m - (q - q^{-1})(b_2 q^{-j+1} + b_3 q^{j-1}) \sum_{m=1}^{N} z_m + aq^{2j} + dq^{-2j}$$
$$\tag{6.3}$$

Finally, annihilation of poles at $z = z_m$ gives the following Bethe Ansatz equations

$$\frac{d(z_l)}{a(z_l)} = q^{4j} \prod_{m=1, m \neq l}^{2j} \frac{q^2 z_l - z_m}{z_l - q^2 z_m}, \quad l = 1, ..., 2j. \tag{6.4}$$

The Bethe equations is a system of $2j$ algebraic equations. It has exactly $2j+1$ solutions. In the case (i) all of them are polynomials of one and the same degree $2j$.

Similar arguments are applicable in the case (ii) . The difference is that the l.h.s. of (6.1) is now regular at $z = \infty$ from the very beginning and the condition of vanishing of (6.2) does not bring a restriction on degree of the polynomials. The Bethe equations valid for any $N < 2j + 1$

$$\frac{d(z_l)}{a(z_l)} = q^{2N} \prod_{m=1, m \neq l}^{N} \frac{q^2 z_l - z_m}{z_l - q^2 z_m}, \quad l = 1, ..., N. \tag{6.5}$$

Apparently, for each degree there is exactly one such polynomial. This means that for each $N < 2j + 1$ the Bethe equations (6.5) must have exactly one solution. This fact is far from obvious when we look at (6.5). In conclusion, we derived equations (6.5) for zeros of q-orthogonal polynomials. It seems to be an interesting to review the theory of orthogonal polynomials from this point of view.

The main property of the q-hypergeometric equation (the case (ii)) that makes it "solvable" is triangularity of the matrix connecting the original basis z^m ($m = 0, 1, ..., 2j$) with the basis formed by eigenfunctions of the operator G. In particular, this leads to a very simple structure of the spectrum of G:

$$E_N = aq^{2N-2j} + dq^{2j-2N}, \quad N = 0, ..., 2j. \tag{6.6}$$

7 Diagonalizing the Hofstadter Hamiltonian

As we have shown in the section 4.,the Hamiltonian of the Bloch particle in a magnetic field can be expressed as a polynomial form in the quantum group operators only in the midband point of the spectrum., namely for $\Lambda = 0$. Moreover, only in the isotropic case $t_x = t_y$ the representation of the quantum group is regular. In this case we may use the functional Bethe Ansatz described in the previous section to algebrize the spectral problem. Let us first consider the representation in which the Hamiltonian is a linear form(4.7). Let us plug the the functional realization (3.7) in the Hamiltonian (4.7). We get a difference equation for a meromorphic function $\Psi(z)$

$$i(z^{-1} + qz)\Psi(qz) - i(z^{-1} + q^{-1}z)\Psi(q^{-1}z) = E\Psi(z) \tag{7.1}$$

This equation can be easily obtained directly from the original Harper equation in the LC gauge (2.9) without a reference to the quantum group. Indeed, Eq.(7.1) is an extension of the (2.9) at $\vec{p} = (\frac{\pi}{2}, \frac{\pi}{2})$ the whole complex plane, provided

$$\chi_n = \Psi(q^n) \tag{7.2}$$

The point is the from the theory of representation of the quantum group we know that the function $\Psi(z)$ is a polynomial of degree $Q - 1$:

$$\Psi(z) = \prod_{m=1}^{Q-1} (z - z_m) \tag{7.3}$$

Note that this trick is impossible in an arbitrary gauge, in particular in the Landau gauge. The similar arguments work in the case of the ML gauge. The functional realization of the quantum group Hamiltonian (4.10) gives the difference equation

$$z^{-1}\Psi(q^2 z) + q^{-2}z\Psi(q^{-2}z) - (z + z^{-1})\Psi(z) = E\Psi(z) \tag{7.4}$$

with a polynomial solution. It is an extension of the Harper equation in the ML gauge (2.4) for $\vec{k}' = (0, \pi)$ can be extended to the whole complex plane with

$$\phi_n = \Psi(q^{2n}) \tag{7.5}$$

Now we can easily apply the method and the results of the previous section to write the Bethe equations for roots of the polynomials (7.3). As a result we obtain the Bethe equations (1.9) and the energy spectrum (1.8) for the advertised in the Introduction . Here we write them in a more conventional form. Let $z_l = \exp(2\varphi_l)$

$$\frac{\cosh(2\varphi_l - i\frac{\Phi}{4})}{\cosh(2\varphi_l + i\frac{\Phi}{4})} = \pm \prod_{m=1, m\neq l}^{Q-1} \frac{\sinh(\varphi_l - \varphi_m + i\frac{\Phi}{4})}{\sinh(\varphi_l - \varphi_m - i\frac{\Phi}{4})} \tag{7.6}$$

The Bethe equations for the realization of the Hamiltonian as a quadratic form (4.10), (i.e. algebraization of the eq.(7.4)) can be borrowed directly from the previous section. They are

$$z_l^2 = q^Q \prod_{m=1, m\neq l}^{Q-1} \frac{q^2 z_l - z_m}{z_l - q^2 z_m}, \quad l = 1, ..., Q - 1 \tag{7.7}$$

The energy spectrum is again proportional to the sum of roots:

$$E = -q(q - q^{-1}) \sum_{l=1}^{Q-1} z_l. \tag{7.8}$$

Inspire of the apparent difference eqs. (1.9) and (1.8) must be equivalent to the eqs. (7.7) and (7.8).Recalling the connection between LC and ML gauges via the Fourier transformation (A.3) one may say that the two forms of the Hamiltonian (4.7) and (4.10) are Fourier dual to each other.

It seems it is easier to deal with Bethe equations in the form (1.9) then with (7.7). Indeed, the roots (1.9) are either real or appear in complex conjugate pairs . The distribution of roots of (7.7) seems to be more complicated due to the different behaviour of both sides of (7.7) under the complex conjugation.

At the end of this section let us note that the difference equation of the form (7.1) can be obtained also from a *quadratic* form

$$H = CA - BD + qBA - qCD$$

for the quantum group $U_q'(sl_2)$ with $q' = q^2$.

8 q-Analog of Orthogonal Polynomials as Exact Zero Mode Wave Functions

There is an intriguing connection between the wave function (7.3)and q-generalization of the classical orthogonal polynomials [29]. The most general deformation of classical orthogonal polynomials contains 4 parameters (except q). These polynomials (so called Askey-Wilson polynomials) satisfy a q-analog of the differential hypergeometric. It is a difference equation

$$A(z)P_n(q^2z) + A(z^{-1})P_n(q^{-2}z) - (A(z) + A(z^{-1}))P_n(z) \tag{8.1}$$
$$= (q^{-2n} - 1)(1 - abcdq^{2n-2})P_n(z) \tag{8.2}$$

where
$$A(z) = \frac{(1-az)(1-bz)(1-cz)(1-dz)}{(1-z^2)(1-q^2z^2)} \tag{8.3}$$

and a, b, c, d are parameters and n is degree of the polynomial. Note that P_n is a *Laurent* polynomial in z and usual polynomial in $z + z^{-1}$ of degree n. The Askey-Wilson polynomials are known to be closely related to the "q-harmonic analysis" providing the most general family of spherical functions on the quantum $SL(2)$ group.

Choosing $c = -d = q, a = b = 0$ we arrive at the equation for the continuous q-Hermite polynomials [29] $H_n^{(q)}$

$$H_n^{(q)}(q^2z) - z^2 H_n^{(q)}(q^{-2}z) = q^{-2n}(1 - z^2)H_n^{(q)}(z).$$

(They are called continuous because of their orthogonality on the unit circle with a continuous measure).

It is clear from (7.4) that for an odd Q at $n = (Q-1)/2$ this yields its zero energy solution:

$$\Psi^{(E=0)}(iz) = z^{(Q-1)/2}H_{(Q-1)/2}^{(q)}(z). \tag{8.4}$$

The explicit form of the q-Hermite polynomials is

$$H_n^{(q)}(z) = \sum_{m=0}^{n} \frac{(q^2; q2)_n}{(q^2; q^2)_m (q^2; q^2)_{n-m}} z^{2m-n} \tag{8.5}$$

where the standard notation

$$(a;q)_n = \prod_{l=0}^{n-1}(1 - aq^l)$$

is used.

Another choice is $c = -d = q$, $a = -b = q$ and then the replacement of q by $q^{1/2}$ gives the q-Legendre equation

$$\frac{1-qz^2}{1-z^2}P_n^{(q)}(qz) + \frac{q-z^2}{1-z^2}P_n^{(q)}(q^{-1}z) = (q^{-n} + q^{n+1})P_n^{(q)}(z) \tag{8.6}$$

Then, comparing with the Eq.(7.1) we conclude that the zero mode solution is given by the continuous q-Legendre polynomial

$$\Psi^{(E=0)}(iz) = z^{(Q-1)/2}P_{(Q-1)/2}^{(q)}(z)$$

Their explicit form is

$$P_n^{(q)}(z) = \sum_{m=0}^{n} \frac{(q;q)_n(q^{1/2};q)_m(q^{1/2};q)_{n-m}}{(q;q)_m(q;q)_{n-m}}z^{2m-n} \tag{8.7}$$

9 Future Problems

The almost immediate and the most interesting task now is to solve the Bethe Ansatz equations in the limit $P,Q \to \infty$ when the flux $\Phi/2\pi$ is irrational. In all previous examples of integrable systems it was always possible to derive an integral equation for a distribution function of roots z_l. We hope that all this would be indeed possible and after all would allow one to obtain fractal properties of the spectrum analytically.

Another very important problem is to manage to apply the Bethe ansatz technique to the whole Hilbert space of the model and obtain the algebraic Bethe equations determining the spectrum in arbitrary points of the bands.

ACKNOWLEDGMENTS

We would like to thank P.G.O.Freund, A.Abanov, A.Gorsky, A.Kirillov, E.Floratos, J.-L.Gervais and V.Pasquier for interesting discussions.

P.W. acknowledges the hospitality of Weizmann Institute of Science and Laboratoire de Physique Théorique de l'Ecole Normale Supérieure where part of this work was done. A.Z. is grateful to the Mathematical Disciplines Center of the University of Chicago for the hospitality and support. He also would like to thank Prof.A.Niemi for the warm hospitality in the Institute of Theoretical Physics at Uppsala University where this work was completed. This work was supported in part by NSF under the Research Grant 27STC-9120000 and NSF-DMR 88-19860

A Gauges

1. The connection Landau gauge and the modified Landau gauge (2.2) is very simple

$$\phi_n(\vec{k}') = \exp(\frac{i\Phi}{2}n(n-Q))\psi_n(\vec{k}) \tag{A.1}$$

$$k'_x = k_x + \frac{i\pi P(Q-1)}{Q}, \qquad k'_y = k_y \tag{A.2}$$

Both k'_x and k_x are defined modulo Φ .

2.The relationship between Landau gauge and the Chiral gauge is a bit sophisticated[d]. They connected by the Fourier transformation

$$\tilde{\phi}_m = \sum_{n=0}^{Q-1} e^{i\Phi nm}\phi_n \tag{A.3}$$

where ϕ_n is the wave function in the ML gauge.spectrum. A new function $\tilde{\phi}_n$ obeys the equation

$$(e^{ik'_y} + e^{i\Phi n - ik'_x})\tilde{\phi}_{n+1} +$$
$$+(e^{-ik'_y} + e^{-i\Phi n + i\Phi + ik'_x})\tilde{\phi}_{n-1} = E\tilde{\phi}_n \tag{A.4}$$

Compare with (2.11) we can identify momentum

$$k'_x = -p_y - \frac{1}{4}\Phi, \qquad k'_y = p_x - \frac{1}{4}\Phi \tag{A.5}$$

B Quantum Integrable Models with Non-Periodic Boundary Conditions and Hofstadter Problem

A general quadratic form in quantum group generators is related to a quantum magnetic chain with a non-periodic boundary conditions , namely to a chain with two sites.

To begin with, let us give a brief summary of the formalism treating the integrable systems with boundaries. The boundary conditions of an integrable model determined by c-number 2×2 matrices $K_+(u)$ and $K_-(u)$ depending on spectral parameter and satisfied the "reflection equations" [22].

$$R(u/v)(K_-(u) \otimes 1)R(uvq^{-1})(1 \otimes K_-(v)) = (1 \otimes K_-(v))R(uvq^{-1})(K_-(u) \otimes 1)R(u/v)$$
$$R(v/u)(K_+^t(u) \otimes 1)R((uvq)^{-1})(1 \otimes K_+^t(v)) = (1 \otimes K_+^t(v))R((uvq)^{-1}) \times$$
$$\times (K_+^t(v) \otimes 1)R(v/u)$$

(t means the transposition) and R and L satisfy to the basic intertwining relation (3.3) Each solution of "reflection equations" specify a boundary condition consistent with integrability. Solutions for K_+ and for K_- are related

$$K_+(u) = K_-^t(u^{-1}) \tag{B.1}$$

---

[d]We are indebted to A.Abanov who found this connection

The transfer matrix for an integrable model with non-periodic boundary conditions is a *quadratic* form of the transfer matrix $L(u)$ of a model with periodic boundary conditions [23].

$$T(u) = L(u)K_-(u)\sigma_2 L^t(u^{-1})\sigma_2 \tag{B.2}$$

The trace of the transfer matrix

$$\tau(u) = Tr(K_+(u)T(u)) \tag{B.3}$$

forms a commutative family $[\tau(u), \tau(v)] = 0$.

It is known that for the models with the trigonometric R-matrix There is a 3-parametric family of boundary K-matrices [25], [26]

$$K_-(u) = \begin{pmatrix} \alpha(q^{-1}s^{-1}u - qsu^{-1}) & \beta(q^{-1}u^2 - qu^{-2}) \\ \gamma(q^{-1}u^2 - qu^{-2}) & -\alpha(su - s^{-1}u^{-1}) \end{pmatrix} \tag{B.4}$$

$$K_+ = \begin{pmatrix} \lambda(qtu - q^{-1}t^{-1}u - 1) & \mu(qu^2 - q^{-1}u^{-2}) \\ \nu(qu^2 - q^{-1}u^{-2}) & -\lambda(t^{-1}u - tu^{-1}) \end{pmatrix} \tag{B.5}$$

where α, β, γ, s and λ, μ, ν, t are arbitrary parameters characterized the boundary conditions. Substituting the L-operator (3.5) into (B.2) we find matrix elements of the transfer matrix

$$T_{11}(u) = \frac{\alpha(q^{-1}s^{-1}u - qsu^{-1})}{(q - q^{-1})^2}(k^2 + k^{-2} - u^2A^2 - u^{-2}D^2)$$
$$+ \frac{q^{-1}u^2 - qu^{-2}}{q - q^{-1}}(\gamma ku^{-1}CD - \gamma k^{-1}uCA - \beta kuAB + \beta k^{-1}u^{-1}DB)$$
$$+ \alpha(su - s^{-1}u^{-1})CB \tag{B.6}$$

$$T_{12}(u) = \frac{q^{-1}u^2 - qu^{-2}}{q - q^{-1}}\left(\frac{\beta(k^2A^2 + k^{-2}D^2 - u^2 - u^{-2})}{q - q^{-1}} - \gamma(q - q^{-1})C^2\right.$$
$$\left. + \alpha(sk^{-1}DC - s^{-1}kAC)\right) \tag{B.7}$$

$$T_{21}(u) = \frac{q^{-1}u^2 - qu^{-2}}{q - q^{-1}}\left(\frac{\gamma(k^2D^2 + k^{-2}A^2 - u^2 - u^{-2})}{q - q^{-1}} - \beta(q - q^{-1})B^2\right.$$
$$\left. + \alpha(skBD - s^{-1}k^{-1}BA)\right) \tag{B.8}$$

$$T_{22}(u) = \frac{\alpha(s^{-1}u^{-1} - su)}{(q - q^{-1})^2}(k^2 + k^{-2} - u^2D^2 - u^{-2}A^2)$$
$$+ \frac{q^{-1}u^2 - qu^{-2}}{q - q^{-1}}(\beta ku^{-1}BA - \beta k^{-1}uBD - \gamma kuDC + \gamma k^{-1}u^{-1}AC)$$
$$- \alpha(q^{-1}s^{-1}u - qsu^{-1})BC \tag{B.9}$$

Summing diagonal elements (B.6) and (B.9) we find the trace of the transfer matrix $\tau(u)$

$$\tau(u) = \frac{(u^4 + u^{-4} - q^2 - q^{-2})}{(q - q^{-1})^2}[(\mu\gamma k^{-2} + \nu\beta k^2 - \alpha\lambda ts^{-1})A^2$$
$$+ (\mu\gamma k^2 + \nu\beta k^{-2} - \alpha\lambda t^{-1}s)D^2$$
$$+ (q - q^{-1})(-(tk^{-1}\lambda\gamma + q^{-1}s^{-1}k\alpha\nu)CA + (q^{-1}t^{-1}k^{-1}\lambda\beta + sk\alpha\mu)BD$$
$$+ (t^{-1}k\lambda\gamma + qsk^{-1}\alpha\nu)CD - (qtk\lambda\beta + s^{-1}k^{-1}\alpha\mu)BA)$$
$$- (q - q^{-1})^2(\mu\beta B^2 + \nu\gamma C^2)] + const \tag{B.10}$$

(here we replace the Casimir element (3.2) by it c-number value).

The trace of the transfer matrix $\tau(u)$ is a general quadratic form in A, B, C, D since it depends on 7 parameters - 3 in each K-matrix and rapidity k. Indeed, the total number of coefficients of a general quadratic form is 10 but one of them is a common multiplier and another two contribute to a *const* in(B.10) due to the two central elements ($AD = 1$ and the Casimir operator).

As we showed previously the Hamiltonian of the Bloch particle in magnetic field is a particular quadratic form of the quantum group generators and therefore can be consider as an integrable model.

References

[1] J.Zak, *Phys. Rev.* **134** 1602 (1964)

[2] M.Ya.Azbel, *Sov. Phys.JETP* **19** 634 (1964)

[3] G.H.Wannier,*Phys.Status Solidi* **88**, 757 (1978)

[4] D.R.Hofstadter,*Phys. Rev. B* **14** 2239 (1976)

[5] D.J.Thouless, M.Kohmoto, P.Nightingale, M.den Nijs, *Phys.Rev.Lett* **49**,405 (1982)

[6] S.Aubry and G.Andre *Ann.Israel Phys.Soc.* **3**, 131 (1980)

[7] D.J.Thouless, *Phys. Rev.* **28** 4272 (1983)

[8] H.Hiramoto, M.Kohmoto, *Int.J.Mod.Phys. B* **6**, 281 (1992)

[9] P.W.Anderson in Proc. of Nobel Symp. 73 *Physica Scripta* **T27**, (1988)

[10] P.B.Wiegmann in Proc. of Nobel Symp. 73 *Physica Scripta* **T27**, (1988)

[11] P.P.Kulish, N.Yu.Reshetikhin *Zap.nauch.semin.LOMI* **101**, 112 (1980)

[12] V.G.Drinfeld , *Dokl.Acad.Nauk* **283** 1060 (1985)

[13] M.Jimbo *Lett.Math. Phys.* **10** 63 (1985)

[14] N.Yu.Reshetikhin, L.A.Takhtadjan, L.D.Faddeev,*Algebra i Analiz* **1**, 178 (1989)

[15] P.Roche, D.Arnaudon *Lett.Math. Phys.* **17**, 295 (1989)

[16] E.K.Sklyanin *Func.Anal.Appl.* **17**, 273 (1983)

[17] E.G.Floratos *Phys.Lett* . **B233**, 235 (1989)

[18] L.D.Faddeev *The Bethe Ansatz*, Andrejewski lectures, preprint SFB 288, No 70 (1993)

[19] E.K.Sklyanin *Zap.nauch.semin.LOMI* **134**, 112 (1983)

[20] A.V.Turbiner *Comm.Math. Phys.* **118** 467 (1988)

[21] A.Ushveridze *Sov.Journal Part.Nucl.* **20**, 185 (1989), *ibid* **23**, 25 (1992)

[22] I.Cherednik *Theor.Mat. Fys.* **61** 35 (1984)

[23] E.K.Sklyanin *J.Phys. A* **21**. 2375 (1988)

[24] V.V.Bazhanov and Yu.G.Stroganov *J.Stat.Phys.* **59**. 799 (1990)

[25] M.Gaudin *La Fonction de Bethe*,Masson,1983

[26] H.J. de Vega, A.Gonzalez-Ruiz, preprint LPTHE-PAR 92-45, To appear in J.Phys. A

[27] M.A.Shifman *Int.Journal of Mod.Phys.A* **4** 2897 (1988)

[28] O.V.Ogievetsky and A.V.Turbiner*Preprint CERN-TH*:6212/91 (1991)

[29] see e.g.G.Gasper and M.Rahman, *Basic Hypergeometric Seriaes* (1990), Cambridge Univ.Press and N.Vilenkin, A.Klimyk, *Representation of Lie Groups and Special Functions*,vol.3 (1992) Kluwer Acad.Publ.

Index

One-dimensional, 234
Polyakov, 73, 142
topological, 285
Superconformal symmetry, 261
Supercurrents
extended supersymmetry, 92
Supergravity, 97
Superspace
analytic, 93
Supersymmetry, 238
$N = 2$, 335, 353, 361
breaking, 349, 367
super self-duality, 87

Tachyon, 77, 100
TBA, *see* Thermodynamic Bethe
Ansatz
Temperley-Lieb algebra, 341
representation, 341
Thermodynamic Bethe Ansatz, 273,
see Y-system, 363
magnonic, 276
on $\mathcal{G} \times \mathcal{H}$, 277
pseudoenergy, 276
universal form, 276
Three-point function, 380
Toda hierarchy, 138
Topological
A and B models, 262
conformal field theory, 25
correlation functions, 24
cosets, 305
field theory, 261, 263
gauge theory, 288
gravity, 26, 242
lattice models, 339
model, 23, 24
rigid string, 153
sigma model, 32, 262, 286
sigma models, 153
string theory, 23, 24, 34, 264,
285
theories, 338, 353
theory, 234, 235, 237, 286

Topological theories, *see* Twisted
algebra
Trace anomaly, 2
Transfer matrix, 12, 114
Triangulated surfaces, 72
Turaev-Viro invariant, 51
Turbulence, 194
Twisted algebra, 352
Twisted cohomology, 60
Twistor transform
supersymmetric, 93
Twistors
and harmonic spaces, 87
Two-matrix model, 81

Verma module, 61
Vertex models, 78, 344
Vertex operators, 380
Coulomb gas representation, 380
Vertex/IRF correspondence, 347, 366
Virasoro algebra, 137, 305
Vorticity, 194

W-algebra, 59, 305
W-gravity, 59
Ward identity, 265
Wess-Zumino action, 2

Y-system, 278
periodicity, 281
Yang-Baxter equation, 12, 46, 356
Yang-Mills theory
in 2 dimensions, 119
self-duality equations, 87
symmetry transformations, 94
topological, 152
two dimensional, 152
Yangian, 12, 13

MIX
Papier aus verantwortungsvollen Quellen
Paper from responsible sources
FSC® C105338
FSC
www.fsc.org

If you have any concerns about our products,
you can contact us on
ProductSafety@springernature.com

In case Publisher is established outside the EU,
the EU authorized representative is:
**Springer Nature Customer Service Center GmbH
Europaplatz 3, 69115 Heidelberg, Germany**

Printed by Libri Plureos GmbH
in Hamburg, Germany